Biotransformations:
Bioremediation Technology for Health and Environmental Protection

Biotransformations:
Bioremediation Technology for Health and Environmental Protection

Edited by

Ved Pal Singh
Department of Botany,
University of Delhi, Delhi, India

Raymond D. Stapleton Jr.
Merck & Co. Inc.,
Elkton, Virginia, USA

progress in industrial microbiology

2002

ELSEVIER

Amsterdam – London – New York – Oxford – Paris – Shannon – Tokyo

ELSEVIER SCIENCE B.V.
Sara Burgerhartstraat 25
P.O. Box 211, 1000 AE Amsterdam, The Netherlands

First edition 2002

Library of Congress Cataloging in Publication Data
A catalog record from the Library of Congress has been applied for.

ISBN: 0-444-50997-6
ISSN: 0079-6352

⊗ The paper used in this publication meets the requirements of ANSI/NISO Z39.48-1992 (Permanence of Paper). Printed in The Netherlands.

PREFACE

For many years, the discipline of microbial transformations was regarded as one of the important components of Applied Microbiology and Biotechnology, which could serve as the basis for understanding the basic concept of bioremediation technology and its implications in the health and environmental protection programmes all over the world. A variety of pollutants of anthropogenic origin are released as organochemicals as well as agricultural, industrial, and municipal wastes, which pose potential threat to both human and animal health as well as the environment. Under such a serious situation, there is an immediate requirement to look for the possibilities for tackling this problem of increased agro-industrial wastes generated by fast-increasing global population.

Amongst living organisms, microbial systems occupy a key position in health and environmental protection programmes, owing to their biotechnological potential in degrading and eliminating the hazardous compounds. The earlier book (edited by VPS and titled "Biotransformations: Microbial Degradation of Health-Risk Compounds" which was published by Elsevier Science B.V.), having been out of print, there was an urgent need for bringing out a consolidated account in the form of another book, covering all these aspects, and hence this endeavour.

"Biotransformations: Bioremediation Technology for Health and Environmental Protection" provides a clear understanding how microbes, following their degradative processes, contribute maximally to the benefit of mankind through biotransformations of waste materials as well as a wide variety of health-risk compounds.

The book contains twenty four chapters contributed by leading scientists from different parts of the world, covering various aspects of bioremediation of xenobiotics such as toxic, carcinogenic, teratogenic, and mutagenic compounds, which include halogenated aromatics, derivatives of heavy metals, microbial toxins, tannins, dyes, sulfur compounds of coal and petroleum and pesticides. The bioremediation of agricultural residue, industrial as well as municipal wastes, fuel oils, lubricants, natural rubber products, and other synthetic polymers, which pollute the environment substantially, also constitutes an important component of the book. All biotechnological aspects of microbial transformations pertaining to biodegradation/bioremediation of hazardous wastes, ranging from screening methods for microbes with degradative potential, processes of degradation, strain improvement for enhanced biodegradation and elimination of xenobiotics of health and environmental concern have been dealt with.

The book intends to widen the scope of Applied Microbiology and Biotechnology in general and biotransformations in particular. It will provide an opportunity for scientists in the areas of biochemistry, food industry, environmental science and engineering and their implications in technologically feasible, environment friendly and economically viable bioremediation options. Also, it forms an interface between agro-industrial establishments and the academic world and will generate new thought provoking ideas for scientists of future generations for the safeguard of both human and animal health as well as the environment.

We are thankful to all the scientists, who accepted our invitation to contribute their valuable review articles for this book, and especially to Professor Dr. Fred Stutzenberger for kindly agreeing to write an Overview on the theme of the book, forming Chapter 1, and Professor Dr. Wolf-Rainer Abraham and Dr. J.L. Faull for contributing lead articles, forming Chapters 2 and 3, respectively. We are greatly indebted to our colleagues and friends, especially Professor S.S. Bhojwani, Professor S.N. Raina, and Professor A.K. Bhatnagar of the Department of Botany, University of Delhi, who have helped us in various ways. VPS is extremely grateful to his mentor Professor Abbas Musavi of Aligarh Muslim University and the fellow scientists Professor R.K. Saxena, University of Delhi South Campus and Professor T.N. Lakhanpal, H.P. University, Shimla for their encouragement and support. RDS would like to thank his mentors Drs. G.S. Sayler, A.V. Palumbo, and J. Zhou for their support.

Thanks are also due to Mr. S.K. Dass, Mr. Suresh Vig, Mr. Ram Pal Giri, Mr. Satish Chand Sharma, Mr. Ganga Prasad, Mr. Surendra Thakur, and Mr. Bhagwat Giri for their help in photography, drawing, postage, communication and other related works.

The help rendered by Mr. Ashok Datta and Ms Reema of LaPrints in preparing the manuscript of the book is gratefully acknowledged.

We are extremely grateful to the Publisher, Elsevier Science B.V., The Netherlands, its Publishing Director, Dr. Bas van der Hoek and especially to Ms Ana Bela Sa Dias, the Associate Publishing Editor, her Boss Mr. Adriaan Klinkenberg, the Senior Publishing Editor, and her Assistant/Secretary Miss Caroline ten Wolde as well as other Office Staff for having taken keen interest in our proposal for this book and keeping us well informed about its status from time to time.

The family members of VPS: his mother, Anandi Devi, wife, Kusum, daughter, Sandhya, sons, Sudhir, & Hemant, and parents-in-law, Biri Singh and Susheela Devi deserve our warmest appreciation for their help throughout the period of our work on this book. RDS would like to thank his family for their support: parents, Ray Sr. and Saluda, wife, Wendy, and son, Joshua.

VED PAL SINGH
RAYMOND D. STAPLETON, JR.

20 August, 2001

ABOUT THE EDITORS

Dr. Ved Pal Singh is a Reader at the Department of Botany, University of Delhi. He has 20 years of teaching and research experience in the field of Applied Microbiology and Biotechnology, and has about 60 publications to his credit. He has edited International Review Series 'Frontiers in Applied Microbiology' and 'Concepts in Applied Microbiology and Biotechnology' as well as Volume 32 (1995) of the 'Progress in Industrial Microbiology' published by Elsevier Science B.V., The Netherlands. Dr. Singh received Young Scientist Awards from INSA (1985) and UNESCO (1987), and was awarded INSA-COSTED Travel Fellowship to visit Hungary (1985). He was a British Council Visitor to UK and Germany (1987). He represented India in the V, VI, and VII International Symposia on Actinomycetes Biology held at Mexico (1982), Hungary (1985), and Japan (1988). He chaired sessions and delivered lectures at a number of symposia/seminars/workshops in India and abroad. He worked as a Commonwealth Academic Staff Fellow at Royal College, Glasgow (1990-1991). He has been honoured by the International Society of Conservators and Explorers of Natural Resources, conferring on him the Founder Fellowship with the title FNRS. Dr. Singh is also a member to various national and international scientific societies. Presently, he is serving as Member, Academic Council, University of Delhi for the second term.

Dr. Raymond D. Stapleton, Jr. received his Ph.D. from The Center for Environmental Biotechnology at The University of Tennessee, Knoxville (USA). He subsequently was the recipient of an Alexander Hollaender Fellowship for post-doctoral studies in the Environmental Sciences Division at Oak Ridge National Laboratory, Oak Ridge, Tennessee (USA). He has spent the last 8 years actively engaged in research in the areas of Microbial Ecology, Biogeochemistry, and Applied Microbiology and Biotechnology, with 14 publications and one patent to his credit. Currently, Dr. Stapleton is employed as a Scientist in the pharmaceutical industry with Merck & Company, Inc. at Elkton, Virginia, USA.

LIST OF ABBREVIATIONS

ADI	:	Arginine deiminase
AF	:	Aflatoxin
ARDRA	:	Amplified rDNAs
B.E.	:	Biological efficiency
BHC	:	Benzene hexachloride
BOD	:	Biochemical oxygen demand
BP	:	Biphenyl
BSA	:	Bovine serum albumin
BTEX	:	Benzene, toluene, ethylbenzene, xylenes
CAM	:	Camphor
CB	:	Chlorobenzene
p-CB	:	*para*-Chlorobiphenyl
CBAs	:	Chlorinated benzoic acids/Chlorobenzoates
4-CC	:	4-Chlorocatechol
CE	:	Coversion efficiency
CFUs	:	Colony forming units
CK	:	Carbamate kinase
COD	:	Chemical oxygen demand
COS	:	Carbonyl sulfide
2,4-D	:	2,4-Dichlorophenoxyacetic acid
DBDS	:	Dibenzyldisulfide
DBPS	:	2′-hydroxybiphenyl 2-sulfinic acid
DBS	:	Dibenzylsulfide
DBT	:	Dibenzothiophene
DBTO	:	DBT5′-sulfoxide
$DBTO_2$	:	DBT5′-sulfone
DCBA	:	Dichlorobenzoate
DCE	:	1,2-Dichloroethylene
DDT	:	1,1,1-trichloro-2, 2′-bis(4-chlorobiphenyl) ethane
DFA	:	Direct fluorescent antibodies
DHA	:	Dihydroxyacetone
DHAP	:	Dihydroxyacetone phosphate
DHBA	:	Dihydroxybenzoate
DHP	:	Dehydrogenative polymerizate

DIDT	:	5,6-dihydro-3H-imidazo(2,1-C)-1,2,4-dithiazol-3-thione
DMS	:	Dimethylsulphide
DMSO	:	Dimethylsulphoxide
DMSP	:	b-Dimethylsulphonioprionate
2,4-DNT	:	2,4-Dinitrotoluene
L-DOPA	:	L-Dihydroxyphenyl alanine
DPHP	:	Diphenylphthalate
DTHT	:	2-n-Dodecyltetrahydrothiophene
EBDC	:	Ethylenebisdithiocarbamate
EBIS	:	Ethylenebisisothiocyanate sulfide
EDA	:	Ethylenediamine
EDAX	:	Energy dispersion X-ray analysis
EDI	:	Ethylene diisothiocyanate
EDTA	:	Ethylenediamine tetraacetic acid
EM	:	Electron microscopy
EP	:	Early phase species
EPA	:	Environmental Protection Agency
EPS	:	Exopolysaccharide/ Extracellular polymeric substances
ESCA	:	X-ray photoelectron spectroscopy for chemical analysis
ETD	:	Ethylene thiuram disulfide
ETM	:	Ethylene thiuram monosulfide
ETU	:	Ethylenethiourea
EU	:	Ethyleneurea
EXAFS	:	Extended X-ray absorption fine-structure
FAO	:	Food and Agricultural Organization
GAC-FBR	:	Granular activated carbon-fluidized bed reactor
GC-MS	:	Gas chromatography-mass spectroscopy
GC-AES	:	Gas-chromatography-atomic emission spectroscopy
GC-SCD	:	Gas-chromatography-sulfur chemiluminiscence detection
GDH	:	Glycerol dehydrogenase
GGE	:	Guaiacol glyceryl ether
GT	:	Glucosyl transferase
2-HBPS	:	2-Hydroxybiphenyl 2-sulfonic acid
HCH	:	Hexachlorocyclohexane
HDC	:	Histidine decarboxylase
HDS	:	Hydrodesulfurization

HFBT	:	3-Hydroxy-2-formylbenzothiophene
HMX	:	Octahydro-1,3,5,7-tetranitro-1,3,5,7-tetraazocine
3-HPA	:	Hydroxy-3-propionaldehyde
IRMS	:	Isotope ratio mass spectrometry
IS	:	Inner salt
LAS	:	Linear alkylbenzene sulfonate
LiP	:	Lignin peroxidase
LP	:	Late phase species
LPS	:	Lipopolysaccharide
MCA	:	Metabolic control analysis
MCBP	:	Monochlorobiphenyl
MDS	:	Microbial desulfurization
MGP	:	Manufactured gas plant
MLF	:	Malolactic fermentation
MMO	:	Methane monooxygenase
MMPA	:	3-Methylmercaptopropionate
MNC	:	4-Methyl-5-nitrocatechol
MnP	:	Mn(II)-dependent peroxidase
MPA	:	3-Mercaptopropionate
MTBE	:	Methyl tert-butyl ether
NAH	:	Naphthalene
NR	:	Natural rubber
OCT	:	Octane
ODC	:	Ornithine decarboxylase
ORFs	:	Open reading frames
OS	:	Orientational spectra
OTC	:	Ornithine transcarbamylase
PAHs	:	Polycyclic aromatic hydrocarbons
PCA	:	Protocatechuic acid
PCBs	:	Polychlorinated biphenyls
PCE	:	Perchloroethylene
PCP	:	Pentachlorophenol
PCs	:	Phytochelatins
PEI	:	Polyethyleneimine
PFGE	:	Pulse field gel electrophoresis
PHS-TVA	:	Public Health Service-Tennessee Valley Authority

PNP	:	*p*-Nitrophenol
PP	:	Polypropylene
PUF	:	Polyurethane foam
RBCs	:	Rotating biological contactors
RFLP	:	Restriction fragment length polymorphism
RDX	:	Hexahydro-1,3,5,7-trinitro-1,3,5-triazine
RT	:	Reverse transcriptase
SAL	:	Salicylate
SBR	:	Styrene butadiene rubber
SCLM	:	Scanning confocal laser microscopy
SEM	:	Scanning electron microscopy
SLF	:	Sanitary Landfill
SRA	:	Substrate-dependent respiratory activity
SRS	:	Savannah River Site
SVI	:	Sludge volume insex
2,4,5-T	:	2,4,5-Trichlorophenoxyacetic acid
T2C	:	Thiophene-2-carboxylic acid
TCA	:	Tricholoroethane
TCBA	:	Trichlorobenzoate
TCE	:	1,1,2-Trichloroethylene
TDC	:	Tyrosine decarboxylase
THC	:	Total hydrocarbon concentration
THTA	:	2-Tetrahydrothiopheneacetic acid
THTC	:	2-Tetrahydrothiophenecarboxylic acid
TLC	:	Thin layer chromatography
TMP	:	Trimetaphosphate
TNT	:	Trinitrotoluene
TOC	:	Total organic carbon
TOL	:	Toluene
TSS	:	Total suspended solids
USAF	:	US Air Force
VC	:	Vinyl chloride
WET	:	Whole effluent toxicity
WSCs	:	Water soluble carbohydrates
WSPs	:	Water soluble phenolics
ZSV	:	Zone settling velocity

LIST OF CONTRIBUTORS

Chapter numbers are shown in parentheses following the address of each contributor

Wolf-Rainer Abraham, GBF - National Research Center for Biotechnology, Chemical Microbiology, Mascheroder Weg 1, 38124, Braunschweig, Germany (2)

Prerna Ahuja, Department of Microbiology, University of Delhi South Campus, New Delhi - 110 021, India (8)

M. Archana, Department of Food Microbiology, Central Food Technological Research Institute, Mysore - 570 013, India (10)

G. Baggi, Department of Food Science & Microbiology, Università di Milano, Via Celoria, 2, 20133 Milan, Italy (6)

R.L. Bezbaruah, Biochemistry Division, Regional Research Laboratory, Jorhat - 785 006, India (19)

A.K. Bhatnagar, Environmental Biology Laboratory, Department of Botany, University of Delhi, Delhi - 110 007, India (20)

Kartiki Bhatnagar, Department of Botany, University of Delhi, Delhi - 110 007, India (17)

Robin Brigmon, Environmental Biotechnology Section, Westinghouse Savannah River Co., Aiken, SC, USA (1)

V.D. Bunin, Institute of Applied Microbiology, Obolensk 142289, Russia (18)

Dwight Camper, Department of Plant Pathology and Physiology, Clemson University, Clemson, SC, USA (1)

J.L. Faull, Biology Department, Birkbeck College, Malet Street, London WC1E 7HX, UK (3)

B.K. Gogoi, Biochemistry Division, Regional Research Laboratory, Jorhat - 785 006, India (19)

Jose M. González, Department of Marine Sciences, University of Georgia, Athens, GA 30602, USA (12)

Rani Gupta, Department of Microbiology, University of Delhi South Campus, New Delhi - 110 021, India (8)

O.V. Ignatov, Institute of Biochemistry and Physiology of Plants and Microorganisms, Russian Academy of Sciences, 13 Pr. Entuziastov; and Saratov State University, 83 U1, Astrakhanskaya, Saratov 410005, Russia (18)

V.V. Ignatov, Institute of Biochemistry and Physiology of Plants and Microorganisms, Russian Academy of Sciences, 13 Pr. Entuziastov, Saratov 410015, Russia (18)

Archana P. Iyer, Centre for Advanced Study in Botany, University of Madras, Chennai - 600 025, India (13)

Samantha B. Joye, Department of Marine Sciences, University of Georgia, Athens, GA 30602, USA (12)

Inderdeep Kaur, Environmental Biology Laboratory, Department of Botany, University of Delhi, Delhi - 110 007, India (20)

Ronald P. Kiene, Department of Marine Sciences, University of South Alabama, Mobile, AL 36688, USA (12)

T.N. Lakhanpal, Department of Biosciences, Himachal Pradesh University, Shimla - 171 005, India (4)

A. Lonvaud-Funel, Faculté d'Oenologie – Université Victor Segalen Bordeaux2, 351, Cours de la Liberation – 33405, Telence Cedex, France (9)

A. Mahadevan, Centre for Advanced Study in Botany, University of Madras, Chennai - 600 025, India (13)

Harapriya Mohapatra, Department of Microbiology, University of Delhi South Campus, New Delhi - 110 021, India (8)

Mary Ann Moran, Department of Marine Sciences, University of Georgia, Athens, GA 30602, USA (12)

S. Nicklin, Bioremediation Group, DERA, Fort Halstead, Sevenoaks, Kent, TN14 7BP, UK (3)

Anthony V. Palumbo, Environmental Sciences Division, P.O. Box 2008, Oak Ridge National Laboratory, Oak Ridge, TN 37830-8002, USA (23)

Tommy J. Phelps, Environmental Sciences Division, P.O. Box 2008, Oak Ridge National Laboratory, Oak Ridge, TN 37830-8002, USA (23)

R.K. Saxena, Department of Microbiology, University of Delhi South Campus, New Delhi - 110 021, India (8)

Stefan Schmidt, Abteilung Mikrobiologie, Institut für Allgemeine Botanik der Universität Hamburg, Ohnhorststraße 18, 22609, Hamburg, Germany (15)

T. Shantha, Department of Food Microbiology, Central Food Technological Research Institute, Mysore - 570 013, India (10)

S.Yu. Shchyogolev, Institute of Biochemistry and Physiology of Plants and Microorganisms, Russian Academy of Sciences, 13 Pr. Entuziastov, Saratov 410015, Russia (18)

Balwant Kumar Singh, Department of Botany, University of Delhi, Delhi - 110 007, India (22)

Dileep K. Singh, Department of Zoology, University of Delhi, Delhi - 110 007, India (7,24)

Mahendra Nath Singh, Department of Botany, S.S.N. College, Alipur, Delhi - 110 036, India (22)

Ved Pal Singh, Department of Botany, University of Delhi, Delhi - 110 007, India (5,11,17,22)

Dimitry Yu. Sorokin, Institute of Microbiology, Russian Academy of Sciences, pr. 60-letiya Oktyabrya 7, k. 2, Moscow 117811, Russia (12)

Raymond D. Stapleton, Environmental Sciences Division, P.O. Box 2008, Oak Ridge National Laboratory, Oak Ridge, TN 37830-8002, USA (23)

Fred Stutzenberger, Department of Microbiology and Molecular Medicine, Clemson University, Clemson, SC, USA (1)

Y. Tokiwa, National Institute of Bioscience and Human-Technology, Tsukuba, Ibaragi 305, Japan (16)

A. Tsuchii, National Institute of Bioscience and Human-Technology, Tsukuba, Ibaragi 305, Japan (16)

R.S. Upadhyay, Department of Botany, Banaras Hindu University, Varanasi - 221 005, India (14)

S. Wilkinson, Bioremediation Group, DERA, Fort Halstead, Sevenoaks, Kent, TN14 7BP, UK (3)

T. Yamanaka, Soil Microbiology Laboratory, Forestry and Forest Products Research Institute, MAFF, P.O. Box 16, Tsukuba Norin Kenkyu Danchi-nai, Ibaraki 305-8687, Japan (21)

Chuanlun Zhang, Department of Geological Sciences, 101 Geological Science Building, University of Missouri Columbia, MO 65211, USA (23)

Jizhong Zhou, Environmental Sciences Division, P.O. Box 2008, Oak Ridge National Laboratory, Oak Ridge, TN 37830-8002, USA (23)

CONTENTS

Biotransformations: Bioremediation Technology for Health and Environmental Protection
V.P. Singh and R.D. Stapleton, Jr. (Editors)
© 2002 Elsevier Science B.V. All rights reserved.

Bioremediation of compounds hazardous to health and the environment: An overview

Robin Brigmon[a], Dwight Camper[b] and Fred Stutzenberger[c]

[a]Environmental Biotechnology Section, Westinghouse Savannah River Co., Aiken, SC, USA

[b]Department of Plant Pathology and Physiology, Clemson University, Clemson, SC, USA

[c]Department of Microbiology and Molecular Medicine, Clemson University, Clemson, SC, USA

INTRODUCTION

Bioremediation is a biological process by which environmental pollutants are eliminated or converted to less toxic (or even useful) substances. Natural bioremediation is often largely catalyzed by the indigenous microbial or plant populations in soil or aquatic ecosystems. These bioremediative processes (such as microbial degradation of petroleum hydrocarbons at oil seepage sites) have doubtlessly been active for millions of years. Farmers have been recycling animal and plant wastes back to the soil via composting for thousands of years. Yet, it is only in recent decades that detailed studies have begun to probe the underlying molecular mechanisms of the composting process as well as other approaches to bioremediation.

Glazer and Nikaido [1] categorize bioremediation strategies as either *in situ* treatment, composting, land-farming or aboveground reactor systems. This opening chapter will provide an overview of those systems, as well as describe some fundamental microbial mechanisms (oxidative, reductive, and hydrolytic reactions) and associations in plant and microbial populations (competition, cross-feeding and commensalism) which may be essential to their success.

DEGRADATIVE PROCESSES

Chemicals in the soil, water and air are affected by various chemical/ physical and environmental factors. These factors are interrelated and influence the ultimate fate of the chemical as well as the degradative processes themselves. Environmental factors (rain, temperature, ultraviolet irradiation and humidity) affect the soil microbial population and resultant biological activities. Chemical/physical factors of the chemical compound include solubility, vapor pressure and structure. These factors govern the susceptibility to transformation reactions catalyzed by the microbial community. Vegetation can also affect the fate, as absorption and translocation can remove the chemical from soil or water. Once inside the plant, the chemical is then subjected to metabolism, sequestration into the vacuole or incorporation into a plant constituent such as lignin. In some cases, the absorbed chemical can be released via exudation from the root system, thus making it available for microbial or chemical processes. Presence of the chemical in the soil or water may be direct as in pesticide application for pest control, or may be indirect as in spillage or inadvertent contamination.

Bioremediation of chemicals in the soil depends on the activities of microbes in the soil, or in association with the root system. Chemical/physical properties of the chemical itself can influence the availability of the chemical to the microbe, or the susceptibility to degradative processes. Molecular alterations catalyzed by microbial processes include oxidation, reduction, hydrolysis, de-esterification, dehalogenation, dealkylation, conjugation and others. These processes usually result in a non-toxic chemical in the case of a pesticide, or in decreasing contamination levels of hazardous materials. Dissipation of pesticides in the soil involves several processes: volatilization, photodecomposition, leaching, adsorption, and microbial degradation.

Degradation strategies exhibited by microorganisms include: co-metabolism - the biotransformation of a molecule coincidental to the normal metabolic functions of the microbe; catabolism - the utilization of the molecule as a nutritive or energy source; and extracellular enzymes (phosphatases and amidases) - secreted into the soil, which can act on the molecule as a substrate. Three basic types of reactions can occur: degradation, conjugation, and rearrangements, and all of which can be microbially mediated. Complete degradation of a chemical in the soil to carbon dioxide and water involves many different types of reactions; however, usually the first one or two transformations frequently result in loss of biological activity.

Composting

Composting is by far the oldest bioremediation technology. Ideally, it is a controlled *biooxidative process* that evolves through a mesophilic-thermophilic-mesophilic humification to yield carbon dioxide, water, minerals and stabilized compost that is greatly reduced in volume compared to the starting material [2]. Successfully composted material is a stabilized and sanitized product which is non-toxic to plant and animal life. In many cases, it may be used as a supplement on poor soil to increase agricultural productivity.

The basic requirements of successful composting have been summarized by Thomas et al. [3]:

(i) biodegradability of materials (de-watered sewage sludges, food processing wastes, animal manure and municipal solid wastes are some examples of compostable materials);

(ii) adequate moisture (generally in the range of 40-65%, limited by the structural characteristics of the material);

(iii)   suitable carbon/nitrogen (C/N) ratios (optimal range of 15-30); sufficient mass of materials to retain heat even in relatively cold climates;

(iv)   sufficient porosity to allow free exchange of gases in and out of the composting mass.

Aerobic composting systems may be divided into three main catagories:

(i)   open windrow, in which the composting material is gathered into a series of long piles (windrows), which are aerated by periodic manual or mechanical turning;

(ii)   static pile, in which the composting materials are aerated via an air distribution system located in or under the pile;

(iii)   in-vessel (composting chamber) system, in which biodegradable materials are continually or periodically mixed and aerated via a forced air blower.

The open windrow process is most amenable to the bioconversion of municipal and agricultural wastes in areas that have sufficient open space for operation. The U.S. Public Health Service-Tennessee Valley Authority (PHS-TVA) Joint Composting Project (Figure 1) was located on the Watauga River near Johnson City in eastern Tennessee.

Municipal solid waste was sorted to remove non-biodegradables, ground to increase surface area, mixed with concentrated sewage sludge and piled into windrows which averaged about 3 m wide by 1 m high by 85 m long. Windrows were mechanically turned and sampled periodically (Figure 2) to determine a variety of physical, chemical and biological parameters at various depths across the cross-sectional area.

Within 3 weeks of the 7-week composting cycle, freshly cut cross-sectional faces of windrows exhibited a typical white band of thermophilic actinomycete

Figure 1. Aerial view of the U.S. Public Health Service-Tennessee Valley Authority (PHS-TVA) Joint Composting Project in eastern Tennessee. Raw sewage sludge was de-watered at the sewage treatment plant (a) and mixed with municipal solid waste which was then composted in long windrows (b).

Figure 2. Cross-sections of windrows at the PHS-TVA Project were routinely cut to allow taking of samples for microbiological and chemical analyses.

growth (Figure 3) where temperatures varied from 60 to 65°C even during cold winter weather [4].

Figure 3. The white band of actinomycete growth (indicated by arrows) yielded thermophiles with high biodegradative capacity.

The composting process selects for thermophilic microbes that can rapidly degrade the raw materials to usable carbon and energy sources. Composting of materials such as municipal and agricultural wastes establish thermophilic microbial populations that efficiently degrade complex polysaccharides such as starch and cellulose (Figure 4a,b).

Materials which are suitable for bioremediation composting include sewage sludge, soils contaminated with diesel fuel and other petroleum products, mineral oil and munitions residues, as well as pharmaceutical and brewing industrial wastes. The materials which are *recalcitrant* to biodegradation (such as munitions residues) are often mixed with a feedstock of highly biodegradable material (such as canning plant waste) to support the formation

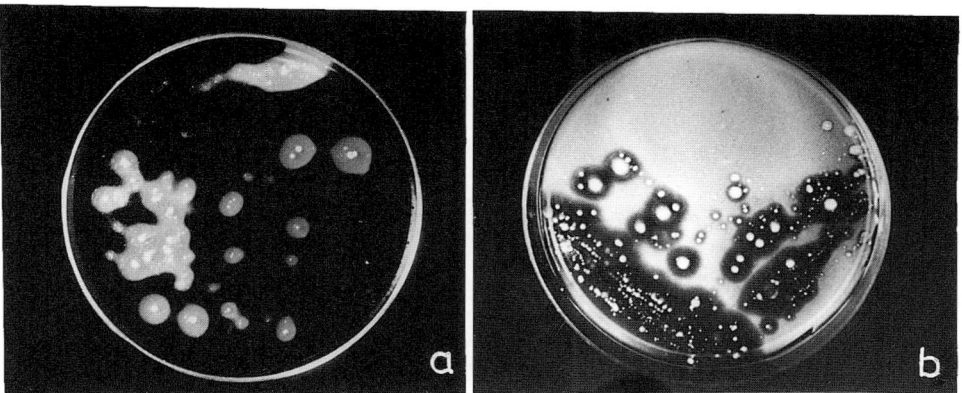

Figure 4. The clear areas of starch degradation (a) around thermophilic actinomycete colonies were defined by flooding with an iodine solution. Similar clearing around colonies growing on cellulose agar (b) indicates cellulase production by isolates from compost at the PHS-TVA Joint Project.

of a dense microbial population, which will cometabolize the hazardous compounds. For example, munitions wastes which contained trinitrotoluene (TNT), hexahydro-1,3,5,7-trinitro-1,3,5-triazine (RDX) and octahydro-1,3,5,7-tetranitro-1,3,5,7-tetraazocine (HMX) were mixed with alfalfa, used animal bedding, horse feed and fertilizer. Comparison of mesophilic (35°C) with thermophilic (55°C) degradation rates demonstrated that TNT, RDX and HMX half-lives were reduced two-fold under thermophilic conditions [3].

The use of thermophiles has potential for application to a wide range of industrial scale processes. Thermophiles isolated from the high temperature composting of municipal solid waste have been mutagenized to generate degradative enzyme *hyperproducers* (Figure 5a) [5] and catabolite repression-resistant strains (Figure 5b) [6]. Whether comparable mutants can establish themselves and benefit the bioremediation process during the composting of other pollutants remains a challenge to be explored.

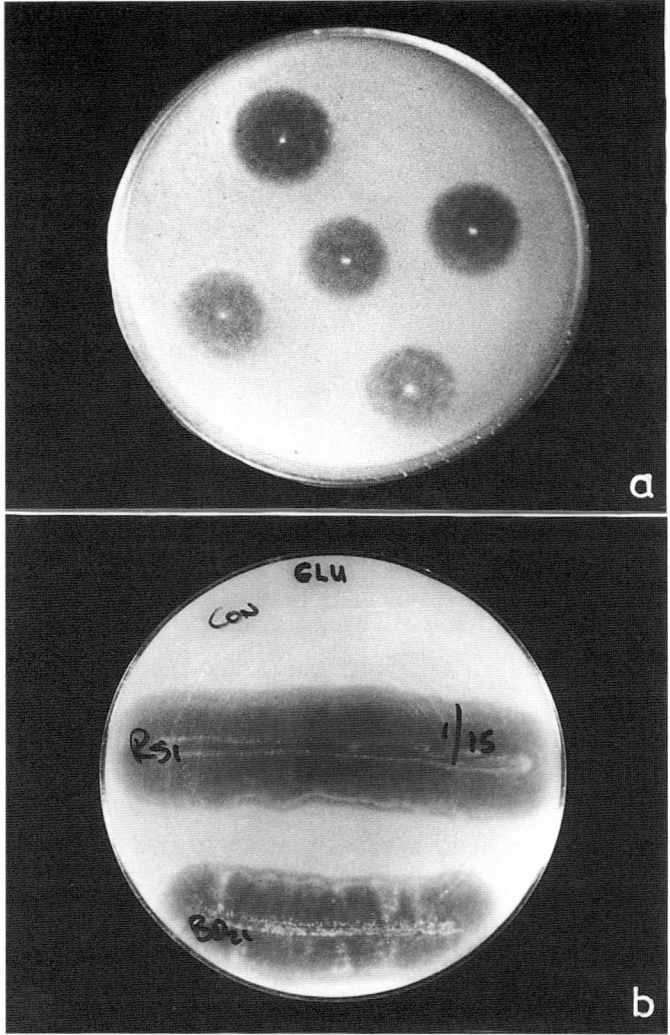

Figure 5. Mutagenesis of *Thermomonospora curvata* isolates from compost yielded cellulase hyperproducers (a) and catabolite repression resistant strains RS1 and BD21 which produced cellulase in the presence of 1.0% glucose, a condition which completely repressed cellulase biosynthesis in the wild-type control (b).

In situ biodegradation

In situ biodegradation is a highly attractive technology for remediation, because contaminants are destroyed in place, not simply moved to another location or immobilized [7]. This results in a decrease in remediation costs, lower contamination risks, shorter restoration time and increased efficiency, as well as enhanced public and regulatory acceptability. Public relations are enhanced because much of the action occurs with minimum above ground activity. Bakst [8] has determined *in situ* bioremediation to be amongst the least costly technologies where its application is feasible.

Demonstrations at the Savannah River Site (SRS), a 320 square mile facility owned by the U.S. Department of Energy and operated by Westinghouse Savannah River Company in South Carolina, have shown *in situ* bioremediation to be cost effective in soil and groundwater contaminated with chlorinated ethenes [9,10,11]. The SRS had been in operation since the early 1950's to produce nuclear materials for national defense, medical research, and space exploration. It contained five production nuclear reactors, one pilot scale reactor and all of the associated construction, fuel fabrication, processing, and waste handling operations. During the first 20 years most of this waste including millions of pounds of solvents was handled via burning rubble pits, evaporation ponds, and waste pits resulting in extensive soil and groundwater contamination.

The deployment of two projects at SRS designed for field demonstrations of *in situ* treatment of groundwater contaminated with chlorinated solvents by gaseous nutrient injection has proven the effectiveness of *in situ* bioremediation. Figure 6 shows a map of SRS and locations of the two project areas in A/M Area and the B-Area Sanitary Landfill.

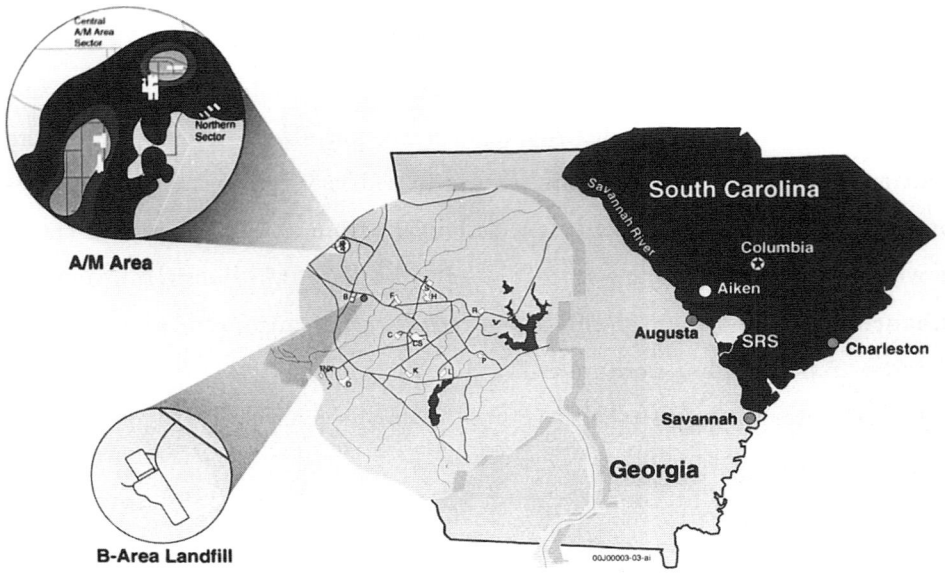

Figure 6. A map of the   Savannah River Site   and locations of two bioremediation projects in the A/M Area and the B-Area Sanitary Landfill.

The first demonstration was near the 300-M Area operations, where fuel and target elements were decreased in processing. An estimated 13 million pounds of solvents were used in this processing from 1952 to 1982. While evaporation was used to reduce much of the solvents an estimated 2 million pounds was released to the M Area Settling Basin [11]. These discharges to the M area settling basin consisted primarily of trichloroethylene (TCE), perchloroethylene (PCE), and trichloroethane (TCA).

The M area *in situ* bioremediation demonstration consisted of two horizontal wells for injection and extraction at the process sewer line leaking PCE and TCE [11]. Figure 7 illustrates a side view of the horizontal wells in relation to the surface nutrient injection and extraction systems. Groundwater, extracted air, and sediment samples were taken before, during and after testing of the system, that ran for over a year.

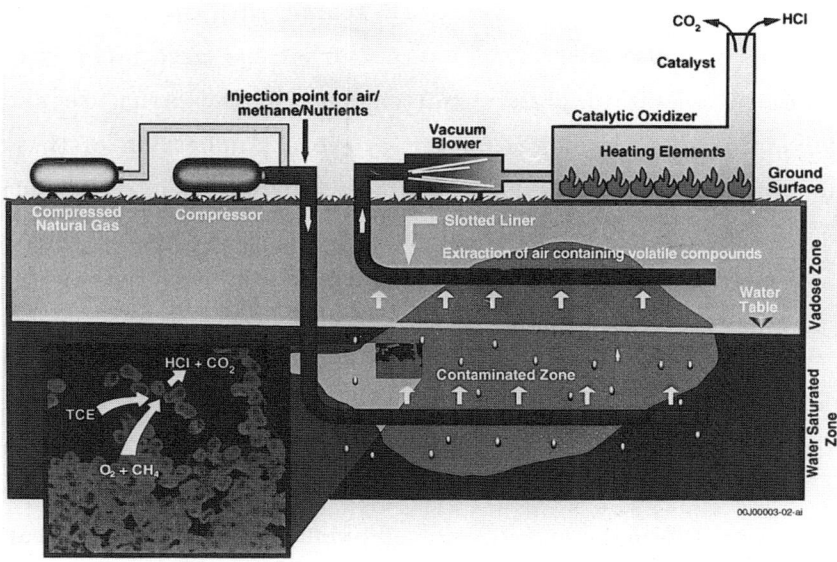

Figure 7. A side view of the horizontal wells in relation to the surface nutrient injection and extraction systems.

Subsurface gaseous nutrient injection, including methane, was found to be an effective *in situ* bioremediation treatment for TCE-contaminated groundwater at the M area site of SRS [11]. The oxidation of methane induces an enzyme, methane monooxygenase (MMO), which is responsible for cometabolizing TCE and its metabolites [12]. Molecular gene probe analysis demonstrated an increased MMO activity, corresponding with significant TCE degradation rates among a high proportion of the indigenous methanotrophic bacteria in the M area groundwater [13]. The biodegradation rate of TCE correlated to specific microbial population growth and associated growth factors, including injected methane and nitrogen [14]. Groundwater monitoring of bacteria and contaminant concentrations provided information on the efficiency of this technology [15]. Research with species-specific molecular probes has demonstrated that perturbations associated with *biostimulation* can result in preferential changes in the structure and physiological status of microbial communities [16]. Application of these methane/air mixtures in this case demonstrated a 3 to 5 order of magnitude increases in TCE-degrading methanotrophic bacteria during the methane- and nutrient-injection. These increases correlated with methane groundwater concentrations and the injection regime.

A second demonstration of *in situ* bioremediation at SRS is at the B-area Sanitary Landfill (SLF). The SLF began receiving solid waste from construction areas, offices, shops, and cafeterias in 1974. During the course of its operation, the Sanitary Landfill received numerous materials that can leach or generate hazardous compounds, e.g., paints, thinners, solvents, batteries, and rags as well as wipes used with organic solvents. Evidence of contaminants in the groundwater includes TCE, PCE, TCA, 1,2-dichloroethylene (DCE), vinyl chloride (VC), and chlorobenzene (CB).

Initially, the SLF bioremediation potential was evaluated using a soil column, which was employed to simulate both vadose and groundwater conditions using SLF sediment and groundwater [17]. An *in situ* optimization test demonstrated that biostimulation by addition of oxygen, nutrients, and methane at the SLF resulted in undetectable levels of contaminants and other organics in both the groundwater and vadose zones [18]. Groundwater samples were taken during nutrient gas injection at the SLF and analyzed for changes in bacterial populations and contaminant concentrations. The groundwater populations of *methanotrophic* bacteria were monitored using direct fluorescent antibodies (DFA) and microbiological techniques [19]. Methanotrophic populations increased significantly in groundwater during the course of gaseous nutrient injections. Figure 8 demonstrates methanotrophic bacteria labelled with DFA from concentrated SLF groundwater. The total number of groundwater microorganisms did not change, indicating a selective stimulation of the methanotrophic bacterial population. A larger scale (400 and 600 ft) set of horizontal nutrient injection wells at the SLF is now being tested for long-term containment of chlorinated ethenes. These demonstrations represent another approach in the development and field deployment of *in situ* bioremediation technologies.

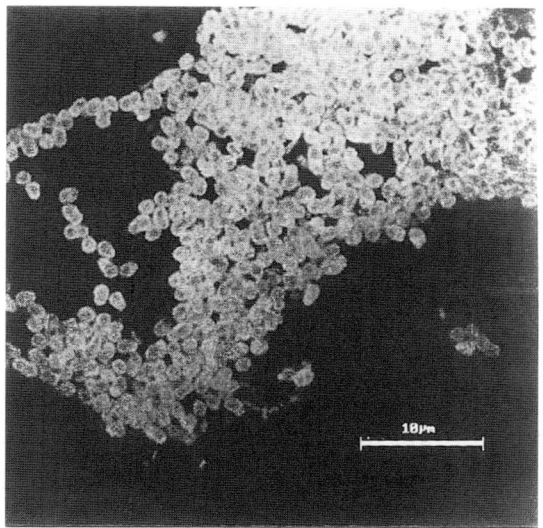

Figure 8. Concentrated groundwater methanotrophic bacteria labelled with fluorescent antibodies.

INTRINSIC BIOREMEDIATION

Intrinsic bioremediation is a risk management option that relies on natural biological, chemical, and physical processes to contain the spread of contamination from a source [20]. Containment includes sorption, destruction, or transformation of contaminants. Comparing rates of contaminant transport to rates of intrinsic bioremediation can quantitatively assess the efficiency of the option to prevent contaminant migration in groundwater systems [21]. If transport rates that are fast relative to rates of contamination removal, contaminants have the potential to reach points of contact with human or wildlife populations. Conversely, if transport rates are slower than the removal rates, contaminant migration will be more confined and less likely to reach a point of contact. Evaluating the factors, mitigating contaminant transport to predetermined points of contact, can assess the efficiency of intrinsic bioremediation. Thus, this assessment includes hydrological (rates of groundwater flow), microbiological (rates of biodegradation), and sociopolitical (points of contact) considerations [22].

The analysis of intrinsic bioremediation at SRS, including the SLF, is continuing to evaluate its effectiveness as a viable addition and/or alternative to engineered remediation. Intrinsic bioremediation, if proven effective in reducing the contaminant concentrations, will be incorporated as part of the *environmental restoration plan* at the SLF. The plan is being addressed in a two-fold manner. Innovative monitoring techniques have been developed to allow the measurement of subtle changes in the subsurface microbial populations and their physiology. Secondly, the targeted SLF has been equipped with key monitoring wells so that the extent and direction of the contaminant plumes can be defined.

Daughter products of TCE biodegradation, including DCE and VC, are present in the SLF groundwater, even though these compounds were never disposed at this site. The presence of these compounds is the evidence of TCE intrinsic bioremediation. Bouwer et al. [23] have reported if the amount of *cis*-DCE is greater than 80% of the total DCE, which is true in this case, it is a biodegradation product of TCE. Intrinsic bioremediation of TCE is associated with the accumulation of daughter products and increase in chloride ions [24]. Similarly, CB groundwater concentrations have significantly decreased in recent years at the SLF with an increase in chloride ions [25]. Chlorobenzene can undergo microbiological dechlorination and the benzene ring can be converted to catechol, followed by ring fission or oxidation of the side chain [26]. In addition, VC, an anaerobic microbiological breakdown product of TCE, is being increasingly detected in groundwater monitoring wells at the SLF. Geochemical data confirm that TCE and CB concentrations are decreasing greater than would be expected due to SLF groundwater transport or dilution. The use of intrinsic bioremediation appears to be an important consideration at this site as part of an environmental restoration program.

Bioreactors

A number of methods have been described where contaminants are treated in bioreactors. Some bioreactors range in complexity from older systems including activated-sludge treatment of wastewater containing contaminants, to the use of specific immobilized microbial cells for degradation of *xenobiotics*. There have been numerous studies on biodegradation of various contaminants using immobilized bacteria in bioreactors [27]. The source of the microorganisms used in bioreactors are often mixtures isolated from soil containing the contaminant of interest and activated-sludge [28,29]. In some cases augmentation of indigenous microbial communities with high-efficiency laboratory strains can enhance bioprocess effectively. This was found to be the case with pentachlorophenol (PCP), where augmentation of a sludge-inoculated bioreactor with *Desulfitobacterium frappieri,* a highly efficient anaerobic dechlorinator, significantly increased PCP removal [30].

Many of these reactors are designed to remove contaminants in waste streams before reaching surface receptors. In Augusta, GA a mobile bioreactor unit was used to treat groundwater pumped out of a contaminated aquifer [31]. A granular activated carbon-fluidized bed bioreactor (GAC-FBR) in operation at a former manufactured gas plant (MGP) consistently removed greater than 99.9% benzene, toluene, ethylbenzene, xylenes (BTEX) and naphthalene from contaminated groundwater at organic loading rates of 4.0 kg COD/m^3-d [31]. Nutrients, including phosphate and nitrogen mixtures, were fed into the reactor along with the groundwater and diffused air. The sole carbon source was the groundwater contaminants. The densities of petroleum-degrading bacteria were as high as 7.58E+07 colony forming units per gram dry weight GAC in the column. Scanning confocal laser microscopy (SCLM) was employed to examine the biofilm that developed on the GAC-FBR. By using SCLM, in conjunction with fluorescent staining techniques, the biofilm was found to be morphologically highly diverse throughout the column (Figure 9).

The biofilm structure near the top or effluent section of the GAC-FBR included large void spaces, large numbers of filamentous bacteria and protozoans. The biofilm composition near the bottom or influent section of the GAC-FBR was composed of primarily smaller bacterial rods. The application of these techniques will allow detailed analysis of biofilm structure and function with respect to contaminant biodegradation.

Biofilters

Biofiltration is an efficient, cost-effective, and safe technique for the destruction of low concentrations of volatile organic compounds and for the elimination of odorous gases from waste air stream mixtures. An increasing number of air purification systems are based on the ability of certain microbial populations to degrade a variety of organic and inorganic compounds. In a biofilter, the gas stream containing the contaminant of interest is forced to diffuse through a layer of biologically active material layered onto a solid support matrix. The microbial consortium forms a biofilm where the biodegradation activity can remove the contaminant. However, as biomass accumulates in the biofilter the available surface area for the contaminant to diffuse into the biofilm can decrease, causing a drop in removal efficiency [32]. Controls must be implemented to monitor the process.

The control of the biofilter efficiency and the influence of factors, including humidity and waste stream contaminant concentrations, such as ammonia, can influence the size and diversity of the microbial communities [33]. The biofilter support matrix can be composed of a number of materials, including poplar wood bark, compost [34], GAC [35], or acrylic materials [36]. Active compost filtration can be an excellent process for bioremediation of gaseous BTEX by biofiltration [34]. For co-oxidation of ethenes from industrial gases, it has been shown that biofilter nitrifying activity is associated with the production of microbial products that can act as co-substrates for heterotrophic bacteria [35].

Land-farming

Land treatment has been proven effective for oil spills and associated oil products in both carefully controlled laboratory experiments and in the field. Effective management of land treatment systems requires optimum design and operation in order to achieve the most efficient degradation of petroleum compounds with minimum environmental impact. Soil characterization and treatment with soil amendments including fertilizer, lime, appropriate moisture levels, and periodic tilling can maximize or biostimulate the bioremediation. After the Gulf War in Kuwait, there was widespread land contamination with oil and sludges. In the Kuwaiti Burgan Oil Fields, where the worst contamination occurred, 16 land-farming plots of 120 m² were constructed [37]. Over an 18-month period BTEX, PAH, and heavy metals were monitored every three months. The results showed over an 80% reduction in contaminants with land-farming after 15 months.

The bioremediation of soil contaminated with petroleum and also PAH compounds is often limited by the low bioavailability of the contaminants. *Bioavailablity* of contaminants is a greater problem in highly consolidated soils that contain a high percentage of clay. Surfactants have been used to enhance mobilization and biodegradation of contaminants in soil [38]. The surfactants act by contaminant transfer to the aqueous phase through emulsification, micellar solubilization, or facilitated transport. These surfactants, common to detergents and of glycolipids, may consist of fatty alcohol ethoxylates, which are homologous in structure to biosurfactants produced by hydrocarbon-degrading bacteria [39]. While the surfactants can increase bioremediation rates several fold, they can also be inhibitory to desorption or toxicity. Therefore, more work is needed to make the application of surfactants a routine method in land-farming.

Rhizosphere

The rhizosphere was described by Hiltner [40] as a zone of unique and dynamic interaction, which occurs between plant roots and soil microorganisms [41]. This specialized region (as illustrated elsewhere) is characterized by enhanced microbial biomass and activity, where the numbers of microorganisms in the rhizosphere are typically an order of magnitude higher than in bulk soils lacking plants.

In agricultural and natural soils inhabited by plants, microbial activity in the rhizosphere is known to affect the fate of added organic chemicals [42-47]. In addition to sustaining higher microbial densities as compared to those in nonvegetated soils, the plant root zone provides a niche which supports a diverse population of microorganisms and facilitates cometabolic transformations [42]. Evidence from degradation studies in a rhizosphere system suggests that a diverse and synergistic microbial community, rather than a single microorganism, is involved in the enhanced degradation of these xenobiotics [48].

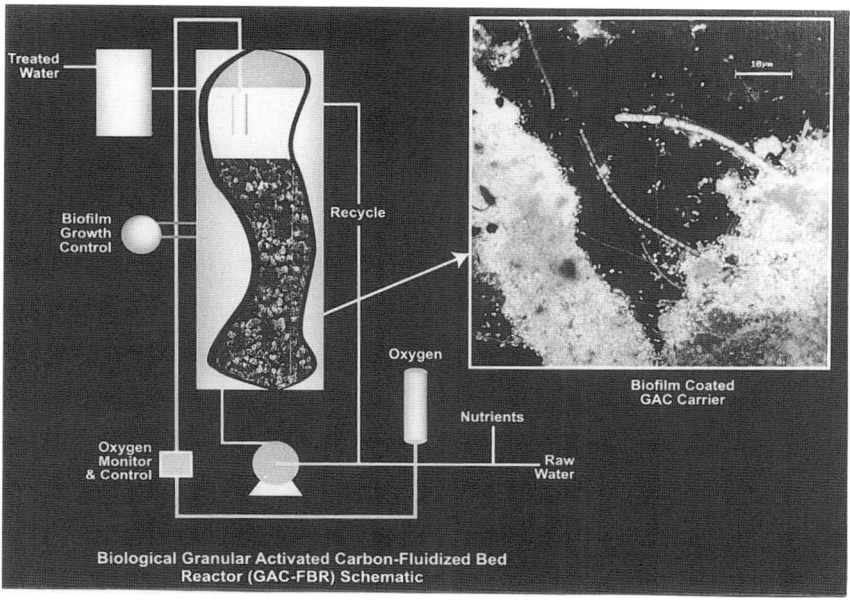

Figure 9. Schematic of a granular activated carbon-fluidized bed bioreactor (GAC-FBR) system with scanning confocal laser microscopy (SCLM) photograph of associated biofilm.

Biodegradation experiments have usually been conducted in a soil matrix in the absence of plants. However, biodegradation studies involving a plant/soil rhizosphere system where microbial activity, diversity and biomass are greater, offer tremendous untapped possibilities of enhanced microbial remediation of a wide variety of pesticides. A number of researchers have reported increased pesticide degradation in the rhizospheres of a variety of plant species [49]. Boyle and Shann [50] showed that, when relatively simple chemicals, such as phenol or 2,4-dichlorophenol, are placed into a rhizosphere or bulk soil system for study, there was no difference in degradation rates. However, when the chemical structure is more complex, as with 2,4-dichlorophenoxyacetic acid or 2,4,5-trichlorophenoxyacetic acid, the degradation rates were approximately 2-times greater in a rhizosphere soil system as compared to bulk soils. They also showed that monocot rhizosphere soils *mineralized* these compounds faster than dicot rhizosphere soils. Seibert et al. (cited in [49]) detected increased degradation of atrazine in the rhizosphere of corn, which was correlated with increased microbial biomass and activity as a function of added compound. Similar results were reported for atrazine and rhizosphere soils collected near the boundaries of a pesticide-contaminated site [51] and for parathion in rice rhizosphere soils [52]. The examples cited above, as well as others, serve to illustrate the influence of rhizosphere soils on the fate of pesticides.

Recent studies have suggested an enhanced degradation of other xenobiotics [44,45,53]. Rasolomanana and Blanandreau [54] observed enhanced degradation of oil by rhizosphere microbes and were able to isolate a *Bacillus* sp. capable of growing on oil residues when rice root exudates were added. A study of four PAHs (benzo[a]anthracene, chrysene, benzo[a]pyrene and dibenz[a,h]anthracene) in the rhizospheres of eight prairie grasses showed a consistently greater degradation than in unvegetated controls [44]. Ferro et al. [46] detected enhanced mineralization of pentachlorophenol in soils containing wheatgrass. Walton and Anderson [53] observed enhanced degradation of 1,1,2-trichloroethylene (TCE) in slurries of rhizosphere soils and mineralization of ^{14}C-TCE in rhizosphere soils; plants represented different varieties. Additional studies showed that a diverse and synergistic microbial community was responsible for TCE degradation [48]. These studies also raised the possibility of *ectomycorrhizal fungi* acting as degraders - enhanced mineralization of ^{14}C-TCE in soil containing loblolly pine (*Pinus taeda*) seedlings [53], degradation of certain congeners of polychlorinated biphenyls (PCBs) [55] and the degradation of a variety of organochlorine compounds including pentachlorophenol, endosulfan, and DDT by *Trichoderma harzianum* [56].

Sorption of a xenobiotic to soil particles or plant surfaces and uptake by the plant are factors that can influence the effect of rhizosphere microbes on degradation. A review of sorption, binding processes and subsequent desorption discusses the resultant effects on bioavailability for subsequent degradation [57]; a few examples will be included herein. Brigmon et al. [31] showed that sorption was the dominant factor in reduction of TCE residues in contaminated soils. Aerobic and anaerobic degradation contributed little to the decrease in TCE residues. Adsorption to the plant root cell wall may decrease xenobiotic availability to the remainder of the rhizosphere. Plant uptake would be a more permanent removal of the xenobiotic from the soil. The absorbed molecule would then be subjected to translocation, sequestration, conjugation or degradation. Thus, the plant can determine xenobiotic availability to rhizosphere microorganisms and subsequent degradation. Hydroponic studies with 2,4-D showed that over 80% of the added radiolabel was found in the plant. Reviews of xenobiotic uptake by plants have been presented by Ryan et al.[58] and Paterson [59]. This feature of plant influence on xenobiotic levels in soils is the basis of the use of plants in phytoremediation experiments, i.e. the use of plants to remediate contaminated sites.

Phytoremediation

Phytoremediation is a process whereby certain plants decontaminate the soils by removal, containment or degradation of pollutants involving physical, chemical and biological mechanisms. This approach is particularly suited to treat large areas of moderately contaminated soils. It can also be applied to contaminated groundwater or wastewater, a practice used for over 300 years [60]. In earlier studies, phytoremediation was applied to decontamination of dredged material slurries [61] and metal-contaminated soils [62]. Plant-based remediation has evolved as a viable method because of new technology and the need for more cost-effective approaches.

Plants remediate organic compounds by three mechanisms:

(i)   direct uptake and subsequent accumulation of the compounds;

(ii)   release of exudates and enzymes that stimulate microbial activity and biochemical transformations;

(iii)   enhancement of transformations in the rhizosphere (the root-soil interface).

A comprehensive review of the literature, pertaining to uptake, accumulation, translocation and biotransformation of organic chemicals by plants, was published by Nellessen and Fletcher [63]. A prerequisite for successful phytoremediation is that the contaminants must be non-toxic to the plant. Direct uptake of a chemical through the roots depends on uptake mechanisms as well as transpiration rates. Transpiration rates depend on plant type, leaf area, soil moisture, and environmental conditions. Other factors, which influence uptake, include the concentration of the chemical in the soil, chemical/physical properties of the chemical, and speciation of the chemical. After uptake, a plant can metabolically convert the parent compound into various products, which may be non-toxic. These products can be stored in plant cells or incorporated into plant constituents, i.e. lignin. The plant may also sequester the parent molecule in the root, thus immobilizing it. Some plants have developed mechanisms to absorb, translocate and conjugate or bind the molecules with plant constituents. Tolerant plants can stabilize the contaminated site against erosion.

Phytoremediation can be used in soils contaminated with heavy metals or various organic molecules. Metal-tolerant plants were used to control erosion on metal mine tailing and waste [64]. In another approach, soil amendments of lime, coarse limestone gravel, manure, and metal-tolerant plants were used in remediation of an upland disposal site. Plants can subsequently be harvested or grazed under controlled conditions. Plants that tolerate high concentrations of heavy metals can absorb, translocate and store the metals; the metals are then non-toxic to the plants. Tolerance mechanisms include metal accumulation in the cell walls, chelation with a plant constituent such as citric acid or malic acid, and binding to specific proteins, termed *phytochelatins*. Differential metabolic activity of different plants is the basis of herbicide selectivity [65] as well as the development of herbicide-resistant plants [66]. This capacity can also be utilized in phytoremediation in that the plant may be able to selectively metabolize the organic contaminant, thus rendering it non-toxic and reducing the potential environmental threat. These capacities are both constitutive as well as inducible [67]. Phytoremediation has much potential for future decontamination of soils and waters, especially when the pollutant is near the surface, relatively immobile in the soil or sediment, and poses little imminent health risk.

This chapter has been an overview of bioremediation processes, together with a few examples to illustrate the principles and efficacy of those processes. In the following chapters, the reader will find expanded descriptions of various bioremediation strategies and their potential for application in a world increasingly in need of the benefits they promise.

REFERENCES

1   Glazer AN, Nikaido H. Microbial Biotechnology. New York: W.H. Freeman & co, 1998; pp 583.

2   Stentiford EI, Dodds CM. In: Doelle HW, Mitchell DA, Rolz CE, eds. Solid State Cultivation, London: Elsevier Applied Science Publishers, 1992; 211-249.

3   Thomas JM, Ward CH, Raymond RL, Wilson JT, Loehr RC. Encyclopidia Microbiol 1992; 1: 369-385.

4   Stutzenberger FJ, Kaufman AJ, Lossin RD. Can J Microbiol 1970; 16: 553-560.

5   Fennington G, Lupo D, Stutzenberger F. 1982. Biotechnol Bioeng 1982; 24: 2487-2497.

6   Fennington G, Neubauer D, Stutzenberger F. Appl. Environ Microbiol 1984; 47: 201-204.

7   Baker K, Herson D. Geomicrobiol J 1991; 8: 133-146.

8   Bakst JS J Indust Microbiol 1991; 8: 13-22.

9   Altman DJ, Berry CJ, Bourquin A, Brigmon RL, Franck MM, Hazen TC, Mosteller D, Washburn FA. Sanitory Landfill Supplemental Test Final Report, WSRC-RP-97-17, Revision 1, April 1, 1998.

10   Brigmon R L, Altman DJ, Franck MM, Hazen TC, Bourquin AW, Fliermans CB. In: Engineered Approaches for In Situ Bioremediation of Chlorinated Solvent Contamination. Columbus: Batelle Press , 1999; 5(2): 107-112.

11   Hazen TC, Lombard KH, Looney BB, Enzien MV, Dougherty JM, Iermans CB, Wear J, Eddy-Dilek CA. In: Gee GW, Wing NR, eds. Proceedings Thirty-Third Hanford Symposium on Health and the Environment: In-Situ Remediation: Scientific Basis for Current and Future Technologies, Columbus: Battelle Press, 1994; 135-150.

12   Fliermans CB, Phelps TJ, Ringelberg D, Mikell AT, White DC. Appl Environ Microbiol 1988; 54: 1709-1714.

13   Bowman JP, Jimenez L, Rosario I, Hazen TC, Sayler GS. Appl Environ Microbiol 1993; 59: 2380-2387.

14   Travis RJ, Rosenberg ND. Environ Sci Technol 1997; 31: 3093-3102.

15   Fliermans CB, Dougherty JM, Franck MM, McKinsey PC, Hazen TC. In: Hinchee RE et al., eds. Applied Biotechnology for Site Remediation. Boca Raton: Lewis Publishers, 1994; 186-203.

16   Brockman   FJ,   Fredrickson JK,   Sun W,   Kieft TL.   In-Situ
        Remediation : Scientific Basis for Current and Future Technologies.
        1994.

17   Enzien MV, Picardal F, Hazen TC, Arnold RG, Fliermans CB. 1994.
        Appl   Environ   Microbiol 1994; 60: 2200-2204.

18   WSRC. Sanitary Landfill In Situ Bioremediation Optimization Test
        Final Report, Savannah River Site. WSRC-TR-96-0065. Westinghouse
        Savannah River Company. Aiken, SC, 1996.

19   Brigmon R L, Franck MM, Bray JS, Lanclos S, Scott D, Fliermans
        CB. J Microbiol Methods 1998; 32: 1-10.

20   Major   D,   Cox E,   Edwards E, Hare P. In: Hinchee RE, Wilson JT,
        Downey DC, eds. Intrinsic Bioremediation, Columbus: Battelle Press,
        1995; 197-204.

21   Rifai HS, Borden RC, Wilson JT, Ward CH. In: Hinchee RE, Wilson
        JT, Downey DC, eds. Intrinsic Bioremediation, Columbus: Battelle
        Press, 1995; 1-30.

22   Chapelle   FH.   1994.   Assessing the Efficiency of Intrinsic
        Bioremediation. In: EPA Symposium on Intrinsic Bioremediation of
        Ground Water, Denver, EPA 1994; 540/R- 94/515.

23   Bouwer E, Durant N, Wilson L, Zhang L, Cunningham WA. FEMS
        Microbiol Reviews 1994; 15: 307-315.

24   iedemeir TH, Swanson MA, Wilson JT, Kampbell DH, Miller RN,
        Hansen E.   In: Hinchee RE, Wilson JT, Downey DC, eds. Intrinsic
        Bioremediation, Columbus: Battelle Press, 1995; 31-52.

25   Brigmon R L, Fliermans CB. Intrinsic bioremediation of Landfills
        Interim Report EM-50 Landfill Stabilization Focus Area SR1-6-LF-52,
        Stabilization/Containment Systems Subtask B: Demo In Situ Intrinsic
        Remediation of Landfills . 1997; WSRC RP-97-323.

26   Gibson DT, Subramanian V. In: Gibson DT, ed. Microbial Degradation
        of Organic Compounds, New York: Marcel Dekker, 1988; 181-193.

27   Cassidy MB, Lee H, Trevors JT. J Indust Microbiol 1996; 16: 79-101.

28   Gisi D, Stucki G, Hanselmann KW. Appl Micribiol Biotechnol 1997; 48: 441-448.

29   Phelps TJ, Niedzielski JJ, Schram RM, Herbes SE, White DC. Appl Environ Microbiol 1990; 56: 1702-1709.

30   Tartakovsky B, Levesque M-J, Dumortier R, Beaudet R, Guiot R. Appl Environ Microbiol 1999; 65: 4357-4362.

31   Brigmon R L, Bell NC, Freedman DL, Berry CJ. J Soil Contam 1998; 7: 433-453.

32   Alonso C, Suidan MT, Kim BR, Kim BJ. 1998. Env Sci Tech 1998; 32: 3118-3123.

33   Malhautier L, Degrange V, Guay R, Degorce-Dumas J-R, Bardin R, Cloirec PL. J Appl Microbiol 1998; 85: 255-262.

34   Matteau Y, Ramsay B. Biodegradation 1997; 8: 135-141.

35   De Heyder B, Van Elst, Van Langenhove H, Verstraetraete W. Biodegradation 1997; 8: 21-30.

36   Elsgaard L. Appl Environ Microbiol 1998; 64: 4168-4173.

37   Al-Awadhi N, Al-Daher, ElNawawy A, Balba MT. J Soil Bioremediation 1996; 5: 243-260.

38   Tiehm A, Stieber M, Werner P, Frimmel FH. Environ Sci Technol 1997; 31: 2570-2576.

39   Madsen T, Kristensen P. Environ Toxicol Chem 1997; 16: 631-637.

40   Hiltner L. Arb DtschLandwirt Ges Berlin 1904; 98: 59-778.

41   Curl EA, Truelove B. The Rhizosphere. Berlin: Springer-Verlag, 1986; 288p.

42   Hsu TS, Bartha H. Appl Environ Microbiol 1979; 37: 36-41.

43   Sandmann ERIC, Loos MA Chemosphere 1984; 13: 1073-1084.

44   April W, Sims RC. Chemosphere 1990; 20: 1551-1557.

45   Knaebel DB, Vestal JR. Can J Microbiol 1992; 38: 643-653.

46 Ferro AM, Sims RC, Bugbee B. J Environ Qual 1994; 23: 272-279.

47 Walton BT, Guthrie EA, Hoylman AM. In: Anderson TA, Costs JR, eds. Bioremediation Through Rhizosphere Technology, ACS Symp No 563, 1994; 11-25.

48 Anderson TA, White DC, Walton BT. In: Singh VP, ed. Biotransformations: Microbial Degradation on Health-Risk Compounds. Amsterdam: Elsevier Science BV, 1995; 205-225.

49 Anderson TA, Guthrie EA, Walton BT. Environ Sci Technol 1993; 27: 2630-2636.

50 Boyle JJ, Shann JR. J Environ Qual 1995; 24: 782-785.

51 Mandelbaum RT, Wackett LP, Allen DL. Appl Environ Microbiol 1993; 59: 1695-1701.

52 Reddy BR, Sethunathan N. Appl Environ Microbiol 1983; 45: 826-829.

53 Walton BT, Anderson TA. Appl Environ Microbiol 1990; 56: 1012-1016.

54 Rasolomanana JL, Balandreau J. Rev Ecol Biol Sol 1987; 24: 443-457.

55 Donnelly PK, Fletcher JS. Abstr Soc Environ Toxicol Chem 1992; 103p.

56 Katayama A, Matsumura F. Environ Toxicol Chem 1993; 12: 1059-1065.

57 Novak JM, Jayachandran K, Moorman TB, Weber JB. In: Skipper HD, Turco RF, eds. Bioremediation: Science and Applications. Madison: SSSA Spec Publ 43. ASA, CSSA and SSSA, 1995; 13-31..

58 Ryan JA. Chemosphere 1988; 17: 2299-2323.

59 Paterson. Chemosphere 1990; 21: 297-331.

60 Hartman WJ Jr. An Evaluation of Land Treatment of Municipal Wastewater and Physical Siting of Facility Installations. US Dept of the Army, 1975.

61   Lee CR, Hoppel RE, Hunt PG, Carlson CA. Techn Rep D-76-4 (June). US Army Engineer Waterways Expt Stn, Vicksburg, MS, 1976.

62   Yamada K, Miyahara K, Kotoyori T. Gamm Ken Nogyo Shienjo Hororu 1975; 15: 39-54.

63   Nellessen JE, Fletcher JS. Environ Toxicol Chem 1993; 12: 2045-2052.

64   Bradshaw AD, Chadwick MJ. The Restoration of Land: The Geology and Reclamation of Derelict and Degraded Land. Berkeley: Univ California Press, 1980.

65   Hatway DE. Molecular Mechanism of Herbicide Selectivity. New York: Oxford Univ Press, 1989.

66   Hart SE, Penner D, Saunders JW. Rev Weed Sci 1994; 6: 251-263.

67   Hatzois KK, Hougland RE. Crop Safeners for Herbicides: Development, Uses and Mechanisms of Action, New York: Academic Press, 1989.

Biotransformations: Bioremediation Technology for Health and Environmental Protection
V.P. Singh and R.D. Stapleton, Jr. (Editors)
© 2002 Elsevier Science B.V. All rights reserved.

Microbial degradation of polychlorinated biphenyls (PCBs) in the environment

Wolf-Rainer Abraham

GBF - National Research Center for Biotechnology, Chemical Microbiology, Mascheroder Weg 1, 38124 Braunschweig, Germany

INTRODUCTION

Polychlorinated biphenyls (PCBs) are a class of compounds consisting of 209 different compounds. These compounds, called congeners, differ by the degree of chlorination and the position of the chlorine atoms at the aromatic rings. They are all poorly soluble in water and the water solubility decreases with increase in the number of chlorine atoms attached to the aromatic nucleus. The high and medium chlorinated congeners are almost insoluble in water and are, therefore, called *"superhydrophobic"*. The poor solubility in water and the reasonably good solubility in organic solvents and in fat are the main factors controlling the distribution of PCB in the environment.

PCB was first produced in the early 1930's for a wide variety of uses that ranged from being an extender in insecticide production to be used as an insulator in transformer production. PCB has a number of useful physical properties which can be amplified by mixing different PCB congeners. It is their unique physical properties that made them attractive compounds for industries. As more uses were found for PCB, their production increased exponentially from 1,000 tons in the 1930's to an estimated high of 200,000 tons in 1975 [1]. The first indication that PCBs may be damaging to human health occurred four decades after PCBs were first introduced into the environment. Preliminary studies suggested that PCB may pose a serious health threat to humans. As more attention was turned towards PCB, it became clearer that PCB was having a negative impact on many biological systems. The excellent physical properties of PCB and its extreme chemical stability were the reasons for its broad industrial applications. However, these properties make PCB a major environmental pollutant as well and allow its enrichment in food chains. As the extent of PCB contamination became more of a public issue, pressure was mounted to ban its production and import. Major producers voluntarily stopped production of open systems containing PCB (insecticides, coatings) but continued with the production of closed system PCB (transformers, capacitors) until government legislation banned all production and import of PCB. Since the mid 1980's the production of PCB was banned worldwide.

It is estimated that about 1.5 million tons of PCBs were produced worldwide [2], and a substantial fraction of it was finally released into the environment. Although some 15 years ago, the production of PCB was stopped worldwide, it is still in use in closed systems. PCB is now widely distributed in environmental samples over the Earth and found even in remote parts of this planet, e.g., in Antarctica or Northern Greenland [3]. In sediment layers of known chronology it can be detected since the 1930's, peaks around 1970 and declines since then [4]. Usually the PCB concentrations are low but due to its extreme hydrophobicity PCB is enriched in fat and concentrated over the food chain. The industrial production of PCB gave complex mixtures of congeners, which are usually grouped according to their content of chlorine atoms and had different trade names like Aroclor (USA), Clophen (Europe) or Kaneclor (Japan). Because of the production of PCB with many congeners (e.g., the PCB mixture Aroclor 1242 has a degree of chlorination of 42% and contains about 40 different congeners in different amounts), its chemical stability and its *bioenhancement*, PCB is an environmental pollutant which is both difficult to quantify and difficult to remediate. Here the main degradation process in the environment, i.e. the degradation by microorganisms, is being discussed.

PCB IN THE ENVIRONMENT

Fate of the PCB in the environment

In a freshwater environment 99.9% of PCB resides in the sediment, and less than 0.01% is dissolved in water and atmosphere combined [5]. Although the majority of PCB is found in lake sediment, there does not seem to be any significant movement of the PCB, once it has entered the sediment. The predominant modes of transport in freshwater systems are the water column, the atmosphere, physical displacement by bottom dwelling organisms, and the resultant movement up the food chain by the accumulation in the body fats. The low vapor pressure and low solubility give PCB a high rate of vapor transport [6]. Since most of the PCB is found in the sediment, the primary method of degradation is by microorganisms; it is however, a slow process due to the effects of both the repartitioning event and microbial availability.

The property of PCB, that makes movement through lake sediment very small, is also dominant in soil environments. In general, sorption increases with increasing chlorination, especially in the *ortho* position, and increasing organic content [7]. Sorption increases significantly with previous contamination, 3.5 times greater if contaminated by oils (based on octanol), and 67 times greater if contaminated previously by PCB [8]. This sorption makes the movement of PCB in soil extremely slow and, as a result, the atmosphere is the predominating avenue for movement. The degradation of PCB is limited by the same constraints as those acting in freshwater systems. Photolysis and chemical degradation are unlikely, and microbial degradation is the only important sink for PCB. In an unsaturated soil environment the transfer between aerobic and anaerobic environments may be easier due to typical micro-habitats around soil particles.

PCB-degrading environmental isolates

Ahmed and Focht reported the first evidence for biodegradation of PCB and isolated two *Achromobacter* strains, growing on biphenyl and 4-chlorobiphenyl [9]. These strains converted several PCB congeners but they did not dechlorinate them.

Baxter et al. [10] measured the PCB degradation by a species of *Nocardia* and a species of *Pseudomonas* and found after 100 days a degradation of 95% of Aroclor 1242 for the *Nocardia* species and 85% for the *Pseudomonas* species. The patterns of the PCB degradation of these two strains were different and up to hexachlorobiphenyls could be degraded to some extent. They observed that the degradation of the congeners depended on the mixture of the PCB. This cometabolism was especially pronounced, when biphenyl was present in the mixture.

Furukawa and Matsumata [11] isolated *Alcaligenes* sp. Y42 from a lake sediment by enrichment with biphenyl. They found that the degradation of PCB ended in chlorobenzoates and that congeners with chlorines in only one ring were degraded more easily than those with chlorines in both the rings. Lower chlorinated congeners were degraded very fast.

Another strain was assigned in the beginning as *Acinetobacter* sp. P6 [12], but later taxonomic studies identified it as the Gram-positive *Corynebacterium* sp. MB1 and finally as *Rhodococcus globerulus* P6 [13]. This strain has been shown to be one of the rather small groups of bacteria, which can degrade highly chlorinated PCB congeners [14]. One explanation for this broad spectrum may be that three different 2,3-dihydroxybiphenyl-1,2-dioxygenases have been found in *Rhodococcus globerulus* P6, presenting a new family of extradiol dioxygenases (see section on degradation pathways).

Techniques for screening isolates were elaborated [15], and in the early 1980's more than 30 isolates were reported to degrade PCB congeneres. The list even contains a *Beijerinckia* sp. strain B1, a genus not very often found in the list of *xenobiotic degraders* [16]. These results clearly demonstrated that bacteria are able to degrade less chlorinated PCB congeners when presented either as single compounds [17] or in PCB mixtures like Aroclor 1221 or 1242 [18]. Throughout all of these studies, it was generally observed that PCB congeners with more than three chlorine atoms were resistant to microbial degradation or only slowly attacked. Testing the PCB degrader *Burkholderia* sp. TSN101, it was found that the degradation pattern of PCB congeners changed with the concentration of PCB in the medium [19].

Bedard and coworkers were looking for bacteria, which were able to degrade PCB congeners with a 2,5-disubstituted phenyl ring. They ended up with the isolation of *Alcaligenes eutrophus* H850 [20]. Extensive characterization of this strain revealed that it could attack congeners, that have no unchlorinated 2,3-sites and could hydroxylate at positions 3 and 4. It produced, in addition to chlorobenzoates, a novel metabolite, 2',3'-dichloroacetophenone [21]. The strain was also unique in its breadth of congener specificity and was able to attack tetra- and pentachlorobiphenyls and several hexachlorobiphenyls. As a result, it is very effective in degrading Arochlor mixtures. The production of chloroacetophenons was more common than just a metabolite of PCB from *Alcaligenes eutrophus* H850 [22]. The analysis of these strains and the degradation pattern reported for other strains led to the grouping of the PCB-degrading bacteria into four classes. These classes were based on their mode of attack and preference for chlorine substitution pattern [23]. As more strains were characterized [24], it now appears that the strains known today exceed these four classes and may represent a continuum of genetic and biochemical variability [25].

At the General Electric Laboratories, a second exceptional strain was isolated and designated as *Pseudomonas* sp. LB400, now *Burkholderia* sp.

LB400 [26]. In terms of breadth of congener specificity and in biochemistry, it is similar to *Alcaligenes* sp. H850. The biphenyl dioxygenase of *Burkholderia* sp. LB400 attacks several di- to tetrachlorobiphenyls with a complex dependence of the regioselectivity and the yield of dioxygenation on the substitution patterns of both the oxidized and the non-oxidized rings. The ability of the enzyme to hydroxylate chlorinated carbons appears to be limited to the *ortho* position and no dioxygenolytic attack, involving chlorinated *meta* or *para* carbons, was found. However, it is not limited to monochlorinated rings, as a 2,4-disubstituted ring was also attacked. The formation of several dioxygenation products was found predominantly with congeners that contain two chlorinated rings, both of which are similarly sensitive to dioxygenation or one is substituted only at carbon 3 [27]. Only two other strains, *Pseudomonas* sp. ENV 307 [28] and *Pseudomonas* sp. ENV 391 [29], have been reported to possess a similar congener degradation spectrum.

MICROBIAL DEGRADATION PATHWAY OF PCB

The metabolism of biphenyl and lower chlorinated PCB congeners proceeds according to a general scheme of degradation of aromatic compounds (Figure 1). The aromatic system is initially activated by the oxidation of a double bond and the introduction of two vicinal hydroxy groups. The resulting diol is further oxidized by a dehydrogenase to the corresponding catechol. Cleavage of the aromatic ring occurs after this activation. By hydrolysis of the ring-cleavage product, benzoate is produced which, in turn, is activated again by the introduction of two hydroxy groups. The other product is an aliphatic carboxylic acid.

The degradation of PCB congeners can be grouped into two steps. One is the attack of the biphenyl ring system and the degradation of one of the rings. This step is called the upper pathway. The second step, the lower pathway, is the degradation of the second ring which usually starts from

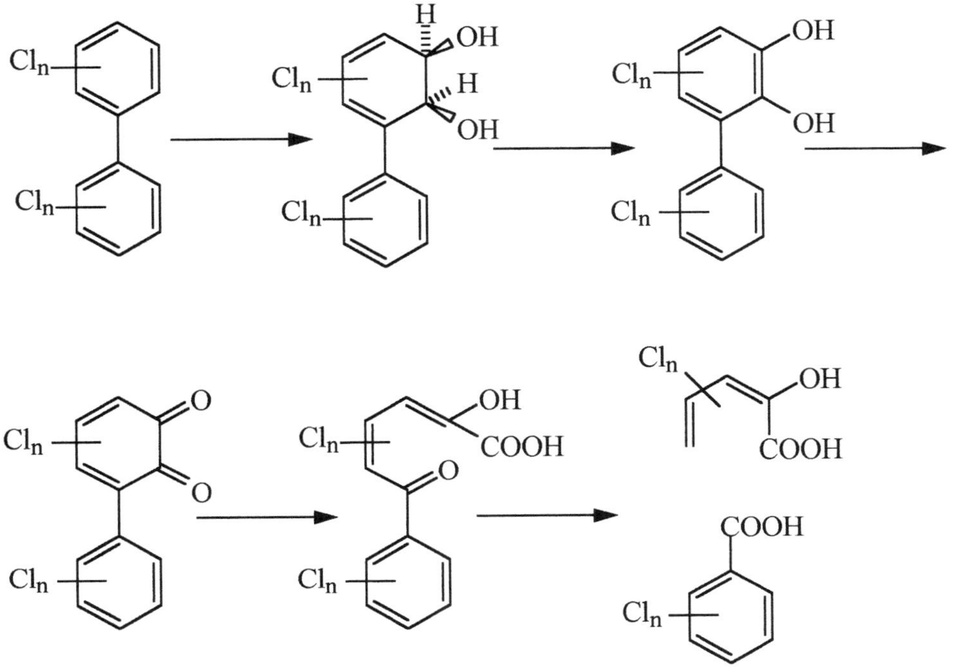

Figure 1. Upper degradation pathway of (chlorinated) biphenyls. The biphenyl ring system is oxidized by a dioxygenase to a *cis*-diol, then dehydrogenated, oxidized to an *ortho*-quinon and then cleaved. The product is finally hydrolyzed.

chlorobenzoates formed after the first step. The reason, why the degradation of PCB congeners was divided into these two steps, came from the metabolic capabilities of the bacteria. Most isolates can carry out only one of the two steps with a fairly broad range of substrates.

As a continuation of the comprehensive and pioneering work on characterization of PCB-degrading microbial strains in the mid 1980's, researchers began to focus on the genes responsible for PCB degradation. The first three genes of a gene cluster in the biphenyl pathway from *Pseudomonas pseudoalcaligenes* KF707 were cloned by Furukawa and

Miyazaki [30]. It contained *bphA* (encoding biphenyl dioxygenase), *bphB* (encoding 2,3-dihydrodiol dehydrogenase), and *bphC* (encoding 2,3-dihydroxybiphenyl dioxygenase). Only the fourth gene, *bphD* (encoding the hydrolase), was missing from the fragment [30]. In the next year the first sequence of a PCB-degradative gene, that being the *bphC* gene of KF707 encoding the 2,3-dihydroxybiphenyl dioxygenase [31], and finally the purification and characterization of the enzyme itself was described by Furukawa and Arimura [32]. The next step was the cloning and sequencing of another *bphC* gene, from *Sphingomonas paucimobilis* Q1, which allowed the comparison of these two genes. These showed only moderate homology to each other and were the first indication of the genetic variability of PCB-degrading strains that exist in nature. Indeed, polyclonal antibodies raised against the protein from *S. paucimobilis* strain Q1, failed to cross react with the previously isolated 2,3-dihydrobiphenyl dioxygenase from *Pseudomonas pseudoalcaligenes* KF 707 [33]. The *bphABCD* genes from *Pseudomonas putida* KF 715 were also cloned and characterized, extending the *genetic library* of these genes further [34]. Experiments with the 2-hydroxy-6-oxo-(phenyl/ chlorophenyl)hexa-2,4-dienoic acid hydrolase, the product of the *bphD* gene of *Comamonas testosteroni*, suggested that an amino acid Ser^{112} is involved in the catalytic activity [35].

To understand what the breadth in sequence variations is, Furukawa and coworkers compared fifteen PCB-degrading strains with gene probes. They used probes derived from *Pseudomonas pseudoalcaligenes* KF707 and *Sphingomonas paucimobilis* Q1. The KF707 probe demonstrated almost identical homology between seven strains, three other strains were homologous but not identical, and five other strains showed no significant *genetic homology* with KF707. However, the Q1 probe lacked genetic homology with any of the other 15 biphenyl degraders tested. Also unique in this respect was *Arthrobacter* sp. M5 (now *Rhodococcus globerulus* P6), which showed no homology to any other strains as well. From this study at least three different types of PCB-degradative genes could be identified, which were in good agreement with the degradation pattern for PCB congeners oberseved with these strains [36].

Both *Burkholderia* sp. LB400 and *Alcaligenes* sp. H850 have a unique PCB-degrading competence. The cloning and expression of the genes from the upper pathway from LB400 made these genes accessible for further studies. Mondello achieved here the first isolation of a whole gene set for upper pathway metabolism of PCB. He cloned them in *Escherichia coli* and got a recombinant strain, which displayed the same exceptional characteristic congener specificity of LB400, but did not require growth on biphenyl for full induction of the PCB pathway [37]. A probe derived from the LB400 genes was used to compare homology of this strain with seven other PCB degraders and only demonstrate significant homology to *Alcaligenes eutrophus* H850 [38]. This finding was consistent with the biochemical and congener specificity, as described previously.

The comparison of the sequences from LB400 and KF707 shed some light on the connection between gene sequences and congener specificity. From the published data on DNA sequences, it could be shown that despite the very significant differences in congener specificity, there were only slight differences in nucleotide sequences. As a consequence, it was postulated that single amino acid substitutions at two positions in the KF707 sequence could play the determining role in substrate specificity for each enzyme [39]. A consequence of this study is an attempt to alter the sequences by *site-directed mutagenesis*. The exchange of four amino acids in the *bphA* gene product of LB400 brought a broader congener specificity with increased activity against several congeners [40]. This work underlined the extremely subtle differences within the initial biphenyl dioxygenase. It also nurses hopes that protein engineering could expand the range of congeners degradable by a single strain leading finally to even more effective biocatalysts for bioremediation of PCB-contaminated sites [41].

Timmis and his group achieved the cloning of the seven structural genes involved in PCB degradation in LB400 and the complete DNA sequences for *bphB*, *bphC* and *bphD*. Together with the sequence information for *bphA* [42], they produced the first DNA sequence for all genes encoding the metabolism from biphenyl to benzoate, the so called upper pathway [43]. In addition the *bphB*, *bphC*, and *bphD* genes from *Rhodococcus globerulus* P6 were cloned and analyzed as well in order to have genes from complementary PCB degraders in hand. The outstanding position of this isolate was proven by the presence of three *bphC* genes with narrow substrate specificity, as well as the fact that none of these genes showed hybridization with any other *bphC* gene from other previously characterized strains [44]. The *Rhodococcus* genes form a diverse family of *bphC* genes which makes them an interesting target for the construction of hybrid strains [45].

Pellizari and colleagues compared isolates from primarily tropical soils obtained by growth on biphenyl and tested them using a *bphA* probe. The hybridization patterns did not correlate well with the substrates of isolation, suggesting that there is considerable diversity in these genes in nature and that *probe hybridization* is not a reliable indication of catabolic capacity. Although the strains with most extensive PCB degradation capacity strongly hybridize with the gene probe, a few strains that exhibit strong hybridization had poor PCB congeners degrading ability. Most of the isolates identified were members of the *beta* subgroup of the Proteobacteria and none were true *Pseudomonas* species [46].

Environmental isolates degrading metabolites from the upper pathway

The upper pathway of PCB degradation ends in the formation of chlorobenzoates and chlorinated acyclic carboxylic acids. Because most PCB degraders do not possess genes for both the upper and the lower pathway, it was of some concern to identify bacteria which are able to mineralize the metabolites of the upper pathway in order to achieve the goal of complete *mineralization* of PCB congeners.

As model substances for the chlorinated aliphatic C_5-acid produced during the last upper pathway step of PCB biodegradation chlorobutyric and chlorocrotonic acids were used as substrates and the biodegradation of these compounds could be shown. *Alcaligenes* sp. CC1 was able to grow on several α- and β-chlorinated aliphatic C_4-acids [47]. This proved that these chlorinated acids can be substrates for a specialized class of bacteria.

The bacteria, like *Alcaligenes* sp. H850 or *Burkholderia* sp. LB400 can attack a broad range of congeners but their end products are not hydrochloric acid and carbondioxide as required in a complete mineralization except chlorobenzoates. Chlorobenzoates are less toxic than PCB, but they are more soluble in water and, therefore, more mobile in the environment. These chlorobenzoates are released into the medium, where they can be degraded by other specialized bacteria. One highly efficient degrader of the chlorobenzoates is *Pseudomonas* sp. B13, which became one of the prime targets for the construction of genetically engineered PCB degraders [48]. *Pseudomonas* sp. MB86 was isolated from a 4-chlorobenzoate enrichment and was able to grow on 4-chlorobenzoate and 4-chlorobiphenyl as sole carbon and energy sources [49]. *Burkholderia cepacia* P166 can mineralize 4-chlorobiphenyl and is another example of this rare class of bacteria having the genes for both the upper and the lower pathways [50]. Recently, another strain was described belonging to this group. It was designated SK-3 but no clear taxonomical identification was given. These bacteria can grow on 2-, 3- and 4-chlorobiphenyl as well as on 4-chlorobenzoate and chloroacetate. Benzoate grown cells of SK-3 were able to degrade several PCB congeners without the need for a primary substrate or previous growth on biphenyl [51].

Anaerobic dechlorination of highly chlorinated PCB congeners

Sediments in rivers and lakes usually do not contain oxygen beginning from a few centimeters below the surface. It was long assumed that PCB in this anoxic environment is not changed by microorganisms. In 1984 the first report on the change of the congener pattern of PCB in river sediments was published [52]. The authors explained these changes by microbial activity in the sediment. Several attempts were made to explain these changes by non-biological processes; but as more and more reports on the shifts in congener pattern of PCB, both from laboratory experiments and from different environmental samples were published, biological processes have been finally accepted [53]. The observed reductive dechlorination patterns in the Hudson River sediment demonstrated that this process favored the reductive dehalogenation of *meta-* and *para*-chlorines and resulted in the accumulation of *ortho*-substituted PCB congeners. A similar observation of an altered PCB pattern was observed from a site in Pittsfield, Massachusetts, which was contaminated with Aroclor 1260. However, Wu and coworkers reported microbial reductive dechlorination of Aroclor 1260 in anaerobic slurries of estuarine sediments, which resulted in *ortho* dechlorinations. From 2,3,4,5- and 2,3,5,6-tetrachlorobiphenyl after 6 months the *meta* chlorines decreased by 65 and 55% and *ortho* chlorines by 18 and 12% [54].

Experiments in microcosm experiments with river sediments confirmed that the microbial activity was responsible for dechlorination of PCB. The sediment samples were divided and one part was autoclaved to kill all microorganisms and to be used as a reference sample. After PCB was added to both, the other part was incubated for 16 weeks under anaerobic conditions and the PCB congeners analyzed [55]. A large increase in lower chlorinated congeners and a corresponding decrease in higher chlorinated congeners were found. As seen in the Hudson River samples, a clear regiospecificity in the dechlorination pattern was observed. This experiment was repeated with different samples with different PCB mixtures. The differences in dechlorination patterns suggested that different organisms were responsible for the dechlorination at the different sites and that each had its characteristic PCB dechlorination specificity [56]. The use of 2,3,4,5,6-pentachlorobiphenyl was described as an enrichment substrate for PCB-dechlorinating bacteria and applied to Housatonic River sediments [57]. The excreted main metabolites of PCB are *para*-hydroxylated PCB congeners. These congeners can be *ortho*-dechlorinated anaerobically by bacteria, one of them is *Desulfitobacterium dehalogenans* [58]. An excellent and very comprehensive overview over the microbial anaerobic dechlorination of PCB was published in 1995 [59].

Using molecular methods, based on 16S rRNA gene sequences, it was possible to identify bacteria responsible for the dechlorination of PCB. A *Clostridium* sp. was identified in anaerobic communities dechlorinating PCB congeners [60]. Interestingly, a polymerase chain reaction product using primers for the 16S rRNA genes was obtained with archaebacterium primers, suggesting that archaebacteria were involved along with *Clostridium* spp. [61]. Some of these bacteria, including *Clostridium* strains, were recently identified. Nine *Clostridium* species were identified based on their ability to dechlorinate *meta*- and *para*-PCB contaminated sediments. The phylogenetic relatedness of these PCB-degrading *Clostridium* species was studied, using ribosomal RNA genes. Two clones form a phylogenetically related cluster, closely affiliated with *Clostridium hydroxybenzoicum* strains. Six other clones also belong to the genus *Clostridium*, but they represent separate species. Furthermore, a close affiliate to *Bacteroides forsynthus*, a *meta*-PCB dechlorinator, was found. The *Clostridium hydroxybenzoicum* strains are primarily *para*-PCB dechlorinators and are the most common. Some less prevalent strains are also mostly *para*-PCB dechlorinators. Other *Clostridium* species, such as *Cl. beijerinckii, Cl. intestinalis* and *Cl. Thermolacticum,* are primarily *meta*-PCB dechlorinators. *Cl. paraputrificum* and *Cl. cellulosi* were less prevalent in the total consortium, but they could dechlorinate both *para*- and *meta*-PCBs. Although a few less prevalent *Clostridium* species can degrade both *para*- and *meta*-PCBs, this study confirmed that *para*- and *meta*-PCB dechlorinating species are generally phylogenetically different [62]. The composition of other PCB-dechlorinating mixed communities was analyzed by restriction fragment length polymorphism (RFLP) of PCR amplified rDNAs (ARDRA) and partial sequencing of 16S rRNA genes amplified from PCB-degrading enrichments. Restriction analysis confirms that the 16S rRNA genes amplified from PCB-dechlorinating communities vary, depending on the PCB congener dechlorinated. Comparison of 16S rRNA sequences to published ones from *ribosomal databases* indicates that the two most abundant species appear to be the species of the genus *Clostridium* [63].

The predominant role of the genus *Clostridium* in the dechlorination of PCB, however, is obviously not invariant. Holoman and coworkers were able to discern between the different groups of bacteria present in an anaerobic PCB-dehalogenating community by combining selective enrichment with molecular monitoring of total community genes coding for 16S rRNA. In enrichment cultures, undescribed species in the *delta* subgroup of the class Proteobacteria, the *Bacillus-Clostridium* subgroup, the Thermotogales subgroup, and a single species with sequence similarity to the deeply branching species *Dehalococcoides ethenogenes* were predominant during active dechlorination of the PCB as well as species with high sequence similarities to Methanomicrobiales and Methanosarcinales archaeal subgroups. Although PCB-dechlorinating cultures were methanogenic, inhibition of methanogenesis and elimination of the archaeal community by addition of bromoethanesulfonic acid only slightly inhibited dechlorination, indicating that the archaea were not required for *ortho* dechlorination of the congener. Deletion of *Clostridium* spp. from the community profile by addition of vancomycin only slightly reduced dechlorination. However, addition of sodium molybdate, an inhibitor of sulfate reduction, inhibited dechlorination and deleted selected species from the community profiles of the class Bacteria. With the exception of one 16S rDNA sequence that had the highest sequence similarity to the perchloroethylene-dechlorinating *Dehalococcoides*, the 16S rDNA sequences associated with *ortho*-PCB dechlorination had high sequence similarities to the *delta*, *Bacillus-Clostridium* and Thermotogales subgroups, which all include sulfur-, sulfate- or Fe^{3+}-respiring bacterial species [64].

FROM THE BENCH TO *IN SITU* BIOREMEDIATION OF PCB-POLLUTED SITES

Although it was a major achievement to show that bacteria can degrade PCB congeners, it was only the first step on a long way to *in situ* bioremediation of PCB pollution in the environment. The next step was the isolation of several degraders and the elucidation of the degradation pathways. A consequence of this were the efforts to isolate, sequence and characterize the genes involved in PCB degradation and their gene products. Here the huge biodiversity of the bacteria taught us that we were only seeing the tip of the iceberg, concerning the variation in gene sequences and, as a consequence, the variation in protein structure. From here the construction of genetically improved bacteria were started and brought many impressive results. But how must an ideal PCB degrader look like? What is required to make it superior to the other bacteria in the environment, which are also able to attack PCB?

Problems of microbial degradation of PCB in the environment

In general, microorganisms in soil and water are degrading organic materials and are, therefore, responsible for a functioning carbon cycle as well as for the cleaning of the environment. A number of microorganisms are also able to degrade industrial chemicals. Biodegradation of hazardous chloroaromatics can be considered complete only when the chlorosubstituents have been removed and the carbon skeleton has been oxidized to carbondioxide (mineralization). Whereas microorganisms mineralizing aromatic compounds are widespread, those mineralizing chlorosubstituted analogues (xenobiotics) are rare. As reported above in recent years, a large variety of microorganisms capable of mineralizing some chloroaromatics, such as chlorobenzoates or chlorobenzenes, have been isolated. These organisms can be differentiated on the basis of how the substituents are eliminated. Various specialized enzymes have been described, which remove the chlorosubstituent during an early reaction, i.e. before ring-cleavage, and thereby reduce the xenobiotic character [65].

The majority of organisms, able to mineralize chloroaromatics, use a different degradation strategy. The chlorosubstituents remain on the aromatic ring and degradation involves the formation of chlorocatechols. This is analogous to the degradation of naturally occuring aromatics which, in a variety of cases, were degraded via catechols as central intermediates. The ring-cleavage of catechols occurs either between (intradiol-cleavage, 3-oxoadipate pathway), or adjacent (extradiol-cleavage) to the hydroxy groups. In the case of chlorocatechols, the further metabolism is performed by a sequence of specialized enzymes (chlorocatechol pathway), resembling the reactions of the classical 3-oxoadipate pathway. The elimination of the chlorosubstituent occurs after ring-cleavage [66].

The capabilities of microorganisms to degrade chemicals have been used for decades for efficient purification of domestic and industrial wastes and, increasingly, for the purification and bioremediation of contaminated air, soils, and aquifers. Bioremediation can be regarded as a promising and clean biotechnological approach for the enhancement of environmental quality. Whereas bioremediation has been used successfully for the treatment of sites contaminated with easily degradable pollutants of low toxicity, with highly stable and toxic compounds, this procedure faces a number of problems at the level of individual strains.

The first problem is the target compound itself. Despite the toxicities and stabilities of various pollutants such as chlorobenzenes, chlorophenols or PCB, a variety of microorganisms, able to mineralize compounds previously assumed to be non-biodegradable, have been isolated in recent years [67]. These bacteria, however, usually possess only incomplete pathways and a number of attempts had been untertaken to improve them by genetic engineering [68]. Usually, the resulting genetically optimized strains possess complete pathways by transfer of *bph* genes into chlorobenzoate degraders [69]. It can be assumed therefore, that, either by new isolates, cloning of genes from unculturable bacteria or by the *genetic optimization* of available microorganisms, strains degrading major parts of target pollutants can be obtained.

A second problem, based on the properties of the pollutants, is their bioavailability. Biological reactions occur in the aqueous phase or at its interface to the proteins of the bacterial cell. A variety of pollutants, however, are hydrophobic, poorly water soluble and, therefore, not bioavailable. It has now been shown that the degradation rate of hydrophobic pollutants by microorganisms could be significantly enhanced through genetic design. These microorganisms are not only able to degrade the target compounds but, simultaneously, produce surface active compounds, which enhance the bioavailability of hydrophobic substances [70]. Only recently, the role of clay in the transportation of PCB to the microbial cell was discovered. This complete novel mechanism will be discussed later.

A third problem is the incomplete transformation of target compounds into *dead-end* or even toxic products. The formation of protoannemonin, semialdehydes or acylchlorides are some of these poisons [71]. This problem has been overcome by the construction of genetically optimized strains, which possess pathways where no toxic end products are formed [72]. An alternative to the avoidance of toxic metabolites by the individual cells is the sharing of substrates in microbial consortia, which will be described in detail for a 4-chlorosalicylate-degrading consortium.

Finally, a fourth problem is the survival and the functioning of the PCB degrader in the environment. Here the strain has to deal with physical and chemical conditions, which are not usually used in the experiments in the laboratory, like low temperature, low pH, high concentrations of heavy metals or other pollutants not related to PCB. This is the field of microbial ecology, which deals with the interaction of bacteria with each other and their environment. Much has been learned from the study of these interactions and the identification of different mechanisms developed by bacteria to protect themselves against heavy metals, grazers or poisonous intermediates [73].

MICROBIAL COMMUNITY INTRACTIONS: ECOLOGICAL ASPECTS

Carbon sharing in a 4-chlorosalicylate-degrading microbial consortium

Contrary to the situation in the laboratories, the mineralization of pollutants in habitats is often achieved not by a sole bacterial strain but by bacterial consortia. Here the different bacteria share substrates and metabolites and, therefore, can protect the consortia from toxic intermediates. Such a sharing of substrates by bacteria was elucidated in a stable community derived from contaminated sediment. It was able to use 4-chlorosalicylate as sole source of carbon and energy. The consortium was grown in a chemostat and fed with 5 mM 4-chlorosalicylate as the sole source of carbon and energy, with a dilution rate of 0.16 d^{-1} and at 12°C [74]. It consists of four different strains, belonging to the genera *Pseudomonas* (MT1 and MT4), *Empedobacter*

(MT2) and *Alcaligenes* (MT3), that were identified by 16S rDNA sequence analysis [75].

To understand the interactions of bacteria within the consortium, it was necessary not only to identify different bacteria but also to determine their relative abundances in the consortium and to monitor the composition of the microbial consortium over a period of time. To achieve this goal, rabbit antisera have been produced against all four isolates for immunochemical analysis of the microbial population and have been tested by different immunological methods. Applying these antibodies in the immunofluorescence, the relative abundances of the strains in the chemostat were determined to be 84 ± 3% for *Pseudomonas* sp. MT1, 1% for *Empedobacter* sp. MT2, 8 ± 4% for *Alcaligenes* sp. MT3, and 8 ± 4% for *Pseudomonas* sp. MT4. Furthermore, it was found that the composition of the mixed culture remained constant over a time period of at least 3 years.

The question remained, why is this microbial community so stable and why is none of the strains replaced by a strain already present in the consortium? The stability of the consortium with regard both to its composition and to the relative abundances of the four strains pointed to a sharing of different tasks within the community. The predominant strain in this consortium, i.e. *Pseudomonas* sp. MT1, grew on 4-chlorosalicylate alone tolerating up to 1 mM, above which cell death occurred. Incubating *Pseudomonas* sp. MT1 in a chemostat, under conditions of the consortium chemostat, led to incomplete mineralization of 4-chlorosalicylate and the excretion of degradation products (metabolites) into the culture medium. The accumulation of 4-chlorocatechol, 3-chloro-*cis*,*cis*-muconate, and protoanemonin was detected. Protoanemonin, first isolated from Ranunculaceae [76] and later observed as a dead-end metabolite of 4-chlorocatechol [77], is known for its *antibiotic* activity [78]. Its formation as a metabolic intermediate represents a severe danger of poisoning for all members of the microbial consortium [79]. Increasing the dilution rate to 0.20 d^{-1} in the *Pseudomonas* sp. MT1 chemostat resulted in the accumulation of 3-chloro-*cis,cis*-muconate to 4.5 µM and that of protoanemonin to 42.7 µM after 50 d. Particularly, the concentration of protoanemonin was critical and caused cell death of *Pseudomonas* sp. MT1, expressed by rapidly decreasing colony forming units (CFUs), optical density and substrate conversion. However, contrary to *Pseudomonas* sp. MT1 the consortium was able to endure much higher dilution rates, with a maximum rate of 0.80 d^{-1}.

This result translates as to why are there four different strains into the question: Who is detoxifying the toxic intermediates produced by *Pseudomonas* sp. MT1 and how is this done? To trace the fate of the substrate and its metabolites into the cell mass of the different strains, a novel technique was used. The addition of very small amounts of nutrients enriched with stable isotopes to the microbial community, followed by a determination of the labelling by means of isotope ratio mass spectrometry (IRMS) into the biomarker of the individual members of the consortium can give important informations about the members of the community, most actively involved in the metabolism of these nutrients and about those bacteria catabolizing the metabolites and debris of the primary degraders [80]. This technique does not require the culturing of isolated microorganisms on laboratory media and gives direct informations on the nutrient flux within the community. Using this approach with metabolites labelled with stable isotopes, it was possible to identify the primary degraders of the intermediates in the consortium and to understand the role of the individual strains in the community.

Addition of very small amounts of ^{13}C-labelled 4-chlorocatechol, a toxic intermediate of the 4-chlorosalicylate pathway, to the community and separation of the individual strains by *immunocapture,* using the strain specific antibodies allowed to determine the activity of the individual strains.

The analysis of the lipid *biomarkers* revealed that this intermediate was preferentially used by *Alcaligenes* sp. MT3, demonstrating that this strain is responsible for detoxification of this metabolite protecting the community from poisoning. This isotope technique revealed further that *Pseudomonas* sp. MT1 degrades the substrate via a novel pathway. This results in the secretion of protoanemonin by this strain, which was previously regarded as a toxic dead-end product. By adding very small amounts of ^{13}C-labelled protoanemonin, we could identify *Pseudomonas* sp. MT4 as degrader of this toxin, which protects the community from this metabolite. *Empedobacter* sp. MT2 finally does not actively degrade the substrate, but obviously lives on debris of the primary degraders [81]. With this new technique, it was possible for the first time to analyze the function of individual members of the community and to analyze the carbon flux within the community (Figure 2). Furthermore, it allows the identification of novel degradation pathways, because the isotopic label is added only in very small amounts preventing it from being toxic for the community.

From this study a dense interaction pattern between the four strains emerged, where sharing of metabolites as substrates is the basis for the stability of this microbial consortium. The consortium solved the problem of incomplete pathways present in the individual strains and their sensitivity against metabolites by sharing functions within the community. This enables them to deal with changing environmental conditions, a situation frequently found in the environment. It appears that a consortium can react with more flexiblility against changes in the environment than a single strain.

Microbial ecology of a PCB-polluted site

In order to understand the interaction of pollutant and autochthonous bacteria the microbial ecology of a PCB-polluted Moorland site in Germany was studied which was a former Russian missile base. The soil consisted of

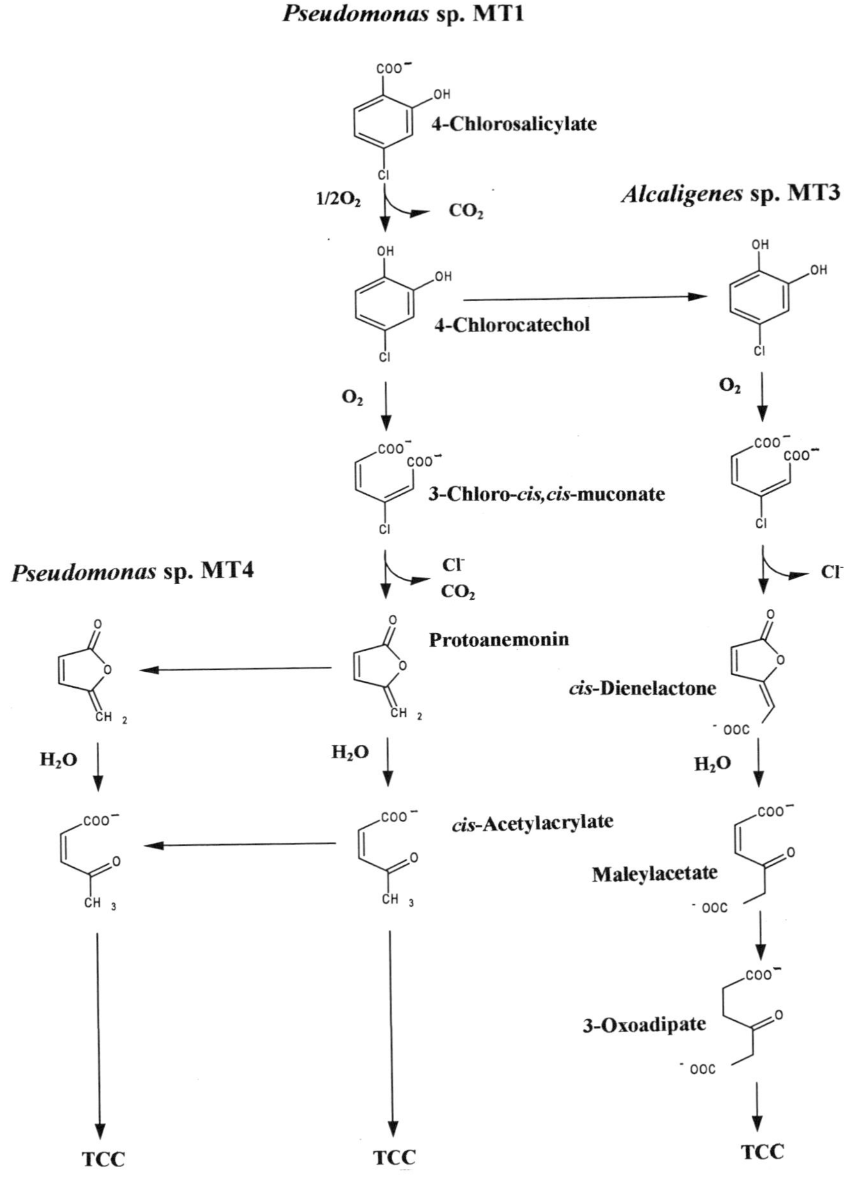

50

Figure 2. Carbon sharing in a 4-chlorosalicylate-degrading microbial consortium. The substrate is degraded by the prime degrader *Pseudomonas* sp. MT1. However, this strain secrets the toxic metabolites 4-chlorocatechol which is degraded by *Alcaligenes* sp. MT3 and protoanemonin which is consumed by *Pseudomonas* sp. MT4.

sand with some clay and a pH of 4.0-4.5. The total organic carbon was low and usually in the range of 1-3 g kg^{-1} dry soil. The PCB concentrations observed varied in a broad range and peaked at 24,000 ppm PCB, making PCB the main carbon source in this soil.

Because the majority of the bacteria, occurring in the environment, cannot be cultivated by methods currently available, a multitude of culture-independent approaches have been developed to study them [82]. Usually the first aim in the study of a microbial community is to know about its presence there. Most methods to study the structure of microbial communities focus on the comparison of 16S rRNA gene sequences. The 16S rRNA is part of the ribosome which functions as the protein factory of the cell. Because the ribosome has such an essential function, a large part of it is highly conserved and very similar or even identical for all bacteria. This makes it to be a very good yardstick for comparing bacteria. The more *phylogenetically* related the bacteria are the more similar are the 16S rRNA gene sequences. The usual approach starts from the extracted total DNA from which the 16S rRNA genes are amplified by PCR. However, because special interest was on the metabolically active bacteria, a different approach was taken. The 16S rRNA was extracted and transcribed by reverse transcriptase into DNA (RT-PCR technique). Because the most active bacteria also possess the highest number of ribosomes, this approach allows the detection of metabolically most active members of the community.

Based on this RT-PCR technique, clone libraries from samples with medium and high PCB pollution could be generated, showing a high microbial diversity despite the low pH and high PCB concentrations [83]. From a sample with a PCB concentration of 10,000 ppm, 34% of the analyzed clones belong to the *alpha*-Proteobacteria lineage, 33% to the *beta*-Proteobacteria and 7% to the *gamma*-Proteobacteria. To the *Holophaga-Acidobacterium* phylum belonged 14% of the clones and 9% to Actinobacteria.

Analysis of the clone libraries revealed an unusual high number of sequences from only two genera of the *beta*-Proteobacteria: *Burkholderia* and *Variovorax*. Three other frequent types of 16S rRNA sequences were similar to those of the species of *Sphingomonas*, members of the *Rhodopila globiformis* group, and the *Acidobacterium* phylum. Some of the 16S rRNA sequences are similar to those of bacterial strains capable to degrade a number of pollutants, including PCB. Because PCB is the main carbon source at this site, it is reasonable to assume that the bacteria identified in these *clone libraries* are those community members which use directly or indirectly PCB as a carbon source. In agreement with this the majority of the strains isolated from this site by enrichment on biphenyl, low chlorinated PCB congeneres and di- and trichlorinated benzoates belong to the genus *Burkholderia*. Most of them are closely related to *Burkholderia* sp. LB400.

A group of cloned sequences clustered within the *Holophaga-Acidobacterium* phylum. Similar sequences were retrieved from different non-polluted soils, i.e. from a peat bog soil in Germany, from Mount Coot-tha soil in Australia, from Drentse A grassland soil in The Netherlands and from an acid mine drainage site in California. The detection of several 16S rRNA sequences from the *Acidobacterium* phylum in the PCB-polluted site reflects one of the still unresolved problems in microbial ecology. Although this phylum is widespread in nature, only very few strains, all belonging to *Acidobacterium capsulatum,* have been isolated and characterized [84]. Therefore, nothing is known about the ability of *Acidobacterium* to degrade PCB or some of its metabolites. This question can only be answered if culture-independent methods like the labelling with stable isotopes of *biomarker molecules*, as described above, are applied.

Microbial formation of clay aggregates in PCB-polluted soil

Scanning electron microscopic analysis of biofilms, grown on sterilized surfaces exposed to different soil fractions of a PCB-polluted site always showed the presence of clay aggregates. Transmission electron microscopy of these ultrathin sectioned biofilms revealed bacteria often to be associated with these clay aggregates. The role of these aggregates was analyzed with respect to the PCB degradation at this site in order to know as to how clay aggregates interact with PCB-degrading bacteria. This question is especially important with regard to the application of genetically modified bacteria for bioremediation. To study this phenomenon in detail, a number of microcosm experiments were performed with medium PCB-polluted soil samples [85]. After the addition of water to five-fold field capacity to the soil samples and incubation for several weeks, individual clay aggregates were formed at the water contact face of the floating substratum. That this process is an active biological one, can be shown by prior autoclaving the soil. After killing the bacteria in this process, no clay aggregates were formed. These clay aggregates were rather homogeneously distributed and, by scanning electron microscopic analysis, only few bacteria could be observed (Figure 3A). Only inspection by transmission electron microscopy revealed the presence of bacteria within the aggregates (Figure 3B). The formation of the clay aggregates can repeatedly be observed over several months. Obviously the bacteria first attach to the substratum, followed by picking up clay leaflets from the bulk water, thus successively building the clay aggregates.

The collection of the clay particles does not seem to be random. Element mapping by electron energy loss spectroscopy in the electron microscope showed, beside clay leaflet, particles with a high content of iron oxide. Characteristically very small particle, with high iron oxide content, were found in direct contact with the bacterial cells. Because this was observed in many of these bacterial clay aggregates, it could be assumed that bacteria collect in this way the micronutrient iron essential for their metabolism. Analysis of the clay of the bacterial aggregates showed that the clay was loaded with PCB. To understand the function of the bacterial clay aggregates,

54

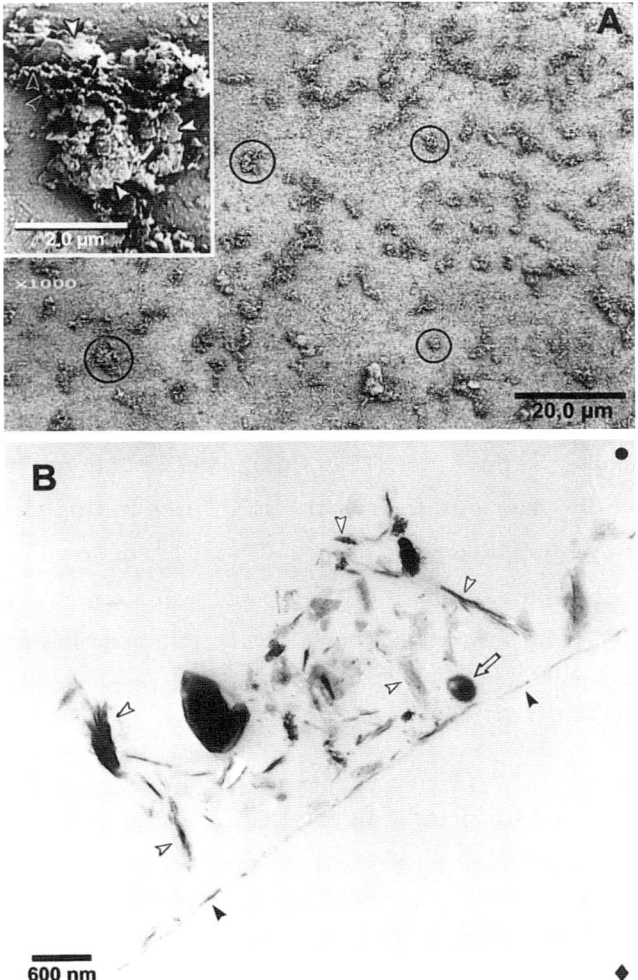

Figure 3. Ultrastructure of clay hutches. A: Scanning electron microscopy reveals a lawn of clay aggregates. Individual aggregates are encircled. In the inset one of these aggregates is enlarged and extracellular polymeric substances (EPS) (open arrow heads) and clay leaflets (filled arrow heads) are shown. B: Transmission electron microscopy of ultrathin sectioned clay hutches demonstrates a cross-section of a clay aggregate. Black arrowheads indicate the substratum surface and white arrowheads point to diverse phyllosilicates, loosely aggregated. An individual bacterium, linked by EPS-fibrils to the substratum surface and clay minerals, is shown (arrow).

another observation is important, i.e. some of the bacteria within the aggregates show electron translucent cytoplasmic inclusions, which are indicative of organic storage material such as polyhydroxybutyric acid. This means that the bacteria are not starving although the soil was poor in carbon, and the weakly bioavailable PCB constitutes the principal source of carbon in the system. Taking all these pieces together, the bacterial clay aggregates are more like *"clay hutches"*. They supply micronutrients like iron and water insoluble organic compounds like PCB to the bacteria. Moreover, they protect the bacteria from grazers like protozoa. Obviously, the bacteria found a way how to handle the water insoluble substrate PCB by using clay as a transport shuttle for it. This is a completely new alternative to the production of biosurfactant used for some bioremediation procedures, and it surely deserves further investigation for bioremediation purposes.

All bacteria observed in these clay hutches are of the Gram-negative type. The structure of the bacterial community of the clay hutches was analyzed using a techniques called T-RFLP. Here PCR of the extracted total DNA was performed using eubacterial primers, which were labelled with two different fluorochromes at the 3'- and the 5'-ends, respectively. The PCR products were then digested with endonucleases and the resulting DNA fragments separated. Only those terminal fragments carried the fluorochromes and could be detected. From the 16S rRNA gene sequence library, the fragments for different genera could be predicted and compared with the results obtained. A number of bacterial genera were identified in the clay hutches, including strains belonging to the genera *Sphingomonas* and *Burkholderia* [86].

Experiments for *in situ* bioremediation of PCB-polluted sites

There are different approaches which can be taken to achieve a clean-up of a PCB-polluted site using bacteria. These approaches can be grouped according to the amount of time, technical effort and money spent to clean such a site.

The easiest and least invasive way is intrinsic bioremediation or natural *attenuation*. Here the site is monitored to make sure that the pollution is not transported to unpolluted site and to record the progress of bioremediation. This approach is usually taken for a site, which only slightly polluted and which does not influence the flora and fauna, including human activities of the adjacent area. Another requirement for this approach is that an indigenous microbial community exists which demonstrates that natural attenuation is already occurring and continuing. This strategy has been taken in the case of PCB pollution with anaerobic dechlorination and even under aerobic conditions [87].

The second choice in biological treatments is the biostimulation of indigenous microbial populations to degrade the target compounds. This alternative is chosen when a natural population exists at the site which has the potential to degrade the chemicals, but is actually lacking oxygen, nitrogen or other nutrients to degrade it. The missing component can then be introduced into the system and the degradative activity of the microbial community can be induced and enabled. For an active bioremediation, this method is usually the most cost-effective one [88]. This approach has been taken for PCB as well [89].

If the intrinsic bacteria do not exist at the polluted site, which are able to degrade the pollutant, the biological approach of *bioaugmentation* has to be taken. In this case, the bacteria, with the necessary catalytic activities and other required characteristics, can be injected directly into the polluted site usally together with nutrients. This approach can be necessary as well in cases where bacteria with the required catalytic activity are present at the site but their activity is too slow or incomplete, and it is desirable to take the bioaugmentation choice. This is the area of the *genetically engineered* and optimized bacteria, but it is also the most expensive alternative method of *in situ* bioremediation.

To achieve the goal of *in situ* bioremediation, a series of scale-ups is usually necessary. From the flask the next step is the integration of the selected strain into a microbial community. Usually this is done in microcosm experiments. Here samples from the environment are brought into the lab and treated under different, but controlled, conditions with the special degraders added. The optimal conditions for the bioremediation can be determined and limiting factor identified. From the lab, the classical next step can be that of a mesocosm. Here the treatment is performed on site but separated from the surrounding area. In these experiments, the performance of the bioremediation strategy can be tested under *close-to-nature* conditions. After that the *in situ* experiment will be done. This is the general course of a scale-up which of course can be modified according to the special needs [90].

The first hurdle, a PCB-degrading strain has to take, is the survival and the functioning with other bacteria. These additional bacteria usually do influence the performance of the degrading strain and it is neither always predictable how this interaction will be, nor what the results will be. Often a bacterial consortium can be isolated, but its individual strains fail to give good results [91].

Even more complicated and less predictable is the situation in microcosms, where a huge number of different bacteria, many of them not at all involved in the degradation of PCB, exist. Therefore, the first experiments in microcosms are generally directed to the monitoring of the survival of the PCB degrader [92]. Microcosm experiment with Hudson River sediments provided evidences that the naturally occurring microorganisms have the potential to attack

PCB (Aroclor 1242) at rapid rates both aerobically and anaerobically. These results are helping our understanding of the fate of released PCB in river sediments [93].

The first field demonstration for PCB bioremediation was in New York in 1987. It was done at a former race track which was contaminated with PCB soil. It was performed in the way of *land-farming* where the soil was tilled with PCB-degrading bacteria and nutrients. During this study, the performance of the highly competent aerobic PCB degrader *Burkholderia* sp. LB400 was monitored over a four-month period. The test demonstrated up to 20% biodegradation from a starting concentration, which was between 50 and 525 ppm. From an adjacent control plot there was no indication of any PCB biodegradation. Both the rate and the extent was about half of that seen in laboratory experiments. This was explained by insufficient temperature and moisture control in the field [94].

A team from General Electric conducted a large field experiment with Hudson River sediment. Over a period of 73 days aerobic PCB biodegradation was stimulated by the addition of inorganic nutrients, biphenyl and oxygen. This resulted in a 37-55% loss of PCB and a production of chlorobenzoates. Interestingly, the use of *Alcaligenes* sp. H850, one of the most competent PCB degraders known at this time, did not improve the result as compared to that observed with the indigenous strains. The probable reason was the poor survival of *Alcaligenes* sp. H850 in the sediment. The fact that even the less chlorinated PCB congeners were not much degraded was explained by poor bioavailability of the PCB in the sediment [95].

In 1985, a five year field study was started on sewage sludge, containing approximately 50 ppm PCB. Two different PCB concentrations, 25 and 75 ppm were tested. The sludge was applied by subsurface injection and commercial fertilizer was added. The result showed 85% degradation of the lower chlorinated congeners and only 6% loss of *biorecalcitrans* PCB marker (2,5,2',5'-tetrachlorobiphenyl). In total 43 PCB congeners showed a measurable degradation. The soil half-life for the dichlorinated congeners was 7-11 months, for the trichlorinated congeners 5-17 months, for the tetrachlorinated congeners 11-58 months, and 19 months for total PCB. The results of this study support land treatment of PCB-polluted sludge as an effective management approach for sludge containing low levels of PCB [96].

Instead of adding only one very efficient PCB degrader to the polluted soil, it is sometimes more effective to use two of them or a consortium. One attempt aiming into this direction was the application of the two complementary PCB degraders ENV 360 and ENV307 in a field study. In addition to the strains, substrates had been added to the site which are expected to support growth or induce the PCB-degradating enzymes. The field had been tilled and the experiment monitored over several months. After two years, the PCB levels have been reduced by 44% [97].

CONCLUSION

It took many years from the start of the PCB production in 1929 to the realization of toxicity and its resistance in the environment and finally to its ban in any industrial applications. At the end at least 1.5 million tons PCB were produced worldwide, from which a considerable percentage was released into the environment. Here it was enriched and concentrated over the food chain because of its solubility in fat.

After the damage was done, the search for solutions of the environmental problem began. This was not an easy task because of the complex mixture of congeners present in industrial PCB products. Each of these congeners has its individual toxicity, *bioaccumulation* and biodegradation behavior. The first isolation of strains, capable of the degradation of some of these congeners, was a huge step forward. More strains were isolated and the understanding of the biochemistry of the degradation of PCB congeners emerged. Isolating the genes, elucidating their sequences and studying the encoded enzymes were another main step forward. A logical consequence was the construction of better, i.e. faster and more universal strains for the mineralization of PCB by genetic engineering.

As more and more strains and their PCB-degradative genes became available, the insight into the biodiversity of the PCB degraders developed. This process is still progressing because most of the bacteria from the environment cannot be cultivated by todays techniques. A way out may be the cloning of environmental DNA and the use of the soil *"metagenome"*. Results from microbial ecology, however, may provide some hopes for the one strain able to deal with all or most PCB degradation problems. It has been shown that a very complex web of microbial interactions exists in the environment; some of these interactions are very different from those usually seen in the flasks and fermenters in the laboratory. Some of these interactions involve the sharing of substrates, which enables the microbial community to tolerate environmental changes and to react very quickly to environmental stresses. It still remains to be determined whether a consortium of different bacteria is more flexible in this respect than a single strain. Other bacterial interactions exist with the surrounding inorganic material. One example is the clay hutches where bacteria use the affinity of clay to organic compounds to solve the problem of transport of the water insoluble PCB. We do not know anything about the interactions between humic acids, PCB, and bacteria. We do not know either what the interactions between the PCB-degrading bacteria and the PCB-tolerating bacteria are. The PCB-metabolizing activity of fungi in PCB-polluted soil and their interaction with the PCB-degrading bacteria is another unknown relationship which should be known for a knowledge-based bioremediation strategy.

The bioremediation of PCB is one of the most difficult tasks in the area of environmental biotechnology. The last decades brought an immense progress in biochemistry and genetics of PCB congener degradation. The stage is now set for the exploration and use of the immense microbial *biodiversity*, concerning PCB-degrading bacteria. In connection with the study of the activities and interactions of microbial communities using the tools developed in microbial ecology, it should be possible to identify the critical ecological parameters controlling the degradation of PCB in the environment. This knowledge-based understanding will improve considerably the *in situ* bioremediation of PCB-polluted site.

ACKNOWLEDGEMENT

Heinrich Lünsdorf is thanked for many inspiring discussions and for supplying some of his excellent electron microphotographs and Peter Wolff for the skilfull drawing of the figures. The support of this work by a grant of the German Federal Ministry for Science, Education and Research (Project No. 0319433C) is gratefully acknowledged.

REFERENCES

1   Fischbein A, Rizzo JN. Mount Sinai J Med 1987; 54: 332-336.

2   Rantanen J. Organohalogen Compounds 1992; 10: 291-294.

3   Riseborough RW, Rieche P, Peakall DB, Herman SG, Kirven MN. Nature 1968; 220: 1098-1102.

4   Venkatesan MI, de Leon RP, van Geen A, Luoma, SN. Marine Chem 1999; 64: 85-97.

5   Eisenreich S, Looney B. Physical Behavior of PCB's in the Great Lakes: Evidence for the Atmospheric Flux of PCB's to Lake Superior. Ann Arbor Science Publishers, Ann Arbor, Michigan, 1983.

6   Tofflemire T, Shen E, Buckley E. Physical Behavior of PCB's in the Great Lakes: Volitilization of PCB from Sediment and Water: Experimental and Field Data. Ann Arbor, Michigan: Ann Arbor Science Publishers, 1983.

7   Paya-Perez A, Riaz M, Larsen B. Ecotoxicology and Environmental Safety: Soil Sorbtion of 20 PCB Congeners and Six Chlorobenzenes. Academic Press Inc, 1991.

8   Sun S, Boyd S. Sorbtion of PCB Congeners by Residual PCB-Oil Phase in Soils, Journal of Environmental Quality, 1991.

9   Ahmed M, Focht DD. Can J Microbiol 1973; 19: 47-52.

10   Baxter RA, Gilbert RE, Lidgett RA, Mainprize JH, Vodden HA. Science of the Total Environment 1975; 4: 53-61.

11   Furukawa K, Matsumata F. J Agric Food Chem 1976; 24: 251-256.

12   Furukawa K, Tonomura K, Kamibayashi A. Appl Environ Microbiol 1978; 35:223-227.

13   Asturias JA, Moore E, Yakimov MM, Klatte S, Timmis KN. Syst Appl Microbiol 1994; 17: 226-231.

14   Kohler H-PE, Kohler-Staub D, Focht DD. Appl Environ Microbiol 1988; 54: 1940-1945.

15   Bedard DL, Unterman R, Bopp LH, Brennan MJ, Haberl ML, Johnson C. Appl Environ Microbiol 1986; 51: 761-768.

16   Kim E, Zylstra GJ. J Bacteriol 1995; 177 : 3095-3103.

17   Tulp MTM, Schmitz R, Hutzinger O. Chemosphere 1978; 1: 103-108.

18   Kaiser KLE, Wong PTS. Bull Environ Contam Toxicol 1974; 11: 291-296.

19   Mukerjee-Dhar G, Hatta T, Shimura M, Kimbara K. Arch Microbiol 1998; 169: 61-70.

20   Bedard DL, Brennan MJ, Unterman R. In: Addis G, Komai R, Palo Alto, eds. Proceedings of the 1983 Polychlorinated Biphenyl Seminar, 1984; 101-108.

21   Bedard DL. Haberl ML, May RJ, Brennan MJ. Appl Environ Microbiol 1987; 53:1103-1112.

22   Barton MR, Crawford RL. Appl Environ Microbiol 1988; 54: 594-595.

23   Bedard DL, Haberl ML. Microb Ecol 1990; 20: 87-102.

24   Hernandez BS, Arensdorf JJ, Focht DD. Biodegradation 1995; 6: 75-82.

25   Unterman R. In: Sayler GS, Fox R, Blackburn JW, eds. Environmental Biotechnology for Waste Treatment. New York: Plenum Press, 1991; 159-162.

26   Bopp LH. J Ind Microbiol 1986; 1: 23-29.

27   Seeger M, Zielinski M, Timmis KN, Hofer B. Appl Environ Microbiol 1999; 65: 3614-3621.

28   Sharma A, Chunn CD, Rothmel RK, Unterman R. Abstr 91st Gen Meet Am Soc Microbiol. 1991; 284.

29   Rothmel RK, Shannon MJR, Unterman R. Abstr 93rd Gen Meet Am Soc Microbiol. 1993; 374.

30   Furukawa K, Miyazaki T. J. Bacteriol 1986; 166: 392-398.

31   Furukawa K, Arimura N, Miyazaki T. J. Bacteriol 1987; 169: 427-429.

32   Furukawa K, Arimura N. J. Bacteriol 1987; 169: 924-927.

33   Taira K, Hayase N, Arimura N, Yamashita S, Miyazaki T, Furukawa K. Biochemistry 1988; 27: 3990-3996.

34   Hayase N, Taira K, Furukawa K. J. Bacteriol 1990; 172: 1160-1164.

35   Ahmad D, Fraser J, Sylvestre M, Larose A, Khan A, Bergeron J, Jateau JM, Sondossi M. Gene 1995; 156: 69-74.

36   Furukawa K, Hayase N, Taira K, Tomizuka N. J Bacteriol 1989; 171: 5467-5472.

37   Mondello FJ, J Bacteriol 1989; 171: 1725-1732.

38   Yates JR, Mondello FJ. J Bacteriol 1989; 171: 1733-1735.

39   Gibson DT, Cruden DL, Haddock JD, Zylstra GJ, Brand JM. J Bacteriol 1993; 175: 4561-4564.

40   Erickson BD, Mondello FJ. Appl Environ Microbiol 1993; 59: 3858-3862.

41   Timmis KN, Steffan RJ, Unterman R. Ann Rev Microbiol 1994; 48: 525-557.

42   Erickson BD, Mondello FJ. J Bacteriol 1992; 174: 2903-2912.

43   Hofer B, Eltis LD, Dowling DN, Timmis KN. Gene 1993; 130: 47-55.

44   Asturias JA, Timmis KN. J Bacteriol 1993; 175: 4631-4640.

45   Asturias JA, Eltis JD, Prucha M, Timmis KN. J Biol Chem 1994; 269: 7807-7815.

46   Pellizari VH, Bezborodnikov S, Quensen JF, Tiedje JM. Appl Environ Microbiol 1996; 62: 2053-2058.

47   Kohler-Staub D, Kohler H-PE. J Bacteriol 1989; 171: 1428-1434.

48   Mokross H, Schmidt E, Reineke W. FEMS Microbiol Lett 1990; 71: 179-186.

49   Barton MR, Crawford RL. Appl Environ Microbiol 1988; 54: 594-595.

50   Arensdorf JJ, Focht DD. Appl Environ Microbiol 1995; 61: 443-447.

51   Kim S, Picardal FW. FEMS Microbiol Lett 2000; 185: 235-239.

52   Bopp RF, Simpson HJ, Deck BL, Kosryk N. Northwestern Environ Science 1984; 3: 180-184.

53   Brown JF, Bedard DL, Brennan MJ, Carnahan JC, Feng, H, Wagner RE. Science 1987; 236: 709-712.

54   Wu Q, Sowers KR, May HD. Appl Environ Microbiol 1998; 64: 1052-1058.

55   Quensen JF III, Boyd SA, Tiedje JM. Science 1988; 242: 752-754.

56   Quensen JF III, Boyd SA, Tiedje JM. Appl Environ Microbiol 1990; 56: 2360-2369.

57   Bedard DL, van Dort H, May RJ, Smullen AL. Environ Sci Technol 1997; 31: 3308-3313.

58   Wiegel J, Zhang X, Wu Q. Appl Environ Microbiol 1999; 65:2217-2221.

59   Bedard DL, Quensen JF. In: Young LY, Cerniglia C, eds. Microbial Transformation and Degradation of Toxic Organic Chemicals. New York: Wiley-Liss Division, John Wiley & Sons, 1995; 127-216.

60   Davenport GJ, Dutta SK. Abstr 98th Gen Meet Am Soc Microbiol 1998; 452.

61   Davenport GJ, Champine JM, Dutta SK. Abstr 97th Gen Meet Am Soc Microbiol 1997; 479.

62   Hou L, Dutta SK. Lett Appl Microbiol 2000; 30: 238-243.

63   La Montagne MG, Davenport GJ, Hou LH, Dutta SK. J Appl Microbiol 1998; 84: 1156-11562.

64   Holoman TR, Elberson MA, Cutter LA, May HD, Sowers KR. Appl Environ Microbiol 1998; 64: 3359-3367.

65   Fetzner S, Lingens F. Microbiol Rev 1994; 58: 641-685.

66   Reineke W, Knackmuss H-J: Ann Rev Microbiol 1988; 42: 263- 287.

67   Hardman DJ. Crit Rev Biotech 1991; 11: 1-40.

68   Furukawa K. Biodegradation 1994; 5: 289-300.

69   Focht DD. Curr Opin Biotechnol 1995; 6: 341-346.

70   Golyshin P, Fredrickson HL, Giulliano L, Rothmel R, Timmis KN, Yakimov MM. Microbiol 1999; 22: 257-267.

71   Erb RW, Eichner CA, Wagner-Döbler I, Timmis KN.   Nature Biotechnology 1997; 15: 378-382.

72   Hofer B, Blasco R, Mallavarapu M, Seeger M, McKay D, Wittich R-M, Pieper DH, Timmis KN. In: Nakazawa T, Furukawa K, Hass D, Silver S, eds. *Pseudomonas*: Molecular Biology and Biotechnology. AS, Washington, D.C., 1996; 121-131.

73   Lünsdorf H, Brümmer I, Timmis KN, Wagner-Döbler I. J Bacteriol 1997; 179: 31-40.

74   Faude, UC. PhD Thesis, University of Leipzig, Germany, 1995.

75   Moore ERB, Mau M, Arnscheidt A, Böttger EC, Hutson RA, Collins
      MD, Van de Peer Y, De Wachter R, Timmis KN. System Appl Microbiol
      1996; 19: 478-492.

76   Seegal BC, Holden M. Science 1945; 101: 413-414.

77   Blasco R, Wittich RM, Mallavarapu M, Timmis KN, Pieper DH. J Biol
      Chem 1995; 49: 29229-29235.

78   Didry N, Dubreuil L, Pinkas M. Pharmazie 1991; 46: 546-547.

79   Blasco R, Mallavarapu M, Wittich RM, Timmis KN, Pieper DH. Appl
      Environ Microbiol 1997; 63: 427-434.

80   Abraham W-R, Hesse C, Pelz O. Appl Environ Microbiol 1998; 64:
      4202-4209.

81   Pelz O, Tesar M, Wittich R-M, Moore ERB, Timmis KN, Abraham W-
      R. Environ Microbiol 1999; 1: 167-174.

82   Amann RI, Ludwig W, Schleifer K-H. Microbiol Rev 1995; 59: 143-169.

83   Nogales B, Moore ERB, Abraham W-R, Timmis KN. Environm Microbiol
      1999; 1: 199-212.

84   Barns SM, Takala SL, Kuske CR. Appl Environ Microbiol 1999; 65:
      1731-1737.

85   Lünsdorf H, Erb RW, Abraham W-R, Timmis KN. Environ Microbiol
      2000; 2: 161-168.

86   Lünsdorf H, Strömpl C, Osborn AM, Moore ERB, Bennasar A,
      Abraham W-R, Timmis KN. Methods Enzymol 2000, in press.

87   Flanagan WP, May RJ. Environ Science Technol 1993; 27: 2207-2212.

88   Smith JR, Tomicek RM, Swallow PV, Weightman RL, Nakles DV,
      Helbling M. In: Kostecki PT, Calabese EJ, Bonazountas M, eds.
      Hydrocarbon Contaminated Soils, Vol 5. Amherst, MA, USA: Amherst
      Scientific Publ, 1995; 531-572.

89   Harkness MR, McDermott JB, Abramowicz DA, Salvo JJ, Flanagan
      WP, Stephens ML, Mondello FJ, May RJ, Lobos JH, Carroll KM,
      Brennan MJ, Bracco AA, Fish KM, Warner GL, Wilson PR, Dietrich
      DK, Lin DT, Morgan CB, Gately WL. Science 1993; 259: 503-507.

90   Shannon MJR, Rothmel R, Chun CD, Unterman R. In: Hinchee RH,
      Leeson A, Simprini L, Ong SK, eds. Bioremediation of Chlorinated
      and PAH Compounds. Boca Raton, Fl: Lewis Publishers, 1994; 354-
      358.

91   Clark RR, Chian ESK, Griffin RA. Appl Environ Microbiol 1979; 37:
      680-685.

92   Havel J, Reineke W. Appl Microbiol Biotechnol 1992; 38: 129-134.

93   Fish KM, Principe JM. Appl Environ Microbiol 1994; 60: 4269-4296.

94   McDermott JB, Unterman R, Brennan MJ, Brooks RE, Mobley DP,
      Schwartz CC, Dietrich DK. Environ Prog 1989; 8: 46-51.

95   Harkness MR, McDermott JB, Abramowicz DA, Salvo JJ, Flanagan
      WP, Stephens ML, Mondello FJ, May RJ, Lobos JH, Carroll KM,
      Brennan MJ, Bracco AA, Fish KM, Warner GL, Wilson PR, Dietrich
      DK, Lin DT, Morgan CB, Gately WL. In: Hinchee RE, Leeson A,
      Semprini L, Ong SK, eds. Bioremediation of Chlorinated and Polycyclic
      Aromatic Hydrocarbon Compounds. New York: Lewis Publishers, 1994;
      368-375.

96   Gan DR, Berthouez PM. Water Environ Res 1994; 66: 54-69.

97   Shannon MJR, Rothmel R, Chun CD, Unterman R. In: Hinchee RH,
      Leeson A, Semprini L, Ong SK, eds. Bioremediation of Chlorinated
      and PAH Compounds. Boca Raton, Fl: Lewis Publishers, 1994; 354-
      358.

Biotransformations: Bioremediation Technology for Health and Environmental Protection
V.P. Singh and R.D. Stapleton, Jr. (Editors)
© 2002 Elsevier Science B.V. All rights reserved.

Biodegradation of fuel oils and lubricants: Soil and water bioremediation options

S. Wilkinson[a], S. Nicklin[a] and J.L. Faull[b]

[a]Bioremediation Group, DERA, Fort Halstead, Sevenoaks, Kent, TN14 7BP, UK

[b]Biology Department, Birkbeck College, Malet Street, London WC1E 7HX, UK

INTRODUCTION

Biodegradation may be defined as the breakdown of organic compounds, in natural or man made environments, by bacteria, actinomycetes and fungi. Recent interest in biodegradation has focussed on the possibility of using these natural processes for the decontamination of soil or water contaminated with complex hydrocarbon mixes.

Accidental crude and fuel oil spillage continues to occur despite many precautions and large scale oil spillages have lead to spectacular pollution events. For example, in 1978 the Amoco Cadiz released 0.2 megatons of crude oil onto the Brittany coastline, in 1989 the Exxon Valdiz released 41,000 cubic metres of crude oil into Prince William Sound in Alaska and in 1993 the Braer released 0.08 megatons of crude oil off the coast of the Shetland Isles [1]. However, it is not only major incidents that contribute to the problem. It has been estimated that amounts similar to those of major oil spills are released yearly in the form of many smaller spills during tanker refuelling, and from urban and *industrial runoff* [2]. Moreover, it is estimated that there will be thousands of newly contaminated sites annually in the USA alone [3]. The problems appear unlikely to diminish as domestic and military consumption of fuel oils continues to increase.

Biodegradative clean up of oil contaminated soils, sediments and waters has only recently been considered a viable clean up option. The detailed studies of the effects of different types of clean up strategy on the coastal regions around Prince William Sound, contaminated by the Exxon Valdiz crude oil spill in 1989, revealed that of all the approaches the best results were obtained where natural processes were allowed to occur [4]. However, the rates of oil removal were slow and in some cases could have lead to an extended exposure of marine species to toxic hydrocarbons and intermediates of biodegradation. A controlled or directed bioremediation process seeks to provide an accelerated natural *decomposition*, and in this review we consider best practice to attempt to achieve this goal.

CRUDE OIL AND FUEL OIL

Crude oil is a natural product, comprising a complex mixture of various hydrocarbons, created by the decomposition of plant remains from the carboniferous period under high temperature and pressure. It is a mixture of aliphatic saturated compounds, including *n* alkanes, branched *n* alkanes and cyclo-alkanes; aromatics, including naphthalene, toluene, xylene and benzene; asphaltanes, including phenols, fatty acids ketones, esters and porphyrins; resins, waxes and high molecular weight tars, including pyridines, quinolines, cardaxoles, sulphonates and amides [5]. Crude oils from different wells differ greatly in their composition. Distillation of the crude oil will yield different fractions which will vary in size, complexity and volatility from the petroleum gases with a boiling point of 30°C to fuel oils residues with a boiling point of over 350°C (Table 1).

PROBLEMS ASSOCIATED WITH FUEL OIL SPILLS

There are a number of problems associated with fuel oil spills. Initially, when the oil still contains many of the lighter fractions, there are safety

Table 1
Crude oil fractions of Iranian oil, Agha Jari by distillation

Generic fraction name	Boiling point (°C)	Molecular size (C N°)	Uses	Volume (%)
Petroleum gases	30	3-4	Solvents	30.5
Light gasoline	30-140	4-6	Petroleum vehicles/ Petrochemical	9.5
Naphtha	120-175	7-10	Petrochemical industry	14.5
Kerosene	165-200	10-14	Aircraft	17.5
Gas oil/diesel	176-365	15-20	Vehicles heating	
Fuel oil/residues	350+	20+	Bunker fuel in ships and power stations and tars	25

Source: Clark [6].

considerations. The volatility of these compounds often leads to an *explosion hazard*. Many of the compounds are acutely toxic to higher animals and inhalation or contact with them should be avoided. Acute human exposure to petroleum distillates causes headaches, central nervous system depression, nausea, respiratory tract irritation and mental confusion [7]. For an ecological community, exposure to oil spills can lead to reduced numbers of species [a loss of species diversity] and in aquatic communities acute toxic effects on shellfish leads to the collapse of various food chains and eventually the death of higher carnivores [8]. Aquatic environments may become eutrophic, leading to oxygen depletion and anaerobiosis. In such situations whole communities may die.

In the median term, but no less important, contamination of the water table may occur as the spill forms a downward moving slick within soil. Groundwater contamination affects palatability and potability of abstracted

72

water. The most frequent source of exposure of animals and man to these hydrocarbons is by the consumption of contaminated drinking water [9].

Over the longer term, soil contamination with hydrocarbons and products of partial degradation prevents sale and development of land and may adversely affect the local economy.

EXTENT OF THE PROBLEM

World wide, many sites are contaminated with a variety of organic compounds such as hydrocarbons, chlorinated compounds and nitro-aromatic compounds. HAzNEWS [10] cites that in the US, the Air force (USAF) has identified about 5,000 sites with some degree of on-site contamination.

Leaks and spills during the delivery and processing of crude oil are inevitable. Between 1992 and 1994, US Air Force had 169 reportable spills amounting to 114,000 gallons [11]. In Germany the number of suspected sites is between 140,000-250,000 [10]. Analysis of contaminant data indicates

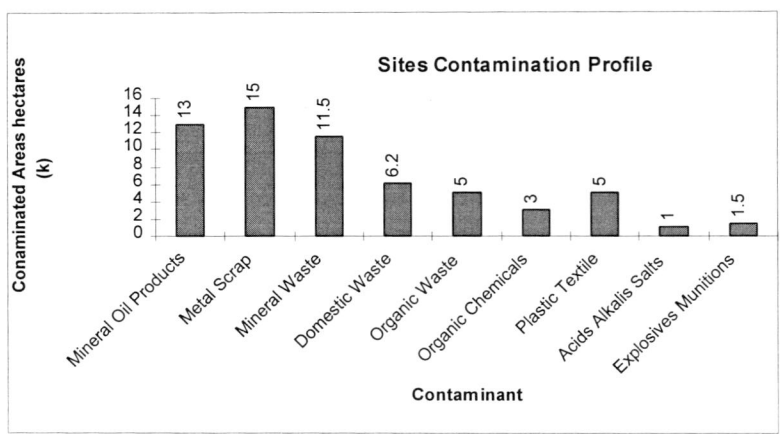

Figure 1. Extent of the waste problem in hectares of contaminated land. Source: Haznews [10].

that mineral oil products represent 57% of the contamination found in soil from fuels, lubricants and oils from leaking tanks, pipelines and spillage's. Figure 1 illustrates the extent of the waste problem expressed in hectares of contaminated land.

MICROBIAL DEGRADATION OF HYDROCARBONS

In recent years the oil industry has shown considerable interest in the use of microorganisms, especially for controlling and dispersing oil spills using surfactants, bioremediation and oil recovery. The term *"hydrocarbonoclastic"* has been used to describe hydrocarbon utilising microorganisms [12]. This specifically relates to microbes, that are capable of degrading hydrocarbons, and all of which share some of the following characteristics [13,14]:

- They are able to extensively degrade partially or fully petroleum based compounds.

- They have a capable and efficient hydrocarbon uptake system.

- They have receptor sites for binding hydrocarbons.

- They are capable of producing of surfactants.

- They are well adapted to the environment, genetically stable with rapid reproduction rates.

- They have been selected for their environment.

- They must not be pathogenic or produce toxic metabolic end-products.

- They must have group-specific oxygenases to introduce molecular oxygen into the hydrocarbon and, with relatively few reactions, generate intermediates that subsequently enter common energy-yielding catabolic pathways.

Microbes capable of degrading hydrocarbons first came to light in the 1950's after a series of air accidents revealed that microbial biomass was capable of blocking fuel lines [15] (Table 2). In most cases the microorganism was identified as the filamentous fungus *Hormoconis (Cladosporium) resinae*. Bacteria were also found to grow in jet fuel, including *Pseudomoas fluorescens, Ps. malophora* and *Alcaligenes* sp. Their activities increased fuel viscosity, blocked filters and produced acidic metabolic byproducts that caused corrosional damage.

Since many hydrocarbons are naturally occurring complex mixtures of organic compounds, derived by biosynthesis, it is not surprising that microorganisms have adapted and evolved the ability to utilise these compounds. The effects of natural selection mean that for every compound there is at least one microorganism able to at least partially degrade it, if the environmental conditions are favourable. Therefore, when oil is lost to the environment the indigenous population will most likely contain a small population of microorganisms capable of degrading the contaminating hydrocarbons [17]. The microorganisms degrade oil and produce intermediate products including alcohol's, phenols, esters, aldehydes, ketones and fatty acids. These in turn are converted into CO_2, water and microbial biomass. This process results in complete mineralization of the pollutant and is clearly the ultimate goal of any bioremediation process.

Because of the presence of hydrocarbon-degrading microorganisms, areas already heavily contaminated with oil are more likely to manage with a further input of hydrocarbon pollutants. Normal populations of hydrocarbon utilizing microorganisms account for 0.1% of the population but may reach

Table 2
Genera of microorganisms reported to utilize petroleum fractions for growth

Bacteria	Yeasts	Filamentous fungi	Actinomycetes	Algae
Acinetobacter	*Candida*	*Absidia*	*Actinomyces*	*Chlorella*
Achromobacter	*Cryptococcus*	*Acremonium*	*Endomyces*	*Prototheca*
Actinomyces	*Debaryomyces*	*Aspergillus*	*Nocardia*	
Alkaligenes	*Endomyces*	*Botrytis*		
Arthrobacter	*Hansenula*	*Cephalosporium*		
Bacillus	*Mycotorula*	*Chaetomium*		
Corynebacterium	*Pichia*	*Chloridium*		
Flavobacter	*Rhodotorula*	*Cladosporium*		
Methanomonas	*Saccharomyces*	*Colletotrichum*		
Micrococcus	*Torulopsis*	*Cunninghamella*		
Micromonospora	*Trichsporon*	*Fusarium*		
Mycobacterium		*Gliocladium*		
Nocardia		*Graphium*		
Pseudomonas		*Monilia*		
Streptomyces		*Helicostylum*		
		Helminthosporium		
		Mucor		
		Paecilomyces		
		Penicillium		
		Rhizopus		
		Scolecobasidium		
		Spicaria		
		Syncephalastra		
		Trichoderma		

Source: Miller and Litsky [16].

100% under selective pressure after a spill or prolonged chronic discharges, returning to background levels after the pollutant is removed. Higgins and Gilbert [18] found that in unpolluted water hydrocarbon oxidizing bacteria were low in number but that their number increased by two orders of magnitude in polluted waters. Environmental factors were found to be important in influencing the prevalence of oil-degrading microorganisms. Atlas [12], Baker et al. [19] and Leahy and Colwell [5] found that hydrocarbon utilizers were more prevalent in summer, at times of high nitrogen and phosphorous, in well aerated water, and in shipping lanes where the autochthonous population had become acclimatised to chronic pollution.

Filamentous fungi, because of their mycelia growth habit, can spread between food bases, penetrate oil droplets and are perhaps more adapted to environments where they grow as a film on a fixed surface. Bacteria are able to utilize a broader range of target structures and provide a crucial role in initiating hydrocarbon biodegradation. The limited available evidence does not suggest an ecologically significant role for algae and protozoan in the degradation of hydrocarbons. However, algae and protozoan populations may be directly affected by hydrocarbon concentration. They may suffer direct toxic effects, or, as hydrocarbon degrading bacteria flourish, protozoa population expand as they graze on the bacteria. However, protozoa population explosions have also been shown to significantly reduce the number of bacteria available for hydrocarbon removal so their presence in a biodegradation system may not always be beneficial [20].

FACTORS AFFECTING BIODEGRADATION

A number of factors affect biodegradation of oils and fuels, and these may be summarized as follows:

Physical state of hydrocarbon/oil

The physical and chemical nature of the oil pollution is a critical factor in determining rates of biodegradation. Access to the pollutant is a major consideration when trying to degrade hydrocarbons which are, to a great extent, hydrophobic. The dispersion of crude oil as an oil-in-water emulsion will increase the surface area available for microbial attack and thus increasing the rate of biodegradation. However, water-in-oil, or *"mousse"*, emulsions can form, creating a low surface area to volume ratio, inhibiting biodegradation. A matter of a few hours in turbulent sea can change an oil-in-water emulsion to a mousse, and render the subsequent cleanup considerably more problematic [21].

Similarly *'tar balls'*, which are large aggregates of weathered un-degraded oil, restrict access by microorganisms because of their limited surface area. Auto-oxidation, photo-oxidation and the removal of low molecular weight hydrocarbons by microbes all aid their formation and tar balls may take thousands of years to degrade [22].

Although mousses do not occur in jet fuel contaminated soils, the hydrophobic nature of jet fuel is a critical consideration in soil decontamination. Air stripping and *mineralization* of volatiles and low molecular weight compounds in the top few inches of a contaminated soil, will leave behind residual oils and waxes. Due to their hydrophobic nature, these are relatively resistant to degradation and will reduce water infiltration and air circulation in the topsoil, thus depriving indigenous bacteria of two key requirements. A lack of infiltration may increase runoff and erosion. The development of preferential flow lines down wax coated cracks in a soil, will mean that some areas of a soil will become deficient in nutrients, water and oxygen. Products such as biodiesel, modified vegetable oil in the form of methyl esters, can be used to dissolve tar-like residual hydrocarbons. By dissolving aggregates, naturally occurring bacteria can degrade the components [23]. Biodiesel is itself biodegradable and less toxic than current alternatives of detergents and dispersants mixed with water.

Clearly, the physical state of the polluting hydrocarbon in soil or water will affect the uptake and utilization of the substrate, and ultimately the speed of *breakdown* of the pollutant. The oil may be dissolved in the aqueous phase and therefore be available for uptake. Alternatively, microbes may directly adhere to large oil drops at aqueous-hydrocarbon interface. There may also be direct contact with pseudosolubilized oil as fine or sub-micron size droplets. There may also be enhanced uptake of oil as a result of natural microbial production of *biosurfactants* or emulsifiers that increase the apparent aqueous solubility of the hydrocarbon [17].

Important environmental factors

Hydrocarbon degradation is also influenced by a number of environmental factors, which include the following:

- Solubility, concentration and state of hydrocarbon in water or soil.

- Temperature and pH — low temperatures have a retarding influence as do extreme pH environments

- Oxygen availability — oxygen is required for respiration and oxidation reactions.

- Nitrogen and phosphorous macronutrients necessary for cell metabolism and, in some cases, to act as electron acceptors.

- Micronutrient availability is also very important. There is often sufficient trace elements in soil, although iron has been found to be limiting in some circumstances.

- Composition of *microfauna* — in chronically polluted areas there will be adapted degradation specialists able to multiply enormously when confronted with fresh inputs.

- Soil composition, clay and organic content are important as they influence bioavailability.

Solubilization

Most hydrocarbons are insoluble. Work has been carried out to establish the relationship between growth rates and solubility of solid hydrocarbons [24]. The solubility of napthalene, phenanthrene, anthracene, and naphthacene in water is 98, 9.0, 0.45, and 0.07 µM respectively. Attempts to isolate bacteria for the most insoluble of these, naphthacene were unsuccessful, thus leading to the suggestion that bacteria only utilize solid aromatic hydrocarbons in the dissolved state.

However, when crude oil is mixed with water approximately 0.02% is extracted into the aqueous phase. This fraction contains the most toxic of the hydrocarbon constituents. This soluble fraction may include phenols, anilines and alkylated derivatives of benzene and naphalene. A number of these fractions such as *n*-hexane, *n*-octane, benzene, and naphthalene are soluble in water as well as compounds shown in Table 3. It is because of their water solubility that these compounds are the greatest immediate threat to groundwater [24].

However, although these may be toxic in saturated concentrations, in low concentration they are readily biodegradable.

Pseudosolubilization

Pseudosolubilization involves the uptake in the bulk phase of fine droplets less than 1 µm in diameter by pinocytosis, resulting in the development of

Table 3
Solubility of some hydrocarbon fractions in water

Compound	Aqueous Solubility mg L^{-1}
Butane (C_4H_{10})	61.4
Decane ($C_{10}H_{22}$)	0.052
Hexadecane ($C_{16}H_{34}$)	0.0063
Benzene (C_6H_6)	1,780
Naphthalene ($C_{10}H_8$)	30
Anthracene ($C_{14}H_{10}$)	1.29
Benzo[a]pyrene ($C_{20}H_{10}$)	0.003
Biphenyl ($C_{12}H_{10}$)	7.5
Arochlor 1221 (mixture, 21% Cl)	0.59
Arochlor 1254 (mixture, 54% Cl)	0.057

Source: Miller [24].

inclusions in the cell membrane (e.g., hexadecane, heptadecane and hexadec-1-ene) in *Acinetobacter* spp. Pseudosolubilized oil is more susceptible to microbial degradation while larger aggregates may persist for long periods of time due to the low surface area available for contact with microorganisms [24].

Temperature

Biodegradation is possible across a range of temperatures (< 0-70°C+), as psychrotrophic, mesophilic, and thermophilic hydrocarbon utilizing microorganisms have been isolated. Substrate utilization at different temperatures is often expressed in terms of Q_{10} values, denoting the factor by which the rate of substrate utilization decreases for each 10°C fall in temperatures [17]. Therefore higher temperatures increase the rates of hydrocarbon metabolism to a maximum, typically in the range 30°C to 40°C, above which the membrane toxicity of the hydrocarbons is increased. Warmer environments generally have greater degradation rates and increased volatilization of low molecular weight hydrocarbons. For this reason there is increasing concern about the *recalcitrance* of such pollutants in arctic and sub-arctic climates [17].

Increased research and recreational activities in these cooler environments, and the associated use of petrol and diesel driven machines, have increased the likelihood of spillage in this environment. Temperature will clearly show its impact on the evaporation of lower molecular weight components of jet fuel [25]. However, Margesin and Schinner [26] found that reduced temperatures (10°C) were not a problem for bioremediation of diesel-oil-contaminated alpine soil. Concentrations of 4000 mg/kg soil were reduced to 380-400 mg/kg after 155 days when inorganic fertilizer was added. A third of this reduction was by abiotic processes, but two thirds of the removal was due to biodegradation of the diesel. The addition of cultures of cold adapted diesel-degrading bacteria had only limited impact and *in situ* bacteria were equally capable of degrading the oil after an acclimation period.

Oxygen

Oxygen is probably the most important limiting nutrient in most biodegradation operations [27]. The rates of degradation decline with reduced O_2 concentrations. Ward et al. [28], working on the Amoco Cadiz spill found that anaerobic degradation caused 5% loss in 233 days, whereas under aerobic conditions there was 20% loss in 14 days. The degradation pathways of aliphatic, cyclic, saturated and aromatic hydrocarbons, by bacteria and fungi involve oxygenases and molecular O_2. Oxygen availability in the oil/water/soil matrix can thus be a rate-limiting variable.

Anaerobic degradation of oil is negligible, but when it does occur nitrate and sulphate can act as surrogate electron acceptors. The byproducts of anaerobic degradation are hydrogen sulphide (H_2S) and methane, both of which can be problem in offshore oil production and soils contaminated with *jet fuel. Desulfovibrio desulfricans* is the only significant anaerobic hydrocarbon- degrading bacterium [19].

Availability of nutrients

The inorganic nutrient requirement is often based on total hydrocarbon concentration (THC), not on the total available hydrocarbon concentration. This is because at any point in time, only a fraction of the soil hydrocarbon content is available for degradation. These are the portions present at the water-hydrocarbon interface, with most hydrocarbon adsorbed in the soil peds acting as a reserve pool. Thus, the instantaneous requirement for nutrients will be considerably less than the calculated requirement for total hydrocarbons. Additionally, there will be a requirement for the addition of nutrients dependent on the rate of hydrocarbon utilization and concurrent nutrient deficit. Similarly, the more water soluble fractions will be degraded more quickly, leaving behind the more recalcitrant fractions. These may require alternative nutrients in varying concentrations before being degraded.

Crude oil varies in composition depending on the location, age and strata of the surrounding rock. Consequently the distillate fractions/cuts derived from a crude oil will contain varying trace quantities of nitrogen and phosphorous. This has important implications for degradation processes since some fuels will contain quantities of N and P whilst others will not.

Micronutrients are also important for the degradation of hydrocarbons. Iron is an important trace element in the degradation of toluene under anaerobic conditions. Iron in the form of iron octoate [ferric (2-ethyl hexanoate)$_3$] or ferric ammonium citrate has been found to be effective at stimulating biodegradation of oil in certain aquatic environments under conditions where the water was supplemented with nitrogen and phosphorous [12].

Ferrous iron may also enhance biodegratation, particularly in soils and sediments containing sulphate-reducing bacteria [29]. In this environment iron prevents sulphide toxicity by the precipitation of ferrous sulphide. The addition of Fe^{2+} is an effective non-toxic approach to sulphide toxicity, in many ways better than the addition of chelating agents or vapour stripping.

Soil characteristics: Clay content

Aelion [30], found that there was a positive correlation between residual gasoline based fuel contamination of a coastal soil and the clay content of the sediment. Clay has a very high surface area to volume ratio due to the fact that stacked silicate plates generate a complex three-dimensional structure. Clay has a great sorption capacity for contaminants, but can easily become anaerobic if saturated by fuels, oils or water.

Some fuels are more viscous than others, and thus less likely to enter the smaller openings of a clay structure. Such fuels remain in discrete hydrocarbon complexes on the outside of peds where oxygen water and nutrients are more readily available. Gasoline is less "oily" and able to penetrate the tertiary structure of the clay where it is either less available to bacteria or where local anaerobic conditions may exist preventing biodegradation taking place. Addition of clay to the soil samples has been shown to reduce evaporation as the components adsorbed to clay particles due to the high cation exchange capacity of many clays. It is also known that as clays become wet they swell and prevent water movement through them. This not only has an impact on the aeration of a clay soil, but also on the supply of nutrients to the soil/

bacteria. A combination of these effects are probably responsible for the reduction in biodegradation in clay soils seen by Aelion [30]. From this data it can be concluded that a clay environment has many more different chemical, and physical properties than a sandy environment (e.g., sorptive and moisture content) which may impact microbial processes [30].

Soil characteristics: Organic content

Monroe et al. [31] found that the incorporation of organic matter into soil material in an Alaskan spill site increased the extent of Prudhoe Bay crude utilized. This may be a function of the increased surface area available for biodegradation created by the addition of the partially degraded moss layer. However, Dean-Ross et al. [25] state that the higher organic contents of a sediment appears to reduce the biodegradability of hydrocarbon components by adsorbing the hydrocarbons thus rendering them unavailable to microorganisms for biodegradation. The difference in the oil type may explain this contradictory evidence. Crude oil, due to its viscosity would remain homogenous and spread over the organic material, whereas because of the solubility of some components of fuel oil it is likely that these would adsorb to the organic and humic material thus rendering them less accessible to microbial attack.

MICROBIAL UPTAKE OF HYDROCARBONS

Clearly, the *bioavailability* of hydrophobic compounds is determined by their sorption characteristics, and chemical properties, all of which dictate their dissolution or partitioning rates and influence uptake and transport into the microbial cell [32].

Direct adhesion

Direct adhesion of microbes onto oil droplets allows exposure to the greatest surface area and clearly facilitates more rapid degradation. The less soluble fractions are usually of chain length C_{10} and greater. *Acinetobacter* spp., grown on pure *n*-alkane or on crude oil, demonstrated direct contact, densely covering the hydrocarbon spheres. There is also a positive correlation between drop size and the growth rate, with direct adhesion on larger droplets (<20 mm diameter) or *pseudosolubilization* on an increased surface area (<1 mm diameter) offering optimal growth [12].

Surfactant formation

Oil-in-water emulsions may be formed in high turbulence resulting in the creation of large phase interfaces in the aqueous medium. A variety of microorganisms bring about emulsion formation with the release of extracellular emulsifying agents. Often referred to as *biosurfactant*, these agents can be classified into several broad groups including glycolipids, lipopeptides, lipopolysaccharides, phospholipids and fatty acids/neutral lipids (Table 4). The production of surfactants thus enhances hydrocarbon uptake by increasing direct contact between the microorganism and its substrate.

Table 4
Production of emulsifying factors by microorganisms

Microorganisms	Excreted emulsifying factor
Candida petrophilium	Peptide and fatty acid moieties
Arthrobacter & *Brevibacterium*	Trehalose lipid
Corynebacterium & *Nocardia*	2 mol a branched β-hydroxyle fatty acid
Torulopsis gropengiesseri	Saphorose acid
Pseudomonas aeruginosa S7B1	Rhamnolipid
P. aeruginosa KY4025	Rhamnolipid

Source: Gutnick and Rosenburg [14].

Surfactant production effectively increases the utilizable surface area of the oil droplet and acts as a protective barrier against other microorganisms competing on the same oil droplet. The fungus, *Hormoconis (Cladosporium) resinae* has a highly hydrophobic cell surface to make intimate contact with the substrate whilst inhibiting other microorganisms from exploiting the same niche [33].

MECHANISMS OF BIOLOGICAL BREAKDOWN

The main pathway of breakdown of alkanes, particularly those with 11 to 19 carbon atoms, involves a monooxygenase (hydroxylase)-mediated oxidation of the C_1 methyl group (Figure 2). This monoterminal attack usually forms a primary *n*-alcohol. This is dehydrogenated to the corresponding aldehyde and fatty acid (monocarboxylic acid), and incorporated into central pathways of metabolism via fatty acid β-oxidation pathways to acetyl groups.

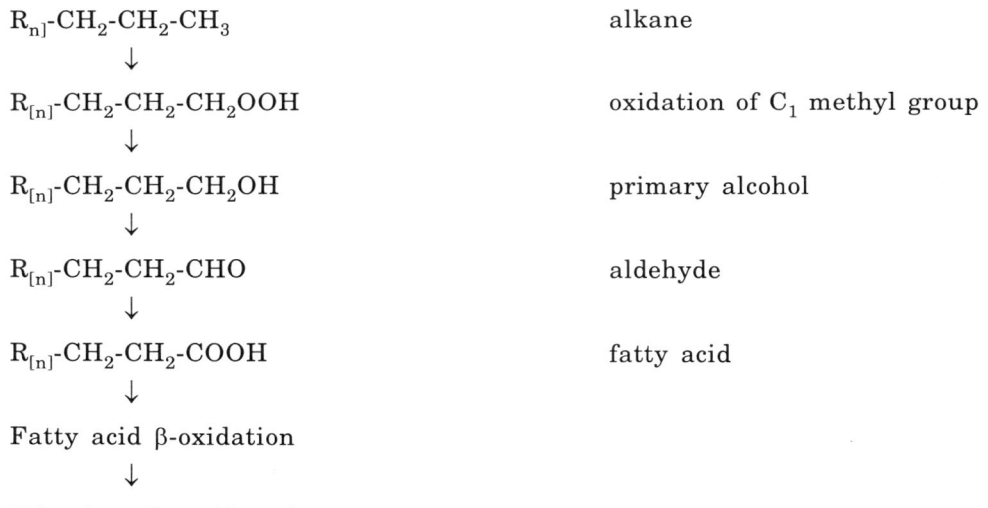

$R_{[n]}-CH_2-CH_2-CH_3$	alkane
↓	
$R_{[n]}-CH_2-CH_2-CH_2OOH$	oxidation of C_1 methyl group
↓	
$R_{[n]}-CH_2-CH_2-CH_2OH$	primary alcohol
↓	
$R_{[n]}-CH_2-CH_2-CHO$	aldehyde
↓	
$R_{[n]}-CH_2-CH_2-COOH$	fatty acid
↓	
Fatty acid β-oxidation	
↓	
Tricarboxylic acid cycle	

Figure 2. Metabolism of alkanes.

In this cyclic β-oxidation pathway, hydrocarbons are degraded to acetyl-CoA, which can be fed into the TCA cycle or used in biosynthesis.

If a secondary alcohol is formed (sub-terminal attack), a ketone is formed which may be converted to acetate and a long-chain primary alcohol. This is then degraded as described above.

Branched alkanes are utilized by fewer organisms but with greater difficulty. The assumption is that the methyl branches either impede uptake into the cell or make the hydrocarbon less compatible with the enzymes involved in β-oxidation. The cyclic alkanes (Figure 3) are also initially attacked by a hydroxylase enzyme producing a cycloalkane and leading to direct insertion of an additional oxygen molecule in the ring. The resulting lactone in eventually cleaved to produce a dicarboxylic acid [34].

Aromatic hydrocarbons are also degraded to intermediates by central metabolic routes, dioxygenation and ring scission, but are broken down more slowly than n-alkanes. Aromatics are attacked by both mono- and dioxygenases (requiring molecular oxygen) producing an initial dihydroxy compound that can subsequently be attacked by a ring-opening oxygenase. The resulting product, often a dicarboxylic acid or semialdehyde, is further metabolized to

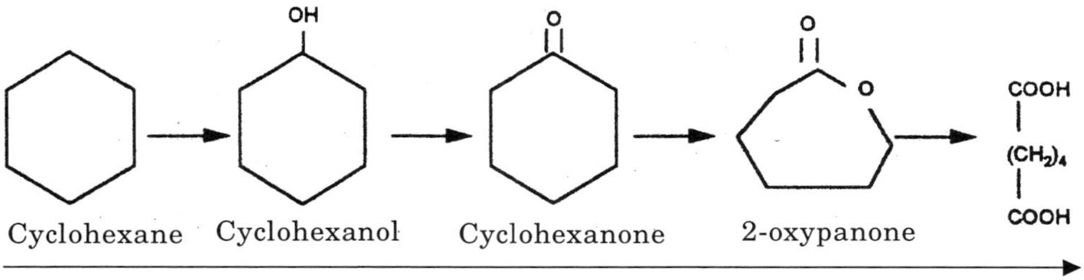

Figure 3. Cycloalkane degradation pathway.

produce intermediates that can be routed into the intermediary metabolism of the cell.

Intermediates of these central metabolic routes are partly oxidised to carbon dioxide and water and partly converted into biomass resulting in the growth and replication of the organism [33].

Molecular weight greatly affects whether a compound is volatilized, biodegraded or recalcitrant. Dean-Ross [25] demonstrated that for hydrocarbons with a molecular weight equivalent to decane ($C_{10}H_{22}$) and lower, evaporation was the main removal process. Above decane, for the compounds undecane ($C_{11}H_{24}$), dodecane ($C_{12}H_{26}$) and hexadecane ($C_{16}H_{34}$), no biodegradation took place. However for tridecane ($C_{13}H_{28}$), tetradecane ($C_{14}H_{30}$) and pentadecane ($C_{15}H_{32}$) biodegradation took place and was the dominant process for removal. Biodegradation is thus incredibly compound specific.

REMEDIATION STRATEGIES AVAILABLE

There are a number of approaches to remediation of a given oil polluted site, but in general there are two options, to treat the polluted soil, sediment or water *in situ* or to remove and treat *ex situ*.

In situ

Engineered *in situ* bioremediation requires that the environment in the soil is maintained in order that degradation can take place. Effectively moisture, oxygen, nutrients, pH and temperature must all be maintained within certain parameters in the soil. Often this means that bioremediation *in situ* is more complex to control and susceptible to small changes to the above parameters [35].

In situ treatment aims to enhance the activity of native subsurface bacteria by the addition of sufficient oxygen in the form of dissolved oxygen, hydrogen peroxide or nitrate. Suitable nutrients may also be needed for degradation to take place depending on the contaminant under consideration. Anaerobic degradation may also be stimulated if necessary by the addition of a starch based compound, i.e. potato wash water.

The choice of system for the bioremediation of an organic compound in a soil environment will depend upon many factors including contaminant solubility, soil type, hydrology, soil water mobility, risk to health, risk to groundwater, depth of contamination and time constraints and available funds.

Oxygen treatments

Due to the low water solubility of oxygen various methods have been investigated as a means of elevating the oxygen levels in oil-contaminated soils. Hydrogen peroxide is the oxygen source most commonly used in biological treatment of subsurface contamination. It can provide a reservoir of available oxygen in the groundwater at much higher concentration than water dissolved oxygen. However, hydrogen peroxide is highly reactive and has a tendency to *"off gas"* at the point of injection or prematurely decompose.

Aggarwal et al. [27] carried out work to improve the effectiveness of enhanced biodegradation methods to decontaminate jet fuel spills. This study was concerned with improvements in nutrient formulation, reducing *in situ* precipitation and the testing of potential stabilizers for hydrogen peroxide.

Nutrient addition

The very low concentration of nitrogen and phosphorous in unpolluted seawater are known to limit the rate of microbial growth on hydrocarbons.

Atlas [17] found that increasing concentration of nitrogen and phosphorous in Kuwait oil at 14°C with 3.0 mg ammonium nitrogen and 0.6 mg phosphorous in 70 mg/l oil, attained maximum biodegradation. Atlas and Bartha [12] also demonstrated the addition of 1 mg N and 0.07 mg P to 8 g/l oil supported maximum degradation of Sweden crude oil. Similarly, Reisfield et al. [36] found that optimal degradation of Iranian crude in the Mediterranean were achieved with 11 and 2 mg/l of N and P for 1 g/l oil in water.

Atlas [17] also suggests that nutrients are rate limiting in free oil-in-water, but less so in soluble hydrocarbons since the solubility is so low that available nutrient concentrations are adequate to support hydrocarbon degradation. Fertilizers such as paraffinized urea and octylophosphate in ratios C/N and C/P, of 10:1 and 100:1 (C:N:P = 100:10:1) respectively have been suggested for optimal growth. Dibble and Bartha [37] suggest ratios of 60:1 C/N and 800:1 C/P, thus illustrating the nutrient requirement specificity of oil-in-water mixes and the need for individual consideration of any acute or chronic discharge.

However, nutrient formulations used to enhance *in situ* biodegradation of fuels pose two major problems, namely: (i) plugging of the aquifer due to excessive precipitation of phosphate, and (ii) insufficient delivery of oxygen due to rapid decomposition of hydrogen peroxide.

Precipitation of phosphates can be reduced by the use of polyphosphate as opposed to orthophosphates. Trimetaphosphate (TMP) was also found by Aggarwal et al. [27] to be a suitable alternative. Furthermore, citrate was shown to suppress peroxide degradation by enzymatic and inorganic catalysts in the laboratory. However these results were not confirmed in the field. Spain et al. [38] found that bacterial catalase is responsible for the premature and rapid decomposition of H_2O_2 at a jet fuel site undergoing enhanced biodegradation.

Before considering the addition of nutrients or oxygen as a means to enhance or facilitate decay the following points should be considered:

- Determine limiting factor(s) and formulate to supply adequate, but not necessarily excess, nutrients. Do not supply unneeded chemicals.

- Determine the solubility limitation of any nutrient salts to be added in groundwater. That is to say add no more nutrients than can be solubilized and transported into the aquifer.

- If H_2O_2 is utilized as an oxygen source, determine the site-specific H_2O_2 stability and the real effectiveness of stabilizers utilized. Do not add stabilizers that are not effective or insufficiently soluble to be transported in groundwater.

- Assess non-biological oxygen demand. This oxygen demand must be considered when determining oxygen requirements.

- The oxygen demand of any nutrient or H_2O_2 stabilizer must be taken into account.

- The effects of increasing the redox potential of injection waters must be considered, e.g., if groundwater containing ferrous iron is re-injected, the ferrous iron will both exert an oxygen demand and form a precipitate potentially causing plugging problems.

- The toxicity of any additive must be considered to both microorganisms and groundwater.

Protozoan populations can be suppressed by pulse dosing nutrients, and this approach also reduced formation of waxes that restricted oxygen and water penetration of soils [39]. There is also some evidence that excess addition of inorganic nutrients may reduce biodegradation [40,41].

Denitrifying conditions

Under certain conditions nitrate can act as an electron acceptor and this type of metabolism can results in anaerobic biodegradation of organic compounds via the process of nitrate reduction and denitrification. This process is much slower than aerobic degradation but under certain conditions may be an acceptable alternative. Nitrate is less expensive and more soluble than oxygen in water and thus more economical for the restoration of fuel contaminated aquifers. Reports on the extent and ability for microorganisms to degrade under denitrifying conditions has lead to the conclusion that biodegradation is influenced by several undefined factors, and that site specific studies should be carried out prior to implementing a nitrate-based remediation for fuel contaminated aquifers [42].

Soil venting

Many remediation strategies have involved injecting oxygen or hydrogen peroxide in water or an alternative electron acceptor (nitrate) and nutrients into contaminated soil. However, the amount of oxygen required for mineralization of hydrocarbons makes water soluble oxygen an unfeasible option and the use of H_2O_2 uneconomical. An alternative strategy is to pump air through the unsaturated soil profile to create aerobic zones and induce biodegradation.

In work by Arthur et al. [43], it was found that soil venting considerably increased biodegradation rates over static soils and offered an effective means of biodegradation in the vadose zone. The study concludes that oxygen (or another suitable electron acceptor) is often the main limiting factor for *in situ* biodegradation.

Hinchee et al. [44] carried out a study on the effect of soil venting on *in situ* biodegradation Findings indicated that forced aeration, coupled with additions of nutrients and moisture, stimulate hydrocarbon degrading microorganisms and presented a feasible approach to bioremediation management.

Surfactant enhancement

Studies investigating the addition of various surfactants to enhance biodegradation of oil in a soil under simulated conditions of soil venting found that surfactant plus soil venting failed to enhance biodegradation of fuel oil contaminated soil when compared to soil venting alone [43]. This may have been due to the surfactant acting as an alternative and preferred carbon source for microbial metabolism. Various bacteria are able to produce surfactants to aid in the biodegradation of fuels. Indeed in a study by Espeche et al. [45] the degradation of n-hexadecane in media was concurrent with the production of an extracellular surfactant by an *Acinetobacter* strain B2-2. This surfactant decreased the surface tension of the culture media and helped to disperse the oil improving access to the biodegrading microorganisms. The optimal conditions influencing n-hexadecane degradation were identified as pH 7.5 and temperature 20-25°C. Variations in the concentrations of substrate up to 2.0% had no inhibitory effect and the presence of an alternative carbon source was not utilised by the n-hexadecane specific bacteria.

Ex situ

Ex situ treatment of contaminated soil requires large scale soil removal using heavy machinery and the creation of extensive, contained land farm areas where treatments similar to the *in situ* treatments can be applied. As a rule this approach is often prohibitively expensive, and where the level of toxicity of the pollutant is intrinsically low, as is the case with hydrocarbons, uneconomic. However, treatment of contaminated surface and groundwater is more easily achieved *ex situ*.

Groundwater

Two approaches to bioremediate soil and groundwater are:

- to pump the groundwater to the surface or excavate soil for treatment in a bioreactor,

- to remediate the aquifer or soil *in situ* via the injection of nutrients oxygen, etc. [35].

Natural attenuation of the contaminant may be considered if there is no threat to abstractable water, no movement of the contaminant and biodegradation is taking place naturally. This approach requires that the contaminant plume is monitored carefully until an satisfactory end point is reached.

Pump and treat bioreactors

In a conventional well designed to recover contaminated groundwater, a pump at the base of the well creates a cone of depression in the groundwater surrounding the well, and the hydraulic gradient aids the flow of contaminant into the immediate area. A second *"oil skimmer"* pump is used to abstract the mixture of hydrocarbon/water for separation, treatment and or recovery above ground.

Alternatives have been developed, including a *"Bioslurper"* system that has a single suction tube inside, positioned at the surface of the free product floating on the groundwater surface [10]. Airflow through the upper soil layers ensures that vacuum-enhanced pumping creates a pressure gradient which induces flow into the well. As a result, free product, water and air are drawn up the tube for separation and treatment above ground. This limits the smearing of free product on the soil as well as requiring one pump for contaminant recovery.

Ex situ treatment may not recover all of the contaminant, is costly and there is the need to dispose of large quantities of sludge from diffused air bioreactors. There is also the problem of volatilization of the lower molecular weight hydrocarbons.

For water contaminated with gasoline, methyl tert-butyl ether (MTBE) is a very water soluble fuel additive and is causing significant problems in the US. MTBE is an oxygenate added to gasoline fuel to improve the octane rating and aid cleaner burning and less air pollution. It is very mobile in soil as it is ten times more water soluble that any other constituent of petrol. It is highly resistant to biodegradation although susceptible to photo-oxidation. The compound in groundwater at tens of ppb concentrations and is detectable to the palate as a turpentine taste. It is also considered a carcinogen and can penetrate through the skin and accumulate in the fatty subcutaneous tissue. The fact that it is so water soluble makes it treatable, as it can be extracted through injection well pump and treat. Mo et al. [46] found that oxygenates, such as methyl tert-butyl ether, ethanol, methanol, ethyl tert-butyl ether, and tert-amyl methyl ether can be degraded by acclimated consortia in lab and field studies. It is envisaged that the combination of biological and photocatalytic treatment technologies could be used in series, for the treatment of difficult or mixed waste streams from gasoline contaminated produced waters.

Phytoremediation/reed beds treatment

Reed beds are also efficient at removing oil from polluted water, the roots, rhizomes and growth media act to trap oil and sediment particles [47]. The growth media can be organic and clay based soils, sand, gravel, and crushed stone. This provides a large surface area for bacteria to adhere to. Although not widely used in the UK, reed beds offer a practical method of jet fuel contaminated water treatment.

The main disadvantage to the use of reed beds in bioremediation is the time to maturity of the reed beds of up to 3 years and thus as an emergency treatment is not practical. However it is envisaged that the development of trailer transported systems for long term pump and treat systems is a practical application.

CONCLUSION

Microorganisms are capable of utilizing hydrocarbons as a carbon and energy source. This has been demonstrated in the lab and at field scale studies. In order to implement the correct remediation strategy for the particular type of fuel it is essential that lab-based investigations are carried out for the site under study. Often in the past methods applicable to one site have been transferred to another site with little regard to the changes in variables, such as soil structure, soil chemistry, soil hydrology, nutrient requirements, and the presence of indigenous hydrocarbon-utilizing population.

A lack of knowledge of these site-specific characteristics has meant that all to often the incorrect remediation strategy has been applied or even worse a detrimental remediation strategy adopted. Field studies and implementation of remediation technology have progressed faster than research into biochemistry, enzymology and microbial ecology. There is still a requirement for basic data on toxicity, hydrocarbon/water hydrogeology and soil sorption of hydrocarbons.

Each contaminated site will be different, and there will be different requirements for nutrient inputs, aeration systems, and inoculum injection.

A suitable approach to remediation can be devised only after standard simulated lab-based investigations are carried out.

It is unlikely that *in situ* remediation will often be an option as the requirements for controlling the soil environment are to great. However, *in situ* remediation may be an option immediately after a *spillage* of hydrocarbon whilst it is still in the aerated topsoil. Additionally where contaminated sites are covered by buildings or the contaminant is too deep for extraction, *in situ* remediation may, in some cases, be the only option available.

Ex situ remediation of contaminated soil and water is the more appropriate remediation strategy. However, the application of excessive nutrients, cofactors and aeration is not only wasteful but may be *detrimental* to remediation of this type.

REFERENCES

1   Swannell RPJ, Lee K, McDonagh M. Microbiol Rev 1996; 60: 342-365.
2   Cooper PF, Job G D, Green MB, Shutes RB. Wastewater Treatment Swindon: WRC, 1996.
3   Aelion MC. J Cont Hydrol 1996; 22: 109-121.
4   Hinton SM, Minak-Bererno, V, Keim LG. In: Skipper HD, Turco, RF, eds. SSSA Bioremediation, Science and Applications Special Publications 1995; 43: 33-54.
5   Leahey JG, Colwell RR. Microbiol Rev 1990; 54: 305-315.
6.   Clark RB. Marine Pollution. Oxford: Oxford Scientific Press, 1989.
7   Howard PH. Handbook of Environmental Fate and Exposure Data. Chelsea, Mich, USA: Lewis Publishers, 1990.
8   Atlas RM, Bartha R. Microbial Ecology. Benjamin Cummings, 1994; 533p.

9   Yucom PS, Irvine RL, Bumpus JA. Water Research 1995; 67: 174-180.

10   HAZNEWS 1995; 84: 17.

11   Baker JA. Evaluation of the Natural Biodegradation of Jet Fuel JP-8 in Various Soils Using Respirometry. Masters Thesis for the Air Force Institute of Technology, Wright Patterson AFB, Ohio USA, 1995.

12   Atlas RM, Bartha R. Adv Microbial Ecol 1992; 12: 287-338.

13   O'Malley S. Microbial Degradation of Oils. BSc Dissertation, Dublin City University, 1991.

14   Gutnick DL, Rosenburg E. Ann Rev Microbiol 1977; 31: 379-396.

15   Hettige G, Sheridan J. In: Garg KL, Gars N, Mukerji KG, eds. Recent Adv Biodet Biodeg 1994; 2: 17-29.

16   Miller BM, Litsky W. Industrial Microbiology 1976; 391-398. Macgraw Hill.

17   Atlas RM. Microbiol Rev 1981; 45: 180-209.

18   Higgins IJ and Gilbert PD. In: Chater K W and Somerville HJ, eds. The Oil Industry and Microbial Ecosystems. Heydon & Sons, 1978; 82-117.

19   CONCAWE The Environmental Impact of Refinery Wastes 1979; 8/79.

20   Sinclair JL, Kampbell DH, Cook ML, Wilson JT. Appl Environ Microbiol 1993; 59: 467-472.

21   Horizon 1997. http//www.bbc.co.uk/horizon/oilspill.shtml.

22   Butler, JN, Morris, BF, Sass, J. Bermuda Biological Station Special Publication 1973; 10.

23   Mudge S. Laboratory News, Jan 15, 1998.

24   Miller RM. In: Skipper HD, Turco RF, eds. SSSA Bioremediation, Science and Applications Special Publications 1995; 33-54.

25   Dean-Ross D. Bull Environ Contam Toxicol 1993; 51: 596-599.

26   Margesin R, Schinner F. Appl Microbiol Biotechnol 1997; 47: 462-468.

27   Aggarwall PK, Means JL, Hinchee RE, Headington GL, Gavaskar AR. Methods to Select Chemicals for in situ Biodegradation of Fuel Hydrocarbons, Ohio,USA: Battelle Columbus Division, 1990.

28   Ward DN, Douglas AG, Porters RJ, Taylor J, Ciger W, Shafner C. Ar. Poll Bull 1983; 14: 103-108.

29   Beller HR, Reinhard M. Microb Ecol 1995; 30: 105–114.

30   Aelion MC, Brabley PM. Appl Environ Microbiol 1991; 39: 57-63.

31   Monroe EM, Lindstrom JE, Brown EJ, Braddock JF. 96th ASM General Meeting Abstract Q, 1996; 309: 439.

32   Sticher Patrick P, Jaspers MCH, Stemmler K, Harms H, Zehnder AB, Van der Mer JR. Appl Environ Microbiol. 1997; 63: 4053-4060.

33   Lindley ND, Heydeman MT. FEMS Microb Ecol 1985; 307-310.

34   Prichard PH, Mueller JG, Lantz SE and Santavy DL In: Allsopp D, Colwell RR, Hawksworth, eds. Microbial Diversity and Ecosystem Function. CABI Bioscience Int 161-185.

35   Holliger C, Gaspard S, Glod G, Heijman C, Schumacher W, Schwarzenbach RP, Vazquez D. FEMS Microbiol Rev 1997; 20: 517-523.

36   Reisfield A, Rosenberg E, Gutnick D. Adv Appl Microbiol 1972; 14: 93-122.

37   Dibble JT, Bartha R. Appl Environ Microbiol 1979; 37: 729-739.

38   Spain JC, Milligan JD, Downey DC, Slaughter JK. Groundwater 1989; 27:163-167.

39   Bergeron L. New Scientist 1997; 11.

40   Morgan P, Watkinson RJ. Wat Sci Technol 1990; 22: 63-68.

41   Moller J, Gaarn N, Steckel T, Wedebye EB, Westerman P. Bull Environ Contam Toxicol 1996; 54: 913-918.

42   Hutchins SR, Downs WC, Wilson JT, Smith GB, Kovacs DA, Fine DD, Douglass RH, Hendrix DJ. Groundwater 1991; 29: 571-580.

43   Arthur MF, O'Brien GK, Marsh S and Zwick, NCEL Report 1992; CR
        92.004.

44   Hinchee RE, Arthur M. Appl Biochem Biotechnol 1991; 28/29: 901-
        906.

45   Espeche ME, MacCormack W, Fraile ER. Int Biodet Biodeg 1994; 33:
        187-196.

46   Mo K, Lora C, Javanmardian M, Yang X, Kulpa CF. 96th ASM General
        Meeting Abstracts 1996.

47   Cooper PF, Green MB. Wat Sci Technol 1995; 32: 317-327.

Biotransformations: Bioremediation Technology for Health and Environmental Protection
V.P. Singh and R.D. Stapleton, Jr. (Editors)
2002 Elsevier Science B.V.

Bioremediation technology for environmental protection through bioconversion of agro-industrial wastes

T.N. Lakhanpal

Department of Biosciences, Himachal Pradesh University, Shimla - 171 005, India

INTRODUCTION

All of us are witnessing periodic sharp increases in the price of food and energy. There has also been observed an increase in the generation of waste materials, including the agricultural residue, through various activities. The accumulation of such material causes pollution, posing danger to health and hygiene. It has become necessary to make efforts to utilize the wastes for generating energy and/or recycle them through same *bioagents* so that they, not only relieve the pressure on the environment, but also help in alleviating health hazards. In this context, mushroom cultivation is one biotechnological process, which can convert the waste plant residues, rich in lignocellulosics into protein rich foods. Mushroom cultivation infact is the only microbial process or product system which can bioconvert all of the major plant polymers, such as lignin, cellulose, and hemicellulose.

The conversion on naturally occurring complex lignocellulosic materials into simpler substances by living substances is termed as bioconversion. It is one of the major, important, and significant processes, which puts the otherwise waste materials into appropriate use. Different types of waste materials are generated daily or else seasonally as crop residues. One such material is cereal straw, which is generated in large quantities every year. Lignocellulosics comprise mainly of cellulose, hemicellulose, and lignin.

It is difficult to exactly estimate the quantities, nature, and disposal of plant wastes because of their varied and scattered availability. However, as per Food and Agricultural Organization (FAO) estimates [1], the total annual production of cereal lignocellulosic straw is 2669 mT, of which 1135 mT is produced in Asia alone. The residue of the agriculture is usually left in the field. If it remains unutilized, it decomposes and harbours microbes, which cause putrifaction and environmental pollution. If it is burnt in the field, as is usually done by farmers, it becomes a source of pollution causing bronchial diseases. Hence, it is essential that same profitable methods for its disposal are developed, whereby pollution is checked and material is properly utilized. Therefore, by growing mushrooms, such materials can be converted into protein-rich food and also good feed.

Depending upon their origin, wastes are classified as crop wastes, forestry wastes, *agro-industrial wastes*. Paddy and wheat straw, sugarcane trash, peels and vines of vegetables, fruits stalks, straw or sticks of maize, jute, cotton, etc. are crop wastes. Sawdust, bark clippings underutilized or non-edible plants constitute the forestry waste. Rice husk, bagasse, molasses cotton linters bran, etc. are examples of agro-industrial wastes (Table 1) [2].

Bioconversion of wastes through mushroom cultivation is an outstanding example of simple and low cost bioconversion technology. Because mushrooms can easily utilize the cereal straw and other lignocellulosic wastes, mushroom cultivation has picked up worldwide. In bioconversions the enzyme complexes of mushrooms act upon the lignocellulosic material degrading and utilizing it to produce mushroom fruiting bodies with high value food protein, which can be directly consumed by human beings. The residues can also be used as an animal feed or as an effective soil fertilizer and/or conditioner. The biodegradation of lignocellulose into soluble products is generally through extracellular enzyme. Among lignocellulosic materials, the degradation of lignin is not easy but cellulose is easily degradable. Lignin is selectively

Table 1
Quantity of agricultural residue/waste/byproducts

Agricultural residue/ waste/byproduct	Estimated quantity/ year (million tons)
I Crop residue	
Wheat straw	31.80
Paddy straw	88.00
Maize stalk	14.25
Maize cobs	6.85
Sorghum sticks	32.80
Pearl millet straw	8.80
Finger millet straw	10.50
Barley straw	4.40
Gram straw	10.80
Pigeon pea stable	3.64
Groundnut shell	5.75
Sugarcane trash and dry leaves	1.77
Jute sticks	2.05
Jute bark and leaves	2.05
Cotton sticks	10.05
Mango peel and kernel	1.25
Pineapple wastes	0.02
Citrus fruit peel, pomace, red, etc.	0.03
Banana pseudostems	0.2
Pea shells and vines	0.01
Tomato seeds and pomace	0.08
Coconut shell	0.88
Coconut water	0.50
Coconut husk	0.06
Arecanut husk	0.10

Table 1 continued

Agricultural residue/ waste/byproduct	Estimated quantity/ year (million tons)
Arecanut leaf sheaths and leaves	0.90
Cashew apple	30.00
Cashew testa	0.04
Cashewnut shell liquid	0.01
Tea fluffs, stalks and sweepings	0.45
Coffee residue	0.62
Tobacco leaf scrap and stalks	0.62
Tobacco seeds	0.15
Rubber wood	0.25
Bark from industrial wood	0.30
II Agro-industrial residue/waste/byproducts	
Rice husk	18.0
Rice bran	3.0
Deoiled rice bran	2.1
Jute mill waste	0.5
Cotton dust	0.3
Groundnut deoiled cakes	2.5
Deoiled cakes of other major oilseeds such as rapeseed and mustard, linseed, seasame and castor	2.0
Deoiled cake of non-edible oilseed, such as Kusum, Neem, Karanza, etc.	0.17
Sugarcane bagasse	5.25
Molasses	1.70
Press mud	0.20
Sawdust	2.00

Source: Srivastava and Maheshwari [2].

degraded by organisms, which secrete lignin peroxidase or laccase secreting organisms like bacterial species of *Nocardia, Streptomyces, Bacillus, Pseudomonas* and species of mushrooms *Agaricus, Plenritus,* etc. [3-5].

Cellulose is degraded by bacteria, e.g., *Cellulomonas, Cytophaga, Polyangicun, Streptomyces* and species of fungi, e.g., *Trichoderma, Chaetomium, Fusarium, Aspergillus, Penicillium, Agaricus,* and *Pleurotus* etc. [6-8]. White-rot fungi, including mushrooms upgrade digestibility and nutritive value of agricultural wastes (Table 2).

Table 2
Comparison of 1990 and 1994 world production of some cultivated *edible* mushrooms (Unit: 1000 metric tons)

Species	1990		1994		% Increase
	Fresh wt.	%	Fresh wt.	%	
Agaricus bisporus	1424.0	37.8	1846.0	37.6	29.6
Lentinus edodes	393.0	10.4	826.2	16.8	110.2
Pleurotus spp	900.0	23.9	797.4	16.3	11.4
Volvariella volvacea	207.0	5.5	298.8	6.1	44.3

Bassham [9] estimated that net productivity of biomass due to photosynthesis by plant on earth is around 155.2 billion tons/year, nearly two third of this is produced on land and one third in the oceans. Dunlap and Chiang [10] estimated that of the total land biomass about 1.25% is used for human food, about 9% is lost as waste during harvesting or processing and rest is available as lignocellulosic wastes. Eveleigh [11] estimated the annual production of biomass to be around 1.8×10^{12} tons, of which about 150×10^{16} tons is used by pulp and paper industry. Lutzen et al. [12] put the estimate of the lignocellulosic wastes at 50% of the total biomass of the world.

Ghose and Ghose [13] estimated that nearly 25 million tons of cellulosic agricultural by-products are produced in India. Garcha [14] pointed out that in 1979, 20.8 million tons of wheat and paddy straw was available for mushroom cultivation. Swaminathan [15] estimated that 269 million tons of paddy as surplus. Chadha and Sharma [16,17] pointed out that around 300 million tones of agricultural waste is annually available in India, and half of the residue remains unused. Chakravarty [18] estimated that in India about 20 million tons of rice husk and chaff and 4 million tons of rice bran are produced annually. Kuhad and Singh [19] estimated that around 30-50 × 10[7] tons lignin waste is produced for pulp and paper industry every year in India. Therefore, various estimates are available for the annual lignocellulosic waste generated. Table 1 summarises all types of waste materials generated in the country.

From amongst the 2000 species of mushrooms reported to be edible from all over the world, technology is available for the commercial cultivation of about 14 species [20]. There are, however, a few species, which account for all but a few per cent of the total mushroom production. They belong primarily to genera *Agaricus, Lentinus, Flammulina, Volvariella,* and *Pleurotus.* Among these the species *Agaricus* and *Pleurotus* rank first and third in world mushroom production (Table 2), whereas second, fourth, fifth, and sixth place is occupied by *Lentinus edodes, Auricularia* spp., *Volvariella volvacea,* and *Flammulina velutipes,* respectively (Table 3).

The white button mushroom (*Agaricus bisporus* or *A. bitorquis*) contributes about 37.9%, having the largest share in the world production. USA, Chine, France, and Holland are the major contributors of button mushroom to the world. However, its production in Asia has shown an overall decline during the last decade, particularly in countries like Taiwan, Korea, and China. On the other hand, countries like Indonesia and Thailand have shown increase in their production capacity. India has also shown increasing trends in the

Table 3
Cultivated mushrooms in the world with available production technology

---

Agaricus bisporus (Lange) Imbach

Agaricus bitorquis (Ouebt) Sacc.

Pleurotus spp.

(*P. ostreatus, P. sajor-caju, P. eryngii, P. cornucopiae, P. fossulatus, P. flabellatus, P. opuntae, P. eitrinopileatus, P. membranaceus, P. platypus, P. eous, P. sapidus, P. cystidions, P. columbinus, P. pulmonarius*)

Volvariella volvacea (Bull ex Fr.) Singer

Stropharia rugosonnulata Forlow ex Murril

Auricularia auricula Auct.

Pholiota aererita Brig.

Pholiota mutabilis (Scheeff. ex Fr.) Onebt

Lentinus edodes (Berk.) Singer

Armillariella mellea (Vahl. ex Fr.) Richen

Coprinus fimetarius Fr.

Coprinus comatus Mill. ex Fr.

Flammulina velutipes (Curb. ex Fr.)

Tremella fuciformis Berk.

---

production and productivity of white button mushroom during the last few years, but its total production is still too meagre to make significant contribution to the world trade of mushrooms. However, during the past few years, India has made considerable progress in the export of processed mushrooms. The world trade in mushrooms, which runs into several thousand crores of rupees, is still concentrated in six countries viz. US, Germany, U.K., France, Italy, and Canada. These countries are also the greatest consumers, consuming 85% of the mushrooms produced. Europe and North America also account for 55% and 27% of the world production, respectively. Eastern Asia contributes meagre 14% of the total output. For canned-mushroom export, Netherland occupies the top position with 36%, followed by China, France, and Spain. The largest importers are Germany and USA. The per capita consumption of mushrooms is also high in Europe and America. Some important mushrooms grown in India with available production technology are given in (Table 4).

LIGNOCELLULOSIC MATERIALS AND THEIR CONVERSION THROUGH CULTIVATION OF DIFFERENT MUSHROOMS

Cultivation of white button mushroom (A. bisporus)

Cultivation of white button mushroom is a complex, technical process, the preparation of compost requires special skills and handling. The success of crop depends upon the preparation of the compost. In the beginning horse

Table 4
Important mushrooms grown in India with available production technology

	Scientific name	Common name
(i)	*Agaricus bisporus*	Button/European/Temperate mushroom
(ii)	*A. bitorquis*	Button/Edulis/Hot weather mushroom
(iii)	*Plenrotus sajor-caju, P. flabellatus P, ostreatus, P. florida, P. Citrinopileatus, P. cornucopiae, P. sapidus, P. membranaceous, P. eryngii, P. fossulatus, P. eous* and *P. platypus*	Oyster mushroom
(iv)	*Volvariella volvacea* and *V. diplasia*	Paddy straw/Chinese/Tropical mushroom
(v)	*Auricularia polytricha*	Black ear mushroom
(vi)	*Colocybe indica*	White milky mushroom
(vii)	*Stropharia ruguso-annulata*	Brown cap or Giant mushroom
(viii)	*Lentinus edodes*	Shiitake

manure was used for making compost. Compost prepared from horse manure is called natural manure, while the compost prepared from cereal straw or other wastes is termed as synthetic compost. Depending upon the availability of the raw materials, different formulations have been proposed and used by different workers and different organizations, from time to time. Two main methods are used for preparing compost in India — the long method, which does not require pasteurization, and the short method, where pasteurization facilities are available. For spawn run or vegetative growth, a temperature range of 22-28°C is required whereas for fruiting body formation it requires a temperature range between 15-18°C, Humidity range of 85-95% if essential along with proper ventilation. The following three basic steps are involved in the cultivation of this mushroom.

(i)   Production of spawn.

(ii)   Preparation of compost.

(iii)   Crop production.

The composting material consists basically of two types of ingredients. For convenience they are categorized into i) base and ii) supplementing materials. The base materials have straw of cereals, maize stalks, hays, sugarcane bagasse or any other material rich in cellulosic contents. The straw should be fresh; older straw does not compost well. The size of straw also has a proper role in composting, the desirable size is 5-8 cm, smaller than this clumps and compacts, not allowing air circulation in the heap; larger than this usually makes the heap loose, causing improper fermentation. The cellulose, hemicellulose, and lignin in the straw provide the fungus with carbon source, and also gives proper texture to the heap so that aeration is proper. Some cereal straws, like that of barley and rice, are soft and degrade quickly. Whenever these are used care has to be taken while wetting and turning them during composting.

The base materials are usually deficient in the requisite amounts of nitrogen and other components required for the fermentation process. Therefore, materials rich in these substances are added as supplements. The various materials used for this purpose include animal manures, animal feed, nitrogenous fertilizers and various carbohydrate sources. Animal manures from horse, chicken, pig, and sheep are good in nitrogen content, which varies from 1-5 per cent. Cattle manure is not suitable for compost preparation. These manures also provide little carbohydrate. Molasses, potato waste, apple and grain pomace, wet brewers grain and malt sprouts provide carbohydrates, which are essential for maintaining proper C/N ratios and also for establishment of bacterial flora in the compost [21]. Nitrogen is also readily supplied with compost by using nitrogenous fertilizers, like ammonium sulphate, calcium ammonium nitrate, and urea. Animal feeds, such as wheat or rice bran, seed meal cotton dried brewers grain soya, castor, mustard and linseed, supply both nitrogen and carbohydrate. Nitrogen content may vary from 3-12 per cent [22]. Various formulations are used for compost preparation. Some of these are given below.

METHODS OF COMPOSTING

Long method

It takes 26-28 days. During the period the compost heap is turned 7-8 times at varying intervals. This method is practiced where pasteurization facilities are not available. This method is adopted by small growers lacking boiler facilities. The compost prepared by this method usually gives comparatively low yield of mushrooms. It is more liable to attacks by pests and disease during cropping; the source of wheat are, otherwise, eliminated during *pasteurization*. The fertilizers, bran and moist straw are thoroughly mixed in heap (1.75 m × 1.5 m × any length). Dry patches are watered lightly while making the heap. The heap is pressed from the top. On the second day, the stack temperature will rise to 65-70°C. On the sixth day, the stack is broken and made looser for proper aeration and remade using wooden boards (first turning), second turning is given on the tenth day, chalk powder (calcium carbonate) is added (if required), watering is done as required to keep it moist, and the heap is slightly pressed. Third turning is given on sixteenth day, if required, and cotton meal is added. The heap is then remade and pressed. On 19th day, fifth turning is given. The stack is remade but not pressed. Sixth turning is given on the 20th day and watering is done, if required.

By this time, the compost is generally free of ammonia, a requisite for good composting. The compost is then spread on the floor for sometime and seventh turning is given on the 25th day of stacking. On 26th day BHC or Lindane dust (5%) is mixed and the compost is filled in trays, one more turning is given if the compost still has ammoniacal smell. A good compost will be dark brown, free of ammonia, lacking greasiness with a moisture content of 65-70%. The pH is neutral or near neutral, i.e. pH 7-8.

Short method

This method takes about 18 days and is followed by growers having pasteurization and boiler facilities. It involves two phases, one out doors (8-12 days) and the other involves a boiler/pasteurization chamber (3-7 days).

In phase I, preliminary stacking is done on a concrete floor. Four days before composting, straw and chicken manure are placed in layers, and sufficient water is added to wet the straw. The stack is usually made of the size as for long method. On the first day, stack is opened, urea added and water sprinkled on dry patches and stack is remade. On the second day, stack is sprinkled with water, if necessary, and pressed hard. Second turning is given on the 4th day and gypsum is added. On 6th day, third turning is given; on 8ᵗʰ day, it is filled in trays for pasteurization and BHC or lintex or lindane dust (5%) is added. During turning, proper reorientation of different portions in the stack is done. The central portion is placed at the bottom of the new stack, the top and sides in the centre, and bottom part is placed on the top and sides. The temperature in the compact heap will rise, which may even go upto 80°C. Maintenance of temperature is necessary at different points for good composting. Over-composted substrate becomes deficient in certain nutrients and vitamins, while under-composted substrate retains ammonia and encourages *Coprinus* infection during spawn run and cropping.

In phase II, which is carried out in insulated room, the compost is loosely filled in trays to a depth of 15 cm. The trays are arranged in tiers, leaving a vertical space (about 10 cm) between trays, each side of the room (30 cm) in the centre (75 cm) and 5 cm between stacks for circulation of the hot air or steam. Avoid cooling of compost during filling, which is possible by heating the room after stacking. On the first day, the door ventilators are closed, the room temperature is brought to 48-49°C by dry heat. Fresh air is inducted with a reverse mounted exhaust fan for 2 hours. Then the temperature of the compost is raised to 54.4°C by injecting steam on the second day, temperature is recorded every hour, but proper aeration is maintained. On the third day fresh air is circulated for 15 minutes, and the bed temperature is raised gradually to 58-60°C by steam after closing all vents. Keep the temperature at 60°C for one hour to bring the bed temperature below 57.6°C and air temperature to 46-49°C. This temperature is maintained for almost an hour, then fresh air is introduced on the fourth day, the *fermentation* of the compost will bring the temperature to 53-55°C. When fresh air is introduced, the temperature of the compost gradually drops on the fifth day. By further aeration, the temperature on 6ᵗʰ day drops down to 35-37°C. On the 7ᵗʰ day, the trays are shifted to the spawn room and the temperature is allowed to drop down to around 29°C. The compost of the stack has 60-70% moisture content.

The spawn is prepared either on sorghum or wheat grains. Fresh spawn is always used. Rate of spawning is important in getting optimum yield. It is generally calculated per unit bed area or according to the quantity of compost to be spawned. Generally, 250 g of spawn is enough to "seed" two trays of the size 90 × 50 × 15 cm or 1 m^2 bed area with 32 kg compost. This comes to less than 0.5% of compost (4.5-5 kg/ton compost). After spawning, the compost is pressed hard to make it compact. The trays are then arranged in the cropping room in tires and covered with newspaper sheets sprayed with 2% formalin. The temperature during spawn run is maintained at 20-25°C. The newspapers are sprinkled with water at least twice a day to ensure 90-95% RH in the room. Within 15-20 days, the compost gets impregnated with mycelial threads completely. The dark brown colour of the compost changes to light brown. Little ventilation is required during spawn run.

The newspapers are removed and after spawn run of compost (3-4 cm in depth) is covered with a layer of sterilized casing soil constituents of either garden soil and sand mixture (4:1) or decomposed cow dung manure and loam soil (1:1) or spent compost (2 years old), and lime (75 kg + 1 kg). Afterwards the cased trays are lightly sprayed with benomyl or carbendazin solution (1 g/l). The humidity in the cropping room is maintained at 95% and temperature at 20-24°C. Not much aeration is required during the first week. The beds are kept moist.

One week after casing, the room temperature is lowered to 18°C. When pinheads start appearing, the temperature is further lowered and maintained at 14-15°C and RH at 90-95%. Good ventilation is required at this stage. The carbon dioxide concentration in the room is kept below 0.1% and 2-4 air changes per hour are required. More watering is required during the production of flushes. Mushrooms are harvested while still in button stage. The compost formulations are given in Table 5.

The cost of cultivation varies from place to place, depending upon the price of raw material, labour wages, environmental conditions at the site and marketing opportunities [23].

Table 5
Different compost formulations

Materials	Types of formulations				
	I	II	III	IV	V
Wheat straw	1000kg	1000kg	1000kg (Paddy straw + 6kg cotton seed meal)	1000kg -	500kg -
Calcium ammonium nitrate	30kg	30kg	30kg	-	-
Calcium carbonate	-	10kg	100kg	-	-
Superphosphate	10kg	10kg	10kg	-	-
Urea	13.3kg	13.3kg	15kg	14.5kg	7kg
Wheat bran/brewer's grain	100kg	100-130kg	100kg	72kg	60kg
Rice bran	-	-	167kg	-	-
Sulphate of potash	10kg	10kg	-	-	-
Gypsum	100kg	100kg	40kg	30kg	-
Molasses	16.6kg	16.6kg	-	-	-
Di-bromo-chloro propane	123ml	123ml	-	-	-
Lindane	8.33g	8.33g	-	-	-
Chicken manure	-	-	-	400kg	300kg
Horse manure	-	-	-	-	1000kg

CULTIVATION OF *AGARICUS BITORQUIS*

Tiwari and sohi [24] first attempted the cultivation of this mushroom in India. Gularia [25] worked on various aspects of cultivation. Dhar [26-29] provided a package of practices for its cultivation during summer months in hills. The production technology of this mushroom is on the same line as that of A. *bisporus,* with a difference that it has high temperature optima. Because of the higher temperature requirements, long method of cultivation without pasteurization is not successful. Therefore, compost for A. *bitorquis* is prepared by short method. The base materials mainly used are wheat and/or paddy straw, supplemented with poultry manure (to every tonne of base material, heap stock of poultry manure is added), breweris grain (200 kg/ton), and urea (16 kg/ton). The C/N *ratio* of the composting material is 35:1. Composting is done in two phases as in A. *bisporus,* outdoors and then pasteurization. The compost should be dark brown in colour with 68-70% moisture and 7.2-7.6 pH. The grain spawn is mixed with the compost at 0.5 to 0.7 per cent of compost weight. Seeded compost is filled into bags or placed in beds and slightly compressed, but not to the extent as in A. *bisporus.* For spawn run, the compost is incubated at 28-30°C. The spawn run takes 14-16 days and is slower than that of A. *bisporus.* The compost trays or bed is covered with newspaper sheets and sprinkled with water. No watering is needed in bags; however, when the compost becomes completely impregnated with spawn, casing is done. No spraying of water is required as in A. *bisporus* at this stage. Ventilation for the supply of fresh air is also not required during spawn run, as CO_2 accumulation increases the rate of spawn run. The RH during spawn run is maintained at 95% supplemented with soybean meal (1 kg/1 kg compost) or casing has been reported to increase the yield. Ruffling of spawn run compost has also been reported to induce leaf flushes. The pinheads start appearing within 10-12 days. The mushrooms are harvested when cap diameter is 3-5 cm and fruit body pileus has not opened [30].

CULTIVATION OF OYSTER OR TROPICAL MUSHROOM (*PLEUROTUS* SPECIES: DHINGRI)

Many species of *Pleurotus*, e.g., *P. sajor-caju, P. sapidus, P. cystidions, P. ostreatus, P. eryngii,* and *P. flabellatus* have been exploited for commercial cultivation. A variety of substances, like wheat and paddy straw, saw dust, maize stalks, dried leaves, waste from industries, synthetic compost (used for button mushroom), dried logs of soft weed trees and roots of water hyacinth, paper waste, and shelled maize cobs, have been used as substrates. Different species have different temperature requirements and hence they are ideally suited for cultivation at different times of the year in various regions. It grows at a temperature higher (20-30°C) than required for button mushroom and hence it is more suitable for Indian conditions, and moreover technology for its cultivation is simple and cheap. Further, it can be dried and stored easily.

The technology for its cultivation involves the following steps.

The paddy and wheat straw should preferably be fresh, should not be more than a year old and cut into 3-6 cm pieces. The wheat straw is kept in hot water at about 60°C for 4-6 hours. For paddy straw, overnight dip even in cold water for 8-10 hours is sufficient. When the temperature of the straw develops to around 25-30°C, it is mixed with spawn @ 2% weight of the substrate (i.e. 200 g in 10 kg). The spawn should be fresh or not more than a month old. Then it is filled into polythene bags (30-40 cm). The bags so filled are perforated at regular intervals (i.e. almost 15-20 cm apart) for allowing aeration. These are placed on a raised platform in the dark, and are incubated at 20-30°C with 70-85% RH for spawning.

When the substrate is covered by the fungus, the polythene bags are gently removed without damaging the compact block of straw. These are again arranged on a wooden platform and watered twice a day or as required to ensure 70-80% RH. Walls and floor of the room are also sprinkled with water. The room is aerated regularly for 1-2 hours daily.

Mushrooms appear within 7-10 days. These need to be harvested prior to the upcurving of the margin of the pileus. Younger fruit bodies have longer shelf life with lesser moisture content. Three to four flushes appear within a period of 4-5 weeks. Extra produce can be sun-dried or dehydrated at 50-55°C and stored in sealed/closed poythene bags for later use. The yield per unit substrate and area is also good, which varies from 2.5-3.0 kg per bag [31,32] (Table 6).

Table 6
Conversion efficiency (CE) and yield (g) of oyster mushroom on different substrates

Substrates	Mean weight of fruiting (g)	Total yield/kg dry wt. (total of 3 flushes)	Biological efficiency (per cent)
Paddy straw	5	620	62
Dried banana leaves	7.1	760	76
Sawdust	2.2	230	23
Oil palm refuse (pericarp)	2.1	110	11
Oil palm bunch refuse	2.5	220	22
Dry grass	3.1	320	32
Coir dust	2.1	210	21
Coir refuse	1.5	310	31
Arecanut husk	2.5	30	30
Areca leaf sheath	1.1	20	0.2
Coconut leaves	20	350	35

Source: Ramesh et al. [31].

The nitrogen contents of the common substrates for growing oyster mushrooms are given in Table 7 [32]. Also, the yield performance and growth conditions of summer cultivated and winter cultivated species of *Pleurotus* are given in Table 8 [32].

CULTIVATION OF *VOLVARIELLA* SPP. (PADDY STRAW OR CHINESE MUSHROOM)

Out of the three species, *V. volvacea, V. diplasia,* and *V. esculenta,* only two are cultivated in India on cellulose-rich paddy straw. The cultivation can

Table 7
Nitrogen contents of common materials (substrates and supplements) for growing oyster mushroom

	Materials	Total nitrogen (%)
(A)	Substrates	
	Wheat straw	0.56 – 0.63
	Paddy	0.57 – 0.62
	Barley straw	0.57 – 0.58
	Soybean straw	0.55 – 0.65
	Sugarcane bagasse	0.16 – 0.18
	Millet straw	0.58 – 0.62
	Cotton stems and leaves	1.50 – 2.00
	Sorghum	0.80 – 0.90
	Comcobs	0.35 – 0.37
(B)	Supplements	
	What bran	2.50 – 2.70
	Rice bran	1.90 – 2.00
	Cotton seed meal	5.80 – 6.10
	Ground nut cake	6.30 – 6.80
	Gram powder	3.50 – 3.80

Source: Upadhyay [32].

Table 8
Yield performance and growth requirements of various *Pleurotus* species

Sl. No.	Species	Optimum temp. for mycelial growth on the substrate (°C)	Optimum temp. for fruiting (°C)	Temp. range where it can grow (°C)	No. of days required for substrate colonization at optimum temp. (°C)	No. of days for fruiting at optimum temp.	Yield performance (B.E.%)
Summer cultivated species							
1.	*P. flabellatus*	25-30	22±2	16-28	12-14	18-22	60-90
2.	*P. sajor-caju*	25-30	24±2	17-30	12-14	18-25	50-70
3.	*P. sapidus*	25-30	24±2	17-30	16-18	22-28	40-75
4.	*P. membranaceus*	25-30	27±2	20-30	7-9	8-12	40-70
5.	*P. citrinopileatus*	25-30	26±2	20-30	12-14	20-28	30-60
6.	*P. eous*	25-30	24±2	16-28	20-22	25-30	30-50
Winter cultivated species							
7.	*P. ostreatus*	25-30	Strain I 20-22 / Strain II 12-20	18-25 / 7-22	20-25	30-35	30-50
8.	*P. florida*	25-30	20±2	12-22	16-18	25-30	50-90
9.	*P. cornucopiae*	25-30	20±2	12-25	16-18	25-30	40-70
10.	*P. fossulatus*	18-22	18±2	12-25	50-55	65-70	20-30
11.	*P. eryngii*	18-22	16±2	12-24	55-60	70-75	20-30

Source: Upadhyay [32].

be carried out in rooms, covered sheds or ploythene cages. It requires a temperature between 25-45°C for growth. The best season for growing this mushroom in the plains is from April to October. The humidity required is above 50%. The ideal temperature for spawn run is between 30-35°C and a relative humidity of 80-90%. One crop cycle takes 30-35 days.

The substrate for composting consists of fresh paddy straw. It is tied in bundles of ½ to 1 kg (of dry straw) weight and upto 15 cm in diameter. These bundles are then steeped in clear water in a tank or a drum for about 12 hours. The moistured straw bundles are laid in slightly raised concrete or wooden platform in layers of four bundles each with butts on one side. The first layer is over laid by four bundles with butt ends on the opposite side. In all 32 bundles are placed in eight layers. Loose straw on the sides is cut and removed. Beds are usually of 90 × 90 × 120 cm size.

Spawning of beds is done simultaneously while making them. Generally 15-20 g spawn/kg paddy straw is used. Each layer of bed is spawned after putting the bundles in position. No seeding is done below the upper most layer. Tur or gram daal powder is sprinkled over the spawn at the rate of 200 g per bed to enhance growth. The beds are subsequently covered with a polythene sheet in order of raise the temperature and also to maintain high humidity. In case the prevailing temperature and humidity are suitable, there will be no need to cover the beds. In no case should the polythene sheet, used for covering the beds, come in direct contact with the compost. If the beds are not covered light sprinkling of water may be necessary once or twice daily. Even the floor and walls of the building may need to be sprinkled with water, if the prevailing humidity is low. Fresh and good quality spawn (3-4 weeks old) may be used to get optimum yield [23].

Good spawn run usually takes 8-10 days at 30-35°C, once the substrate gets impregnated with mycelial strands; polythene sheets, if used, should be removed. 'Pin heads', which appear soon after, gradually turn into egg shaped buttons. The mushrooms are removed gently so as not to harm the buttons still developing. The mushrooms are removed gently so as not to harm the buttons. Light sprinkling of water is essential to keep the bed wet. In 25-30 days, usually two flushes of fruit bodies appear. The beds are then dismantled and a new cycle begins. The yield from 2-3 kg/heap is obtained. The fruit bodies can be harvested and dried for future use. The protein content on fresh weight basis is around 4.9%.

Cultivation of *Calocybe indica* Purkayastha and Chandra

Cultivation of *Calocybe indica* is an edible mushroom described from India by Purkayastha and Chandra [33]. It is a robust mushroom milky white in colour and, therefore, attempts were made right from the time when it was first described to cultivate it. Purkayastha and Chandra [34] were able to induce *in vitro* fructifications of *C. indica*. Subsequently, Purkayastha and Nayak [35] evaluated various substrate formulations and reported the suitability of maize meal and wheat bran for the production of fructifications. In 1981, Chakarvarty et al. [36] attempted to grow *C. indica* on composted and non-composted substrates. The composted substrate was prepared with rice straw, using long method of composting (refer to button mushroom). The non-composted substrate consisted of chopped and pasteurized paddy straw mixed with 4% NPK fertilizer. Later, they reported soil : sand (2:1) mixed with 50% maize meal as the most suitable substrate. Wheat grain spawn was reported to be better for *C. indica* yield. Doshi et al. [37] reported suitability of sorghum grain as spawn substrate. Among the substrates, wheat straw was reported to be the best. Pandey and Tiwari [38] reported cultivation of *C. indica* on paddy straw, using sorghum spawn and sandy loam soil (40-50% sand) as the casing medium, with *biological efficiency* (B.E.) of 60-100%. Later on, Pandey [39] described the cultivation techniques of *C. indica* on six different agrowastes namely: wheat straw, paddy straw, ragi straw, reddish waste, sawdust, and coir waste. Krishnamoorthy et al. [40] evaluated nine substrates viz. paddy straw, sorghum stalks, maize stalks, sugarcane bagasse, vativer grass, groundnut, palmrosa grass, soybean hay, and black gram hay for the cultivation of *C. indica* (strain APK$_2$).

Composting of the substrate

Various types of substrates, listed above, are chopped into small pieces and soaked in clean water for 10-12 hours. After draining excess of water, the substrates are filled in polypropylene (PP) bags (30 × 40 cm), 2 kg wet substrate (bag) tied and sterilized at 15 lb/inch^2 pressure for 1 hour in steam sterilizer. After cooling the bags, the bags are spawned with 4% grain spawn in asceptic environment. After inoculation, the bags are incubated in the spawn running chamber, where the temperature of 28-32°C is maintained.

The composting experiments have revealed that there is significantly higher yields after 6 days of composting the straw. The yields obtained on fresh straw or straw compost for more than 6 days are lower. This shows that the chemical state and physical properties of the substrate, in term of aerobic conditions, pH and good drainage, are also most suitable after 6 days of composting.

Spawn run

A temperature of 30°C has been observed to be the best for spawn growth (4.75 mm/day). There was absolutely no growth at 10-20°C. The growth at 40°C was very slow (0.34 mm/day). The growth at 45°C is very much affected. The growth of spawn initially kept at 10-20°C and 40°C could be revived at 30°C. The initial exposure at 40°C acted as stimulant, increasing the growth when brought back to 30°C [39]. Wheat, bajra (boiled for 30 minutes), and jower (boiled for 30-60 minutes) have been reported to be good substrates, for spawn production [41]. Spawn can be stored upto 8 weeks. The yield decreased after 2-3 weeks of storage. A spawn dosage of 4% gave optimum yield.

After complete spawn run, the bags are cased (3-4 cm) with red sandy soil. Peat soil, clay loam, sand, *biogas slurry*, farmyard manure, and composted coir pith have also been used [40]. The cased bags are kept in cropping room when a temperature of 28-32°C and RH of 80-90% and 12 hours of light are maintained. Supplementation with 100% rice bran or 50% oat meal and others, like maize grain, sorghum grain, etc., show an increase in yield.

Pandey [39] obtained biological efficiency of 81.87%, whereas Krishnamoorthy et al. [40] reported 117% biological efficiency with an improved strain (APK$_2$). It seems that this indigenous mushroom is catching up in India quickly.

CULTIVATION OF SHIITAKE MUSHROOM: *LENTINUS EDODES* (BERK) SINGER

L. edodes, a white rot wood decay fungus inhabits the wood of many hardwood tree species in Asia, especially those belonging to family Fagaceae, e.g., oaks. *Castanopsis* (Shii) Chesnut, hornbeam, beach and also grows on pasania (*Lithocarpus*) in Japan [42]. It ranks second among the cultivated mushrooms in the world from the stand point of total production [43]. Because Japan has been leading in world production of Shiitake, therefore the mushroom gets its name Shiitake or '*Japanese mushroom*'. However, historical accounts reveal that it was perhaps China that edibility of the mushroom was first recognized, and its initial cultivation technology developed.

MUSHROOM CULTIVATION ON WOODEN LOGS

For cultivation of the suitable mushroom on natural wood logs, approximately 84 hardwood tree species have been used. But most commonly used species are: *Quercus rerrata* and *Q. aevtissima*. For Shiitake cultivation, instated of beds, the logs of certain woods are used for mushroom growing, and these logs are called *"bed logs"*. Trees for cultivation are felled in autumn from coppices. Time of cutting is important; if a cut is made at the wrong time, the bark strips off easily, and contamination with weed fungi takes place before the Shiitake mycelium could have had a chance to become established. More important is the sugar content of the wood, because the mycelium grows faster after inoculation when the percentage of easily available carbohydrates in the wood is high. After trees are felled, the trunks remain in the forest. Just before inoculation, they are cut into logs measuring 1-15 m in length and 5-15 cm in diameter; the thicker branches are also used.

Inoculation is done with solid wood pieces or sawdust. If solid places of inoculum are used, these are drilled into the log holes. In case of sawdust spawn, the incisions for holes are more or less of the same size. These are used with wood chip spawn. The holes and incisions are painted with hot wax in order to prevent evaporation. The number of holes or incisions is calculated to correspond to about 1 holes per 1000 sq cm.

CULTIVATION ON SAWDUST

Lentinus has been traditionally cultivated on hardwood logs outdoors or under *greenhouse* conditions. It takes quite a lot of time to find suitable species and size of wood for Shiitake culture. In order to find a way to overcome these difficulties, an alternative and modern technique for cultivation by using sawdust (mixed with various ingredients) to grow Shiitake mushroom in plastic bags under fully controlled conditions has been developed. Many farmers in China, Taiwan, USA, Korea, Canada, Singapore, Holland, Philippines, Finland, Thailand, and Belgium are now producing Shiitake on supplemented sawdust [43].

The medium for growing Shiitake in the plastic bag is sawdust with 5-10% of rice bran to enrich the substrate. Sawdust stacked out doors for 5-10 days is mixed with rice bran and is adjusted to a measurable water content i.e., 60-70%. The mixed sawdusts are packed into plastic bags (15-18 cm × 10-12 cm) and pressed to form a cylindric cake. A plastic ring is wrapped at the top end to the bag to form a bottleneck with a cotton plug in it. The filled bags are sterilized in a special steaming room or an autoclave at 90-100°C for 240 minutes.

The bags are inoculated after cooling and transferred to a growing house maintained at 28°C or lower for 40-50 days, until the surface of the bagged sawdust becomes dark brown in colour and acquires a hard texture [44,45].

CONCLUSION

Utility of mushroom for bioconversion of wastes into useful fructification is well known, in addition to those for which production technology has been described in detail. Other mushrooms have also been cultivated, but their commercial cultivation is limited only to a few, which are relished by people. Most of these possess good cellulolytic and hemicellulolytic systems, and a few like *Agaricus, Pleurotus,* and *Lentinus* are also strongly lignolytic. This bioconversion is achieved through extracellular enzyme production. Detailed studies on enzyme production are needed. *Strain improvement* has been done, but further improvement will probably depend upon DNA technology and use of protoplasts, combined with incorporation of novel genetic material into suitable commercial strains of mushrooms. In case of mushrooms, land is not a limiting factor, and in a small space mushrooms can be produced repeatedly during a season. The mushroom cultivation helps in recycling of agro-industrial wastes, converting them into protein-rich fruitbodies, thereby reducing environmental pollution and *health hazards*. Moreover, cultivation of mushroom gives good economic returns.

REFERENCES

1   FAO. Production Year Book Vol. 41. Rome: Food and Agriculture Organisation, 1989.

2   Srivastava PK, Maheswari RC. Utilization of Agricultural Wastes and Byproducts in India. Paper presented in XVIII Annual Convention of ISAE, held at CSSRI, Karnal on Feb. 26-28, 1981.

3   Knapp JS. In: Moo-Young M, ed. Comprehensive Biotechnology. 1985; 835-847.

4   Ericsson KEL. J Biotechnol 1993; 30: 149-158.

5   Singh MP, Kaushal SC. In: Fourth Chandigarh Symposium on New Biology, 6-7th Feb Chandigarh, Abstract 1996; 47.

6   Michalski KJ, Beneke ES. Mycologia 1969; 61: 1041-1047.

7   Tangu SK, Blanch HW, Charles R. Biotechnol Bioeng 1981; 23: 1837-1849.

8   Messner R, Kubicek CP. Enzyme Microb Technol 1990; 2: 685-690.

9   Bassham JA. In: Wilke CR, ed. Cellulose as a Chemical and Energy Resource. Biotechnol Bioeng, New York: Interscience, 1975; No. 5: 9.

10   Dunlap CE, Chiang LC. In: Shular ML, ed. Utilization and Recycling of Agricultural Waste and Residues, Florida: CRC Press, 1980; 19-65.

11   Eveleigh DE. In: Honghton DR, Smith RN, Eggins HOW, eds. Biodeterioration. London: Elsevier Applied Science, 1987; 315-324.

12   Lutzen NW, Nielson MH, Oxibobell KM, Schiilen M, Olesson BS. Philos Trans Royal Soc London 1988; 300: 283.

13   Ghose TK, Ghose P. J Appl Chem Biotechnol 1978; 28: 309-320.

14   Garcha HS. Mush Sci 1981; 11: 245-254.

15   Swaminathan MS. Curr Sci 1982; 51: 13-24.

16   Chadha KL, Sharma SR. eds. Advances in Horticulture — Mushroom, Vol 13, New Delhi: Malhotra Publ House, 1995.

17   Chadha KL, Shram SR. In: Chadha KL, Shrama SR, eds. Advances in Horticulture – Mushroom. New Delhi: Molhotra Publ House, 1995; 13: 1-33.

18   Chakravarty DK. In: Biotechnology and Other Alternative Technologies, New Delhi: Oxford and IBH Publ 1989; 3-17.

19   Kuhad RC, Singh A. Crit Rev Biotechnol 1993; 13: 151-172.

20   Chang ST, Tan YH. Mush Sci 1989; 12: 761-767.

21   Hayes WA. Mush Sci 1969; 7: 173-186.

22   Gupta Y, Vijay B. Mush Res 1992; 1: 115-117.

23   Sohi HS. Cultivation of Edible Mushroom. Extension Bull 1, Solan: NCMRTC, 1990.

24   Tiwari RP, Sohi HS. Proc. Ist Symp on Survey and Cultivation of Edible Mushrooms in India, RRL, Srinagar, J&K, 1976.

25   Gularia DS. Studies on *Agaricus bitorquis* (Ovd.) Sacc. Ph.D. Thesis, HPKVV College of Agriculture, Solan, HP, 1985.

26   Dhar BL. Annual Report, Solan: NCMRT 1998; 21-26.

27   Dhar BL. Annual Report, Solan: NCMRT 1989; 18-22.

28   Dhar BL. Annual Report, 1991-92 Solan: NCMRT; 1991; 30-36.

29   Dhar BL. Mush Res 1992; 1: 19-29.

30   Dhar BL. In: Chadha KL, Sharma SR, eds. Advances in Horticulture — Mushroom. New Delhi: Malhotra Publ House 1995; 13: 99-108.

31   Ramesh CR, Ansari MM, Bandhopadhyay AK. In: Chadha KL, Sharam SR, eds. Advances in Horiculture — Mushroom, New Delhi: Molhotra Publ House, 1995; 13: 399-404.

32   Upadhyay RC. Cultivation of Oyester Mushroom, Tech Bull 1, Solan: NCMRT, 1990; 1-28.

33   Purkayastha RP, Chandra A. Trans Brit Mycol Soc 1975; 62: 415-418.

34   Purkayastha RP, Chandra A. The Mushroom Journal 1976; 40: 1-2.

35   Purkayastha RP, Nayak D. The Mushroom Journal 1977; 52: 1-3.

36   Chakravarty DK, Sanker BB, Kundu BM. Curr Sci 1981; 50: 550.

37   Doshi A, Sidana N, Chakravarty BP. Mushroom Sci 1989; 12: 395-400.

38   Pandey M, Tiwari RP. Mushroom Information 1993; 5: 5-10.

39   Pandey M. Cultivation Technology of Specially Mushroom (*Calocybe indica*) Ph.D. Thesis, H.P. Univ, Shimla, 1998.

40   Krishnamoorthy AS, Muthuswami MT, Nakkeeran S. Indian J Mush 2000; 18: 19-23.

41   Pandey M, Lakhanpal TN, Tiwari RP. Indian J Mush 2000; 18: 15-18.

42   Chang ST, Hayes WA. The Biology and Cultivation of Edible Mushrooms. New York: Academic Press, 1978.

43   Chang ST, Miles PG. Mushroom J Tropics 1987; 7: 31-37.

44   Kaur MJ. Studies on Cultivation of Shiitake Mushroom *Lentinus edodes* in India. Ph.D. Thesis, H.P. Univ, Shimla, 1994.

45   Kaur MJ, Lakhanpal TN. Indian J Microbiol 1995; 35: 339-342.

Biotransformations: Bioremediation Technology for Health and Environmental Protection
V.P. Singh and R.D. Stapleton, Jr. (Editors)
© 2002 Elsevier Science B.V. All rights reserved.

Enzymatic transformations of xenobiotics of health and environmental concern

Ved Pal Singh

Department of Botany, University of Delhi, Delhi - 110 007, India

INTRODUCTION

The present-day industrialized world with fast-increasing global population, by virtue of being involved in generating vast variety of pollutants in the environment through various developmental activities, is now feeling threatened with far reaching hazards to both human and animal health as well as the environment. There are various methods by which different governmental agencies are setting up norms to undertake the task of protecting the environment, so as to safeguard the human civilization against the health hazards. Agricultural and industrial wastes introduce a great variety of *health-risk compounds* in the biosphere and pollute it. Therefore, there is an urgent need to look for the possibilities to tackle this situation of increased agro-industrial wastes of anthropogenic origin. Owing to their biotechnological potential in degrading and eliminating hazardous organochemicals, microorganisms occupy a key position in health and environmental protection programmes. Degradation of hazardous organic compounds by rhizosphere microbial communities has been studied in great detail by Anderson et al. [1]. *Bioprotection* of environment through potential use of microorganisms and their products, including enzymes, is of great significance in the present day context by implicating them in the degradation of health-risk compounds [2]. Thus, various microbial metabolic processes might be implicated in the biodegradation of a variety of xenobiotic compounds. Some of the processes associated are: fermentation, some types of anabolic metabolism, chemotrophic metabolism, and exoenzyme metabolism.

The present article deals with the enzymatic degradation of wide variety of *xenobiotics*, such as toxic, carcinogenic, teratogenic, and mutagenic compounds, which are of health and environmental concern.

NITROGENOUS COMPOUNDS

Nitroaromatics

Various nitrogenous xenobiotics, particularly nitroaromatics are degraded by microorganisms both under aerobic and anaerobic conditions [3]. Nitroaromatic compounds are produced industrially on a large scale. They are highly toxic to both man and animals. Two major *catabolic pathways* are involved in the degradation of nitroaromatics. In the first pathway, the nitro group is reduced to an aniline intermediate, which is further degraded to ammonium ion and catechol.

The reduction of nitro substituent under both aerobic and anaerobic conditions seems to be a common enzymatic mechanism in the environment. The microbial degradation of nitroaromatics to catechol involves a series of reductions and oxidations, generally catalyzed by reductases and oxygenases. A *Pseudomonas* sp. has been found to degrade aerobically, a *priority pollutant* 2,4-dinitrotoluene (2,4-DNT), using the latter as the sole source of carbon and energy. 4-Methyl-5-nitrocatechol (MNC) accumulated transiently when cells grown on acetate were transferred to medium containing 2,4-DNT. The conversion of 2,4-DNT to MNC was catalyzed by dioxygenase.

The pathway for degradation of nitrotoluenes, which are largely used in manufacture of azo and sulphur dyes, and in the production of *explosives*, is located on TOL-plasmid. In fact, TOL-encoded toluene monooxygenase enzyme biotransformed 3-nitrotoluene and 4-nitrotoluene into their corresponding benzyl alcohol and benzaldehyde, but not 2-nitrotoluene [3]. Various

Pseudomonas spp., a *Rhodococcus* sp., a *Flavobacterium* sp., and *Morexella* have been shown to implicate their enzymes, particularly oxygenases, in the degradation of nitrobenzenes and nitrophenols.

Nitriles

Nitriles, which are cynaide-substituted carboxylic acid with general formula R-CN, are used industrially in benzonitrile herbicide as organic solvents and in the synthesis of polymers, plastics, synthetic fibres, resins, and dye stuffs. These are toxic, mutagenic, and carcinogenic compounds. Microbial enzymes, which participate in the degradation of nitriles are nitrilases from *Pseudomonas* sp., *Nocardia* sp. strains NCIB 11215 and NCIB 11216, *Fusarium solani, Arthrobacter* sp., *Escherichia coli*, transformed with a *Klebsiella ozaenae* plasmid DNA, *Rhodococcus rhodochrus* and *Alcaligenes faecalis* JM3.

Similarly, aliphatic nitrile hydratase, that catalyzed the hydration of nitriles to amides, was purified and characterized in *Arthrobacter* sp. J1, *Brevibacterium* R312, and *Rhodococcus* sp. N774. In the first strain, the activity of an amidase, which forms acetic acid and ammonia, was also detected by Mukai and Doi [3]. Biotransformation of dinitrile to mononitrile, catalyzed by nitrile hydratase and amidase has been obtained from *Corynebacterium* sp. C5, and both these enzymes are synthesized constitutively in the microbial cells [3].

CARBAMATES, PHENYLUREAS AND ANILIDES

Among carbamates, carbofuran - an insecticide, which is extensively used to control corn rootworm, has been shown to be partially degraded by soil microbes. Recently, it has been reported that an *Achromobacter* sp. WM111, isolated by soil enrichment cultures, is capable of hydrolyzing carbofuran at an exceptionally rapid rate [4]. This microorganism catalyzes the degradation of other N-methylcarbamate insecticides (Carbaryl, Aldicarb, Baygon), but is ineffective in the degradation of acylamide or urea pesticides. These results indicated that the hydrolytic enzyme of this strain is specific for phenol-carbamate ester linkages. In another report, Karns et al. [5], by studying plasmid and chromosomal DNA of *Achromobacter*, found that the strain harboured a plasmid, which encoded for hydrolase activity.

The phenylurea *herbicides* are sensitive to degradation by *Bacillus sphaericus*. An amidase, named arylacylamidase, hydrolyzed acylanilide herbicides, and this enzyme was induced by different acylanilide herbicides, acylanilide fungicides and by phenylcarbamate herbicide (Propham); however, the maximum enzymatic activity was revealable with Linuron as inducer. Similarly, anilide degradation has been demonstrated, implicating acylanilide hydrolases isolated from strains of *Fusarium* and *Bacillus*. On the other hand, an aniline oxidase and a peroxidase from the fungus *Geotrichum candidum* could dimerize different anilines except nitroanilines.

In a variety of microorganisms, 4-chloroaniline was found to be degraded by inducible oxygenase to 4-chlorocatechol, which was further degraded via a modified *ortho*-cleavage pathway. The enzyme oxygenase exhibits a broad substrate specificity.

HALOGENATED AROMATIC COMPOUNDS

Among these compounds, polycyclic aromatic hydrocarbons (PAHs) are important as they are ubiquitous environmental contaminants, which occur as natural constituents and combustion products of fossil fuels. It is well known that microorganisms can degrade naphthalene and other PAHs. Grund et al. [6] have demonstrated various microbial enzymatic reactions, that occur during aerobic breakdown of PAHs, which include the oxidative catabolism of the compounds, such as naphthalene. It has already been reported that "*cis* - naphthalene dihydrodiol" is the first metabolite in the bacterial metabolism of naphthalene and the enzyme which catalyzes this reaction is naphthalene dioxygenase. As such, dioxygenases are the enzymes that activate molecular oxygen and introduce two hydroxy groups. However, naphthalene dioxygenase enzyme accepts indole as a substrate, which leads to the formation of indigo. All known bacterial species, that can be grown on naphthalene as sole source of carbon and energy, can degrade salicylate as well. In *Pseudomonas* sp. NCIB 9816, catechol has been found to be metabolized by the enzymes of the *ortho*-pathway or by the *meta*-pathway, and it was quite evident that a number of variants of NCIB 9816 were in circulation, which are probably due to the presence of a plasmid in *Pseudomonas putida* strain [6].

The key enzymes for the *meta*- and *ortho*-pathways of catechol are the ring-opening dioxygenases. Similarly, different bacterial dioxygenases have been implicated in the biodegradation of dibenzo-*p*-dioxin and polychlorinated biphenyls (PCBs). Chlorobenzoates or chlorinated benzoic acids (CBAs) are often dead-end products of the major steps for the degradation of CBAs.

The enzymatic breakdown of CBAs via chlorocatechol and subsequently the dehalogenation of nonaromatic intermediates are quite common in Gram-negative bacteria, especially *Pseudomonas, Alcaligenes,* and *Acinetobacter* strains. Oxidative dehalogenation takes place by dioxygenases, whereas, hydrolytic dehalogenation occurs by dehalogenase enzyme of *Pseudomonas* sp. CBS3. Also, various strains of rhizobia have been found to degrade a variety of aromatic compounds [7].

Microbial degradation of chlorinated polycyclic hydrocarbons, such as PCBs, *p*-chlorobiphenyl (*p*-CB), 1,1,1-trichloro-2,2-bis(4-chlorobiphenyl) ethane (DDT) and other polycyclic compounds, has been studied by Bhat and

Vaidyanathan [8]. They have further shown that various microbial enzymes have been implicated in the degradation of a variety of halogenated aromatic compounds, such as polychlorinated phenols, chlorobenzenes, chlorobenzoates, chlorotoluene, 4-chlorophenylacetate, and chlorophenoxyalkanoic acids, including 2,4-dichlorophenoxyacetic acid (2,4-D) and 2,4,5-trichlorophenoxyacetic acid (2,4,5-T), which pose potential threat to both human and animal health as well as environment [8].

The genetic studies, related to biodegradation of chlorinated aromatic compounds, have revealed that a gene cluster (*bph ABC* operon), cloned by Furukawa and his group, as reported by Bhat and Vaidyanathan [8], encodes biphenyl/PCB-degrading enzymes from chromosomal DNA of *Pseudomonas pseudoalcaligenes* KF707. Subcloning, deletion analysis and *transposon mutagenesis* of the cloned *bph* genes revealed that the biphenyl/PCB catabolic *bph* operon of *Ps. pseudoalcaligenes* KF707 is very well organized.

Azo Dyes

About one-half of all known dyes are azo dyes, thus making them the largest group of synthetic colourants, but have relatively small contribution to the total mass of pollutants released into the environment [9]. Until recently, azo dyes were not thought to be degraded by microorganisms unless an aerobic step was included in the process. The discovery that the white-rot fungus, *Penicillium chrysosporium*, which degrades a wide variety of azo dyes, provides an entirely new approach to the study of azo dye degradation. Possibly, the most important finding is that the initial degradation of many azo dyes, in this system, is not a reduction of the azo linkage, but rather an oxidation reaction, mediated by lignin peroxidases or Mn peroxidases that are secreted by the fungus during idiophase metabolism. The ability of white-rot fungi (*Myrothecium* spp. and *Ganoderma* spp.) to degrade azo dyes has also been reported by Bumpus [9].

Role of microsomal cytochrome P-450 monooxygenase and microsomal epoxide hydroxylase in the degradation of styrene [10] and the involvement of the monooxygenase as well as ethene epoxide dehydrogenase in ethene-utilizing *Mycobacterium* strain E3 in the enzymatic transformation of vinyl chloride are well documented [11].

Tannins

Tannins are the major tannery effluents, which constitute the most complex industrial wastes that are toxic to plants, animals and soil microorganisms [12]. Tannins are defined as water soluble polyphenols which differ from other natural polyphenolic compounds in their ability to precipitate proteins, such as gelatin. These are most abundant plant constituents, following cellulose, hemicellulose, and lignin [13]. The enzyme, which degrades tannins is a tannin acylhydrolase, that is commonly known as tannase [13]. This enzyme has been characterized from various microorganisms, such as *Aspergillus niger, Penicillium* sp., and *Candida* sp. Tannase from *A. niger* was *inducible* and hydrolyzed the ester linkages of gallic acid. But according to a report, *Aspergillus* sp. and *Penicillium* sp., isolated from soil, were able to produce tannase even in its absence, thereby suggesting that inducer is not required for tannase production. A detailed account of tannin degradation by microorganisms has been presented in this book elsewhere [Chapter 11].

HETEROCYCLIC COMPOUNDS

Heterocyclic compounds, which are released into the environment and which are toxic and mutagenic, are also prone to degradation by microbes. For example, indole and its derivatives from a class of toxic recalcitrant compounds released into the environment through *cigarette smoking*, coal-tar and sewage, which are responsible for toxicity and mutagenicity in man and animals. A strain of *Aspergillus niger*, which cometabolized indole in the presence of glucose and nitrate, monohydroxylated this compound into indoxyl.

This mechanism is considered prevalent in fungi and other higher organisms as well as in anaerobic bacteria, which hydroxylate indole into oxindole, whereas in aerobic bacteria, unhydroxylated aromatic compounds are attacked by deoxygenase. Indoxyl was further connected to N-formylanthranilate by a dioxygenase, but this activity was not demonstrated for instability of the substrate. In the cytosolic fraction, however, N-formylanthranilate deformylase, anthranilate hydroxylase, dihydroxybenzoate (DHBA) decarboxylase, and catechol deoxygenase, induced by growth on the glucose plus indole, were detected. The enzyme DHBA decarboxylase has been found only in fungal systems.

Another group of heterocyclic compounds, S-triazines and particularly diamino-S-triazines are used as herbicides. Cook and Hutter [14] reported that diethylsimazine was quantitatively utilized as a nitrogen source by a strain of *Rhodococcus corallinus*, yielding ethylamino-dihydroxy-triazine, which is utilized in co-culture with a *Pseudomonas* to yield cell material. They had further confirmed that the initial reaction is a quantitative hydrolytic ring dechlorination by two isofunctional, but different, hydrolases [15].

MYCOTOXIN BIOTRANSFORMATIONS

Owing to the potential deleterious effects of mycotoxins on human and animal health, the understanding of their metabolism in terms of enzymatic detoxification becomes very much important. In the following text we would be dealing with the aspects of biotransformations of some of the important mycotoxins, which are known to cause potential hazards to both human and animal health and which have been well characterized biochemically.

Aflatoxin detoxification

Among aflatoxins, aflatoxin B_1 is considered to be a model secondary metabolite, produced by *Aspergillus flavus* which occupies a central position in the understanding of aflatoxin metabolism in general. In animal systems, aflatoxin B_1 is transformed in liver into three important hydroxylated metabolites, namely aflatoxin R_0 (aflatoxicol), aflatoxin Q_1, and aflatoxin P_1 which can form conjugated molecules with polar compounds. The resulted products are then rapidly eliminated (or excreted out) from the body of the animals. The other parameters such as physical, chemical, and biological methods of aflatoxin detoxification have been dealt with in detail elsewhere [16-18]. Biological aspects of *detoxification* (or degradation) of aflatoxins are of great significance as most living organisms exclude them through degradation with the help of their own metabolic machinery. Various microorganisms have also been found to play role in aflatoxin degradation, when cultivated with the *toxigenic* fungi. *Flavobacterium aurantiacum* removed aflatoxin B_1 from liquids [19,20]. *Dactylium dendroides, Absidia repens*, and *Mucor grieseocyanus* converted aflatoxin B_1 to aflatoxin R_0 [21]. Biotransformation of aflatoxin B_1 to as yet uncharacterized compounds has also been reported with bacteria, including *Corynebacterium rubrum* and *Lactobacillus* spp; with fungi, *A. niger, Trichoderma viride, Mucor ambigus, M. alternans, Helminthosporium sativum, Rhizopus arrhizus, R. oxyzae* and *R. stolonifer*, and the protozoan, *Tetrahymena pyriformis* [22,23]. Bol and Smith [24] reported about 87% degradation of aflatoxin B_1 by food-grade *Rhizopus* strains on agar media. Cole and Kirksey [25] have demonstrated the conversion of aflatoxin G_1 to aflatoxin B_3 (an intermediate metabolite reported in *A. flavus* and as parasiticol in *A. parasiticus*) by various *Rhizopus* spp. Complete elimination of aflatoxin B_1 from peanut meal by *R. oryzae* NRRL 395 has been reported by Knol et al. [26] under solid state process. Cuero et al. [27] have also shown about 40%, 70%, and 75% reduction in

138

aflatoxin concentrations when *A. flavus* (toxigenic strain) was co-cultivated with *Fusarium graminiarum, A. oryzae,* and *Penicillium viridicatum* on maize. Microbial transformations of aflatoxins have also been reported by Banwort [28], Bhatnagar et al. [29], Bol and Smith [24], and Faraj [30]. Even the toxigenic strains of aspergilli can degrade their own aflatoxins both *in vitro* as well as *in vivo* [2,30-35]. However, the degradation of alfatoxins by the

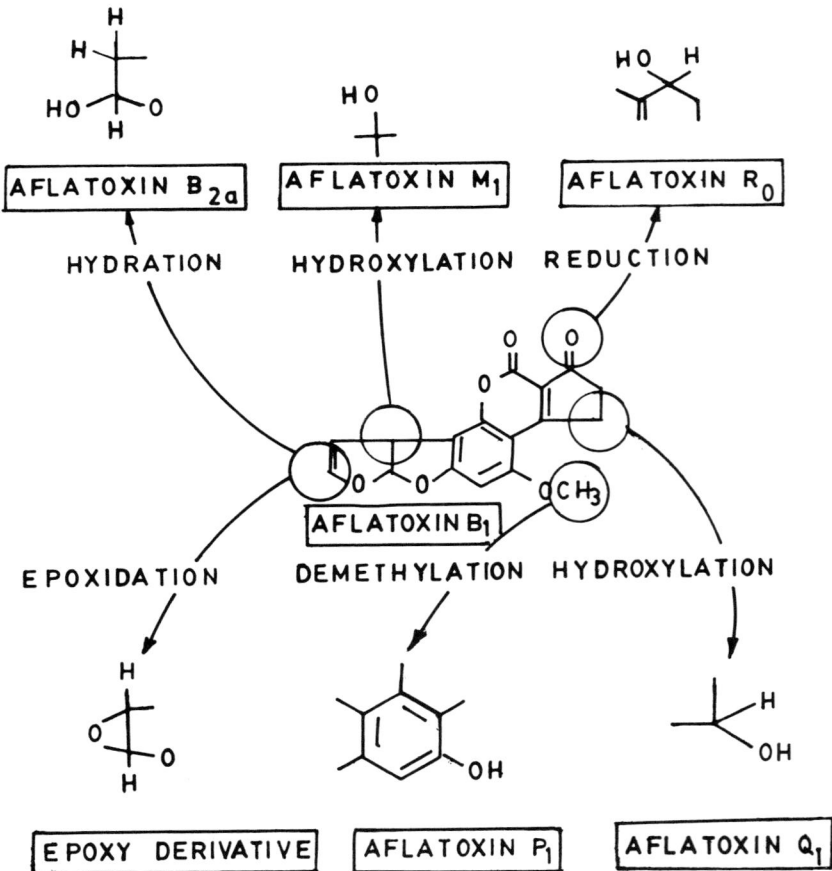

Figure 1. Metabolic biotransformation products of aflatoxin B_1.

toxigenic fungi requires certain optimal conditions, such as medium composition, age of culture, proper aeration, pH, temperature, etc. [24,31,32]. The role of enzymes, such as cytochrome P-450 monooxygenase [33] and microsomal peroxidase [32,34-36] in aflatoxin degradation has been well documented. The mixed function hydroxylases as well as other *microsomal enzymes* of livers have also been implicated in aflatoxin detoxification in animal systems [37]. Figure 1 shows various metabolic routes of aflatoxin biotransformations, pertaining to conversion of most hepatotoxic, carcinogenic and mutagenic aflatoxin B_1 to relatively less toxic products.

Sterigmatocystin detoxification

Detoxification of sterigmatocystin has been well understood only in the animal systems. The possible reactions, which can occur in liver, are demethylation and conjugation, which are coupled with microsomal functions. These two reactions are shown in Figure 2, representing glucuronides of sterigmatocystin, which appear in the urine of treated vervet monkeys [38]. The reactions are considered to be the possible routes to detoxification, as they generate more polar and more readily *excretable forms of toxin*.

Ochratoxin detoxification

The main route for detoxification of this microbial peptide toxins is the enzymatic hydrolysis to generate a naturally occurring amino acid phenylalanine and biologically inactive ochratoxin (*alpha*) in the intestine as well as in the liver of animals [39]. The hydrolytic enzymes, which participate in detoxification of ochratoxins A and B are carboxypeptidase A and chymotrypsin, that break the amide bonds; and the products, thus generated, are much less toxic than the ochratoxin A and B. This suggests that whole molecule of ochratoxin is required for developing toxicity, and any change in its structure leads to detoxification. The major route to ochratoxin A detoxification has been depicted in Figure 3.

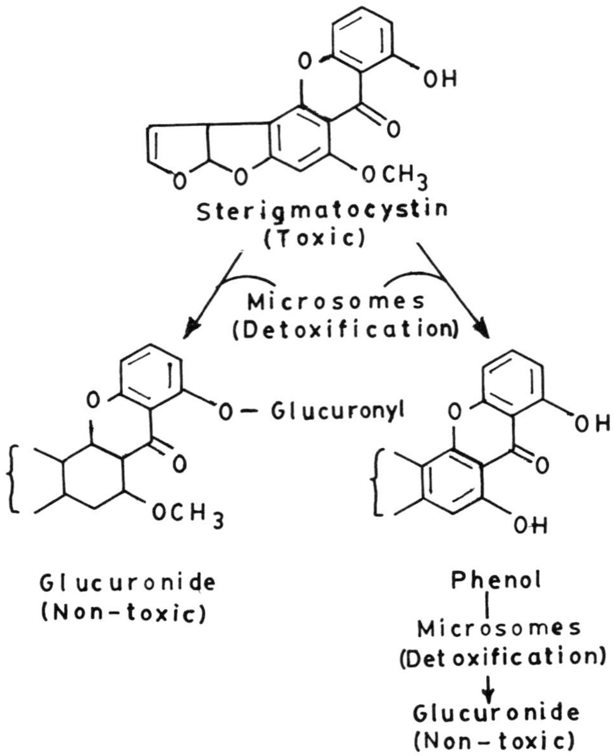

Figure 2. Sterigmatocystin metabolism in liver.

Rumen protozoa and bacteria of sheep and cattle have been found to degrade ochratoxin A to ochratoxin (*alpha*) and phenyl alanine [40]. Earlier, Hult et al. [41] also found these metabolites of ochratoxin A, when this mycotoxin was incubated with contents from 3 out of 4 compartments of stomachs of cow tried. More recently, Westlake et al. [42] have demonstrated that ochratoxin A was degraded, to some extent, by *Butyrivibrio fibrisolvens* without affecting the specific growth rate of this rumen bacterium, thus

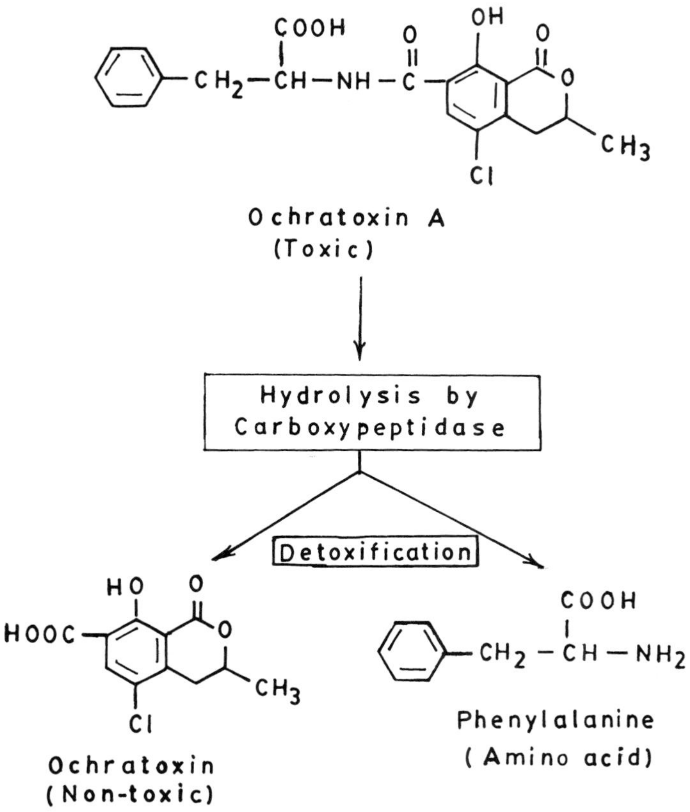

Figure 3. Ochratoxin A metabolism in the intestine and in the liver.

making ruminants more resistant to food poisoning by this toxin. Ochratoxin B was about 7 to 8 times more sensitive than the ochratoxin A with respect to carboxypeptidase-dependent hydrolysis [43].

Trichothecene detoxification

Trichothecenes constitute a broad group of toxins produced by several groups of fungi, including *Fusarium, Trichothecium, Myrothecium*, and

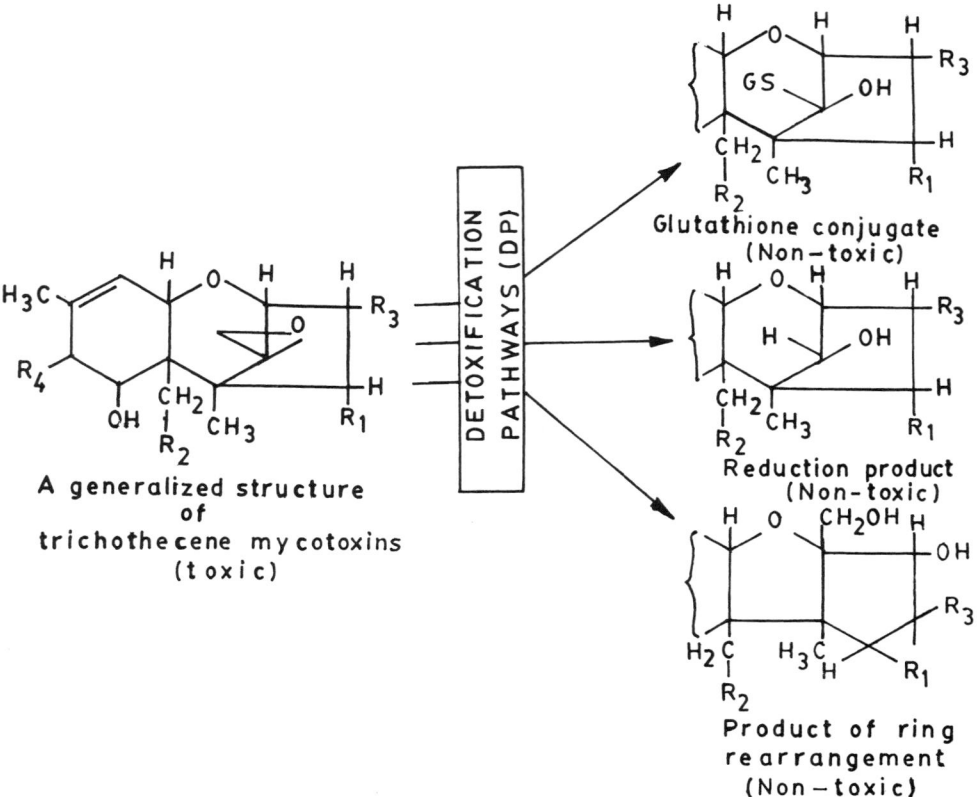

Figure 4. Three possible pathways for trichothecene detoxification. R_1, R_2, R_3 and R_4 may represent hydrogen, hydroxyl, esterified hydroxyl, keto (R_4 only), epoxide (replaces R_4 and adjacent OH group), or a macrocyclic bridge may link R_1 and the adjacent CH_3 group. The OH group to adjacent R_4 is replaced by hydrogen in some toxic molecules.

Stachybotrys spp. [37,44], but not by *F. moniliforme* [45]. They exhibit necrogenic action on the skin. Epoxide being the active centre of trichothecene molecule, any modification of epoxide results in detoxification (Figure 4). There is evidence that T-2 toxin is hydrolyzed by liver tissues *in vitro* to the HT-2 toxin. It has been established that C-4 acetyl group of T-2 toxin is selectively deacetylated *in vitro* and *in vivo* by microsomal nonspecific carboxyesterase of livers of rats, rabbits and other animals [29]. The hydroxylated thrichothecene mycotoxins are excreted as glucuronides [37]. The chemical or biological reduction of trichothecenes with lithium aluminium hydride or enzymatic modification with microsomal epoxide hydrase or the soluble glutathione S-epoxide transferase in crude liver homogenate has been found to destroy the toxic properties of T-2 toxin, scirpene triacetate, and diacetoxyscirpenol [37]. Kiessling et al. [40], while studying the effects of rumen bacteria and protozoa from sheep and cattle on 6 mycotoxins (aflatoxin B_1, ochratoxin A, zearalenone, T-2 toxin, diacetoxyscirpenol, and deoxynivelenol) demonstrated that trichothecenes, such as diacetoxyscirpenol and T-2 toxin, were deacetylated to monoacetoxyscirpenol and HT-2 toxin, respectively. Rumen protozoa were more efficient in such biotransformation of these toxins than the bacteria. Deoxynivalenol was not affected by any of these microbes. Westlake et al. [46] have shown the presence of a T-2 toxin-degrading membrane protein fraction of 65,000 dalton molecular weight, which also exhibited nonspecific esterase activity in it. They have also shown that *B. fibrisolvens* CE51 degraded T-2 toxin to HT-2 toxin (22%), T-2 triol (3%), and neosolaniol (10%). The deacetylation of T-2 toxin appeared to be mediated by the esterase activity of the membrane fraction of *B. fibrisolvens* (42,46). On the other hand, *Anaerovibrio lipolytica* and *Selenomonas ruminantium* have been found to degrade this toxin to HT-2 toxin (22 and 18%, respectively) and T-2 triol (7 and 10%, respectively).

Zearalenone detoxification

The information available on zearalenone detoxification is very scanty, as little is known about the metabolism of this fusarial toxin. The possible role of a microsomal enzyme – glucuronyl transferase has been implicated in the formation of glucuronide conjugates of this mycotoxin (Figure 5). Another possible metabolite of zearalenone contains hydroxyl group [37]. Kallela and Vasnius [47] studied the degradation of zearalenone by bovine rumen fluid.

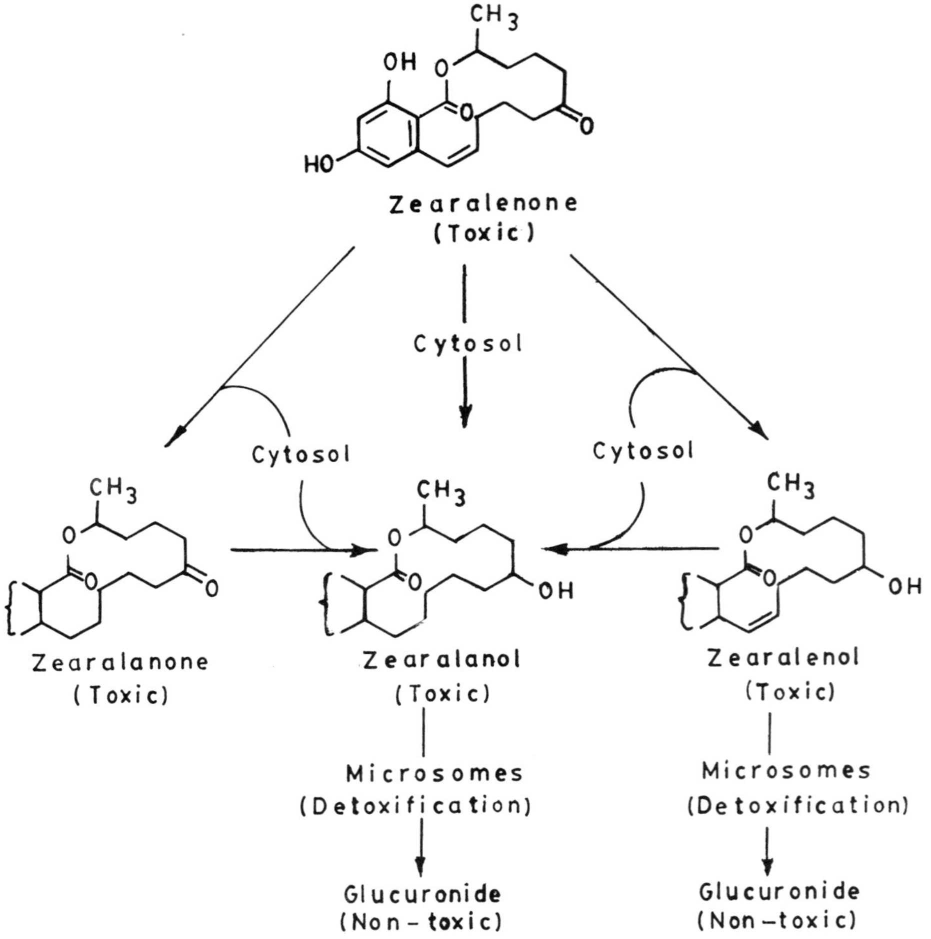

Figure 5. Pathways for zearalenone biotransformation in liver.

They observed that after 48 h of incubation, the zearalenone was degraded by an average of 37.5%. However, Kiessling et al. [40] showed that protozoa and bacteria from ruminants, such as sheep and cattle, transformed zearalenone by reducing to α-zearalenol, and to some extent to β-zearalenol. Zearalenone has been found to be degraded by *Butyrivibrio fibrisolvens* into as yet unidentified metabolites [42]. Cuero et al. [27] have shown the decreased levels of zearalenone in cultures of *Fusarium graminearum* when cultivated in the presence of *A. flavus* at 16°C, suggesting that co-cultivation of zearalenone-producing fungus with other organism could help in detoxification of this mycotoxin.

CONCLUSION

Microbes are *omnipresent* and their role in health and environmental protection is well understood. However, the understanding of the microbial enzymatic mechanisms as the detoxification strategies, which form the basis for bioremediation of wastes and biological control of potentially hazardous compounds, is of tremendous importance. Microbial degradation of various xenobiotics of health and environmental concern, such as nitrogenous compounds, carbamates, phenylureas, anilides, halogenated aromatic compounds, azo dyes, tannins and heterocyclic compounds has been described. Mycotoxins in general and aflatoxins in particular have been found to be degraded by microsomal *cytochrome P-450 monooxygenase* and peroxidase enzymes. A microsome technology based model of detoxification strategies for health and environmental protection has been proposed.

REFERENCES

1   Anderson TA, David C, White C, Walton BT. Progress in Industrial Microbiology 1995; 32: 205-225.

2   Singh VP (ed). Progress in Industrial Microbiology. Vol 32, Amsterdam: Elsevier Science BV, 1995; pp. 282.

3   Mukai K, Doi Y. Progress in Industrial Microbiology 1995; 32: 189-204.

4   Karns JS, Mulbry WW, Nelson JO, Kearney PC. Pest Biochem Physiol 1986; 25: 211-217.

5   Karns JS, Mulbry WW, Kearney PC. In: Augustive PL, Danforth HD, Bakst MR, eds. Biotechnology for Solving Agricultural Problems. BARC Symposium 10, Boston: Martinus Mijhoff Publishers, 1986; 339-354.

6   Grund E, Schmitz A, Fiedler J, Gartemann KH. Progress in Industrial Microbiology 1995; 32: 103-123.

7   Waheeta A, Mahadevan A. In: Mukerji KG, Singh VP, eds. Frontiers in Applied Microbiology. Meerut: Rastogi and Company, 1993; 37-56.

8   Bhat MA, Vaidyanathan CS. Progress in Industrial Microbiology 1995; 32: 125-156.

9   Bumpus JA. Progress in Industrial Microbiology 1995; 32: 157-176.

10   Hartmans S. Progress in Industrial Microbiology 1995; 32: 227-238.

11   Hartmans S. Progress in Industrial Microbiology 1995; 32: 239-248.

12   Mahadevan A, Sivaswamy SN. In: Mukerji KG, Pathak NC, Singh VP, eds. Frontiers in Applied Microbiology. Luknow: Print House (India), 1985; 327-347.

13   Saxena RK, Sharmila P, Singh VP. Progress in Industrial Microbiology 1995; 32: 259-270.

14   Cook AM, Hutter R. J Agric Food Chem 1984; 32: 581-585.

15   Cook AM, Hutter R. FEMS Microbiol Lett 1986; 34: 335-338.

16   Lisker N, Lillehoj EB. In: Smith JE, Handerson R, eds. Mycotoxins and Animal Feedstuffs. Boca Raton: CRC Press, 1991.

17   Natarajan KR, Rhee KC, Cater CM, Maittil KM. J Am Oil Chem Soc 1975; 53: 160-163.

18   Samarajeewa U. In: Smith JE, Handerson R, eds. Mycotoxins and Animal Feedstuffs. Boca Raton: CRC Press, 1991.

19   Ciegler A, Lillehoj EB, Peterson RE, Hall HH. Appl Microbiol 1966; 14: 934-939.

20   Lillehoj EB, Stubblefield RD, Shannon GM, Shotwell OL. Mycopath Mycol Appl 1971; 45: 259-266.

21   Detroy RW, Hesseltine CW. Can J Microbiol 1969; 15: 495-500.

22   Detroy RW, Hesseltine CW. Can J Biochem 1970; 48: 830-832.

23   Mann R, Rehm HJ. Eur J Appl Microbiol 1976; 2: 297-306.

24   Bol J, Smith JE. Food Biotechnol 1989; 3: 127-144.

25   Cole RJ, Kirskey JW. J Agric Food Chem 1971; 19: 222-223.

26   Knol W, Bol J, Huis In T, Veld JHJ. In: Zenthen P, Cheftel JC, Erikson C, Gormley TR, Liko P, Panlus K, eds. Processing and Quality of Food. London, New York: Elsevier Applied Sciences, 1990; 2133-2136.

27   Cuero R, Smith JE, Lacey J. J Food Protect 1988; 51: 452-456.

28   Banwort GI (ed). Basic Food Microbiology. Westport, Connecticut: AVI Publishing Co Inc, 1981; pp. 343.

29   Bhatnagar D, Lillehoj EB, Bennett JW. In: Smith JE, Handerson R, eds. Mycotoxins and Animal Feedstuffs. Boca Raton: CRC Press, 1991.

30   Faraj MK. PhD Thesis, University of Strathclyde, Glasgow, UK 1990.

31   Hamid AB, Smith JE. J Indust Microbiol 1987; 2: 137-141.

32   Singh VP, Smith JE, Harran G, Saxena RK, Mukerji KG. Indian J Microbiol 1992; 357-369.

33   Hamid AB, Smith JE. J Gen Microbiol 1987; 133: 2023-2029.

34   Singh VP. Current Science 1997; 73(6): 529-532.

35   Singh VP. J Phytol Res 1998; 11(1): 7-10.

36   Singh VP, Gupta S, Singh I, Gupta R. Indian J Aerobiol 1998; 11(1&2): 6-15.

37   Patterson DSP. Pure Appl Chem 1977; 49: 1723-1731.

38   Thiel PG, Steyn M. Biochem Pharmac 1973; 22: 3267-3273.

39   Chu FS, Noh I, Chang CC. Life Sci 1972; 11: 503-509.

40   Kiessling KH, Petterersson H, Sandholm K, Olson M. Appl Environ Microbiol 1984; 47: 1070-1073.

41   Hult K, Teilling A, Gatenbeck S. Appl Environ Microbiol 1976; 32: 443-444.

42   Westlake K, Mackie RI, Dutton MF. Appl Environ Microbiol 1987; 53: 613-614.

43   Doster RC, Sinnhuber RO. Food Cosmet Toxicol 1972; 10: 389.

44   McCarmick SP, Taylor SL, Plattner RD, Beremand MN. Appl Environ
      Microbiol 1990; 56: 702-706.

45   Mirocha CJ, Abbas HK, Vesonder RF. Appl Environ Microbiol 1990;
      56: 502-525.

46   Westlake K, Mackie RI, Dutton MF. Appl Environ Microbiol 1987; 53:
      587-592.

47   Kallela K, Vasnius L. Nord Veterinaermed 1982; 34: 336-339.

Biotransformations: Bioremediation Technology for Health and Environmental Protection
V.P. Singh and R.D. Stapleton, Jr. (Editors)
2002 Elsevier Science B.V.

Microbial degradation of chlorobenzoates (CBAs): Biochemical aspects and ecological implications

G. Baggi

Department of Food Science & Microbiology, Università di Milano, Via Celoria, 2, 20133 Milan, Italy

INTRODUCTION

Halogens are distributed in soils and waters mostly as inorganic compounds, while more than 200 naturally occurring chloroaromatics were identified in mammals, invertebrates, plants, algae, fungi, and bacteria [1]. After the second world war, synthetic compounds carrying halogen substituents, largely used as solvents, pesticides, lubricants, hydraulic fluids etc., were released in the environment. For their persistence and, in some cases, their high toxicity, they are potent hazards to the environment. Their recalcitrance to microbial degradation depends on the substitution grade, differing in a range from 1 to 12 as in mirex ($C_{10}Cl_{12}$), insecticide totally substituted. The nucleus breakage, in fact, requires the C atoms subjected to introduction of atmospheric O_2, catalyzed by bacterial mono- and dioxygenases, which are free from substituents.

The microbial degradation of halogenated aromatic compounds has been extensively studied for many years in order to elucidate the degradative mechanisms involved and, more recently, to develop adequate technologies for environmental protection. Particularly, the behaviour of chlorobenzoates (CBAs) in soils and waters has been carefully considered, as these compounds are used themselves as *herbicides*, e.g., 2,3,6-trichlorobenzoate (2,3,6-TCBA). They can derive from the catabolism of related xenobiotics (i.e. dicamba) or are cometabolites of ubiquitous pollutants like polychlorinated biphenyls (PCBs) and chlorotoluenes [2,3] (Figure 1). Due to their ubiquitous presence,

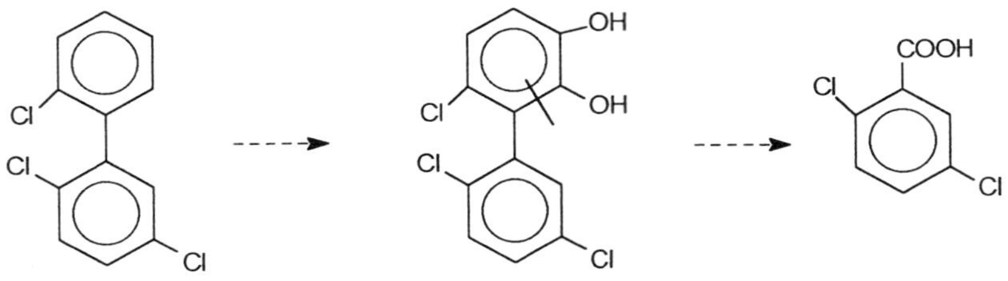

Figure 1. Formation of chlorobenzoic acids from PCBs. The degradation starts from the less substituted ring having free 2.3 sites for dioxygenases.

good water solubility and low toxicity, CBAs have been used for many years as models to study the degradation of more complex chloroaromatics, both in aerobic and anaerobic conditions and to elucidate the microbial strategies implicated in the release of chlorine substituents.

The degradation of synthetic chloroaromatics, whose structure was not found in natural compounds, might be due to: (i) pre-existing enzymes implicated in the degradation of non-substituted analogs, (ii) enzymes specifically involved in the degradation of natural halogenated compounds (e.g., 2,4-dichlorophenol) or, finally, to (iii) the evolution of new catabolic pathways because of the selective pressure exerted by the introduction and the persistence of these xenobiotics in the environment.

The total degradation of CBAs involves the stoichiometric release of Cl$^-$ as inorganic ions, which may occur after the breakage of the aromatic nucleus through spontaneous dechlorination of the intermediate chloromuconolactone. Differently, Cl$^-$ release is due to oxygenolytic mechanisms with the involvement of dioxygenases, more or less specific for the halogenated compounds. In both cases O_2 is required, while a third mechanism, involving hydrolytic reactions with the formation of corresponding hydroxy derivatives, has been reported in aerobic and anaerobic conditions. In anaerobiosis, the dechlorination mostly involves reductive reactions with the detachment, step-by-step, of the ring substituents.

METABOLIC PATHWAYS FOR DEGRADATION OF CHLORO-BEZOATES

Investigations on the microbial degradation of CBAs have allowed the isolation of different microorganisms capable of degrading these compounds. Beside two strains characterized by a strict specificity versus the isomer utilized for growth [4], other strains have exhibited a remarkable versatility, being capable of degrading different isomers mono-, di- or trisubstituents [5,6]. Among the different CBA isomers, those carrying a chlorine substituent in the *ortho* position have been recognized as compounds of environmental concern, as they exhibit herbicidal or fungicidal properties. Unfortunately, these compounds were shown to be more persistent than other isomers because of the steric hindrance plus the effect of halogen atoms on the electron density at the *ortho* position of the benzene ring, hampering the attack of the benzoate 1,2-dioxygenase responsible, in many cases, of the first step in CBA catabolism.

Monochlorobenzoates

Natural isolates of *Flavobacterium, Alcaligenes,* and *Pseudomonas* genera were obtained in pure cultures through enrichment procedures on 2-CBA as unique C source [6-11]. Among these strains, *Ps. aeruginosa* JB2 showed a large versatility in utilizing, apart from 2-CBA, other different CBAs isomers [5] and was utilized for mating experiments to obtain recombinant organisms with the ability of degrading CBAs [12]. A hybrid strain, *Pseudomonas* JH 230, furtherly selected for growth on 2-CBA and capable of degrading also 3- and 4-CBAs plus 2,4- and 2,5-DCBAs, was constructed from the cross between *Psudomonas* sp. strain WR401, which degrades 2-MB via *meta*-cleavage and *Pseudomonas* sp. strain B13 capable of degrading chlorocatechols, and seems to have recruited the genes coding for the formation of 3-chlorocatechol (3-CC) by the first strain and those for the *ortho*-fission of 3-CC by the second one [13].

Sylvestre et al. [8], working with a *Pseudomonas* strain B-300 observed that, in the presence of glucose which represses some enzymes of the pathway, the growth of this strain on 2-CBA allowed to detect the formation of catechol and both chloro- and muconic acids. It suggests the concomitant occurrence of a metabolic pathway where the chlorine release occurred before the breakage of the aromatic ring, giving catechol and muconic acid with another pathway, where the detachment of chlorine takes place after the nuclear fission with the formation of 2-chloromuconic acid [8].

In cell extracts of *Ps. putida* CLB250, the higher activities for 1,2-dihydro-1,2-dihydroxybenzoate dehydrogenase, catechol 1,2-oxygenase and the subsequent enzymes of the *ortho*-pathway of catechol compared with the negligible activity for 3-CB and 4-chlorocatechol (4-CC) found after growth on 2-CB, suggested a degradative pathway involving as the initial step, the oxygenolytic attack of the compound with chemical loss of Cl- and CO_2 to give catechol. As in these extracts, benzoate 1,2-dioxygenase activity was never found, the involvement of a 1,2-dioxygenase which preferentially attacks 2-substituted benzoates, was postulated [9]. Also *Ps. cepacia* 2CBS degrades 2-CBA through the formation of catechol which, in this case, is further metabolized via *meta*-ring cleavage. An *inducible* two-component 2-halobenzoate 1,2-dioxygenase system is purified and characterized with respect to some of its physical and chemical properties [14,15]. Analogously, the activity of a unique broad-spectrum CBA 1,2-dioxygenase was postulated by Fava et al. for the degradation of 2-CBA and 2,5-DCBA by a *Pseudomonas* sp. CPE2 strain, capable of growing on both CBAs as unique carbon sources with the formation of catechol and 4-CC, respectively [16,17].

When Cl⁻ release occurs after the fission of the aromatic ring, the intermediate 3-CC undergoes *ortho*-fission to give 2-chloro-*cis-cis*-muconate. After cycloisomerization, the halide is spontaneously lost with the formation of 4-carboxy-methylen-2-en-4-olide. This pathway was proposed for the above cited hybrid strain, *Pseudomonas*. JH230 on the basis of simultaneous adoption studies, enzymatic assays and co-oxidation of the analog 2-methylbenzoate [13], and for a *Ps. stutzeri* strain isolated by enrichment techniques, capable of utilizing 2-CBA and 2,5-DCBA [18].

On the other hand, *Ps. aeruginosa* BBZA, isolated on 2-bromobenzoate was capable of utilizing 2-CBA and other halogenated CBAs, with the exception of the compounds having *para*-substituents, seems to degrade 2-halobenzoates via the formation of 2-hydroxybenzoic acid mediated by an enzyme of broad specificity which recognizes only the *ortho*-substituent [10]. This result was obtained with resting cells grown on 2-bromobenzoate in the presence of 3-hydroxybenzoate and 3-CC, which were inhibitors of salicylate and catechol metabolism respectively, and showed that the ring position of halogen is more important than size or electronegativity.

The alternative pathways proposed for 2-CBA degradation are presented in Figure 2. Several bacteria belonging to genera *Arthrobacter, Acinetobacter, Alcaligenes, Pseudomonas,* etc., having the ability of using 3- and/or 4-CBA as unique carbon sources, have been isolated [19-25]. Particularly, 4-CBA degradation has been widely studied for many years as this halogenated compound, used for synthesis of dyestuffs, pigments and pharmaceuticals, is also a common metabolite of some PCBs and of the herbicide bisidin [26]. In most of the cases studied, the degradation of 4-CBA occurs via dehalogenation in the initial stage with the formation of 4-hydroxybenzoate. Many authors have reported this pathway for different microorganisms, among which a facultatively alkalophilic *Arthrobacter* sp. strain SB8 and an *Alcaligenes* sp. strain ALP83, carrying the dehalogenase activity on the plasmid pSS70

154

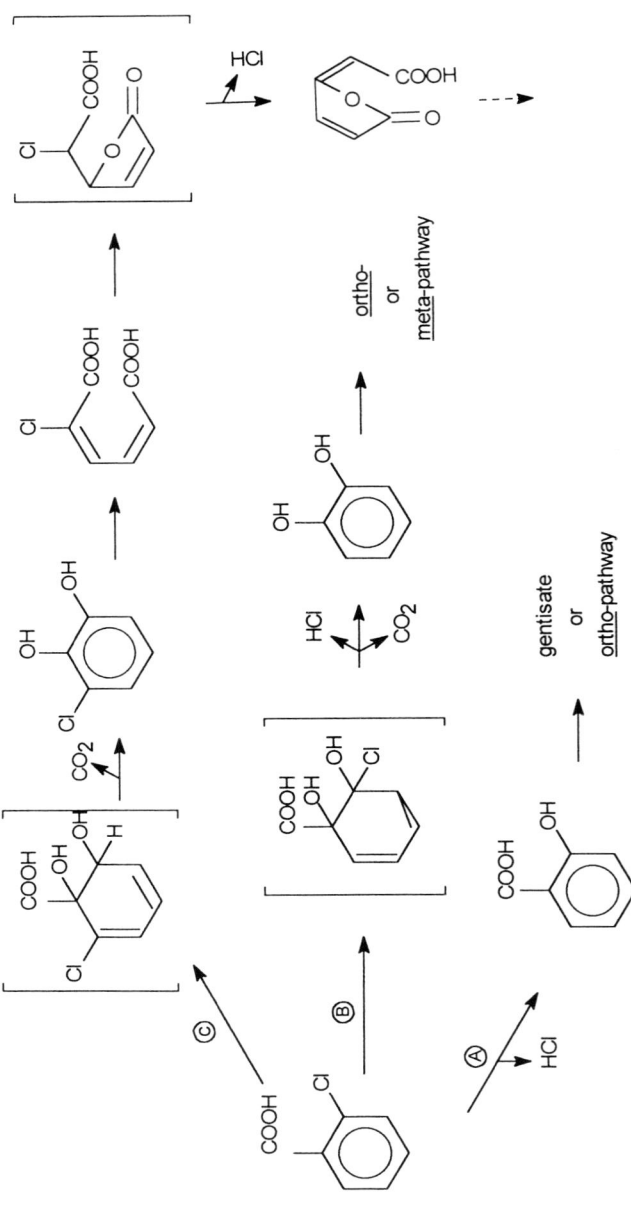

Figure 2. Alternative pathways for the degradation of 2CBA. A): The first step is a hydrolytic dechlorination [10]; B): The first step is catalyzed by a specific 2-halobenzoate 1,2-dioxygenase with the formation of the corresponding diol subsequently converted into a catechol [8,9,16,17]; C): The first step is catalyzed by a 1,6-dioxygenase with the formation of the corresponding diol subsequently converted into 3-chlorocatechol [8].

[27,28]. Differently, in *Pseudomonas* sp. DJ-12 the same hydrolytic activity was specified by chromosomal genes [29]. The Cl⁻ release was believed to be due to a hydrolytic reaction; in fact O^{18} labeling experiments have shown that the oxygen atom of the hydroxyl group is not derived from air and, furthermore, the reaction may proceed also in the absence of oxygen [30-32].

The mechanism of this dehalogenation, studied in cell extracts of the 4-CBA-degrading *Pseudomonas* sp. CBS3 and *Acinetobacter* strain 4CB1, required ATP and CoA and proceeded via the initial formation of 4-chlorobenzoyl CoA which undergoes the detachment of chlorine [33-35]. The dehalogenase of *Pseudomonas* CBS3 was found to be a three-component system in which the first component was highly unstable [36-38] and converted also chlorinated nitroaromatic compounds such as 4-chloro-3,5-dinitrobenzene and 1-chloro-2,4-dinitrobenzene to the corresponding hydroxy derivatives, even at lower activities [39].

A dehalogenation after the breakage of the aromatic ring was found in *Pseudomonas* sp. B13, which, exhibited high activities for halogenated catechols due to the induction of pyrocatechase II, a catechol 1,2-dioxygenase specific for substituted catechols [4]. In successive researches this strain, whose degradative properties were harboured in the plasmid pB13 homologous to pAC25 of a strain of *Ps. putida*, was genetically modified to enlarge its substrate spectrum to 4-CBA and 4-methylbenzoic acid, which were all mineralized via *ortho* pathway [40-43].

Also, *Alc. eutrophus* JMP134 metabolized 3-CBA via 3-CC and 4-CC, which underwent the *ortho*-fission of the aromatic nucleus; whereas 4-CC was efficiently degraded. The degradation of 3-CC caused the accumulation of 2-chloro-*cis-cis*-muconate because of the poor activity of the dichloromuconate cycloisomerase [44]. On the other hand, in *Ps. cepacia* P166, a strain unusually capable of utilizing for growth the three monochlorobiphenyl (MCBP) isomers and the corresponding CBAs' central metabolites of the breakdown of the first compounds, metabolized 3-CC deriving from 2- and 3-chlorobiphenyl (2-CBP and 3-CBP) via *meta*-cleavage, giving a toxic acyl halide which impeded further degradation [45]. On the contrary 4-CBA, resulting from 4-chlorobiphenyl (4-CBP), was transformed to 4-CC, which was also slowly mineralized via *meta*-cleavage [46,47]. Also, a *Pseudomonas* sp. S-47 was found to degrade 4-CBA with the same pathway, as demonstrated by the identification of 5-chloro-2-hydromuconic semialdehide and 2-hydroxy-penta-2,4-dienoic acid. *Ps. cepacia* MB2 was able to grow on 3-chloro-methylbenzoate, which was degraded through the formation of 4-chloro-3-methylcatechol and successively cleaved in *meta* [48,49]. The metabolic pathways for 3- and 4-CBA are presented in Figure 3.

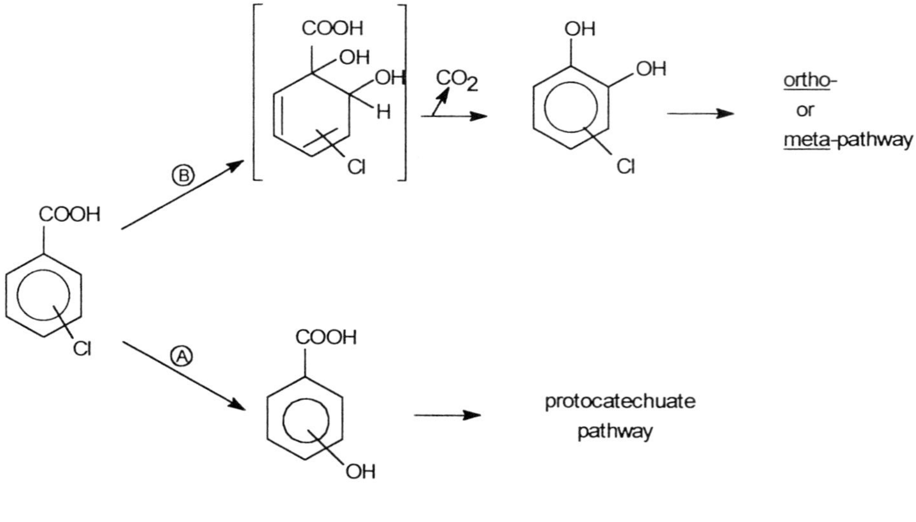

Figure 3. Alternative pathways for the degradation of 3-CBA or 4-CBA. A): The first step is a hydrolytic dechlorination [19-25,27,28]; B): The first step is catalyzed by a dioxygenase with the formation of the corresponding diol subsequently converted into a chlorocatechol [4,45-49].

Dichlorobenzoates

The presence of two chlorosubstituents in different positions of the aromatic ring negatively affects the microbial attack of dichlorobenzoate (DCBA) isomers. In particular, 2,6-DCBA, which presents both substituents in the *ortho* position, did not support the growth of any organism known till now, showing that an unsubstituted carbon next to carboxyl group is important for the attack of the dioxygenase. This compound was only cometabolically degraded at a remarkable extent by a *Ps. stutzeri*, capable of utilizing 2-CB and 2,5-DCBA, while growing on glucose [18]. On the other hand, 2,5-DCBA, which presents only a chlorine in the *ortho* position, is utilized for growth by some microorganisms capable of growing also on 2-CB [7,16,18]. The degradation of 2,5-DCBA was hypothesized to proceed via 4-CC with the probable involvement of a benzoate 1,2-dioxygenase, in accordance with the mechanism proposed for 2-CBA [16].

The degradation of 3,5-DCBA was studied by Hartmann et al. in a *Pseudomonas* strain, which was able to grow on 3-, 4- and 3,5-DCBAs. These authors suggested a pathway proceeding through 3,5-dichlorocatechol (3,5-DCC) formed by a direct dioxygenation, followed by decarboxylation, similar to the metabolism of the analog 3,5-dimethylbenzoate [50]. A *Ps. putida* P111 strain, which was unable to utilize 3,5-DCBA for growth, metabolized it via same pathway through the formation of 1-carboxy-1,2-dihydroxy-3,5-dichlorocyclohexadiene and subsequently converted into 3,5-DCC [6]. 3,5-DCC was also metabolized by cell extracts of *Pseudomonas* strain B13 grown on 3-CBA with the formation of *cis* and *trans* isomers of 2-chloromuconolactone and of chloroacetylacrylate as decarboxylation product of 2-chloromaleylacetate [51]. Another 3,5-DCBA degrader, *Ps. aeruginosa* AC869, was obtained by engineering an environmental strain to utilize this compound; the strain, found to be pathogenic for mice, was attenuated without impairing its degradative ability, but its survival in contaminated soil was drastically reduced [52,53]. Resting cells of *Acinetobacter* 4CB1 utilized 4-CBA anaerobically and converted 3,4-DCBA into 4-carboxy-1,2-benzoquinone, presumably deriving from a double hydrolytic dehalogenation of 3,4-DCBA [54,55] (Figure 4). An *Alcaligenes* strain CPE3, capable of utilizing 3,4-DCBA for growth, degraded this compound via 3,4-dihydroxybenzoic acid, as CPE3 extracts did not exhibit hydrolytic dechlorination activity and metabolized both 3-chloro-4-hydroxy benzoic acid and 3,4-dihydroxybenzoic acid. The involvement of a chlorobenzoate 3,4-dioxygenase was proposed [16,17].

2,4-DCBA degradation was firstly shown in an *Alc. denitrificans* NTB-1 strain, subsequently classified as a coryneform bacterium; this microorganism converted this compound into 4-CBA and used it only as carbon source, through an *inducible reductive dechlorination*, a mechanism usually operating in anaerobiosis. 4-CBA was successively catabolized via 4-hydroxybenzoate, as shown for many other 4-CBA-degrading organisms [56] (Figure 5).

The metabolic pathway of 2,3-DCBA is still unknown; however, two strains of *Alc. denitrificans* BRI3010 and BRI6011, were able to grow on this compound [57]. Dicamba (2-methoxy-3,6-dichlorobenzoic acid), a herbicide used for the control of broad-leaf weeds, was mineralized by *Ps. maltophilia* DI-6 through the formation of 3,6-dichlorosalicylic acid as the first degradation product [58,59]. The involved O-demethylase, separated into its three components, was found to be similar to other mono- and dioxygenase systems implicated in the degradation of other natural and xenobiotic compounds [60].

Trichlorobenzoates

Among trichlorobenzoate (TCBA) isomers, 2,3,6-TCBA, extensively used as herbicide itself or as component of mixtures of herbicides, was shown to be cometabolically converted into 3,5-DCC by a *Brevibacterium* sp. [61].

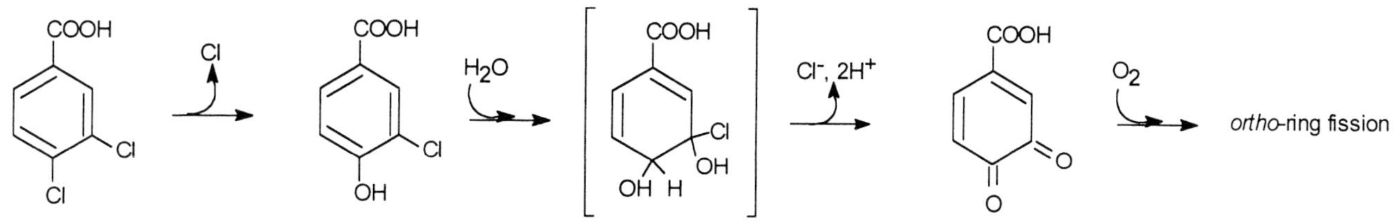

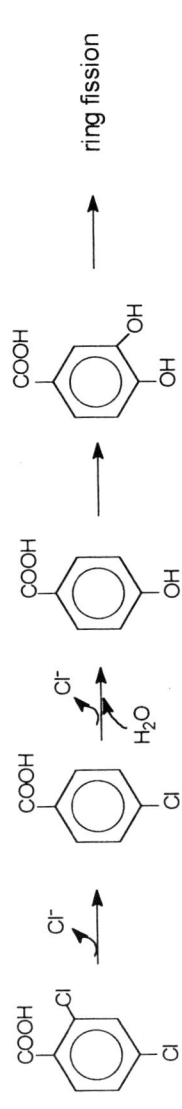

Figure 4. Degradation pathways of 3,4-DCBA by *Acinetobacter* sp. 4CB1. The strain, unable of growing on 3,4-DCBA, cometabolized it to 3-chloro-4-hydroxybenzoic acid the same strain could use as C source [55].

Figure 5. Proposed catabolic pathway for 2,4-DCBA in *Alc. denitrificans* NTB-1. The reductive dechlorination giving 4-CBA is an unusual reaction in aerobic microorganisms [56].

Two *Pseudomonas* strains were shown to be capable of utilizing 2,3,5-TCBA for growth, but the metabolic pathway of this compound has not yet been elucidated [5,6]. Data regarding other TCBA isomers or CBAs, more than three substituents are not available.

ANAEROBIC BIODEGRADATION OF CHLOROBENZOATES

The anaerobic biodegradation of CBAs has been largely ignored until the detachment of halogens bound to the aromatic ring and has been shown to occur through a reductive mechanism [62]. This process was first described in anaerobic consortia of aquatic sediments [63,64]. *Desulfomonile tiedeje* and *Desulfitobacterium dehalogenans* were isolated as anaerobic, dehalogenating bacteria in sulphate-reducing conditions. Besides other chloroaromatic and chloroaliphatic compounds, the first strain was capable of dechlorinating 3-CBA to benzoate [65-67]. *Rhodopseudomonas palustris*, organism versatile in the anaerobic metabolism of various aromatic compounds, was adapted to metabolized 3-CBA completely. Its possible contribution to the decontamination of the aquatic environments polluted by complex mixtures of aromatic substrates, including chlorinated compounds was postulated [68].

However, the anaerobic biodegradation of CBAs with regard to 3-CBA, is still under study, mostly in anaerobic consortia. It is known, in fact, that different microorganisms acting in succession are involved in anaerobic *mineralization* of aromatic compounds [69-71]. In the most recent researches focused on 3-CBA anaerobic biodegradation, the influence of the exposure of the sediments to chlorinated compounds, the effect of the substituents on the aromatic ring, the availability of alternative electron acceptors, and the addition of heavy metal ions were investigated in aquatic sediments and in soil samples by anaerobic consortia, among which two mixed cultures deriving from geothermal and non-geothermal environments were able to grow at 75°C [70,72-77].

The degradation of dicamba, already reported in aerobic conditions, was shown to be performed also by an anaerobic consortium, consisting of a sulphate-reducer, three methanogens and a fermenter, in analogy with the aerobic degradation. The microbial attack starts with an initial demethylation giving 3,6-dichlorosalycilic acid and is followed by a reductive dechlorination with the formation of 6-chlorosalicylic acid without mineralization of the aromatic ring [78]. Finally, 2,3,6-TCBA was reported to undergo reductive dechlorinations by anaerobic enrichments, giving 2,6- and 2,5-DCBAs. As the latter DCBA was used for growth of aerobic bacteria, the alternative of anaerobic and aerobic conditions was proposed to improve the degradation of 2,3,6-TCBA and, more generally, of polychlorinated compounds [79].

Inhibitory effects in the degradation of chloroaromatic mixtures

The contemporary presence of mixtures of compounds, often structurally similar in many cases, produce synergistic or antagonistic interactions on the degradative process. Such interferences, recognized also for ecologically relevant compounds present in mixtures in the environment (BTX , PAH, CBAs isomers, etc.), were ascribed to different causes such as the toxicity of the co-substrate, hindrance in the enzymatic induction, catabolite repression mechanisms, competition at the active sites of the degradative or uptake enzymes etc. In natural polluted ecosystems, these phenomena might compromise with the success of decontamination strategies based on the activities either of indigenous microorganisms or of specialized strains used as inoculum in *"bioaugmentation"* technologies.

Interferences among CBA analogs

Interactions between benzoate and CBAs were first reported in a strain of *Ps. fluorescens* capable of growing on benzoate. The washed cells were able to cometabolize monosubstituted halobenzoates without dehalogenation, but were inhibited in the conversion to benzoic acid. This inhibition was ascribed to a *synergism* between the inhibitor and the repressor of the synthesis of the inducible enzymes involved [80]. On the other hand, similar cometabolic transformations of CBAs, not utilized for growth by different genera of benzoate-degrading microorganisms, were imputated to the lack of specificity of the benzoate-oxidizing system opposite to the substrate specificity of the ring - cleaving dioxygenase [81]. An interesting positive interaction among CBAs, showing a cometabolic-catabolic sequence, was reported for *Acinetobacter* 4CBA, capable of cometabolizing 3,4-DCBA to 3-chloro-4-hydroxybenzoate that the same microorganism could use as growth substrate and then totally degrade [55]. An *Alcaligenes denitrificans* strain BRI6011, able to grow on different CBAs, including 2,4-DCBA and 2,5-DCBA when singly supplied, was inhibited by 2,5-DCBA in the utilization of 2,4-DCBA until the former DCBA was depleted in the growth medium. This inhibition seemed to be due to the different uptake rates of the two compounds, probably mediated by different systems [52]. The growth of another *Alc. denitrificans* strain CB on 4-CB was inhibited by the presence, in mixture together with the growth substrate, of *meta*-substituted DCBAs, In this case, these effects were attributed to interferences in the induction mechanism, as the oxidative activities on 4-CBA by the resting cells, which have pre-synthesized the necessary catabolic enzymes during the growth on 4-CBA, were not inhibited [82,83].

In the above mentioned *Ps. putida* P111, capable of growing on MCBAs, 2,3-, 2,4-, 2,5-DCBAs and 2,3,5-trichlorobenzoate (2,3,5-TCBA), the presence of 3,5-DCBA inhibited the growth of the strain on CBAs substituted in *ortho*. Surprisingly, the growth of P111 and chloride release on 3-CBA or 4-CBA was, viceversa, increased in the presence of 3,5-DCBA which was moreover completely degraded. Obviously, 3,5-DCBA, which the strain could not use as its growth substrate, was metabolized in the presence of 3- and 4-CBAs as their degradation involved the induction of a the dihydrodiol dehydrogenase capable of converting also the chlorodihydrodiol deriving from 3,5-DCBA into 3,5-DCC, which was subsequently degraded totally. On the other hand, the metabolism of the *ortho*-substituted CBAs did not required a functional dihydrodiol dehydrogenase, as the chlorodihydrodiols deriving from these compounds spontaneously dechlorinated into the corresponding (chloro)-catechols [6].

Similar inhibitory effects were observed also with *Burkholderia cepacia* JHR22, a hybrid strain having the ability to degrade different CBAs and monochlorobiphenyl isomers, during the utilization of 2-CBA, 2-, 3-, and 4-chlorobiphenyls, in the presence of 2,3- and 3,4-DCBAs. However, the formation of toxic chlorocatechols, considered in this case as the cause of this inhibition, was never detected in the cultural broths [84] (Figure 6).

The influence of the presence of CBAs in mixture on the abilities of the microorganisms to degrades these compounds, was also demonstrated in experiments conducted with soil suspensions in the presence of the indigenous microflora and/or specialized microbial inocula. Brunsbach and Reineke [85] observed that the indigenous microflora of a soil capable of mineralizing only 4-CBA when supplied as unique contaminant, acquired the ability of cometabolizing other CBAs when present in two compound mixtures. The addition of CBA-degrading organisms, having different degradative spectra, resulted in the rapid degradation of the chloroaromatics singly supplied to the soil; while in the presence of CBA mixtures the degradative abilities of the inoculated strains were differently affected [85] (Figure 3). In similar experiments conducted with a soil having itself 4-CBA-degrading capacity, the inhibitory effects exerted by 3,4- and 2,3-DCBAs were evidenced on the 4-CBA degradation both in the presence of microbial inoculum and/or of the indigenous soil microflora alone [83] (Figure 7).

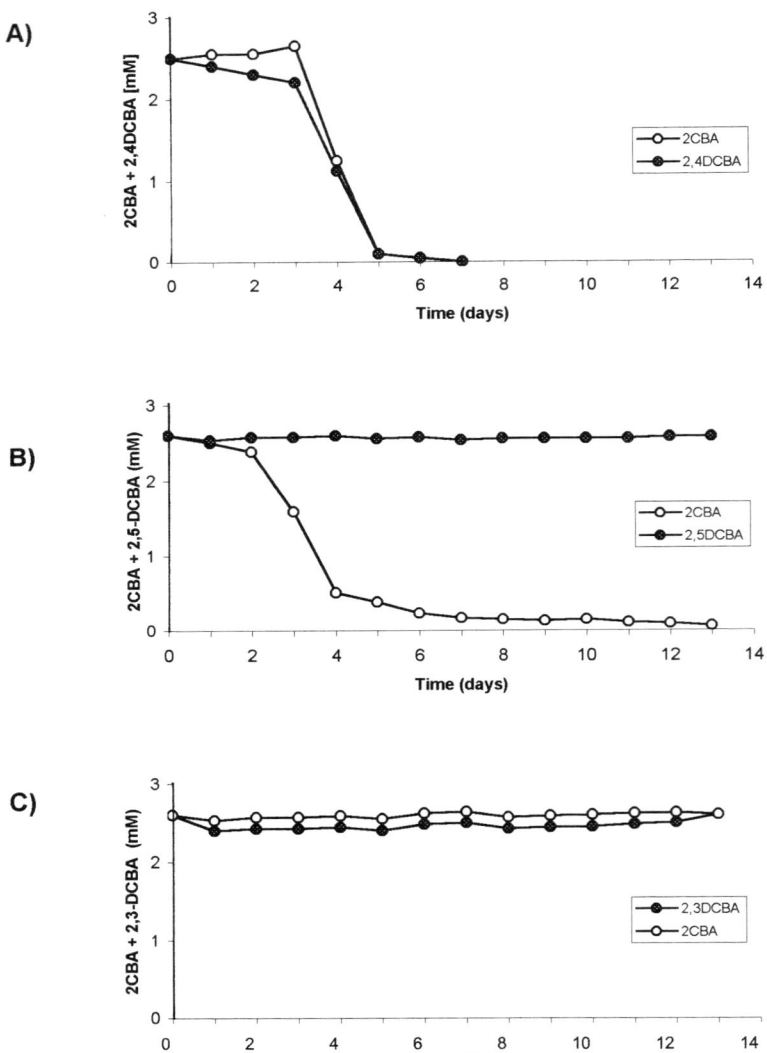

Figure 6. Interactions in the degradation of chlorobenzoate mixtures in *Burkholderia cepacia* JHR22, a hybrid strain capable of degrading different CBAs and monochlorobiphenyls. A): 2CBA and 2,4-DCBA present in mixture are contemporarily degraded; B): The presence of 2,5-DCBA does not hinder 2-CBA degradation; C): The presence of 2,3-DCBA hinders 2-CBA degradation. Data from [84] with kind permission

Interferences between CBAs and PCBs

The inhibition phenomena exerted by CBAs on pure cultures of PCB-degrading microorganisms have a great ecological significance as in sites polluted by PCB congeners, mixtures ofdifferentCBA isomers deriving from

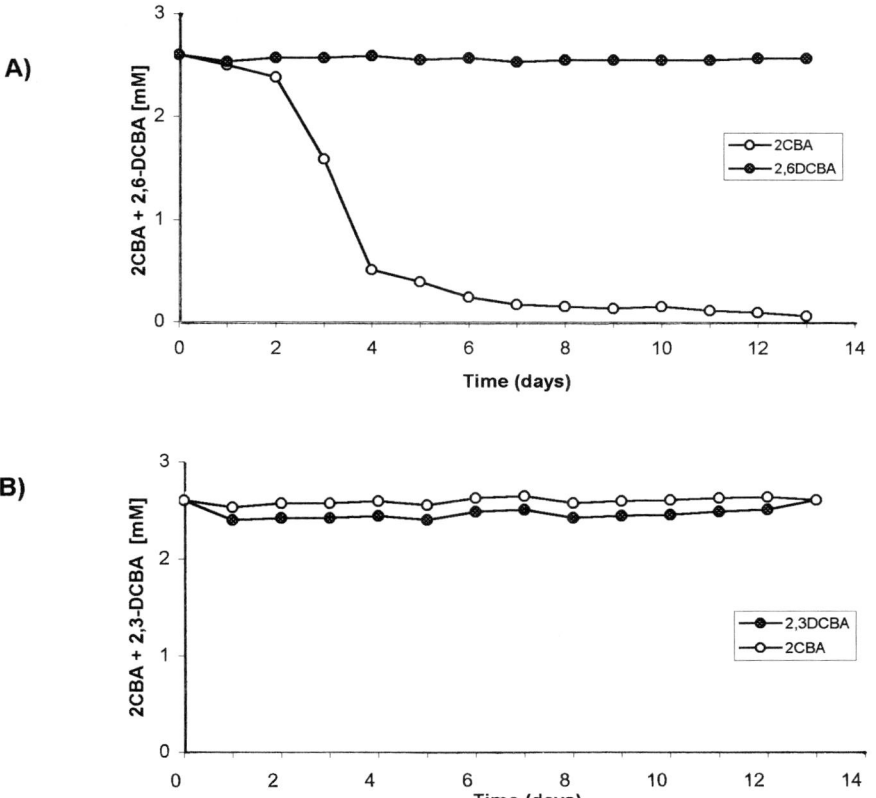

Figure 7. Interactions in the degradation of chlorobenzoate mixtures in soil slurries by the indigenous microflora having intrinsic 4-CBA-degrading capability. A): The addition of 2,6-DCBA does not hinder 4-CBA degradation; b): The addition of 2,3-DCBA exerts an inhibitory effect on 4-CBA degradation. Data from [83]

the cometabolism of these compounds are present. In fact, PCB degraders are usually unable to metabolize CBAs, because of the incompatibility of the pathways of the two classes of the compounds; biphenyl (BP) and PCBs are *meta*-cleaved by a dioxygenase, which gives *"unproductive metabolites"* when it acts on CBA-derived chlorocatechols [45].

This is likely the cause of the inhibition exerted by 3-CBA and *meta*-substituted DCBAs on the degradation of different chlorinated PCB congeners recognized in single strains and in the microbial population of a PCB-contaminated soil [84,86-88]. The same indigenous microflora was shown to be inhibited, even if to a lesser extent than 3-CBA, also by 4-CBA. The pattern of this inhibition has not yet been elucidated. The *meta*-cleavage product of 4-CC did not generate toxic intermediates. However, 4-CBA and/ or its metabolites were shown to inhibit PCB degradation also in *Ps. testosteroni* B356 [87]. The 2-CBA and 2,4-DCBA have been shown to be reasonably the least effective inhibitors, as the metabolic pathways of these compounds never include the formation of *meta*-cleavage products of 3-CC [13,56].

To control the ecological and practical implications due to these interferences, the introduction of CBA-degraders into PCB-contaminated sites was suggested as the prevention of CBA, and/or their metabolite accumulation might enhance PCB mineralization [89-90]. In addition, natural genetic exchanges between CBA utilizers and indigenous PCB degraders could improve the degradative abilities of soil microorganisms [89,91].

Interferences between CBAs and methyl-substituted benzoic acids

The degradation of CBAs can be seriously hampered in the presence of methylsubstituted benzoic acids and viceversa. The usual *meta*-cleavage of the alkylsubstituted compounds generates the toxic metabolites reported above when operating on 3-CC, while the *ortho*-cleavage of 3-methylcatechol gives a methyllactone as dead-end product [45,92]. For these reasons, the addition of mixtures of chloro- and methylaromatics often disrupts bacterial growth, destabilizing microbial consortia in the environment and breaking off mineralization activity. An exception to this rule is the case of *Ps. cepacia* MB2, capable of utilizing 3-chloro-2-methylbenzoic acid through the *meta*-fission pathway [49].

On the other hand, the degradation of mixtures of these compounds might be obtained with the incorporation of new catabolic genes. A genetically modified *Pseudomonas* sp. B13 FR1 SN45P degraded 3-CB, 4-CB and 4-methylbenzoate both as single substrate or as a substrate mixture by solely using the *ortho*-ring cleavage pathway and avoiding the formation of toxic intermediates or dead-end metabolites [93].

CONCLUSION

Chlorobenzoic acids from mono- to tri-substituted compounds, were shown susceptible to *microbial attack*, even if more chlorinated isomers and/or having chlorine atoms in *ortho* position were demonstrated more refractory to biodegradation. CBAs can be totally mineralized with stoichiometric release of the chlorine atoms, both in aerobic and anaerobic conditions, or can undergo only cometabolic transformations giving dead-end products, which the other microorganisms could utilize for their growth.

The bacterial strategies for CBA degradation, elucidated with pure cultures or in microbial consortia, turn around the detachment of chlorine atoms which may occur through: (i) oxygenolytic elimination in an early stage mediated by more or less specific 1,2- or 1,6-dioxygenases leading to the formation of catechol or chlorocatechols; (ii) spontaneous Cl⁻ release at a later stage, by *lactonization* of the *ortho*-ring fission product; (iii) initial dehalogenation through hydrolytic or oxidative reactions with the formation of corresponding hydroxy derivatives; and (iv) reductive dechlorinations, mostly occurring in anaerobic conditions, and on polychlorinated compounds, with the formation of the corresponding derivatives carrying n-1 chlorine substituents.

In natural ecosystems the different pathways may intersect leading to the complete degradation of CBAs through the involvement of different microorganisms and the alternating of different environmental conditions. However, the recent recognition of mixed substrate competition and inhibition phenomena, observed in laboratory experiments with pure and mixed cultures and in soil microcosms, is becoming a potential problem in addressing the efficacy of biodegradation in *polluted ecosystems* both by indigenous microorganisms and specialized microbial inocula for the restoration of contaminated sites.

ACKNOWLEDGEMENT

The skilful help of Maurizio Zangrossi in preparing the manuscript has been highly appreciated.

REFERENCES

1   Siuda JF, De Bernardis JF. Lloydia 1973; 36: 107.

2   Bedard DL. Biotechnology and Biodegradation. The Woodlands, Tex: Portfolio Publishing Company, 1990; 369-388.

3   Lehning A, Fock U, Wittich R-M, Timmis K N, Pieper D H. Appl Environ Microbiol 1997; 63: 1974-1979.

4   Dorn E, Hellwig M, Reineke W, Knackmuss H-J. Arch Microbiol 1974; 99: 61-70.

5   Hickey WJ, Focht DD. Appl Environ Microbiol 1990; 56: 3842-3850.

6   Hernandez BS, Higson FK, Kondrat R, Focht DD. Appl Environ Microbiol 1991; 57: 3361-3366.

7   Baggi G. Ann Microbiol 1985; 35: 71-78.

8   Sylvestre M, Mailhiot K, Ahmad D, Massé R. Can J Microbiol 1989; 35: 439-443.

9   Engesser KH, Schulte P. FEMS Microbiol Lett 1989; 60: 143-148.

10   Higson FK, Focht DD. Appl Environ Microbiol 1990; 56: 1615-1619.

11   Kozlovsky SA, Kunc F. Folia Microbiol 1995; 40: 454-456.

12   Pérez-Lesher J, Hickey WJ. FEMS Microbiol Lett 1995; 133: 47-52.

13   Hartmann J, Engelberts K, Nordhaus B, Schmidt E, Reineke W. FEMS Microbiol Lett 1989; 61: 17-22.

14   Fetzner S, Müller R, Lingens F. Biol Chem Hoppe-Seyler 1989; 370: 1173-1182.

15   Fetzner S, Müller R, Lingens F. J Bacteriol 1992; 174: 279-290.

16   Fava F, Di Gioia D, Marchetti L, Quattroni G, Marraffa V. Appl Microbiol Biotechnol 1993; 40: 541-548.

17   Fava F, Baldoni F, Marchetti L. Lett Appl Microbiol 1996; 22: 275-279.

18   Kozlovsky SA, Zaitsev GM, Kunc F, Gabriel J, Boronic AM. Folia Microbiol 1993; 38: 371-375.

19   Johnston HW, Briggs GG, Alexander M. Soil Biol Biochem 1972; 4: 187-190.

20   Ruisinger S, Klages U, Lingens F. Arch Microbiol 1976; 110: 253-256.

21   Klages U, Lingens F. FEMS Microbiol Lett 1979; 6: 201-203.

22   Klages U, Lingens F. Zbl Bakt Hyg I Abt Orig 1980; C1: 215-223.

23   Keil H, Klages U, Lingens F. FEMS Microbiol Lett 1981; 10: 213-215.

24   Marks TS, Smith ARW, Quirk AV. Appl Environ Microbiol 1984; 48: 1020-1025.

25   van den Tweel WJJ, Ter Burg N, Kok JB, de Bont JAM. Appl Microbiol Biotechnol 1986; 25: 289-294.

26   Kagaku Kogyo Nippo Sya (The Chemical Daily Co, Ltd), No Kagaku Syohin Kagaku Kogyo Nippo, Tokyo: 1988; 10188.

27   Shimao M, Onishi S, Mizumori S, Kato N, Sakazawa C. Appl Environ Microbiol 1989; 55: 478-482.

28   Layton AC, Sanseverino J, Wallace W, Corcoran C, Sayler GS. Appl Environ Microbiol 1992; 58: 399-402.

29   Chae JC, Ahn KJ, Kim CK. J Microbiol Biotechnol 1998; 8: 692-695.

30   Müller R, Thiele J, Klages U, Lingens F. Biochem Biophys Res Commun 1984; 124: 178-182.

31   Marks TS, Wait R, Smith ARW, Quirk AV. Biochem Biophys Res Commun 1984; 124: 669-674.

32   Groenewegen PEJ, van den Tweel WJJ, de Bont JAM. Appl Microbiol Biotechnol 1992; 36: 541-547.

33   Thiele J, Müller R, Lingens F. FEMS Microbiol Lett 1987; 115-119.

34   Löffler FR, Müller R, Lingens F. Biochem Biophys Res Commun 1991; 176: 1106-1111.

35   Löffler F, Müller R. Fed Europ Biochem Soc 1991; 290: 224-226.

36   Elsner A, Löffler F, Miyashita K, Müller R, Lingens F. Appl Environ Microbiol 1991; 57: 324-326.

37   Scholten JD, Chang KH, Babbitt PC, Charest H, Sylvestre M, Dunaway-Mariano D. Science 1991; 253: 182-185.

38   Copley SD, Crooks GP. Appl Environ Microbiol 1992; 58: 1385-1387.

39   Thiele J, Müller R, Lingens F. Appl Environ Microbiol 1988; 54: 1199-1202.

40   Chatterjee DK, Kellogg ST, Hamada S, Chakrabarty AM. J Bacteriol 1981; 146: 639-646.

41   Chatterjee DK, Chakrabarty AM. J Bacteriol 1983; 153: 532-534.

42   Weisshaar MP, Franklin FCH, Reineke W. J Bacteriol 1987; 169: 394-402.

43   Rojo F, Ramos JL, Pieper D, Engesser KH, Knackmuss HJ, Timmis KN. Biotechnology 1988; 2: 65-74.

44   Pieper DH, Knackmuss HJ, Timmis KN. Appl Microbiol Biotechnol 1993; 39: 563-567.

45   Bartels I, Knackmuss HJ, Reineke W. Appl Environ Microbiol 1984; 47: 500-505.

46   Arensdorf JJ, Focht DD. Appl Environ Microbiol 1994; 60: 2884-2889.

47   Arensdorf JJ, Focht DD. Appl Environ Microbiol 1995; 61: 443-447.

48   Seo DI, Chae JC, Kim KP, Kim Y, Lee KS, Kim CK. J Microbiol Biotechnol 1998; 8: 96-100.

49   Higson FK, Focht DD. Appl Environ Microbiol 1992; 58: 2501-2504.

50   Hartmann J, Reineke W, Knackmuss HJ. Appl Environ Microbiol 1979; 37: 421-428.

51   Schwein U, Schmidt E, Knackmuss HJ, Reineke W. Arch Microbiol 1988; 150: 78-84.

52   Chatterjee DK, Chakrabarty AM. Mol Gen Genet 1982; 188: 279-285.

53   Zhou X, George SE, Frank DW, Utley M, Gilmour I, Krogfelt KA, Claxton LD, Laux DC, Cohen PS. Appl Environ Microbiol 1997; 63: 1389-1395.

54   Adriaens P, Kohler HPE, Kohler-Staub D, Focht DD. Appl Environ Microbiol 1989; 55: 887-892.

55   Adriaens P, Focht DD. Appl Environ Microbiol 1991; 57: 173-179.

56   van den Tweel WJJ, Kok JB, de Bont JAM. Appl Environ Microbiol 1987; 53: 810-815.

57   Miguez CB, Greer CW, Ingram JM. Arch Microbiol 1990; 154: 139-143.

58   Cork DJ, Krueger JP. Adv Appl Microbiol 1991; 36: 1-66.

59   Yang J, Wang XZ, Hage DS, Herman PL, Weeks DP. Anal Biochem 1994; 219: 37-42.

60   Wang XZ, Li B, Herman PL, Weeks DP. Appl Environ Microbiol 1997; 63: 1623-1626.

61   Horvath RS. J Agric Food Chem 1971; 19: 291-293.

62   Suflita JM, Horowitz A, Shelton DR, Tiedje JM. Science 1982; 218: 1115-1117.

63   Horowitz A, Suflita JM, Tiedje JM. Appl Environ Microbiol 1983; 45: 1459-1465.

64   Genthner BRS, Price II WA, Pritchard PH. Appl Environ Microbiol 1989; 55: 1466-1471.

65   DeWeerd KA, Mandelco L, Tanner RS, Woese CR, Suflita JM. Arch Microbiol 1990; 154: 23-30.

66   DeWeerd KA, Suflita JM. Appl Environ Microbiol 1990; 56: 2999-3005.

67   Utkin IB, Woese C, Wiegel J. Int J Syst Bacteriol 1994; 44: 612-619.

68   Varsha SK, Wyndham RC. Appl Environ Microbiol 1990; 56: 3871-3873.

69   Shelton DR, Tiedje JM. Appl Environ Microbiol 1984; 48: 840-848.

70   Genthner BRS, Townsend GT, Dalton D. 91st Gen Meet Am Soc Microbiol, Abstract K-1, 1989; p. 214.

71   Dolfing , Tiedje JM. Appl Environ Microbiol 1991; 57: 820-824.

72   Zhang X, Wiegel J. Appl Environ Microbiol 1992; 58: 3580-3585.

73   Haggblom MM, Rivera MD, Young LY. Appl Environ Microbiol 1993; 59: 1162-1167.

74   Kuo CW, Genthner BRS. Appl Environ Microbiol 1996; 62: 2317-2323.

75   Haggblom MM, Rivera MD, Young LY. FEMS Microbiol Lett 1996; 144: 213-219.

76   Townsend GT, Ramanand K, Suflita JM. Appl Environ Microbiol 1997; 63: 2785-2791.

77   Maloney SE, Marks TS, Sharp RJ. Lett Appl Microbiol 1997; 24: 441-444.

78   Taraban RH, Berry DF, Berry DA, Walker HL Jr. Appl Environ Microbiol 1993; 59: 2332-2334.

79   Gerritse J, Gottschal JC. FEMS Microbiol Ecol 1992; 101: 89-98.

80   Hughes DE. Biochem J 1965; 96: 181-188.

81   Horvath RS, Alexander M. Appl Microbiol 1970; 20: 254-258.

82   Baggi G, Zangrossi M. Ann Microbiol Enzymol 1995; 45: 185-189.

83   Baggi G, Zangrossi M. FEMS Microbiol Ecol 1999; 29: 311-318.

84   Stratford J, Wright MA, Reineke W, Mokross H, Havel J, Knowless CJ, Robinson GK. Arch Microbiol 1996; 165: 213-218.

85   Brunsbach FR, Reineke W. Appl Microbiol Biotechnol 1993; 39: 117-122.

86   Bedard DL, Haberl ML. FEMS Microbiol Ecol 1990; 20: 87-102.

87   Sondossi M, Sylvestre M, Ahmad D. Appl Environ Microbiol 1992; 58: 485-495.

88   Guibeault B, Sondossi M, Ahmad D, Sylvestre M. Intern Biodet Biodegr 1994; 33: 73-91.

89   Hickey WJ, Searles DB, Focht DD. Appl Environ Microbiol 1993; 59: 1194-1200.

90   Fava F. Chemosphere 1996; 32: 1477-1483.

91   Focht DD, Searles DB, Koh SC. Appl Environ Microbiol 1996; 62: 3910-3913.

92   Knackmuss HJ, Hellwig M, Lackner H, Otting W. Eur J Appl Microbiol 1976; 2: 267-276.

93   Müller R, Deckwer WD, Hecht V. Biotechnol Bioeng 1996; 51: 528-537.

Biotransformations: Bioremediation Technology for Health and Environmental Protection
V.P. Singh and R.D. Stapleton, Jr. (Editors)
2002 Elsevier Science B.V.

Microbial degradation of insecticides: An assessment for its use in bioremediation

Dileep K. Singh

Department of Zoology, University of Delhi, Delhi - 110 007, India

INTRODUCTION

All pesticides are *biocides*, the chemical substances intended to kill the organisms. They are used for destroying, controlling, preventing, repelling and attracting the pests, such as insects, nematodes, rodents, and fungi and are named accordingly, i.e. insecticides, nematicides, rodenticides, and fungicides. Pesticides find their way in the environment as a result of their application to control the agricultural pests as well as through their use in animal and human health programmes. Manufacturing, storage, processing and transport also contribute to their release in the environment. It has been observed that only 0.1% of pesticides applied reaches the target pests, leaving 99.9% pesticide to affect the environment [1]. It was reported that widespread use of pesticides has resulted in global contamination of the ecosystem [2]. No matter, how and where the pesticides are applied, they finally reach to the soil, which acts as a reservoir for all toxic chemicals in the environment. These pesticides in the soil may get attached to its different components, where they may undergo many intra- and inter- compartmental transport mechanisms as well as transformation processes. The persistence of insecticides and their degradation products depend on how deeply they are mixed into the soil; even the most persistent compounds disappear relatively quickly when on the soil surface, yet when incorporated into the soil they are very persistent [3]. Once an insecticide enters the soil, it undergoes transformation caused by microbial, chemical, and photochemical processes. A significant proportions of insecticides may form bound residues with the soil, which are not readily available to the plants and are often not very toxic to the biota [4]. This may represent a mechanism by which the insecticide residues may persist for longer periods in soil and may be released only very slowly [5-7]; and a relatively small proportion of these may be taken up by the plants [8] and earthworms [9,10] and responsible for contamination of ecosystem.

Generally, insecticide residues will occur in the top 6 inch of soil [11,12]. This is also the region of greatest activity of soil fauna and flora [13], thus setting the stage for interaction of insecticide residues with the fauna and flora of the soil ecosystem. A pesticide loss or persistence in the environment is determined by its resistance to degradation. Most often, pesticides react in the environment with oxygen (oxidation) or water (hydrolysis). In addition, all pesticides are degraded by sun light (photo-degradation). In soil, micro-organisms (bacteria, fungi, etc.) are primarily responsible for pesticide degradation. Thus, the microbial breakdown of insecticides is considered to be the most important catabolic reaction in soil.

PESTICIDE CONSUMPTION PATTERN

A wide range of insecticides has become important in agriculture in both developed and developing countries. About 70% of the pesticides used in world is applied in developed countries, and the remaining 30% is consumed in developing countries. India is predominantly an insecticide user. The annual consumption of pesticides in India has risen from 2000 tones in 1950 to over 80,000 tones today, which includes consumption both in agriculture and public health. Increased consumption of insecticides in India has started creating many environmental and health problems and, therefore, requires immediate attention (Table 1). Use of insecticides in agriculture has resulted in contamination of agricultural soil and pollution of terrestrial and aquatic components of the environment. Insecticide pollution to the environment

Table 1.
Consumption pattern of pesticides in India

Pesticides	Per cent consumption	Group-wise consumption
Insecticides	80	Organochlorine 40%
Fungicides	10	Organophosphorus 30%
Herbicides	7	Carbamates 15%
Others	3	Synthetic pyrethroids 10%
		Others 5%

causes carcinogenic, mutagenic, and teratogenic effects in different organisms and results in the development of resistance in pests. These problems can be more serious, when insecticides will accumulate and persist in the environment, and their effect is cumulative on the organisms. It is reported that microbes present in environment degrade these xenobiotic compounds and use them for their metabolic requirements. This process is known as biodegradation/biotransformation.

INSECTICIDE BIODEGRADATION MECHANISMS

Insecticides can be degraded by microbes, either by using them as a substrate for growth and energy or by cometabolism without supplying energy for the microorganisms. Whether microorganisms can utilize insecticide molecules as the sole carbon source has been the subject of investigation. This biodegradation/biotransformation of pesticides by soil microbes may be utilized for remediation of contaminated sites.

Blackburn and Halker [14] reported that bioremediation of contaminated soil depends on the biological transformation of contaminant to less toxic product (biochemistry), accessibility of contaminant to microorganisms (*bioavailability*) and opportunity for optimization of biological activity (*bioactivity*). Bioremediation of insecticide residues in soil is probably the most effective method for decontamination of soil from these xenobiotics [15] and for restoring the soil health. Bioremediation of DDT and HCH by microbes has been reported by several workers [16,17]. There are many reports indicating that bioremediation can be used as degradation of insecticide residues and in decontamination of soil [18-21]. In our laboratory, it was observed that monocrotophos degraded upto 90% by bacteria.

BIOPROCESS DEVELOPMENT AND BIOREMEDIATION

Biodegradation/biotransformation is a naturally occurring bioprocess, using innumerable strains of microbes that can effectively degrade/transform the toxic contaminants to non-toxic/harmless byproducts. Bioremediation development has made it possible to greatly accelerate natural bioprocesses by selecting, concentrating and acclimating microbes to degrade or transform many xenobiotic contaminants in a few weeks rather than months or years. *Biostimulation* and *bioaugmentation* are important in development of bioprocess and bioremediation of xenobiotic contaminants (Table 2).

Bioremediation is a technology using microorganisms to decontaminate or to clean the contaminated soil biochemically. The technology is based on the activation of microbial degradation of xenobiotic compounds in contaminated sites by optimizing environmental factors, such as nutrient concentration, water content, pH, oxygen supply and availability of contaminants to microorganisms, and temperature [22-24]. The most important advantages of bioremediation are: (i) It is a natural microbial process, (ii) The bioremediation products are usually non-toxic, and (iii) The decontaminated soil can be used for re-cultivation of crops.

Soil microorganisms may also respond to the insecticides as substrates and, thereby, derive energy or utilizable nutrients for their metabolism. Availability, low microbial toxicity and high nutritive value seem to be the properties that enhance biodegradation in the soil. An insecticide may also

Table 2
Bioprocess development and bioremediation

Site inspection

- Level of pesticide contamination

- Contaminating pesticide groups

Sample collection and data generation

- Pesticide residue analysis

- Half-life estimation

Objectives

- Degradation of pesticides

- Mineralization of pesticides

Search for microbes

- Bacteria

- Fungi

- Other microorganisms

Feasibility of study

- Screening technology

- Isolation of microbes

- Identification of microbes

- Use of microbes for degradation/mineralization of pesticides

- Engineered microbes and their efficacy in field

Bioremedial designs

Bioremedial action

National Law: Law on bio-hazards

undergo degradation by analog-induced or constitutive cometabolism, where the insecticide itself does not act as a source of energy [25,26].

Microbes can transform a variety of insecticides and utilize them as a source of energy (Figure 1). Organochlorines have a considerably high persistence in the environment owing to their stable structure and very less biodegradation. Organophosphorus and carbamate insecticides are rapidly degraded chemically as well as by microbes. Synthetic pyrethroids are not recommended for outdoor purposes, as they are highly photolabile and breakdown easily in sunlight.

In the soil, the insecticide undergoes *microbial decomposition* [27]. Leoni et al. [28] detected the formation of 3,5,6-trichloro 2-pyridinol in treated soils and very little residues were detected after 6 months. Insecticides,

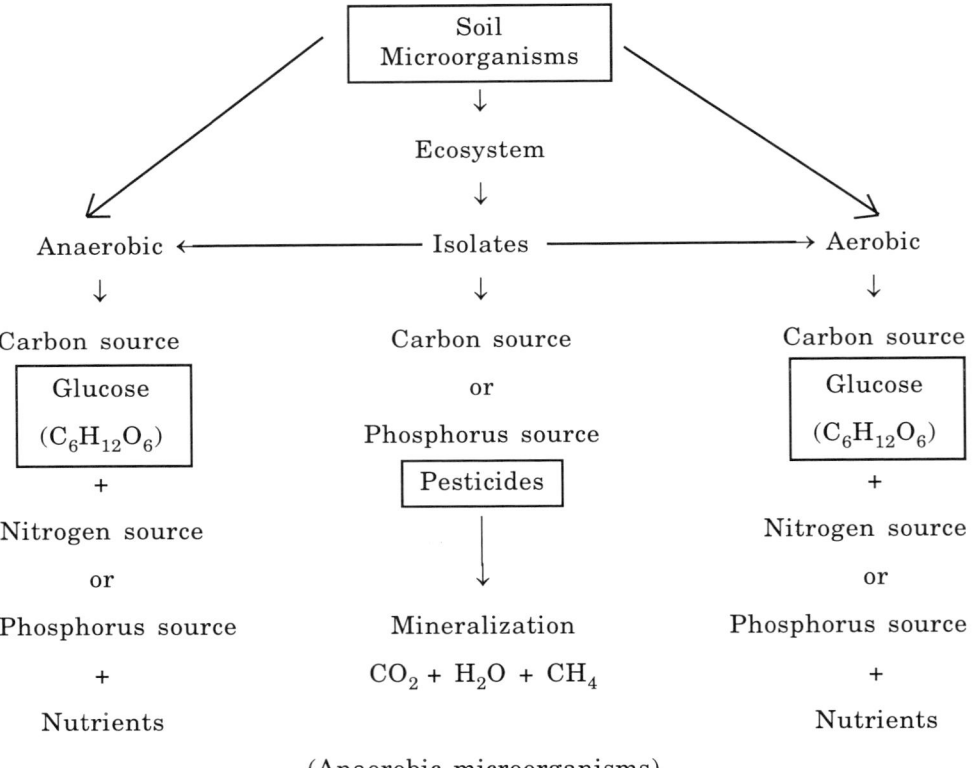

Figure 1. Microbial degradation/mineralization of pesticides.

which are employed for the crop protection, expose the soil in the fields to a heavy insecticide burden. Such effects may get accentuated following repeated insecticide treatments year after year. It is observed that some insecticides have toxic effects on soil microbes viz. bacteria, *Azotobacter*, actinomycetes, fungi, algae and protozoa, affecting their respiration and metabolism. The insecticides used for plant protection may adversely affect the soil enzymes, an indicator of soil health. Hexachlorocyclohexane (HCH) was found to significantly inhibit soil dehydrogenase activity [29]. Hexachlordane and heptachlor decreased the urease and catalase activity in meadow soil [30]. This is an indication that if the enzyme activity will decrease the microbial population will be affected and, in turn, the rate and amount of nutrients available to the plants will be affected. To reduce the toxicity of insecticide residues in soil environment and reclamation of the soil, their degradation to non-toxic products and their fast removal will become an important issue for environmental toxicologists. Here, biodegradation and biotransformation processes become an important tool for reclaimation of contaminated sites.

The half-life of chlorpyriphos in estuarine environment is found to be about 24 days [31]. This shorter half-life of chlorpyriphos may be due to fast degradation of compound by microbes and supporting environmental factors. Bacteria are the most enormously available decomposers and are able to degrade chlorpyriphos [32,33], diazinon [34], methyl-parathion [35] and many other organic compounds. Fungi [36] are the next more important biodegraders of chlorpyriphos in the soil. Omar [37] isolated certain fungi from soil, which could mineralize chlorpyriphos completely even at a concentration of 100 ppm and were utilizing it as a source of sulfur and phosphorus. This ability of microbes can be exploited in development of bioreactors for remediation of contaminated sites. In our laboratory, it was observed that *Pseudomonas aeruginosa* and *Clavibacter michigenense* sub sp. *insidiosum* degraded technical monocrotophos (80% pure) upto 98.9% and 85.88%, respectively. In another study with pure monocrotophos (99.9% pure), the degradation was 78% and 82%, respectively in 24 hours at 37°C. It was observed that both of the strains were using monocrotophos as sole source of phosphorus. Further studies indicated that the enzyme responsible for monocrotophos degradation by hydrolysis was phosphotriesterase.

Bioremediation is probably the safe method here for decontamination of soil and restoration of soil health. Francis et al. [38] reported that *Escherichia coli* was able to degrade γ-HCH in γ-PCCH. Tu [39] isolated thirteen soil microbes able to degrade γ-HCH to γ-PCCH, α-3,4,5,6-TCCH, β-TCCH and PCB. Doelman et al. [40] studied the microbial degradation of α-HCH and β-HCH and observed 40, 80, and 37% degradation of α-HCH by soil microbes. Biodegradation of DDT to different metabolites viz. DDE, DDD, DDMU, DDMS, DDOH and DDA were observed by various workers [41-47]. Many bacteria and fungi were tested for their ability to degrade pesticides to non-toxic metabolites. Strains of *Aspergillus niger, Penicillium brefeldianum, Phanerochaete crysosporium,* and *Pseudomonas acidovorans* M3GY are few of the examples showing capacity to degrade pesticides. Biodegradation of agricultural chemicals is well documented by Stephen et al. [48].

In bioremediation, the ability of microbes to degrade the insecticides are enhanced biotechnologically to achieve the required cleanup target. Possible pathway for microbial degradation of insecticides are dehalogenation, dehydrohalogenation, oxidation, reduction, hydrolysis and conjugation (Table 3). It is reported that *Pseudomonas dehalogenas* and *Arthrobacter* sp. have dehalogenases responsible for degradation of certain halogen compounds. Chlorinated hydrocarbons are reported to be resistant to degradation by microorganisms. Degradation product of cyclodiens are toxic metabolites, i.e. epoxides. The epoxides are persistent in environment and are potent

Table 3.
Insecticide metabolism

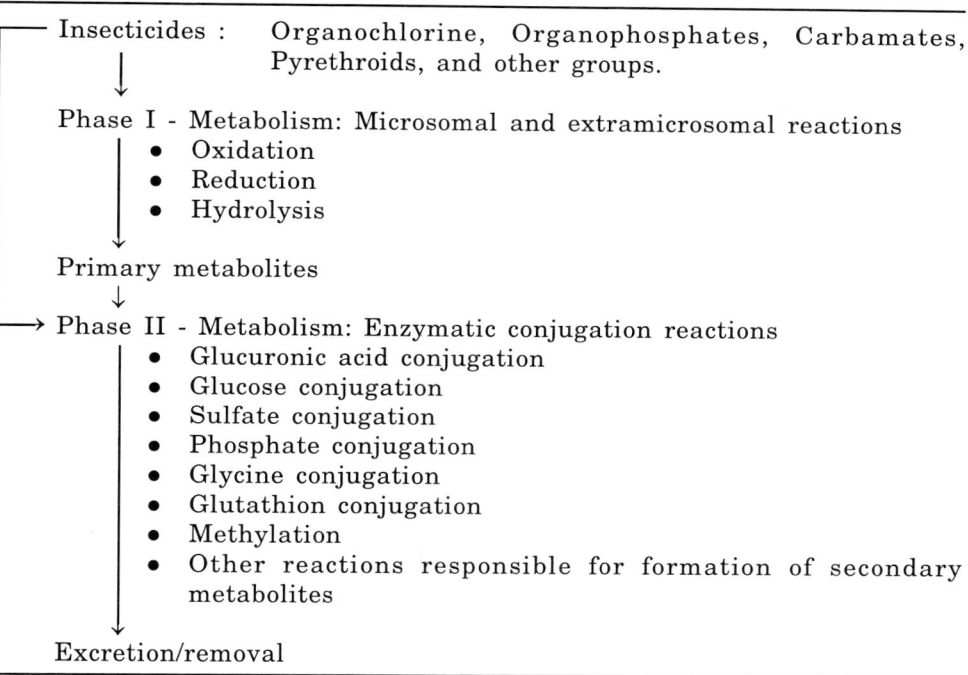

Insecticides : Organochlorine, Organophosphates, Carbamates, Pyrethroids, and other groups.

Phase I - Metabolism: Microsomal and extramicrosomal reactions
- Oxidation
- Reduction
- Hydrolysis

Primary metabolites

Phase II - Metabolism: Enzymatic conjugation reactions
- Glucuronic acid conjugation
- Glucose conjugation
- Sulfate conjugation
- Phosphate conjugation
- Glycine conjugation
- Glutathion conjugation
- Methylation
- Other reactions responsible for formation of secondary metabolites

Excretion/removal

inhibitors of biological activities. DDT is reported to be degraded by certain bacteria to DDD, DDE and DDA. Reductive dechlorination reactions are important in the transformation of halogenated insecticides in anerobic condition. Mohn and Tiedje [49] reported reductive dehalogenation of organochlorine insecticides DDT, γ-lindane, dieldrin and heptachlor. Degradation of organophosphorus insecticides by certain microbes leads to activation of compounds and degradation products are more toxic. Finally, these degradation products may further degrade to non-toxic metabolites by other enzymes present in microbes. Barik and Munnecke [50] and Havens et al. [51] reported that bacteria were involved in hydrolysis of parathion, and paraoxon. Similar reports were available for the degradation of diazinon and coumaphos [52]. Similarly, benzene molecule of carbofuran is hydroxylated by a *Rhodococcus* strain [53]. Conjugation processes, i.e. alkylation (methylation) and acylation are commonly observed [54] (Table 4).

Blackburn and Halker [14] reported the following facts in bioremediation of xenobiotics: (i) The behaviour and fate of the contaminants are site specific and it may be difficult to generalize from one case to other. (ii) The complexity of the hydrocarbon mixture, and perhaps its age, may affect the rate and extent (end point) of treatment. (iii) There is large variation in biological removal of pollutants between large and small scale investigations, and (iv) the effectiveness of bioremediation depends on experimental design and technology.

They also suggested that feasibility of bioremediation depends on the following factors: (i) An ability to achieve required clean-up of target site. (ii) It is acceptable as *cost-effective process* relative to other remediation options. (iii) Risk in residual contamination remaining after remediation. (iv) Favourable public opinion and regulatory perception for bioremediation. (v) Ability to meet time limitations, and (vi) ability to show conformity to space limitation.

CONCLUSION

Microbes are degrading xenobiotic compounds/pesticides in environment and use them for their normal metabolic processes as carbon or phosphorus source or consume the pesticides along with other source of food or energy. This bioprocess of microbes can be utilized for the development of pesticide decontamination and restoration of health of the environment. Hydrolytic enzymes, responsible for degradation of pesticides to non-toxic products in the environment can be developed for the future remedy for toxic compounds — as bioremediation (Table 4).

Table 4
Bioremediation technology used in pesticide decontamination

BIODEGRADATION

DEGRADATION PATHWAY

- Samples taken out every few hours from microbial culture: extracted, purified and analyzed by HPLC/TLC/MS/HPTLC.
- Mass gives the identity of newly formed compounds
- Enzymes responsible for degradation of parent compound and its metabolites
- Identification of the said enzyme or the enzyme systems

ENZYME KINETICS

- Enzyme responsible for the degradation of pesticides
- The step at which the enzyme/enzyme systems are working
- The degradation mechanism of the pesticides
- Optimization parameters for the production of enzyme(s) degrading the said pesticides
- Isolation and purification of the enzyme/enzyme systems
- Characterization of the purified enzyme

MOLECULAR BIOLOGY

- Targeting the gene responsible for the degradation
- Plasmid or the genomic DNA encoded to check the location of the gene responsible for the degradation of pesticide
- Isolating the gene and cloning in expression vectors with strong promoters
- Gives the idea of the localization, size, the sequence and other related aspects of the gene

REFERENCES

1   Pimental D. J Agric Environ Ethics 1995; 8: 17-29.

2   Edwards WM, Glass BL. Bull Environ Contam Toxicol 1971; 6: 81-84.

3   Edwards CA. Res Reviews 1966; 13: 82-132.

4   Kaufman DD. In: Kaufman DD, Still GG, Paulson GD, Bandal SK, eds. Bound and Conjugated Pesticide Residues, ACS Symposium Series 29, Washington DC: American Chemical Society, 1976; 1-10.

5   Khan SU, Ivarson CK. J Agric Food Chem, 1981; 29: 1301-1303.

6   Agarwal HC, Singh DK, Sharma VB. J Environ Sci Health 1994; B29(1): 87-96.

7   Agarwal HC, Singh DK, Sharma VB. J Environ Sci Health 1994; B29(1): 189-194.

8   Khan SU. Pesticides in Soil Environment. Amesterdam: Elsevier Science Publication Co, 1980; 24.

9   Yadav DV, Pillai MKK, Agarwal HC. Bull Environ Contam Toxicol 1976; 16(50) 541-545.

10   Fuhremann TW, Lichtenstein EP. J Agric Food Chem 1978; 25: 605-610.

11   Lichtenstein EP, Muller CH, Myrdal GR, Schulz KR. J Econ Entomol 1962; 55: 215.

12   Harris CR, Sans WW. Proc Entomol Soc Ontario, 1969; 100: 156.

13   Alexander M. Introduction to Soil Microbiology. New York: John Wiley and Sons, 1961.

14   Blackburn JW, Halker WR. Trends in Biotechnology, 1993; 11(8,115): 328-333.

15   Raghu K, Sethunathan N, Agarwal HC, Naidu Ravi. 2nd International Conference on Contaminants in the Soil Environment in the Australasia-Pacific Region, New Delhi, India, 1999; 12-17: 99-100.

16   Mitra J, Raghu K. Environ Technol Lett 1988; 9: 847-852.

17   Sethunathan N, Sahu SK, Raghu K. In: Seeking Agricultural Produce Free of Pesticide Residues, ACIAR Proceedings No 85, 1998; 325-333.

18   Singh N, Sethunathan N. J Agri Food Chem 1992; 40: 1062-65.

19   Wahid PA, Sethunathan N. J Agri Food Chem 1980; 28: 623-625.

20   Sahu SK, Patnaik KK, Bhuyan S, Sethunathan N. Soil Biol Biochem 1993; 387-391.

21   Bhuyan, Sreedharan B, Adhya KS, Sethunathan N. Pestic Sci 1993; 38: 49-55.

22   Wolf K, Van der Brink WJ, Colon FJ. Contaminated Soil 88. Second International TNO-BMFT - Conference on Contaminated Soil. Dordrecht: Kluwer Academic Publishers, 1988.

23   Mueller JG, Lantz SE, Ross D, Colvin RJ, Middlaugh DP, Pritchard PH. Environ Sci Technol 1993; 27: 691-698.

24   Skladany GJ, Metting FB Jr. In: Metting FB Jr, ed. Soil Microbial Ecology. New York: Marcel Dekker, 1993; 483-513.

25   Roath FW. Rev Weed Res 1986; 2: 45-46.

26   Kaufman DD, Kalan JJ, Edwards DF, Jordan EJ. In: Hilton JL, ed. Agricultural Chemicals of the future, New Jersey: Roman & Alanheld, 1985; 437-451.

27   Racke KD, Coats JR, Titus KR. J Environ Sci Health 1988; B23: 527-539.

28   Leoni V, Hollick CB, Deluca D'A lessandro E, Collison RJ, Merolli S. Agrochimica 1981; 25: 414-426.

29   Tyuryukanona. In: Singh Dileep K, Agarwal HC. (1994-97), ICAR Adhoc Research Project on "Impact of Repeated Insecticide Usage on Soil Health in Groundnut (Arachis hypogaea L.) Field.", 1980.

30   Tsirkov YI. Pochv Agrokhimiya, 1970; 4: 85-88.

31   Schimmel S, Garnes RL, Patrick Jr JM, Moore JC. J Agric Food Chem 1983; 31: 104-113.

32   Tu CM. J Environ Sci Health 1991; B 26: 557-573.

188

33   Malik K, Bharati K, Banerji A, Shakil NA, Sethunathan N. Bull Environ Cantam Toxicol 1999; 62(1): 48-54.

34   Sethunathan N, Yoshida T. Can J Microbiol 1973; 19: 873-875.

35   Mishra D, Bhutiyan S, Adhya TK, Sethunathan N. J Environ Sci Health 1992; B 18: 705-712.

36   Bumpus JA, Kakar SN, Coleman RD. Appl Biochem Biotechnol 1993; 39-40: 715-726.

37   Omer SA. Biodegradation 1998; 9(5): 327-336.

38   Francis AJ, Spanggord RJ, Ouchi GL. Appl Microbiol 1975; 29: 567-568.

39   Tu CM. Arch Microbiol 1976; 108: 259-263.

40   Doelman P, Haanstra L, Vos A. Chemosphere 1988; 17: 489-492.

41   Castro TF, Yoshida T. Soil Sci Plant Nutr 1971; 20: 363.

42   Wedemeyer G. Science 152, 1966; 647.

43   Wedemeyer G. Appl Microbiol 1967; 15: 1494-1495.

44   Wedemeyer G. Appl Microbiol 1967; 15: 569-574.

45   Pfamder FK, Alexander M. J Agric Food Chem 1972; 20: 842-846.

46   Middledrop PJM, Jhaspers M, Zehnder AJB, Schraa G. Environ Sci Technol 1996; 30: 2345-2349.

47   Quensen JF, Mueller SA, Jain MK, Tiedje JM. Science 1998; 280: 722-724.

48   Stephen NB, Terry BH, Ronald LC. In: Manual of Environmental Microbiology. 1997; 815-821.

49   Mohan WW, Tiedje JM. Microbiol Rev. 1992; 56: 482.

50   Barik S, Munnecke DM. Bull Environ Contam Toxicol 1982; 29: 235.

51   Havens DL, Rase HF, Tedder DW, Pohland FG. ACS Symposium Series No 468, 1991; 11: 261-268.

52   Smith JM, Payne GF, Lumpkin JA, Karns JS. Biotech Bioeng 1992; 39: 471-752.

53   Behki RM, Topp E, Blackwell B. J Agric Food Chem 1994; 42: 1375.

54   Bollag WB, Bollag JM. Encyclopedia of Microbiology 1992; 1: 269-275.

Biotransformations: Bioremediation Technology for Health and Environmental Protection
V.P. Singh and R.D. Stapleton, Jr. (Editors)

189

Microbial variables for bioremediation of heavy metals from industrial effluents

Rani Gupta, R.K. Saxena, Harapriya Mohapatra and Prerna Ahuja

Department of Microbiology, University of Delhi South Campus, New Delhi - 110 021, India

INTRODUCTION

The use of heavy metals dates back to the metal age. However, metal's use by man began to affect the environment only with the advent of industrial revolution. Today nearly 200 years later, we are opening the doors to the age of metal removal. These metals are dispersed in the earth's elements viz. soil, air and water. In this chapter the main focus will be on the heavy metal removal from wastewater and industrial effluents.

The incomplete 'd' orbital of these metals endows upon them the ability to form complex compounds, which may or may not be redox active, thus conferring toxicity to the living cell. Beyond a permissible concentration, these heavy metals form unspecific compounds inside the cells, thereby, affecting the cellular processes. Thus, the removal of heavy metals is of utmost importance for the survival of the living beings.

Being non-biodegradable and persistent, the only alternative resort for removing heavy metals from the wastewater is by immobilizing them. The conventional methods adapted earlier for this purpose included chemical precipitation (as hydroxides/sulphides), chemical oxidation, chemical reduction, filtration, electrochemical treatment, evaporation, adsorption (on activated carbon, peat, zeolite, clay, and minerals) and ion-exchange resins. These methods suffered from having significant disadvantages such as incomplete metal removal, especially when the metals are in solutions containing 1 to 100 mg/l of dissolved metals, high energy requirements and production of toxic sludge, thus transforming a liquid waste disposal problem into a solid waste disposal one [1]. Though the ion-exchange resins stood apart from all others being the most effective and opted ones, they lacked from being cost-effective and *eco-friendly*. In such a scenario, the biological materials emerged as the choice of all. Among the biological materials, peat was the forerunner, but it had to succumb to its non-uniform distribution. Also, the concept of phytoremediation (use of higher plants in bioremediation) was not acceptable at large as the time required was more. Thus, came up the idea of microbial metal removal.

The concept of microbial metal removal brought into focus the metal resistance (natural/induced) among the microbes for metal removal. It was observed that these metal resistant strains had the ability to actively accumulate heavy metals into the living cells, the process designated as *"bioaccumulation"*. The process was dependent on metabolic activity of the cell, which in turn was found to be significantly affected by the metal ions. Thus, metal tolerance reflected the ability of an organism to survive in an environment with a high concentration of metals or to accumulate high concentrations of metal without dying. However, the prediction of the effectivness of bioaccumulation by the living cells in a natural water body, characterized by varied physio-chemical properties was a difficult task. Hence search was directed towards the microorganisms, which could remove metals from their growth environment without making them available for their own metabolic processes. This, to a great extent, would involve the external components, which is independent of the cells' metabolic activities and is strictly a function of chemical makeup of the cells' outer protective envelope, the cell wall. This *"passive"* physico-chemical uptake of heavy metals takes place to different degree of affinity resulting in different types of binding between the metallic species or its ionic forms and the *"active site"* of a particular molecular structure of the cell wall. This phenomenon was termed as "bioadsorption" or simply *"biosorption"* [1].

In 1986, a one day meeting organized by "The Solvent Extraction And Ion-Exchange Group of the Society of Chemical Industries" in U.K regarded biosorption as an emergent technology [2]. The meeting was solely devoted to the recovery and removal of the metals by biosorption. Since then a number of centres all over the world have been engaged in various areas of biosorptive processes such as: (i) examination of biosorptive potentials of naturally occurring biomasses, (ii) engineering improved biosorbents from natural materials, (iii) the elucidation of the mechanism of metal ion biosorption, (iv) comparing biosorption to adsorption by ion-exchange resins, and (v) assessment of commercial potentialities of biomasses.

The biosorptive methods offer several advantages such as use of renewable biomaterials, which reduce, production costs, fast adsorption kinetics, high selectivity of biosorbents (possibility to recover valuable metals, separation of mixtures), cleansing of aqueous solutions with low metal concentration, high capacity by small equilibrium concentration, low investment and operation costs, easy desorption of metals by pH swing and low affinity with competing cations (calcium and magnesium).

The role of various groups of microorganisms in the removal and recovery of heavy metal(s) by biosorption has been well reviewed several times. A large number of microorganisms belonging to various groups viz. bacteria, fungi, yeasts, cyanobacteria, and algae have been reported to bind a variety of heavy metals to different extents.

Amongst the microorganisms, fungal biomass offers the advantage of having a high percentage of cell wall material which shows excellent metal-binding properties· Many fungi and yeast have shown an excellent potential of metal biosorption, particularly the genera *Rhizopus, Aspergillus, Streptoverticilium,* and *Saccharomyces.* Among bacteria, *Bacillus* sp. has been identified as having a high potential for metal sequestration and has been used in commercial biosorbent preparation. Besides there are reports on the biosorption of metal(s) using *Pseudomonas* sp., *Zoogloea ramigera,* and *Streptomyces* sp. Among photo-autotrophs, marine algae became the candidate of interest due to bulk availability of their biomass from water bodies. *Sargassum natans* and *Ascophyllum nodosum* in this group have shown very high biosorptive capacities for various metal(s). Besides marine algae, there are quite a few reports on binding of heavy metal(s) to cyanobacteria and a green alga *Chlorella* sp.

EVALUATION OF BIOSORBENT PERFORMANCE

The biosorption isotherms

The metal binding has been conventionally quantitatively evaluated from experimental biosorption isotherms, similar to those used for activated carbon. Biosorption isotherms are derived from results of relatively simple contact experiments usually done using small Erlenmeyer flasks (250 ml) with 50-100 ml of metal bearing solution (volume, V) of known initial metal concentration (C_i). A known amount of biomass (M) is suspended and placed on a shaker, allowing enough time for development of sorption equilibrium. The contents of each flask are filtered and the filtrate is collected and analyzed for final/residual/equilibrium concentration (C_f). The uptake (U) of the metal in milligrams of metal per gram of biomass is calculated using the formula:

$$U = V \, (C_i - C_f) \, / \, M$$

where: V = volume of metal solution

C_i = initial metal concentration

C_f = final/residual concentration

M = amount of biomass

The final concentration measured usually differs according to the amount of biosorbent initially added. The resulting uptake data is plotted against the C_f, representing the biosorption isotherm. The maximum sorption uptake (loading capacity) is important to characterize the performance of a given sorbent. The shape of the biosorption isotherm, which is steep from the origin at low residual concentrations of the sorbate, is usually desirable because it indicates a good affinity of the sorbate for the sorbent species.

The biosorption isotherm represents an equilibrium process where the metal bound to the biosorbent is in the state of equilibrium with the metal dissolved in the solution, and the sorption process requires some time to reach the equilibrium. The initial concentration of the metal (sorbate) does not play a significant role in these equilibrium characterizations, but the final or residual concentration is important [1].

There are two widely accepted models of single layer adsorption viz. Langmuir model [3] and the Freundlich model [4]. The Langmuir model is based on the assumption that maximum adsorption occurs when a saturated monolayer of solute molecules is present on the adsorbent surface, the energy of adsorption is constant, and there is no migration of adsorbate molecules in the surface plane. A general form of the Langmuir model equation is:

$$q = q_0 bC \ / \ 1 + bC$$

Where: q = uptake of species

q_o = maximum uptake

b = constant related to the energy of adsorption

C = equilibrium/final concentration in solution

The Freundlich equation is a special case of heterogeneous surface energies and does not become linear at low concentration but remains concave to the concentration axis; also it does not show a saturation or limiting value. The general form of the Freundlich equation is:

$$Q = KC^{1/n}$$

This can be linearized by taking natural logarithm of both sides of the equation, which can be given as follows:

$$\ln q = \ln k + 1/n \ln C$$

Here, ln k gives the measure of the adsorbent capacity and the slope 1/n gives the intensity of adsorption.

Another model, the BET isotherm [5] represents isotherm with multilayer adsorption at the adsorbent surface and assumes that a Langmuir equation applies to each layer. The BET equation is:

$$q_e = BCq^o \,/\, (Cs\text{-}C) \,[(1+B\text{-}1)\,(C/Cs)]$$

Here it is assumed that each layer has an equal energy of adsorption. Cs is the saturation concentration of the solute, B is the constant related to the energy of interaction with the surface.

In most of the studies, the adsorption data has been fit either into Langmuir or Freundlich models [6-8]. However, both these models may exhibit irregular pattern due to complex nature of both the sorbent material and the varied multiple compounds. Also, the presence of other ions in solution can largely complicate the evaluation of the sorption system depending on the way the new solute species interact with the sorbent and the original one. These simple basic models also do not incorporate the effects of any external variable environmental factor [9].

In *Pseudomonas aeruginosa* and *Bacillus* sp., Cu and Cd biosorption could be described by Freundlich isotherm. However, it was not useful in describing the removal of silver and lanthanide from the solution, as there were a few data points over a wide range of concentrations [10].

Khummongkol et al. [11] developed a model for metal uptake by microorganisms based on surface adsorption; and it has been applied to cadmium uptake by *Chlorella vulgaris*. A linear equilibrium relationship between the metal in the solution and that adsorbed on the cell surface was assumed, which was confirmed at low concentration in short term uptake experiments. This model showed fast initial uptake and a subsequent slow uptake due to simultaneous effects of growth and surface adsorption. Ting et al. [12,13] developed another model, which also postulates an initial rapid binding to cell wall and relatively slow uptake subsequently due to membrane transport. This was applied to cadmium and was also extended for its applicability in *multi-ion situation*.

MICROBIAL BIOSORBENTS: TAXONOMIC CONSIDERATIONS

The role of various groups of microorganisms in removal and recovery of heavy metals by biosorption has been reviewed by many authors [14-17]. There are considerably large number of reports on biosorption of heavy metals, using heterotrophs like fungi and bacteria [1]; and comparatively less work has been done on biosorption of metals using photoautotrophs, such as algae and cyanobacteria.

Of the microorganisms, fungal biomass offers the advantage of having a high percentage of cell wall material which shows excellent metal-binding properties [18-21]. Many fungi have shown an excellent potential of metal biosorption, particularly the genera *Rhizopus*, *Penicillium*, *Aspergillus*, and *Saccharomyces* [22-26]. Among bacteria *Bacillus* sp. has been identified and, as having high potential for metal sequestration, has been used in commercial biosorbent preparation [27]. Besides, there are quite a few reports on the biosorption of metals using *Pseudomonas* sp. [10,28,29], *Streptomyces* sp. [29] and *Zoogloea ramigera* [30,31].

The photoautotrophs are much less explored in the area of biosorption, although they offer many advantages including their low cost of biomass production using minimal medium without sugars. Among photoautotrophs, marine algae became the candidate of interest due to bulk availability of their biomass from water-bodies [1]. *Sargassum natans* and *Ascophyllum nodosum* in this group have shown very high biosorptive capacities for various metals [32-34]. Besides marine algae, there are quite a few reports on binding of heavy metals to green algae viz. *Chlorella* spp. [12,13,35-37]. However, very few reports exist on cyanobacteria in the area of biosorption [35,38,39]. Table 1 gives an extensive list of photoautotrophs involved in metal biosorption.

The heterotraphs involved in metal biosorption are enlisted in Table 2.

MECHANISM OF BIOSORPTION

An understanding of the mechanism by which microbes biosorb metals is important to the development of microbial processes for the removal and recovery of metals from aqueous solution [40]. This would also help in possible manipulation of the biosorbent to optimize its performance [41].

Table 1
Metal biosorption by photoautotrophs (algae and bacteria)

Organism (% dry weight) (1)	Metal biosorbed (2)	Metal bound (3)	Reference (4)
Chlorella vulgaris	Cd	0.20	11
	Pb	8.50	37
	Zn	0.13	12
	Au	10.00	19
Chlorella regularis	U	0.39	78
	Cu	0.40	78
	Zn	2.80	78
	Mn	0.80	80
	Mo	1.32	80
Chlorella salina	Zn	0.02	100
Chlorella sp.	Hg	0.01	78
	U	0.19	78
Scenedesmus obliquus	Cd	0.30	115
Scenedesmus sp.	Mo	2.30	78
	U	0.19	78
Chlamydomonas sp.	U	0.34	78
	Cd	0.35	78
Ankistrodesmus sp.	U	1.00	14
Selenastrum sp.	U	1.00	14
Sargassum natans	Au	25.00	32
	Pb	8.00	32
	Ag	7.00	32
	U	4.50	32
	Cu	2.50	32
	Zn	2.00	32
	Co	6.00	32
	Cd	8.30	82
	Pb	21.10	82

Table 1 continued

(1)	(2)	(3)	(4)
Sargassum fluitans	Pb	21.60	82
Sargassum vulgaris	Pb	14.90	82
Ascophyllum nodosum	Au	4.00	32
	Co	15.00	32
	Cd	10.00	82
	Pb	20.10	82
Palmaria palmata	Au	12.50	32
	Pb	1.10	82
Chondrus crispus	Au	7.50	32
	Co	4.50	32
	Pb	6.50	82
Porphyra tenera	Au	15.00	32
	Co	2.50	32
Halimeda opuntia	Co	8.00	32
	Cd	5.20	82
Vaucheria	Pb	17.40	82
	Cu	3.20	116
Fucus vesiculosus	Pb	20.10	82
Ascophyllum nodosum (cross-linked biomass)	Pb	13.10	82
Fucus vesiculosus (cross-linked biomass)	Pb	19.30	82
Sargassum fluitans (cross-linked biomass)	Pb	14.20	82
Chondrus crispus (cross-linked biomass)	Au	30.00	72
Chlorella vulgaris (cross-linked biomass)	Cd	1.13	117

Table 2
Metal biosorption by heterotrophs (fungi, yeast and bacteria)

Organism	Metal biosorbed	Respective uptake (mg/g)	Reference
Rhizopus arrhizus	Ag, Cd, Cr, Cu,	54, 30, 31, 16,	118
	Hg, Mn, Pb, Zn	54, 12, 91, 20	
	Au,	164	34
	Ni	18	119
	U	220	22
	Th	160	22
Saccharomyces cerevisiae	Ag, Co,	4.7, 4.7	26
	Cu, U, Zn	17-40, 24, 14-40	120
	Th	70	14
Aspergillus niger	Au	176	34
	Cu	1.7	121
	U	12	122
Penicillium chrysogenum	Cd	56	123
	Cr	0.33	21
	Cu, Pb, Zn	9, 122, 6.5	124
	Hg, U	20, 25	125
Candida tropicalis	Cd, Cr, Cu, Ni, Zn	60, 4.6, 80, 20, 30	126
Bacillus subtilis	Au, Fe, Mn, Ni	79, 201, 44, 6	127
	Cd, Cu, Pb, Zn	101, 152, 601, 137	127
Citrobacter sp.	Pb	4000	128
	U	8000	129
Bacillus licheniformis	Au, Cu, Fe, Mn, Ni	59, 32, 45, 38, 29	127

Biosorption of heavy metals is a very rapid phenomenon, not affected by the presence of metabolic inhibitors, hence, is a passive process rather than an *energy-driven* metabolic process [35,42]. Biosorption of metals takes place mainly by cell surface complexation, ion-exchange, and *microprecipitation* [1,16,20].

Microbial biomass due to its complex nature is not readily amenable to instrumental analysis by techniques such as infrared spectroscopy and, consequently, little information is available on mechanism of metal binding, in general [43]. Infrared spectroscopy, energy dispersive X-ray analysis (EDAX), and electron microscopy (EM) techniques have been used to localize heavy metals in the microbial cells. X-ray photoelectron spectroscopy for chemical analysis (ESCA) is a relatively new technique for determination of binding energy of electrons in atoms/molecules, which depend on distribution of valence charge, and thus, gives information about oxidation state of atom/ion [43]. Various chemical modifications of cell wall functional groups have also been used for indirectly deducing the mechanism of biosorption [44].

The metal binding appears to be atleast a two step process, where the first step involves a stoichiometric interaction between metal and reactive chemical groups in the cell wall following which is an inorganic deposition of increased amounts of metals [45]. All the metal ions, before gaining access to the plasma membrane and cell cytoplasm, come across the cell wall. The cell wall consists of a variety of polysaccharides and proteins hence, offers a number of active sites capable of binding metal ions [41,46]. Cell wall is thus regarded as a complex ion exchanger similar to commercial resins. Difference in cell wall composition among different groups of microorganisms viz. algae, bacteria, cyanobacteria, and fungi; and the intragroup differences can thus cause significant differences in the type and amount of metal ions bound to them. Among the photoautotrophs, the eukaryotic algal cell walls are mainly cellulosic. According to Crist et al. [47], the potential metal binding groups in this class of microbes are carboxylate, amine, imidazole, phosphate sulphydryl, sulfate, and hydroxyl. Of these, amines and imidazoles are positively charged when protonated and may build negatively charged metal complexes [38]. The cell walls of brown algae contain fucoidin and alginic acid. The alginic acid offers anionic carboxylate and sulfate sites at neutral pH. The freshwater forms contain galacturonic acid and its polymer pectin, which also has anionic sites to which metals can bind by electrostatic attractions [48].

The amino, carboxyl, nitrogen, and oxygen of the peptide bonds are also available for coordination bonding with metal ions, such as lead (II), copper (II) or chromium (VI). Such bond formation could be accompanied by displacement of protons and is dependent, in part, on the extent of protonation, which is determined by the pH [11,12,49,50].

The groups involved in metal binding have been discerned, using the modification/blocking of groups. Carboxyl groups were suggested to be involved in binding Cu and Al in algal species, as blocking of carboxyl groups by esterification lead to a decrease in metal binding [51]. In *Sargassum natans*, the functional groups viz. carbonyl (C=O) and amine were found to provide binding sites for metals [41].

Cell walls of bacteria and cyanobacteria are principally composed of peptidoglycans, which consist of linear chains of the disaccharide N-acetylglucosamine - β 1,4-N- acetylmuramic acid with peptide chains. Cell walls of Gram-negative bacteria are somewhat thinner than the Gram-positive ones and are also not heavily cross-linked. They have an outer membrane, which is composed of an outer layer of lipopolysaccharide (LPS) and of phospholipids and proteins [52].

Gourdon et al. [42] have compared Cd biosorption capacities of Gram-positive and Gram-negative bacteria, isolated from activated sludge and reported a 20% higher uptake capacity of Gram-positive bacteria. Glycoproteins present on outer side of Gram-positive bacterial cell walls have been suggested to have more potential binding sites for Cd than the phospholipids and lipopolysaccharides and hence, held responsible for the observed difference in capacity. Carboxyl group modification caused a marked reduction in metal uptake by *Bacillus subtilis* [45]. However, amine modification did not affect the metal uptake by the bacterium.

In *Bacillus subtilis* teichoic acid [45] and in *B. licheniformis* teichoic and teichuronic acids [53] were found to be the prime sites for metal binding. In *E.coli* K12, peptidoglycan was found to be a potent binder of most of the metals tested, and the carboxylate groups were the principal components involved in metal binding [53].

The phosphoryl groups of lipopolysaccharides (LPS) and phospholipids have been demonstrated to be the most probable sites for metal cations in *Escherichia coli* outer membrane [54,55]. In purified cell envelopes of *E.coli* K12, most of the metal deposition occurred at the polar head regions of constituent membranes or along the peptidoglycan [56]. In *Streptomyces longwoodensis*, phosphate residues were suggested to be the primary constituents responsible for uranium binding [49]. The metal binding in many studies have been suggested to be ion-exchange phenomenon. Greene et al. [57] demonstrated the binding of Cu, Pb, Zn, Ni, Cd, and Cr to *Spirulina platensis* to be accompanied by the liberation of protons, suggesting an ion-exchange reaction. Similar results were obtained in Cd, Cu, and Zn binding by previously protonated biomass of *Sargassum fluitans,* where the metal binding was coupled with the release of H ions [58].

Crist et al. [59] have reported release of Na, Ca, Mg from the hydroxides in essentially the stoichiometric amounts during the sorption of lead and cadmium on *Rhizoclonium* from solid hydroxides by *Vaucheria* sp. Extracellular polymeric substances have also been shown to bind metal ions selectively [60]. *Zoogloea ramigera*, a bacterium from activated sludge has been known to produce substantial amounts of extracellular negatively charged polysaccharides, which form a matrix around the cells for cation attachment [30,31].

A very unique mechanism of metal uptake in *Citrobacter* sp. was described [61,62]. *Citrobacter* sp. showed removal of U, Cd, Cu, and Pb from solution supplemented with glycerol 2-phosphate. The mechanism of uptake involved a phosphatase-mediated cleavage of glycerol 2-phosphate to release HPO_4^{-2}, which precipitated metal on the surface as insoluble metal phosphate (e.g., $CdHPO_4$). This process appears to have a potential application, where phosphate containing organic substrates is present in metal and/or radionuclides containing effluents.

The localization of metals has mainly been carried out using electron microscopic studies. Electron microscopic observations carried out by Mullen et al. [10] revealed the presence of Ag as discrete particles at or near cell wall of both Gram-positive and Gram-negative bacteria, and the presence of silver was confirmed by energy dispersive X-ray analysis (EDAX). Large particles containing gold were localized in *Sargassum natans* cells by EDAX carried out in conjunction with scanning electron microscopy (SEM) [34]. Scott and Palmer [63] have also used these techniques for the presence of Cd in the cells of *Arthrobacter viscosus* following its exposure. In a study carried out by Golab et al. [64] in which *Streptomyces,* showing lead accumulation in the cell wall structures did not show any specific site for uranium accumulation.

BIOSORPTION: INFLUENCE OF PHYSICO-CHEMICAL ENVIRONMENT

Physico-chemical factors strongly influence the availability, mobility, and binding of the metal. Temperature, pH as well as concentration and age of biomass are among the important factors affecting the process of biosorption.

Hydrogen ion concentration (pH)

Both the cell surface metal binding sites and the availability of metal in the solution are affected by pH [65]. At low pH, the cell surface sites are closely linked to H+ ion, thereby making them unavailable for the other cations. However, with increase in pH, there is an increase in ligands with negative charge, which results in increased binding of cations [21,66].

The uptake of uranium by actinomycetes was maximally obtained at a pH near 6 [36]. Similar results were obtained in a green alga *Chlorella vulgaris* [35]. Killed *Streptomyces* biomass also showed a pH-dependent binding of U and Pb (II), with maximum binding occurring at pH 5.0-6.0 [64]. *Synechocystis* exhibited maximum uranium uptake at pH 5.0 [35].

In brown algae, the optimum pH for biosorption of U, Cd, Cu, Zn, and Co was found to be between pH 4.0-5.0. However, optimum pH for gold biosorption was found to be below 3.0 [32]. At pH 4.0-5.0, almost all the metallic species, except gold, get ionized, as in various cationic species at this pH, the carboxylate groups are expected to be largely dissociated generating charged surfaces. The removal of Cu from an aqueous solution was also more efficient with increasing pH. However, according to Norberg and Rydin [31], at higher pH values only part of copper is removed due to adsorption to bacterial biomass, since there is formation of copper hydroxide above pH 5.0. In *Zoogloea* sp., copper was more effectively adsorbed at low values of pH than Cd, and negligible Cd removal was observed at pH 5.5 [30].

A positive effect of increasing pH for Cu and Cd biosorption was also observed in *Bacillus circulans* [67]. Dave and Patwari [68] achieved the maximal bioremoval of Cd in a buffer of pH 7.0 using different bacterial isolates. *Ectocarpus siliculosus*, a macroalga exhibited only 14.8% Cd removal at pH 1.06, which increased to 91-96% between pH 2.04 and 6.75. However, the rate of adsorption decreased to 83% at pH 12.0 [69]. *Chlorella vulgaris* also exhibited a pH-dependent uptake of lead [37]. The rate of adsorption was maximum (15.4 mg/mg/min) at solution pH 5.0. At pH 2.0, the rate declined to 1.2 mg/mg/min. However, at higher pH values, lead gets precipitated in the adsorption medium. Panchandikar and Das [66] reported an increase in zinc uptake, when the initial pH was increased from 3.5 to 5.5 in *Pseudomonas aeruginosa,* and low pH appeared to be detrimental for metal uptake capacity of biomass. Sampedro et al. [39] showed marked pH dependency for biosorptive uptake of various heavy metals in a cyanobacterium, *Phormidium laminosum*, where an increase in pH led to an increase in biosorption efficiency.

In contrast to these studies, a decrease in adsorption rate was observed in case of Cr with increasing pH [70]. The maximum adsorption rates were obtained at a pH of 1-2 in a comparative study of various types of microbial systems, *Chlorella vulgaris, Cladophora crispata, Zoogloea ramigera, Rhizopus arrhizus,* and *Saccharomyces cerevisiae*. In general, it has been observed that pH of the effluents or aquatic bodies varies between pH 4 and 8, and an increase in pH favors the process of adsorption, but beyond a certain limit, high pH results in precipitation of metal, which makes the metal unavailable to the system [71].

Temperature

Temperature affects a number of factors, which govern metal ion biosorption viz. stability of metal ion in solution, presence of competing ligands, cell wall configuration, stability of cell-metal complexes and the ionization of chemical moieties on the cell [1]. The effect of temperature on biosorption is related to the groups involved in metal biosorption. It has been found that the formation of coordination complexes between transition metal cations and carboxylate ligands is endothermic, whereas amine ligand complex formation is exothermic [72].

For a brown alga, *Ascophyllum nodosum*, there was rise in the capacity of cobalt biosorption between 4-23°C. But between the temperature range 23-40°C, the increase in biosorptive capacity was not much pronounced. However, an appreciable increase in adsorption of lead ions by *Zoogloea ramigera* and *Rhizopus arrhizus* biomass was obtained in the range of 35-45°C [73].

Higher uptake capacities at increased temperature have also been reported for uranium by Shumate et al. [74], Strandberg et al. [28], and Tsezos and Volesky [75], and for copper, by deRome and Gadd [24]. In general, in most of the studies, no change in biosorption is observed over a wide range of temperature. Horikoshi et al. [36] reported a high passive uptake of uranium by *Actinomyces levoris* and *Streptomyces viridochromogenes*, which was not affected by increase in temperature from 0-30°C.

Concentration of metal

Metal ion concentration in solution greatly affects the metal loading of the microbial cell biosorbent. Upon contact between the sorbent material and the solution containing the sorbed species, an equilibrium is established at a given temperature, whereby a certain amount of metal sequestered by the biomass is in equilibrium with the residue left in the solution containing the residual/final/equilibrium concentration of the metal.

In general, an increase in external metal concentration has been observed to cause a rise in metal uptake [26,76]. However, beyond a particular concentration, in most cases, a saturation is attained. Horikoshi et al. [35] reported an increase in uptake of uranium with increase in concentration of uranium in the sea water. The living and scalded cells of *Chlorella vulgaris* showed a rapid increase in manganese uptake with a rise in concentration up to 20 ppm following which a gradual increase was observed up to 100 and 200 ppm for living and scalded cells, respectively. The amount of uranium adsorbed by *Actinomyces levoris* and *Streptomyces viridochromogenes* increased almost linearly with increase in concentration of uranium in the solution up to 8 ppm, and the rate of uptake slowed down thereafter [36]. Brady and Duncan [26] have made similar observations for Cd, Cu, and Co binding in *Saccharomyces cerevisiae*.

Time of contact

The major advantage of the biosorptive metal uptake is its fast *equilibrium kinetics*. Biosorption generally completes within first few minutes of the initial contact of biomass with the metal bearing solution [7,50]. Following this, equilibrium is established between the metal uptake by the biomass and the residual metal in the solution. Rapidity of the process of biosorption seems to be consistent with the mechanism of passive adsorption to the cells rather than being a metabolically active process [14,77].

Horikoshi et al. [35,36] have reported a very fast uptake of uranium by *Chlorella regularis*, which was not affected by the presence of metabolic inhibitors and increase in temperature, suggesting the existence of a passive process. In another time course study, copper uptake by *Zoogloea ramigera* also followed fast kinetics [31]. Rapid adsorption of lead has been observed in *Chlorella vulgaris*, which established equilibrium within 10-15 minutes [37].

In *Chroococcus paris,* approximately 90% Zn and Cu were bound rapidly in one minute [7]. Cd-binding in *Chlorella vulgaris* also occurred rapidly, and within 10 minutes, equilibrium was attained [11]. Sahoo et al. [67] have reported very fast uptake kinetics in *Bacillus circulans* for Cd and Cu during the first 15 minutes of initial exposure. In a macroalga, *Ectocarpus siliculosus,* about 96.3% of available Cd ions were found to be adsorbed within 5 minutes of exposure according to Winter et al. [69]. Similar results have also been reported by Dave and Patwari [68].

Concentration and age of biomass

Nature of biomass determines the availability and the type of binding sites for metal ions. Hence, it is most crucial for the phenomenon of biosorption. It has generally been observed that at high biomass concentrations, although the total metal removal from the solution is high, the specific binding, i.e. the micrograms of metal bound per milligram of dry biomass tends to be low [25,26,35,78-80]. According to Itoh et al. [81], the dependence of adsorption on the cell density may be due to electrostatic interactions, as more cations are adsorbed on the cells when the cell distances are more. Sampedro et al. [39] have reported a similar trend of decrease in sorption yield in heterocystous cyanobacterium, *Phormidium laminosum,* when the biomass concentration was increased. The biomass concentration had a pronounced effect on the amount of copper removed from the aqueous solution by *Zoogloea ramigera* [31]. Higher biomass concentration tends to form larger aggregates, which disturbed the equilibrium within the reactor.

The age of biomass is sometimes a governing factor in biosorption, since it can have some effect on the cell wall characteristics, i.e. the groups involved in metal binding and hence, on the mechanism of uptake. However, variable observations have been made. In some studies, older cultures have given better removal and, in others, the young ones gave better metal removing capacities, depending on the organism involved [35,36,67].

Norberg and Persson [30] obtained maximum removal of cadmium and copper by 6-8 day old culture of extracellular polysaccharide producing bacterium, *Zoogloea ramigera*. Scott and Palmer [63] obtained better cadmium accumulation with the older culture of *Arthrobacter viscosus,* which was probably related to the excretion of biomass, as investigated with cultures of different ages from 24-192 h. The 144 h culture was found to give the best metal removal efficiency [67]. In another study, for different bacterial isolates tested for removal of Cd, 20 and 24 h cultures gave the best removal as compared to the young and older cultures [68]. However, the uranium uptake by *Streptomyces viridochromogenes* did not vary much with the age of culture [36].

PRESENCE OF CO-IONS: CATIONS AND ANIONS

Despite the fact that a single metallic species does not exist in natural and wastewater, very few studies have been conducted related to the interactions of different ionic species. The presence of another solute generally inhibits the biosorption of desired metal by competing for the binding sites at the cell surface. It can complicate the evaluation of sorption system to a greater extent, depending on how the new solute species interacts with the sorbent and with the first solute [1,82].

Cations

A biosorbent which binds weakly to the ions found in hard water viz. Ca, Mg, Na, and K is desirable for commercial applications and provides an advantage over commercial ion-exchange resins. Hence, it becomes important to study the interaction of different ions. Inhibition of Mn uptake by Ca by living *Chlorella* cells was observed by Nakajima et al. [80], which was suggested to be due to competition between Mn and Ca ions for uptake. Sampedro et al. [39] have reported an inhibition of Ni uptake with increase in calcium, but its uptake increased in the presence of Fe, Cd, and Pb. The binding of Cu, Cr, and Zn however, remained unaffected.

K and Ca did not affect the uptake of uranium by *Pseudomonas aeruginosa* [28]. Chang and Hong [8] have reported similar observations where Hg biosorption was not inhibited by the presence of high concentrations of Na, indicating a selective uptake of Hg over Na ions. Comparable studies showed that Hg binding by commercial resin was strongly inhibited by high Na concentration [8]. Greene et al.[57] observed a very weak binding of Ca and Mg to *Chlorella vulgaris, C. pyrenoidosa, Spirulina platensis,* and *Cyanidium caldarium*. This is an added advantage of the algal cells over commercial ion-exchange resins in recovery of heavy metal ions from hard waters, as Ca and Mg can saturate cation-exchange resins and interfere with binding of heavy metal ions [38]. Cobalt uptake by *Ascophyllum nodosum* was unaffected by Fe and Ca. However, K exhibited a positive effect [33]. Nakajima and Sakaguchi [29] studied the interaction of nine metals in solution viz. Mn, Co, Ni, Cu, Zn, Cd, Hg, Pb, and UO_2 for various fungi, yeast, bacteria, and actinomycetes. They showed that out of nine metals UO_2, Hg, Pb, and Cu ions were more readily accumulated. They also showed that the accumulation of cobalt occurred to a large extent, when present alone, but when it was present along with other metals, its accumulation was poor.

Kuyucak and Volesky [33] investigated the interaction of various metal ions with cobalt biosorption by *Ascophyllum nodosum*. At pH 4.5, Cu, Ni, Zn, Cr, Pb, and UO_2 had an appreciable effect on the cobalt uptake capacity of the biomass, regardless of the initial cobalt concentration in the solution. In *Ectocarpus siliculosus*, the interference of cadmium adsorption was found to increase in the following order in the presence of co-ions in the solution: $MnCl_2 < MnSO_4 < MgCl_2 < CaCl_2$ [69].

The chemically treated *Chlorella vulgaris* cells took up the same amounts of uranium from a solution containing equal concentration of copper, cadmium, and uranium, as from a pure uranium solution exhibiting no interference of Cu and Cd on uranium binding, the Cu and Cd uptake was markedly inhibited by the presence of uranium [83]. The presence of Cd ions did not inhibit the Cu biosorption by *Zoogloea ramigera* [30]. In this case, the trivalent aluminium was found to be more effective than the divalent Mg and monovalent Na in decreasing the Cu binding. Thus, it can be concluded that, in general, the monovalent cations are adsorbed to less extent as compared to the polyvalent ones. Hence, it is the ionic radii of the cation that affects the adsorption as well as the ion-exchange phenomenon [33].

Anions

Anions can bind to the metal ions resulting in complexes [84,85], which are either weakly adsorbing, non-adsorbing or strongly adsorbing [44]. Of the anions there are quite a few reports on the interference of biosorption by carbonate ions. The marked inhibition by carbonate ion in *Chlorella* sp. has been suggested to be due to the formation of stable complex ions viz. $(UO_2CO_3)_2^{2-}$ and $UO_2(CO_3)_3^{4-}$ which are not adsorbed [35,78,83].

In *Synechococcus* sp., the uranium uptake was found to be strongly inhibited by carbonate ions. It could take up uranium from artificial decarbonated sea water to the tune of 5450 mg/g dry cell, but only 60.9 mg/ g dry cell of uranium from natural sea water and about 56.5 mg/g of uranium from artificial sea water [35]. Volesky [1] reported a reduced biosorption of uranium by *Rhizopus arrhizus* and *Penicillium chrysogenum* from sea water due to the presence of carbonate ions.

In case of actinomycetes, uranium uptake was hardly affected by 0.3 mM sodium bicarbonate, but when the concentration of the latter was increased from 0.3 mM to 3.0 mM, the amount of uranium adsorbed decreased and was completely prevented at 3.0 mM concentration [36]. The Co^{2+} uptake in *Ascophyllum nodosum* was suppressed by 4-14% and 35%, by carbonate and nitrate ions, respectively. However, phosphate and sulphate ions did not affect the biosorption of cobalt [33].

Nakajima et al. [86] have reported an interference of uranium uptake by *Chlorella regularis* by carbonate and phosphate ions. Uranyl ions form quite stable complexes with dihydrogen phosphate ions and hydrogen phosphate ions, such as $UO_2(H_2PO_4)^{3-}$ and $UO_2(HPO_4)_2^{2-}$, which are not taken up by the cells. Sodium hydrogen carbonate at a concentration of 0.01 M reduced the uranium binding in both living and scalded *Chlorella regularis* cells down to zero.

Phosphate ion concentration was found to have a profound effect on Cd adsorption by *Chlorella vulgaris* [11]. An increase in concentration of phosphate from 0-20 mg PO_4^{3-}/l caused 48% reduction in Cd adsorption; and at a concentration of 350 mg PO_4^{3-}/l, there was virtually no adsorption of cadmium.

According to Greene et al. [87], the binding of U (VI) was not affected by sodium salts of acetate, chloride, sulphate, and nitrate at a concentration of 0.2 M, but bicarbonate and phosphate ions caused a strong inhibition of uranium binding. The bicarbonate interference was found to be pH-dependent because of stable U (VI) carbonate complex formation. At pH 5.0, the bicarbonate interference was virtually eliminated. In contrast, anions could hardly affect the manganese uptake by *Chlorella regularis* [80].

Anions like PO_4^{3-} and thiosulphate reduce metal availability by precipitation, thus cause reduction in biosorption [84,85]. More work at present is being carried out on interaction of metals in biosorption to predict the biosorption performance of new materials and to gain more information on the mechanism of metal biosorption [9].

BIOMASS MODIFICATION/PRE-TREATMENTS

As biosorption process involves mainly cell surface sequestration, cell wall modification can greatly alter the binding of metal ions. A number of methods have been employed for cell wall modification of microbial cells in order to enhance the metal binding capacity of biomass and to elucidate the mechanism of biosorption. These modifications can be introduced at two stages: either during the growth or to the pre-grown biomass.

During the growth of a microorganism the condition, in which it is grown, can determine the biosorption potential, since it affects the cell surface phenotype [88]. A lot of work has been done on the effect of culture conditions of cells on the biosorptive capacity of yeast and fungi. However, not much is known about the changes in biosorptive capacity of algal biomass and cyanobacterium. In *Phormidium laminosum*, biosorption decreased initially with nitrogen starvation and, subsequently, increased until it reached the value of N_2-sufficient cells [39]. Besides culture conditions, the treatment to pre-grown biomass also affects biosorption potential. Pregrowing the bacterial cells at pH 7.4 resulted in increased efficiency of Cd removal than the cells grown at pH 5.0, suggesting an influence of pH on the surface layers of the cells [68].

In the pre-grown biomass, several physical and chemical treatments have been tried to tailor the metal binding properties of biomass to specific requirements. The physical treatments include heating/boiling, freezing/ thawing, drying, and lyophilization. The accumulation of Pb and UO_2 by killed cells was found to be slightly higher than that observed with living cells in case of *Streptomyces* [64]. Similar observations have been reported for algae [29] and bacteria [89]. In *Chlorella* sp., the amount of Mn taken up by scalded cells was far greater (5100 mg/g) than the living cells [35]. Golab et al. [64] have reported considerably high uptake capacities of *Streptomyces* for UO_2 and Pb by both living and killed cells.

Heat killed biomass of *Pseudomonas aeruginosa* showed a high capacity for zinc uptake, particularly during the period of rapid binding, probably due to destruction of cell membrane, with the result exposing intracellular components as well as surface binding. Autoclaving the biomass at 121°C for 15 minutes did not lead to any change in metal binding capacity of *Phormidium laminosum* [39]. Freezing and thawing of biomass did not result in any change in metal accumulating capacity of *Bacillus circulans* [67]. Norberg and Persson [30] have reported similar results for *Zoogloea ramigera*.

The various chemical treatments used for biomass modification include (i) washing the biomass with detergents, (ii) cross-linking with organic solvents, and (iii) alkali or acid treatments. Ross and Townsley [90] observed better Cu sorption in detergent treated mycelium of *Penicillium spinulosum*. Similar reports have been made in *Saccharomyces cerevisiae* and *Rhizopus arrhizus* by Gadd [14]. Processing of biomass by cross-linking have been found to improve the physical properties viz. strength, hardness and swelling characteristics, which are desirable for column applications [46].

Formaldehyde modification is a chemical cross-linking between adjacent chemical groups preferably hydroxyl groups in the sugars of the cell wall. Glutaraldehyde cross-links the chemical groups (mainly amino groups), which are more distant from each other because of the prolonged carbon chain. In the polyethyleneimine (PEI) resin modification, the biomass is embedded in the resin and the free amino groups of this matrix are further cross-linked with glutaraldehyde. Polyethyleneimines are likely to introduce NH_2 groups to the biomass, which alter the sorption behavior of the ionic interactions. Also the particle swelling property is a characteristic feature of each modification. After the modification, the biomass is expected to have a lower expansion during swelling because of the rigid structure [91].

In *Ascophyllum nodosum*, the uptake of cadmium decreased in the order: glutaraldehyde > formaldehyde > PEI, but in *Sargassum natans*, glutaraldehyde cross-linking caused 11% better uptake than the native biomass [91]. The metal uptake was higher for larger particles (0.84-1.00 mm) than the smaller particles (0.195-0.295 mm). Nakajima et al. [83] treated the *Chlorella* cells by hot water-CHCl$_3$-methanol/concentrated alkali and observed relatively less capacity of uranium uptake, showing 73.5, 77.8, and 26.8% respectively than that for the dry cells. This suggests that the cell components, which are extracted with hot water (oligo- and polysaccharides, proteins, and low molecular weight substances) and those extracted with CHCl$_3$ -methanol (lipids) as well as the cell fraction obtained with CHCl$_3$ - methanol/concentrated alkali (mainly the cellulose) are all concerned with uranium binding to the cells.

Alkali pre-treatment increased the biosorption capacity (5-34%) for different metals in *Phormidium laminosum* [39]. However, there was loss in biosorption capacity following acidic pretreatment which has been suggested to be due to competition effect between the bound protons and the dissolved metals, which is higher than that resulting from the alkali and alkaline earth cations, normally present in the untreated biomass. Sampedro et al. [39] also did not observe any improvement in biosorption capactiy of *P. laminosum,* following its washing with 1% v/v Triton-X 100. Alkali pre-treatment has also been shown to increase biosorption of heavy metals by fungal biomass [25,91]. These studies clearly indicate that pre-treatments do modify the surface characteristics (groups) either by removing or masking the groups or by exposing more metal binding sites [21].

Cell immobilization

In order to retain the ability of microbial biomass to adsorb metals during the continuous industrial process, it is important to utilize an appropriate immobilization technique. The free cells can provide valuable information in laboratory experimentation, but are generally not suited for column packing in industrial applications [14,82,93]. The free cells generally have low mechanical strength and small particle size, and excessive hydrostatic pressures are required to generate suitable flow rates. High pressures can cause disintegration of free biomass. These problems can be alleviated by the use of immobilized cell systems [19,94,95]. Immobilized biomass offers many advantages including better reusability, high biomass loading and minimal clogging in continuous flow systems [94]. Besides, the particle size and flow rates can be controlled.

A number of matrices have been employed in order to immobilize the microbial cells. One of matrices that has been used in metal recovery by both viable and non-viable cells is the entrapment in the matrix of insoluble Ca-alginate [96,97]. In this process, commercially available alginic acid is precipitated or gelled by the addition of polyvalent metal cations, such as calcium. Alginic acid is a linear polymer of D-mannuronic and D-glucuronic acid obtained from algae. In such immobilization process, cell slurry is mixed with sodium alginate in a particular ratio. This is then pumped dropwise into calcium chloride solution, forming insoluble alginate beads containing entrapped cells. The insoluble alginate matrix has large enough pores, through which metal ions can diffuse and binding interactions with the cells can occur. Alginate beads show excellent resistance to hydrostatic pressures, when packed into columns [98]. Fluidized beds of Ca-alginate entrapped *Chlorella vulgaris* and *Spirulina platensis* were successfully used to recover gold from a simulated gold bearing solutions, containing $AuCl_4$, $CuCl_2$, $FeCl_2$, and $ZnCl_2$ [57].

Kuhn and Pfister [99] achieved more than 95% removal of Cd, Zn, Mn, Pb, Cu, and Sr by alginate immobilized *Zoogloea ramigera* cells. Alginate beads alone showed some binding but did not reach the level achieved with the alginate immobilized cells showing that the bacterium significantly contributed to the metal binding. The calcium alginate immobilized cells of *Chlorella salina* also showed greater binding of cobalt, zinc, and manganese than the free cells [100,101]. Accumulation was also dependent on cell density in alginate beads with greater uptake of cobalt at the highest cell densities.

Whole cell immobilization within a polyacrylamide gel provides a useful laboratory scale system and have been used to biosorb and recover a number of heavy metals [72]. The polyacrylamide immobilization procedure is based on free radical polymerization of acrylamide in an aqueous solution containing microbial cells. The resulting gel is forced through a sieve to get uniform sized particles. Nakajima et al. [102] observed an improved stability of *Chlorella* cells immobilized in polyacrylamide to remove uranium. Similar results were obtained with polyacrylamide immobilized *Streptomyces albus* cells [29]. The immobilized *Streptomyces* cells lost 2% of dry weight in 5 cycles of adsorption-desorption as compared to 50% reduction in weight by free cells [29]. Pons and Fuste [103] have used this matrix for the removal of uranium, using *Pseudomonas* cells.

The polyacrylamide immobilized cells of *Citrobacter* were capable of high removal of uranium, cadmium, and lead from solutions supplemented with glycerol 2-PO$_4$ [94,104]. Wong and Kwok [105] could achieve significantly high nickel from electroplating effluents using polyacrylamide immobilized *Enterobacter* sp. Another important matrix, which is being used for immobilization for metal removal, is silica. Silica immobilized preparations offer advantage in terms of reusability and stability. The silica immobilized product is mechanically strong and exhibits excellent flow characteristics [57]. Silica immobilized algal preparation commercially used is *AlgaSORB*

(Bio-Recovery Systems, Inc., Las Cruces, NM 88003, U.S.A.). After metal recovery, approximately 90% of original metal uptake efficiency was retained even after prolonged use (> 18 months). In a study conducted by Mallick and Rai [97], among the various matrices tried for immobilization, chitosan and carrageenan immobilized *Anabaena doliolum* and *Chlorella* showed the highest efficiency for uptake of anions and cations.

In a number of studies, the biomass has been immobilized using inert solid supports as biofilms [94]. These inert matrices include those with planar surfaces (e.g., wood shavings, clays, sand, and crushed rocks) and porous materials (foams and sponges). Diels et al. [106] have used ZirFon R membrane composed of ZrO_2 and polysulfone to immobilize *Alcaligenes eutrophus* CH34 cells to induce metal precipitation in a reactor and the metal crystal recovered in the recuperation column. They have been able to achieve 99.9% Cd and Zn and 98% or more removal for Ni, Cu, and Co. Living cells of *Pseudomonas aeruginosa* have been immobilized on particles of polyvinyl chloride (PVC) and used in batch and column systems for simultaneous denitrification and heavy metal removal from contaminated water [107]. Another system has employed a mixed bacterial culture, comprising mainly of *Pseudomonas* sp. for denitrification and uranium removal [74].

ELUTION/RECOVERY OF METAL

Biotechnological exploitation of microbial metal binding depends on efficiency of the regeneration of biosorbent for metal recovery. The method used for elution depends on the element involved and the mechanism of its binding [88]. Biosorption is often reversible by simple non-destructive physical or chemical treatments, whereas intracellular accumulation is often irreversible, requiring drastic treatments like incineration or dissolution in acids or alkali. The *non-destructive recovery* by mild and cheap desorbing agents is desirable for regeneration of biomass for use in multiple biosorption-desorption cycles. In multiple cycles of sorption and desorption, where the biosorbent is being regenerated for reuse, the evaluation of the biosorbent deterioration in terms of weight loss and structure for its effect on metal loading capacity is to be taken into consideration [88,108].

The efficiency of the desorbing agent or the eluant is often expressed by S/L, i.e. solid to liquid ratios. The solid represents the amount of solid sorbent (mg dry wt.) and the liquid represents the amount of eluant applied (ml). High values of S/L are desirable for preferably complete elutions and to make the process more economical [1,109]. The physico-chemically sequestered metal to the cell surface can be easily desorbed by ethylenediamine tetraacetic acid (EDTA) [109]. About 80% of the Mn taken up by the *Chlorella regularis* cells was eluted by EDTA, and the metal taken up into the inner spaces increased with the passage of time as the EDTA washable fraction declined [80]. The uranium adsorbed to the actinomycetes was released by EDTA, suggesting a surface binding to the ligands [36]. Dilute EDTA solutions (EDTA : metal, 2 : 1) could rapidly desorb 98-100% of Cu, Cd, and Zn adsorbed by the sheath forming cyanobacterium *Chroococcus paris* in a few minutes [7]. EDTA released about 60% of the cobalt at pH 2.5 from *Ascophyllum nodosum* [109]. Carbonates and bicarbonates have also been used in non-destructive recovery of uranium and other metals [29,87,110].

It has been observed that for metal ions that show a marked pH dependence in binding to the microbial cells, stripping of bound metals can be accomplished by pH adjustments. The Cu(II), Cr(II), Ni(II), Zn(II), Cd(II), and Co(II) bound to *Chlorella vulgaris* at pH 5.0, could be quantitatively reversed by lowering the pH down to 2.0 [38]. Dilute mineral acids have been used in various studies to remove metals from the loaded biomass [93]. The release of Cu and Cd from *Zoogloea ramigera* was achieved using acid treatment. Increasing the acidity lead to effective removal of these metals from the biomass.

Darnall et al. [72] have used mercaptoethanol to selectively remove Au, Ag, and Hg bound to *Chlorella vulgaris* at pH 2.0. The metal selective elution is important and is desirable for some applications and can be achieved by the basic understanding of the mechanism involved in particular metal sequestration. Volesky and Kuyucak [111] have obtained maximum elution of gold using a mixture of thiourea and 0.2M ferric ammonium sulphate from the biosorbent based on marine alga, *Sargassum natans*.

Kuyucak and Volesky [109] have evaluated the effectiveness of different eluants in stripping cobalt from *Ascophyllum nodosum*; $CaCl_2$ (0.05M) in HCl was found to be the best eluent capable of desorbing about 96% of cobalt at pH 2.0–3.0. $CaCl_2$/HCl also did not result in any alteration in cell architecture and its structural materials, whereas all the other eluants including acids, NH_4OH, $KHCO_3$, KSCN, and EDTA resulted in some changes in cell architecture. Cadmium and uranium could also be desorbed from *Citrobacter* cells using citrate buffer [61,104]. In studies conducted in our own laboratory, 10 mM EDTA eluted 64.72% Zn^{2+} from loaded biomass, while citrate buffer (pH 3.0) and Na_2CO_3 (1.0 mM) was best suited for eluting Cu^{2+} and Co^{2+}, respectively [112-114].

COMMERCIAL BIOSORBENTS

Various types of microbial biomasses have formed the basis of formulation of new and potent metal sequestering biosorbents. This is important in the present day scenario, as there is an increasing need for an effective and economical process to remove metal ions from industrial wastewater and drinking water. A potent algal biosrobent, AlgaSORB™ has been developed using a fresh water alga *Chlorella vulgaris* to treat wastewater [72]. This can efficiently remove metallic ions from dilute solutions, i.e. 1-100 mg/l and reduces the concentration of metals down to 1 mg/l or even below. Calcium and magnesium also do not affect the sorption of heavy metals by

AlgaSORB™. Another metal sorption agent, AMT-BIOCLAIM™ (MRA) has employed *Bacillus* biomass to manufacture granulated material for wastewater treatment and metal recovery. This can accumulate metal cations with efficient removal (>99%) from dilute solutions. It is non-selective, and metals can be stripped from it after loading by H_2SO_4, NaOH or complexing agents and the granules can be regenerated for repeated use [27].

Bio-Fix biosrobent uses biomass from a variety of sources, including cyanobacterium (*Spirulina*), yeast, algae, plants (*Lemna* sp.), and guar gums to give a consistent product and immobilized as beads using polysulfone. Zinc binding to this biosorbent is approximately 4-fold higher than the ion-exchange resins. There is variable affinity for different metals Al > Cd > Zn > Mn and a much lower affinity for Mg and Ca. Metals can be eluted using HCl or HNO_3, and biosorbent can be reused for more than 120 extraction-elution cycles. Two marine organisms, *Sargassum natans* and *Ascophyllum nodosum* have been found to have excellent biosorption capacities for gold and cobalt respectively, among the photoautotrophs [32,111].

CONCLUSION

Biosorption is an economically feasible and technically efficient technology for metal removal/recovery and can comfortably fit into the metal treatment processes and is eco-friendly in nature. Inspite of these advantages why has the biosorption/wastewater treatment remained as an embryonic industry? This is because not all the companies, which generate metal-polluted wastewater, will have the capability or the interest to do anything other than the basic treatment to comply with the legislation. Hence to overcome this, what is needed is a series of specialists, centralized facilities which would be capable of removing metal from wastewater and regenerating or processing the metal-loaded sorbent and then converting the recovered metal into reusable form. Alternatively, if the biosorbent used is a waste product, its incineration could be used to produce metal rich slag.

Looking into the economic feasibility, in terms of scale-up and working efficiency as a technology, the *microbial biosorbents* provide encouraging results to be utilized in wastewater treatment. What is needed is an extra mural input from the industries, generating metal-polluted wastewater to invest into such clean-up technologies before discharging their liquid effluents into the water streams.

REFERENCES

1   Volesky B. Biosorption of Heavy Metals. Boca Raton, Florida: CRC Press, Boca Raton, 1990.

2   Eccles H, Hunt S. Immobilisation of Ions by Biosorption. New York: John Willey & Sons, 1986.

3   Adamson AW. Physical Chemistry of Surfaces. New York: John Wiley & Sons, 1976.

4   Freundlich, H. Colloid and Capillary Chemistry. London: Methuen, 1926.

5   Brunauer S, Emmett PH, Teller E. J Am Chem Soc 1938; 60: 309-319.

6   Horikoshi T, Nakajima A, Sakaguchi T. Eur J Appl Microbiol Biotechnol 1981; 12: 90-96.

7   Les A, Walker RW. Water Air Soil Pollut 1984; 23: 129-139.

8   Chang J, Hong J. Biotechnol Bioeng 1994; 44: 999-1006.

9   Volesky B, Holan ZR. Biotechnol Prog 1995; 11: 235-250.

10   Mullen LD, Wolf DC, Ferris FG, Beveridge TJ, Flemming CA, Bailey GW. Appl   Environ Microbiol 1989; 55(12): 3143-3149.

11   Khummongkol D, Canterford GS, Fryer C. Biotech Bioeng 1982; 24: 2643-2660.

12   Ting YP, Lawson F, Prince IG. Biotechnol Bioeng 1988; 34: 990-999.

13 Ting YP, Lawson F, Prince IG. Biotechnol Bioeng 1991; 37: 445-455.

14 Gadd GM. In: Rehm HJ, Reed G, eds. Biotechnology – A Comprehensive Treatise (Vol 6b) Special Microbial Processes. Germany: VCH VerlagsgesellSchaft Weinheim, 1988; 401-433.

15 Volesky B. Trends Bio Sci 1987; 5: 96-101.

16 Muraleedharan TR, Iyengar L, Venkobachar C. Curr Sci 1991; 61(6): 379-385.

17 Gadd GM, White C. Trends Biotech 1993; 11: 353-359.

18 Rosenberger D. In: Smith JE, Berry DR, eds. The Filamentous Fungi (Vol 2). London: Edward Arnold, 1975; 328-343.

19 Gadd GM. In: Edwards C, ed. Microbiology of Extreme Environment. New York: McGraw Hill Publishing Company, 1990.

20 Gadd GM. Experientia 1990; 46: 834-840.

21 Paknikar KM, Palnitkar US, Puranik PR. In: Torma Ae, Apel ML, Brierley CL, eds. Biohydrometallurgical Technologies (Vol II). Wyoming, USA: TMS Publications, 1993; 229-236.

22 Volesky B, Tsezos M. US Patent 4 320 093, 1981.

23 Galun M, Keller P, Malki D, Feidstein H, Galun E, Siegel S, Siegel B. Water Air Soil Pollut 1984; 21: 411-414.

24 deRome L, Gadd GM. Appl Microbiol Biotechnol 1987; 26: 84-90.

25 Luef E, Prey T, Kubicek CP. Appl Microbiol Biotechnol 1991; 34: 688-692.

26 Brady D, Duncan JR. In: Torma AE, Apel ML, Brierley CL, eds. Biohydrometallurgical Technologies (Vol II). Wyoming, USA: TMS Publications, 1993; 711-723.

27 Brierley JA, Brierley CL, Goyak GN. In: Lawrence RW, Branion RMR, Ebner, HG, eds. Fundamentals and Applied Biohydrometallugy. Amsterdam: Elsevier Science Publishers, 1986; 291-304.

28 Strandberg GW, Shumate SE, Parrott JR. Appl Environ Microbiol 1981; 41: 237-245.

224

29   Nakajima A, Sakaguchi T. Appl Microbiol Biotechnol 1986; 24: 59-64.

30   Norberg A, Persson H. Biotechnol Bioeng 1984; 26: 239-246.

31   Norberg A, Rydin S. Biotechnol Bioeng 1984; 26: 265-268.

32   Kuyucak N, Volesky B. Biotech Lett 1988; 10(2): 137-142.

33   Kuyucak N, Volesky B. Biotechnol Bioeng 1989; 33: 809-814.

34   Kuyucak N, Volesky B. Biorecovery 1989; 1: 189-204.

35   Horikoshi T, Nakajima A, Sakaguchi T. J Ferment Technol 1979; 57(3):
      191-194.

36   Horikoshi T, Nakajima A, Sakaguchi T. Eur J Appl Microbiol Biotechnol
      1981; 12: 90- 96.

37   Aksu Z, Kutsal T. J Chem Tech Biotechnol 1991; 62: 109-118.

38   Greene B, Darnall DW. In: Ehrlich HL, Brierley CL, eds. Microbial
      Mineral Recovery. McGraw Hill Inc 1990; 277-301.

39   Sampedro MA, Blanco A, Llama MJ, Serra JL. Biotechnol Appl Biochem
      1995; 22: 355-366.

40   Shumate SE, Strandberg GW. In: Moo-Young M, Robinson CN, Howell
      JA, eds. Comprehensive Biotechnology, New York: Pergamon Press,
      1985; 235-247.

41   Kuyucak N, Volesky B. Biotechnol Bioeng 1989; 33: 823-831.

42   Gourdon R, Bhende S, Rus E, Sofer SS. Biotech Lett 1990; 12(11): 839-
      842.

43   Kuyucak N, Volesky B. Biorecovery 1989; 1: 219-235.

44   Tobin JM, Cooper DG, Neufeld RJ. Enzyme Microbiol Technol 1990;
      12: 591-595.

45   Beveridge TJ, Murray RGE. J Bacteriol 1980; 141: 876-887.

46   Holan ZR, Volesky B. Biotechnol Bioeng 1994; 43(11): 1001-1009.

47   Crist RH, Oberholser K, Shank N, Nguyen N. Environ Sci Technol
      1981; 15(10): 1212-1217.

48   Crist RH, Oberholser K, Schwartz D, Marzoff J, Ryder D, Crist DR.
      Environ Sci Technol 1988; 22(7): 755-760.

49   Friis N, Myers-Keith P. Biotechnol Bioeng 1986; 27: 21-28.

50   Aksu Z, Kutsal T. Environ Technol 1990; 2: 979-987.

51   Gardea-Torresdey JL, Becker-Hapak MK, Hosea JM, Darnall DW. Environ Sci Technol 1990; 24: 1372-1378.

52   Remacle J. In: Volesky B, ed. Biosorption of Heavy Metals. Boca Raton, Florida: CRC Press, 1990; 83-92.

53   Hoyle B, Beveridge TJ. Appl Environ Microbiol 1983; 46(3): 749-752.

54   Strain SM, Fesik SW, Armitage IM. J Biol Chem 1983; 258: 13466-13477.

55   Ferris FG, Beveridge TJ. FEMS Microbiol Lett 1984; 24: 43-47.

56   Beveridge TJ, Fyfe WS. Can J Earth Sci 1985; 22: 1892-1898.

57   Greene B, McPherson R, Darnall D. In: Patterson JW, Passion R, eds. Metals Speciation Separation and Recovery. Chelsea, MI: Lewis Publishers, 1987; 315-338.

58   Schiewer S, Volesky B. Environ Sci Technol 1995; 29: 3049-3058.

59   Crist DR, Crist RH, Martin JR, Watson JR. FEMS Microbiol Rev 1994; 14: 309-314.

60   Fogg GE, Westlake DF. Proc On Ass Theor Appl Limnol 1955; 12: 219-232.

61   Macaskie LE, Dean ACR. J Gen Microbiol 1984; 130: 53-62.

62   Macaskie LE, Bonthrone KM, Rouch DA. FEMS Microbiol Lett 1994; 121: 141-146.

63   Scott JA, Palmer SJ. Appl Microbiol Biotechnol 1990; 33: 221-225.

64   Golab Z, Orlowska B, Smith RW. Water Air Soil Pollut 1991; 60: 99-106.

65   Roomans GM, Theuvenet APR, Borst Pauwels GWFH. Biochem Biophys Acta 1979; 55(1): 187-196.

66   Panchandikar VV, Das RP. Intern J Environ Studies 1993; 44: 251-257.

67   Sahoo DK, Kar RN, Das RP. Bioresource Technol 1992; 41: 177-179.

226

68   Dave SR, Patwari RA. In: Torma AE, Apel ML, Brierley CL, eds. Biohydrometallurgical Technologies. Wyoming, USA: TMS publications, 1993; 119-124.

69   Winter C, Winter M, Pohl P. J Appl Phycol 1994; 6: 479-487.

70   Nourbakhsh M, Sag Y, Ozer D, Aksu Z, Kutsal T, Caglar A. Process Biochem 1994; 29: 1-5.

71   Rai LC, Gaur JP, Kumar HD. Environ Res 1981; 25: 250-259.

72   Darnall DW, Greene B, Henzl MT, Hosea JM, McPherson RA, Sneddon J, Alexander MD. Environ Sci Technol 1986; 20(2): 206-208.

73   Sag Y, Ozer D, Kutsal T. Process Biochem 1995; 30(2): 169-174.

74   Shumate SE, Strandberg GW, Parrott JR. Biotechnol Bioeng Symp 1978; 8: 13-20.

75   Tsezos M, Volesky B. Biotechnol Bioeng 1981; 23: 583-604.

76   Venkateswerlu G, Stotzky G. Appl Microbiol Biotechnol 1989; 31: 619-625.

77   McHale AP, McHale S. Biotech Adv 1994; 12: 647-652.

78   Sakaguchi T, Tsuji T, Nakajima A, Horikoshi T. Eur J Appl Microbiol Biotechnol 1979; 8: 207-215.

79   Nakajima A, Horikoshi T, Sakaguchi T. Agric Biol Chem 1979; 43(7): 1455-1460.

80   Nakajima A, Horikoshi T, Sakaguchi T. Agric Biol Chem 1979, 43(7): 1461-1466.

81   Itoh M, Yuasa M, Kobayashi T. Plant Cell Physiol 1975; 16: 1167-1169.

82   Volesky B. FEMS Microbiol Rev 1994; 14: 291-302.

83   Nakajima A, Horikoshi T, Sakaguchi T. Eur J Appl Microbiol Biotechnol 1981; 12: 76-83.

84   Gadd GM, Griffiths AJ. Microbial Ecol 1978; 4: 303-317.

85   Beyersmann D. In: Merian E, ed. Metals and Their Compounds in the Environment: Occurrence, Analysis and Biological Relevance. Weinheim, New York: VCH, 1991; 491-509.

86   Nakajima A, Horikoshi T, Sakaguchi T. Agric Biol Chem 1979; 43(3): 625-629.

87   Greene B, Henzl MT, Hosea M, Darnall DW. Biotech Bioeng 1986; 28: 764-767.

88   Gadd GM. In: Lederberg, ed. Encyclopedia of Microbiology. San Diago: Academic Press Inc. Harcourt Brace Javsanovich Publishers, 1992; 2: 351-360.

89   Beveridge TJ, Murray RGE. J Bacteriol 1976; 127(3): 1502-1518.

90   Ross IS, Townsley CC. In: Eccles H, Hunt S, eds. Immobilization of Ions by Biosorption. Chichester : IRL Press, 1986; 49-58.

91   Leusch A, Holan ZR, Volesky B. J Chem Tech Biotechnol 1995; 62: 279-288.

92   Muzzarelli RAA, Tanfani F, Scrapini G, Tucci E. J Appl Biochem 1980; 2: 54-59.

93   Cotoras D, Viedema P, Pimentel J. In: Torma AE, Apel ML, Brierley CL, eds. Biohydrometallurgical Technologies (Vol II). Wyoming, USA: TMS Publications, 1993; 103-109.

94   Macaskie LE, Dean ACR. In: Mizrahi A, ed. Biological Waste Treatment. New York: Alan R Liss, 1989; 159-201.

95   Tsezos M. In: Ehrlich HL, Brierley CL, eds. Microbial Mineral Recovery. New York: McGraw Hill, 1990; 325-339.

96   Mallick N, Rai LC. World J Microbiol Biotechnol 1993; 9: 196-201.

97   Mallick N, Rai LC. World J Microbiol Biotechnol 1994; 10: 439-443.

98   Kennedy JF, Cabral JMS. Appl Biochem Bioeng 1983; 4: 189-191.

99   Kuhn SP, Pfister RM. Appl Microbiol Biotechnol 1989; 31: 613-618.

100   Garnham GW, Codd GA, Gadd GM. Appl Microbiol Biotechnol 1992; 37: 270-276.

101   Garnham GW, Codd GA, Gadd GM. Environ Sci Technol 1992; 26(9): 1764-1770.

102   Nakajima A, Horikoshi T, Sakaguchi T. Eur J Appl Microbiol Biotechnol 1982; 16: 88-91.

103   Pons MP, Fuste MC. Appl Microbiol Biotechnol 1993; 39: 661-665.

104   Macaskie LE, Dean ACR. Biotech Lett 1985; 7(7): 457-462.

105   Wong PK, Kwok SC. Biotech Lett 1992; 14(7): 629-634.

106   Diels L, Van Roy S, Taghavi S, Doyen W, Leysen R, Mergeay M. In: Torma AE, Apel ML, Brierley CL, eds. Biohydrometallurgical Technologies. Wyoming, USA: TMS Publications, 1993; 119-124.

107   Hollo J, Toth J, Tengerdy RP, Johnson JE. In: Venkatasubramanium K, ed. Immobilized Microbial Cells. Washington D.C.: American Chemical Society, 1979; 73-86.

108   Tsezos M. Biotechnol Bioeng 1983; 25: 2025-2040.

109   Kuyucak N, Volesky B. Biotechnol Bioeng 1989; 33: 815-822.

110   Tsezos M. Biotechnol Bioeng 1984; 26: 973-981.

111   Volesky B, Kuyucak N. US Patent 4 769 233, 1988.

112   Ahuja P, Gupta R, Saxena. Curr Microbiol 1997; 35: 1-5.

113   Ahuja P, Gupta R, Saxena. Curr Microbiol 1999; 39: 49-52.

114   Ahuja P, Gupta R, Saxena. Process Biochem 1999; 34: 77-85.

115   Cain JR, Paschal DC, Hayden CM. Arch Environ Contam Toxicol 1980; 9: 9-16.

116   Crist RH, Martin JR, Guptill PW, Eslinger J, Crist D. Environ Sci Technol 1990; 24: 337-341.

117   Fernandez-Pinas F, Mateo P, Bonilla I. Arch Environ Contam Toxicol 1991; 21: 425-431.

118   Tobin JM, Cooper DG, Neufeld, RJ. Appl. Environ. Microbiol 1984; 47: 821-824.

119   Fourest E, Roux JC. Appl. Microbiol. Biotechnol 1992; 37: 399-403.

120   Volesky B, May-Phillips HA. Appl Microbiol Biotechnol 1995; 42: 797-806.

121   Townsley CC, Ross IS, Atkins AS. In: Lawrence RW, Branion RMR, Ebner HG, eds. Fundamental and Applied Biohydrometallurgy. Amsterdam: Elsevier, 1986; 279-289.

122   Khalid AM, Asfaq SR, Bhatti TM, Anwar MA, Shemsi AM, Akhtar K. In: Torma AE, Apel ML, Brierley C, eds. Biohydrometallurgical Technologies Vol 2, Warrendale, PA: Minerals, Metals & Materials Society, 1993; 299-308.

123   Holan ZR, Volesky B. Appl. Biochem. Biotechnol. 1995; 53: 133-146.

124   Niu H, Xu XS, Wang JH, Volesky B. Biotechnol. Bioeng. 1993; 42: 785-787.

125   Nemec P, Prochazka H, Stamberg K, Katzer J, Stamberg J, Jilek R, Hulak P. US Patent 4, 021 368, 1977.

126   Mattuschka B, Junghaus K, Straube G. In: Torma AE, Apel ML, Brierley C, eds. Biohydrometallurgical Technologies Vol 2, Warrendale, PA: Minerals, Metals & Materials Society, 1993; 125-132.

127   Beveridge TJ. In: Ehrlich HL, Holmes DS, eds. Biotechnology and Bioengineering Symposium No 16, New York: John Wiley Interscience, 1986; 127-140.

128   Macaskie LE. J Chem Technol Biotechnol 1990; 49: 357.

129   Macaskie LE. Science 1992; 257: 782.

Biotransformations: Bioremediation Technology for Health and Environmental Protection
V.P. Singh and R.D. Stapleton, Jr. (Editors)

Lactic acid bacteria in winemaking: Influence on sensorial and hygienic quality

A. Lonvaud-Funel

Faculté d'oenologie – Université Victor Segalen Bordeaux 2, 351, Cours de la Libération – 33405 Talence Cedex, France

INTRODUCTION

At the beginning of the 20th century, after the famous « Etudes sur le vin » of Louis Pasteur, various observations showed that several months after alcoholic fermentation, wine acidity spontaneously decreases. This phenomenon was attributed to the transformation of malic acid into lactic acid [1]. At that time, the reason for such change was not clearly established. However, all the defects due to microorganism development were accurately described. Although not really understood, bacteria were shown to be responsible for wine diseases: *"tourne"* — the degradation of tartaric acid, *"bitterness"* — the degradation of glycerol, and *"ropiness"* — the unacceptable increase in wine viscosity. Further studies demonstrated that lactic acid bacteria were the spoilage agents.

On the other hand, the works of Ferre in Burgundy and Ribéreau-Gayon and Peynaud in Bordeaux showed the interest in growth and metabolism of lactic acid bacteria during winemaking, particularly in the step named malolactic fermentation [2]. During this second stage of vinification, following alcoholic fermentation, the main event is the deacidification of wine caused by decarboxylation of L-malic acid. However, many other wine components are metabolized from time to time, depending on conditions. Some are already known such as citric acid or some amino acids, while others are only supposed such as precursor aromas. Some of these metabolites are desirable or undesirable, both from the health and the sensorial point of view.

The increasing interest of wine microbiologists for lactic acid bacteria and especially malolactic fermentation led to extensive work in the 1950-1960s. Bacteria were isolated, then classified and the main growth parameters of enological significance were studied. Temperature, pH, and ethanol content were found to be decisive for the development of bacteria in wine. During vinification, these parameters are at the limit of tolerance of the bacteria and their negative (or positive) effects are additive. Winemakers know that at a temperature of about 18-22°C and a pH above 3.3-3.4, malolactic fermentation will normally start soon after alcoholic fermentation. The latent phase between both fermentations must be as short as possible since wine remains subject to spoilage by different kinds of yeasts and bacteria if it is not stabilized. Sulfur dioxide addition is used to prevent chemical oxidation as well as proliferation of *spoilage microorganisms*, but it cannot be used before malolactic fermentation is totally completed.

In the 1960s, while malolactic fermentation was finally recognized as a necessary step of vinification for almost all red wines and most white wines, increasing knowledge on lactic acid bacteria helped winemakers to control the process. However, very soon it appeared that difficulties often occurred and could not be explained. In spite of optimized physico-chemical factors, such as temperature and pH, bacteria might not grow enough to induce malolactic fermentation. Today, despite much progress in wine lactic acid bacteria, it is still impossible to explain all the cases of recalcitrant malolactic fermentation.

During the early 1970s, the first attempts to inoculate wine with selected strains of lactic acid bacteria were readily given up because of too many drawbacks. It was obvious that strains isolated from fermenting wines lost some of their adaptive capacities when cultivated in laboratory or industrial conditions. They could not survive when added to wine; within a few hours more than 95% of the viability was lost and never retrieved. Meanwhile, the demand of winemakers for better control of malolactic fermentation by inoculation increased, while at the same time the first yeast starters came on the market to improve alcoholic fermentation. Great progress was made when it was found that a preliminary *"reactivation"* of lactic acid bacterial cultures used as starters enhanced survival after wine inoculation [3]. Reactivation, which consists in incubation of the culture for 24-48 hours in a medium composed of grape juice and yeast extract, proved to be satisfactory and efficient for several bacterial strains, and the first malolactic starters were marketed. Even if they represented real progress, such starters were difficult to use in cellar conditions.

More advances in basic research focused on physiology in stress conditions recently led to the production of ready-to-use malolactic starters [4]. Today it is possible to induce malolactic fermentation in most wines with selected strains. The challenge now for these starters is to develop resistance to low pH for very acidic wines and to evaluate their incidence on typicity and the development of aromas. When malolactic fermentation is completed, vinification is over and the growth and survival of bacteria must be prevented. Sulfur dioxide or if necessary physical treatments, such as filtration of heat treatment, can eliminate lactic acid bacteria when they are undesirable.

Thus it seems that winemakers now have the appropriate tools to promote or prevent microorganisms in order to enhance wine quality. However, many defects are still observed. Conventional microbiology has given us the main knowledge but it is now the era of molecular studies; they will give the sound basis to take advantage of bacterial growth and avoid the undesirable side effects of their intervention.

DIVERSITY OF LACTIC ACID BACTERIA IN GRAPE MUST AND WINE

The classification of enological lactic acid bacteria was established by using the conventional methods, which focus on phenotypes. According to morphological, biochemical and physiological properties, wine and grape lactic acid bacteria can be classified as lactobacilli and cocci, being either obligate and facultative heterofermentative or homofermentative.

There are Gram-positive non-motile, non-spore forming bacteria which mainly produce lactic acid from fermentation of carbohydrates. The stereoisomeric form of L- or D-lactic acid, the presence or not of other products from glucose fermentation and the ability to ferment various carbohydrates are the keys used in classification using the Bergey's manual [5]. With this method wine lactic acid bacteria can be classified as shown in Table 1.

However, the most important species and the only one with real enological interest is *Oenococcus oeni*. These bacteria were formerly named *Leuconostoc oenos* and were classified according to their phenotype as heterofermentative cocci in the genus *Leuconostoc*. Their characteristic acidophily was recognized

Table 1
List of principal species of lactic acid bacteria of grape and wine

Lactobacilli

Facultative heterofermentative	*Lactobacillus plantarum, L. casei*
Obligate heterofermentative	*Lactobacillus hilgardii*
	Lactobacillus brevis

Cocci

Homofermentative	*Pediococcus damnosus*
	Pediococcus parvulus
	Pediococcus pentosaceus
Heterofermentative	*Leuconostoc mesenteroides*
	Oenococcus oeni

as an additional character [6]. Recently, phylogenetic studies have been used to classify lactic acid bacteria. Since it appeared that *Leuconostoc oenos* was very distant from the other lactic acid bacteria, so that a new genus was created in 1995, comprising one species *Oenococcus oeni* [7].

In grape must and during the early days of alcoholic fermentation, up to 8 species can be identified. However, at the end of alcoholic fermentation, the very dominant species is *O. oeni*. The other lactobacilli, *Leuconostoc* or pediococci have totally disappeared or at least cannot be detected among the high population of *O. oeni* [8]. A natural selection occurs due to the change in composition of the medium. As yeast ferments grape must, the ethanol level increases, as do other inhibitory yeast metabolites, such as fatty acids and probably others, which have not yet been identified [9]. Moreover, bacteria are transitorily starved of essential amino acids and other growth factors, which are preferably used by yeast. The complex interactions between yeast and bacteria explain why only the most resistant strains survive among the diverse initial bacterial microflora. This is the case of *O. oeni* strains in general and of some strains of heterofermentative *L. hilgardii* and *L. brevis* as well as homofermentative *P. damnosus*. Not all the strains of these species seem to adapt, but only some of them. The factors of adaptation which characterize these strains still need to be identified. However, pH is certainly the major parameter, which determines the survival of lactic acid bacteria in wine, irrespective of the species. The higher the pH, the more easily bacteria grow, despite the other inhibitory factors.

Usually during malolactic fermentation, only *O. oeni* strains can be identified at the level of 10^6 to 10^8 CFU.ml^{-1}. Other species are certainly present but at much lower concentrations. Thanks to molecular methods of DNA analysis, it is now known that several *O. oeni* strains are present. They can be characterized by restriction fragment length polymorphism (RFLP) after DNA restriction by rare cutting enzymes and separation of DNA fragments by pulse field gel electrophoresis (PFGE) (Figure 1). Analysis of phage-sensitive spectra and the restriction profile of prophage DNA have also showed diversity of *O. oeni* strains. The same work proved that due to diversity of bacterial and phage strains, it is almost impossible for malolactic fermentation to be stopped by a massive phage attack. The latter would be the case if only one or a few strains were involved, as in the dairy industry [11].

Finally, the total bacterial population rises and falls according to the phase of vinification. At each moment, strains disappear while others, which may be in the minority for a while, can adapt to the changing conditions and grow. While in the past this diversity was restricted to species, it is now possible using the powerful tools of molecular biology to affirm that wine always contains various strains of each species. This diversity is beneficial

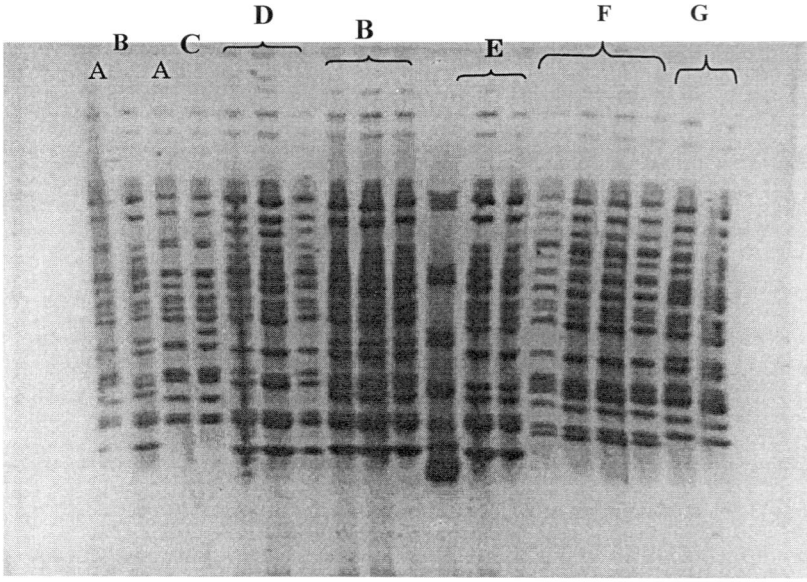

Figure 1. Genomic profile of *O. oeni* strains isolated from wine during malolactic fermentation. (DNA restriction by *Not*I and PFGE [10]).

since a more efficient population can replace a falling one. This is also the reason for the complexity of the unknown transformations of aromas, which are so important in wine typicity. On the other hand, there are drawbacks due to the presence and development of undesirable strains.

Therefore, the intervention of lactic acid bacteria in wine is a balance between the improvement and depreciation of quality. In most cases this intervention is favourable and is promoted by winemakers. The rightful demand of consumers for food and beverages of high quality has now forced wine microbiologists to study more thoroughly the minor metabolism of lactic acid bacteria in addition to malolactic transformation, being the most important one.

POSITIVE EFFECTS OF THE GROWTH OF LACTIC ACID BACTERIA IN WINEMAKING

Lactic acid bacterial growth (*O. oeni*) is needed after alcoholic fermentation to achieve malolactic fermentation. Many wine components are metabolized by *O. oeni* and their variety depends on the strains. However, the common trait is that all strains degrade L-malic acid, citric acid, and residual sugars, which have not been fermented by yeast. Depending on the strain, amino acids are also used. The most striking effect is the deacidification produced by decarboxylation of L-malic acid into L-lactic acid (Figure 2).

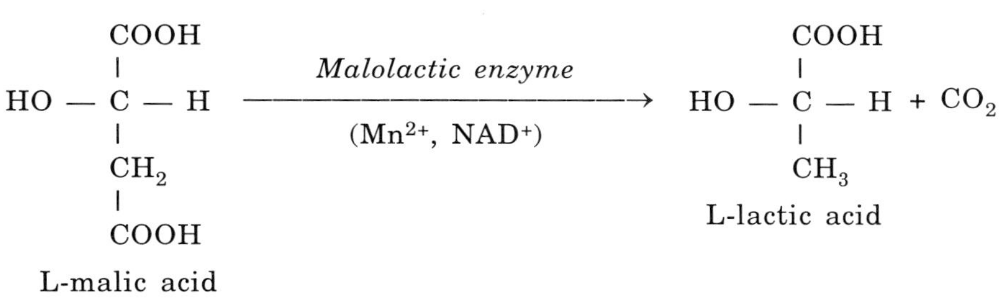

$$
\begin{array}{ccc}
\text{COOH} & & \text{COOH} \\
| & \textit{Malolactic enzyme} & | \\
\text{HO} - \text{C} - \text{H} & \xrightarrow{\hspace{3cm}} & \text{HO} - \text{C} - \text{H} + CO_2 \\
| & (Mn^{2+},\ NAD^+) & | \\
\text{CH}_2 & & \text{CH}_3 \\
| & & \text{L-lactic acid} \\
\text{COOH} & & \\
\text{L-malic acid} & &
\end{array}
$$

Figure 2. The malolactic reaction.

Depending on the initial L-malic acid concentration, pH is on average higher by 0.1 or 0.2 units after malolactic fermentation. Accordingly on the production area, from 2 to 10 g.L^{-1} of malic acid is present after alcoholic

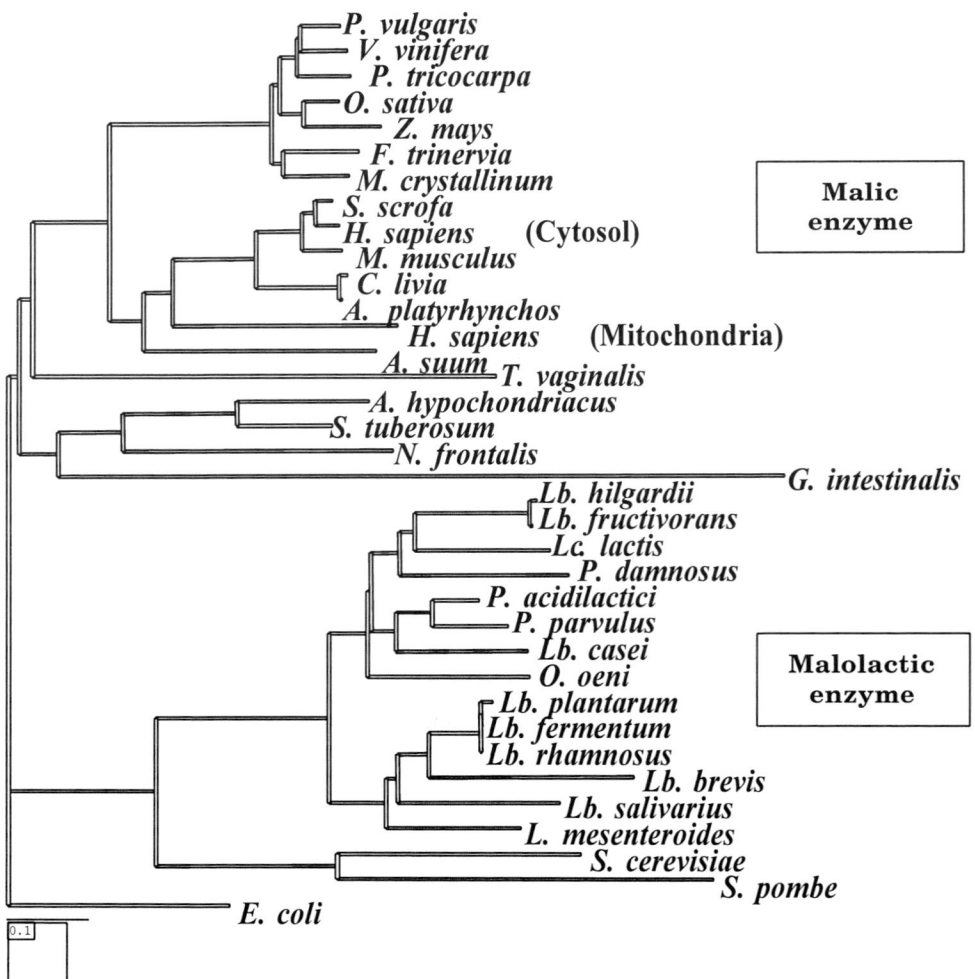

Figure 3. Fitch distance matrix tree of amino acid sequence of malolactic and malic enzymes.

fermentation and it is totally degraded. The result is a loss of the total acidity of wine. In addition, the replacement of the strong green taste of malic acid by the softer lactic acid is the main justification for malolactic fermentation.

L-malic acid is decarboxylated by the malolactic enzyme which was purified from *Lactobacillus plantarum* [12,13] and afterwards from many other lactic acid bacteria, including non-wine lactic acid bacteria. The gene *mleS* of *Lactococcus lactis* and *O. oeni,* which encodes this protein, was then sequenced [14,15]. The deduced amino acid sequence of the *mleS* gene isolated from 13 different species of lactic acid bacteria showed a high amino acid sequence similarity of the malolactic enzyme with the malic enzyme of many other organisms, including microorganisms [16]. Curiously, the malic enzymes of *Schizosaccharomyces pombe* and of *Saccharomyces cerevisiae,* which are wine microorganisms, were similar to the malolactic enzyme (Figure 3).

However, at least for its primary structure, the malolactic enzyme is very close to the malic enzyme in spite of a fundamental difference in the reaction. Indeed, all malic enzymes produce pyruvate by an oxidative decarboxylation, using NAD^+ or $NADP^+$ as coenzyme, while malolactic enzyme produces L-lactate. Nevertheless, the malolactic enzyme also needs NAD^+ as coenzyme, but it is not reduced. Its exact role in the reaction remains unclear. It might participate in the assembly and configuration of the functional protein.

The malolactic activity of the whole bacterial cell is strictly dependent on the integrity of the cell membrane. Only viable cells, which can maintain optimized conditions of pH (optimum pH 5.7), cofactors and the absence of inhibitors inside are able to decarboxylate malic acid. The membrane must fulfill its biological function of a barrier between cytoplasm and wine, which is acidic (average pH ranges from 3.0 to 3.6) and contains strong inhibitors for the malolactic enzyme, such as organic acids and polyphenols.

The membrane is also the site where the malolactic fermenting bacteria can retrieve energy. Malolactic reaction implies an influx of L-malate and an efflux of lactate which, together with the alkalinization of the internal medium due to decarboxylation, creates a proton motive force, used by the membrane ATPase to produce energy [17]. Therefore, even if the malolactic reaction at the enzyme level itself is non-energetic, the transformation provides energy to the cell. This is probably the best explanation for the fact that malic acid enhances bacterial growth. Moreover, the winemakers' conviction that malolactic fermentation helps to stabilize wine is supported. The growth of a high biomass of O. oeni deprives the medium of various energy-providing substrates, one of which is malic acid.

The other important biochemical transformation is degradation of citric acid. Its concentration is much lower than that of malic acid. The optimum is about 250 to 350 mg.L^{-1} after alcoholic fermentation but the impact on growth of lactic acid bacteria and wine quality is obvious. Citric acid is always metabolized by O. oeni during malolactic fermentation but not so fast

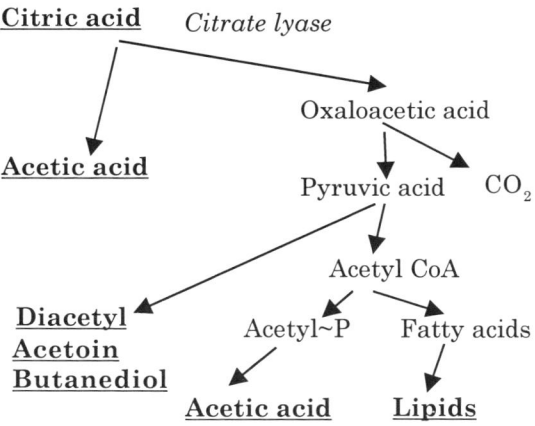

Figure 4. Schematic pathway of citric acid metabolism by O. oeni.

as malic acid. When malic acid has totally disappeared, a few dozens of mg. L^{-1} of citric acid always remain. The pathway is shown in Figure 4.

The interest of this metabolism is two-fold. It generates ATP from acetyl phosphate, and produces acetoinic compounds and lipid precursor acetyl CoA from pyruvate. This means that part of the citric acid is used for growth while the other part leads to aroma compounds, especially diacetyl, which have the strongest flavour influence. According to the conditions of the medium, pyruvate is preferentially funnelled toward lipid synthesis when pH, temperature, etc. are favourable to growth or toward the acetoinic compounds as a detoxifying process under unfavourable conditions [18].

While it is produced, diacetyl is also reduced to acetoin by the diacetylreductase participating in the regeneration of the oxidized cofactor. This enzyme exists not only in bacteria but also in yeasts. Therefore, an extended contact of yeast and lactic acid bacteria lees in wine after vinification lowers the final diacetyl concentration. On the contrary, early racking or clarification of wine enhances its participation in wine aroma [19]. Diacetyl increases the complexity of wine giving it a buttery taste, which is appreciated by most consumers. The average thresholds are 4.5-9.5 mg.L^{-1} in white wines and 12-14 mg.L^{-1} in red wine. Concentrations higher than these values are not appreciated by all people.

In addition, acetic acid is another obligate product of citric acid metabolism. More than one molecule is produced from one citrate molecule. This is particularly important in vinification since high acetic acid concentrations are prejudicial to wine quality. Therefore, throughout wine making, acetic acid production (volatile acidity) must be minimized. Citric acid degradation, which always occurs during malolactic fermentation, participates inevitably in the increase of volatile acidity. However, this involvement is moderate, since the initial concentration of citric acid is low.

Some workers have suggested the use of special strains, which would not degrade citric acid in order to lower the volatile acidity. However, it is not recommended since the participation of diacetyl to wine taste is needed and, on the other hand, the persistence of citric acid in wine makes it microbiologically less stable. Overall it can be considered that, under usual conditions, the metabolism of citric acid is beneficial to wine quality.

Other sensorial descriptions of wine aroma and taste are used when comparing wine before and after malolactic fermentation (Figure 5) [20]. The profile aroma is certainly changed during the process; however, except for diacetyl, no relationship has been established yet with the chemical

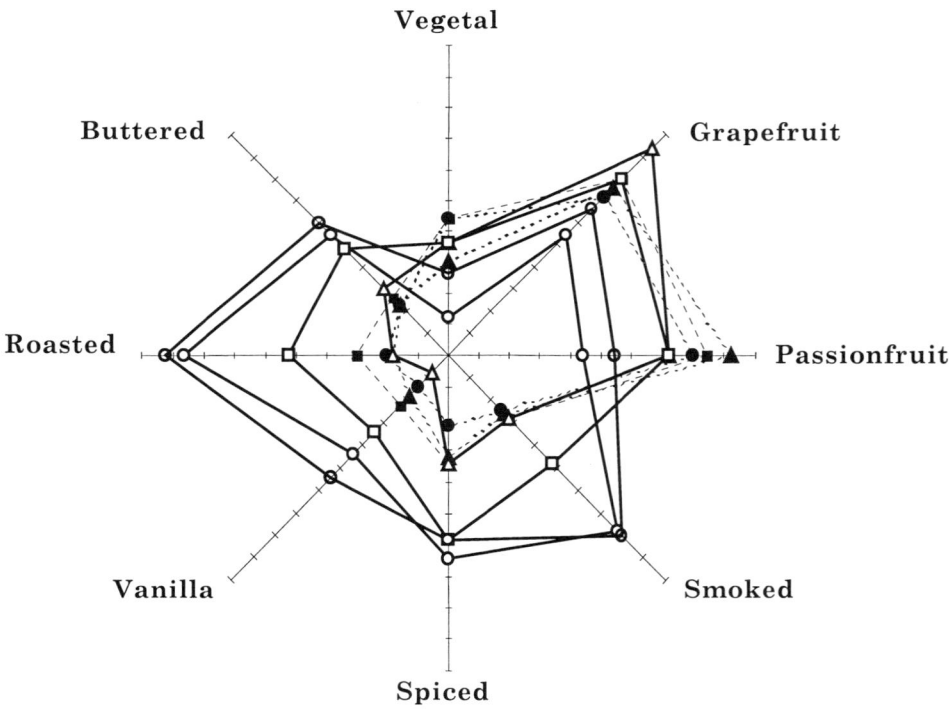

Figure 5. Differentiation of wines before and after malolactic fermentation in different tanks [20].

composition of wine [21]. Roughly malolactic fermentation enhances complexity and fruitiness, while it lowers vegetative aromas. Its impact depends probably on the predominant *O. oeni* strains as well as on the grape variety and on the enological process chosen by winemaker, which defines the type of wine produced. If malolactic fermentation is conducted in barrels, wood components are in higher concentration [20].

Finally another recognized role of malolactic fermentation concerns red wine colour, which becomes darker and more stable. Astringency decreased by the reaction between tannins and anthocyanins which is favoured. Some phenolic compounds precipitate due to structural changes. However, the role of *O. oeni* is completely unknown in the mechanisms involved.

UNDESIRABLE PRODUCTS OF LACTIC ACID BACTERIA IN WINE

Even if winemakers promote growth of lactic acid bacteria to achieve complete malolactic fermentation, they must sometimes face various problems due to the development of undesirable strains or to the growth conditions of normal strains. Undesirable products of lactic acid bacteria are either metabolites, which lead to sensorial defaults such as acetic acid, bitterness, ropiness, and others which are poorly understood, or to components which may represent health-risk compounds, such as acrolein, biogenic amines, and ethyl carbamate.

Until recently, it was thought that wine lactic acid bacteria could be separated into good and bad bacteria. This classification was very simple and *O. oeni* was recognized to be the only desirable species to be promoted. Today the situation has evolved somewhat different. *O. oeni* is still the enological lactic acid bacterium par excellence. However, some strains give poor results and even the best strains may cause wine spoilage, if their growth is not well controlled. Moreover, recent investigations showed that some *O. oeni* could be prejudicial to wine quality due to their capacity to metabolize some amino acids.

Spoilage by lactic acid bacteria in general can be produced either by normal strains, which ferment too much sugar or by special strains, which are characterized by a particular metabolism not represented in the whole species.

Increase in volatile acidity

O. oeni and other species are heterofermentative bacteria, which ferment glucose, fructose, and the pentoses of wine to D-lactic acid, acetic acid, and other minor secondary products. The acetic acid concentration is the major concern of winemakers from harvest to bottling line. Normally *O. oeni* and possibly other species grow after alcoholic fermentation, when yeasts have fermented sugars (hexoses). Nevertheless, at this stage some hundred mg.L^{-1} of hexoses and pentoses are available for lactic acid bacteria which use them together with other energy and carbon sources, to increase their population from about 10^3 CFU.ml^{-1} to 10^6 CFU.ml^{-1} and higher. They produce acetic acid, so the volatile acidity of wine increases but normally in very limited amounts.

In some cases, growth of lactic acid bacteria starts near the end of alcoholic fermentation but before the end of total sugar exhaustion. In such conditions, it is due to favourable conditions, the main one being high pH. The concentration of sugars fermented by lactic acid bacteria is substantial and the volatile acidity becomes unacceptable. Moreover, due to early growth of bacteria, the competition between declining yeast and bacteria turns in favour of bacteria, and yeasts are inhibited. Alcoholic fermentation slows down and sometimes stops, leaving sugar available for bacteria. This is one of the worst incidents during winemaking. It is called *"piqûre lactique"* because the volatile acidity produced by lactic acid bacteria reaches high levels [22].

O. oeni is responsible for piqûre lactique in table winemaking. However, this problem is not rare either in *"fortified"* wines (Port, Sherry, Pineau, etc.) made by the addition of Brandy or Cognac to unfermented or slightly fermented grape must. In this case, the strains of several species such as *O. oeni, L. hilgardii,* and *L. fructivorans,* which have an abnormal ethanol tolerance, survive and grow, thus fermenting sugars and producing volatile acidity.

Production of biogenic amines

Biogenic amines result from decarboxylation of some amino acids. In wine histamine, tyramine, and diaminobutane are the most important and they are present in variable amounts from zero to several mg.L^{-1}. Due to their physiological activity and their possible toxic effects in human metabolism, they are more and more studied nowadays in wine and foods. In general, at the levels encountered in wine, there is no risk for consumers. However, if high amine concentrations are ingested or if the consumer's detoxification system is defective, they can produce hypotension, digestive disturbance and other allergic symptoms.

So far no regulations limit their concentration in wine, but some rough guidelines have been established in commercial exchanges. Therefore, it is important to know more about their presence in wines. Table 2 gives some idea of the concentration of the main amines in different kinds of wines.

Table 2
Concentration of biogenic amines (mg.L^{-1}) in some French wines [23]

Amine	Alsace	Bordeaux	Burgundy	Champagne
Histamine	3.12 ± 0.76	4.91 ± 1.02	9.74 ± 1.26	10.78 ± 1.16
Putrescine	1.95 ± 0.63	4.03 ± 0.68	9.26 ± 2.96	6.83 ± 0.95
Cadaverine	3.41 ± 0.91	0.88 ± 0.28	1.41 ± 0.52	3.82 ± 0.88
Tyramine	6.10 ± 1.59	7.31 ± 1.03	9.68 ± 1.23	13.70 ± 2.36

These amines are respectively produced from histidine, ornithine, lysine, and tyrosine by the corresponding decarboxylase. So far histamine has been the best studied. For a long time, enologists considered that it was mainly produced by undesirable *Pediococcus* strains and that only wines produced in poor conditions of vinification contained histamine and other amines. However, for others the amines were strictly correlated with the completion of malolactic fermentation [24]. An extensive survey of wines finally showed that even *O. oeni* could produce histamine [25]. The highest levels of histamine are reached in the poorest nutritional conditions. Indeed, in such conditions, when the medium is deprived of better nutrients, bacteria can probably derive some energy from this metabolism. In *Lactococcus lactis*, histidine decarboxylation has been shown to provide ATP to the membrane level [17].

In reality, not all *O. oeni* strains can decarboxylate histidine. This depends on the histidine decarboxylase (HDC) activity which is related to the expression of the *hdc* gene. Investigations on the enzyme, cloning and then sequencing of the *hdc* gene of an *O. oeni* strain gave very useful results [26].

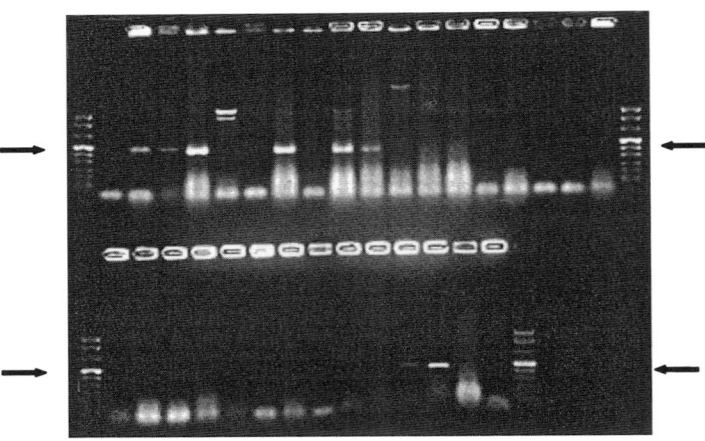

Figure 6. Detection of histamine-producing strains in wine by PCR [26]. The arrows indicate the specific amplicon.

First it was found that some *O. oeni* strains carry the gene, while it is totally absent from others. In addition, for all the strains studied, when the gene was present the strain was active. This naturally led to the design of two kinds of detection tests. The *hdc* gene is used as a DNA probe for DNA/DNA hybridization to assess its presence in bacteria. Hybridization is conducted on colonies in order to appreciate the percentage of histamine-producing strains in a population. To detect low levels of such strains, a PCR test is possible using primers selected from the sequence of the *hdc* gene (Figure 6). Moreover, the primers and the DNA probe may hybridize on histamine-producing strains of other species.

HDC is very stable in the bacterial cell. It is conserved even during the decline phase, while the viable cell population disappears. Thus histamine can accumulate in wine during storage as long as the remaining population retains some activity.

Tyramine, together with histamine, is thought to cause headache and other minor complaints. Its production has been less studied than histamine, and no bacterial tyrosine decarboxylase (TDC) is really known. From wines containing biogenic amines, bacterial strains were isolated, which were able to produce tyramine. So far no *O. oeni* strains have been identified; only *L. brevis* and *L. hilgardii* have been characterized [27]. Some simultaneously produced tyramine and phenylethylamine. Unlike histamine, tyramine is produced in larger amounts when growth conditions are not restrictive. Preliminary results have shown that the TDC of wine lactic acid bacteria is optimal at the end of the rapid growth phase. It decreases during the stationary phase.

After vinification, even if bacteria survive sulfur dioxide addition, in most cases they cannot grow but remain in such a physiological state that they are not capable of high activity. Thus, tyramine production may be limited during wine storage. TDC of wine lactic acid bacteria is very unstable and its purification has not been achieved. Therefore, up till now there is no easy rapid molecular method to detect tyramine-producing strains in wine unlike histamine-producing bacteria.

In general, wines contain either no biogenic amine or all of them (histamine, tyramine, and diaminobutane). Their concentration often increases even after malolactic fermentation. HDC is specific for its substrates and this also seems to be the case for TDC. No result is available yet for wine bacteria on ornithine decarboxylase (ODC). Therefore, this suggests that in such wines a variety of bacteria grow and survive even after malolactic fermentation, and/or that some strains carry several decarboxylase activities producing all the amines. Field observations show that such wines indeed carry a great variety of species and strains. These are generally wines with relatively high pH, above 3.6 before malolactic fermentation (MLF). This confirms the predominant role of acidity in the selection of bacterial microflora.

Another usual finding is that in wines of certain cellars, biogenic amines are always present while in other cellars, they are always absent. The indigenous microflora can be very different from one area to another and it may be that some parameters select the undesirable bacteria.

Besides the nature of bacteria, the prerequisite for the production of high amine levels is the availability of amino acid precursors. Amino acids of wine come from grape must, but above all are determined by yeast. After alcoholic fermentation, amino acid concentrations are enhanced during yeast autolysis. Moreover, an extended contact of wine with yeast lees, as is customary in some wine-producing areas, enriches wine not only in free amino acids but also in peptides and proteins. Lactic acid bacteria use these substrates directly or after hydrolysis by proteolytic activities. This is why wines stored with lees for beneficial purposes, can contain very large concentration of amines.

Since lactic acid bacteria are necessary to obtain malolactic fermentation, winemakers must find a solution for minimizing amine levels, even if indigenous bacteria possess decarboxylating activities. Today, a solution is the use of selected malolactic starters. *O. oeni* strains, free of undesirable activities, are produced and lyophilized in concentrated populations [4]. They are massively added to wine to complete malolactic fermentation. They normally overcome the indigenous bacteria and seem to eliminate them by competition. Then after malolactic fermentation, wine is readily stabilized by sulfur dioxide which eliminates the malolactic starter and the residual wild-type strains. The result on the *biogenic* amine concentration is striking as shown in Table 3.

Table 3
Concentration of biogenic amines $(mg.L^{-1})$ in wines and the same wines inoculated by a malolactic starter (initial concentrations before malolactic fermentation: not detectable)

	Histamine		Tyramine		Diamino-butane	
Time after malolactic fermentation (days)	1	15	1	15	1	15
Wine A	3.6	4.8	3.9	4.3	4.3	6.3
A + starter	ND	ND	ND	ND	ND	ND
Wine B	1.9	2.5	3.0	3.0	3.0	5.0
B + starter	0.5	ND	ND	ND	0.2	ND
Wine C	2.0	4.2	3.3	3.3	3.6	4.6
C + starter	0.6	ND	ND	ND	ND	ND

ND: Not detected.

Bitterness of wine due to glycerol metabolism of lactic acid bacteria

Among the *"wine diseases"* described by Pasteur, one was named bitterness because of the unacceptable taste produced. This was correlated with the presence of acrolein [28], which results from a particular metabolic pathway of glycerol in *Lactobacillus* [29,30]. In fact, glycerol produces hydroxy-3-propionaldehyde (3-HPA), a precursor of acrolein in acidic conditions and in heat treatment. Bitterness results from the combination of polyphenols with acrolein. Normally, lactic acid bacteria can assimilate glycerol by glycerol dehydrogenase to dihydroxyacetone (DHA) and dihydroxy-acetone phosphate (DHAP), which finally reaches the glycolytic pathway (*"oxidative branch"*). However, another enzyme, glycerol dehydratase, is present in some strains. It produces 3-HPA which is mainly reduced to 1,3-propanediol, the *"reductive*

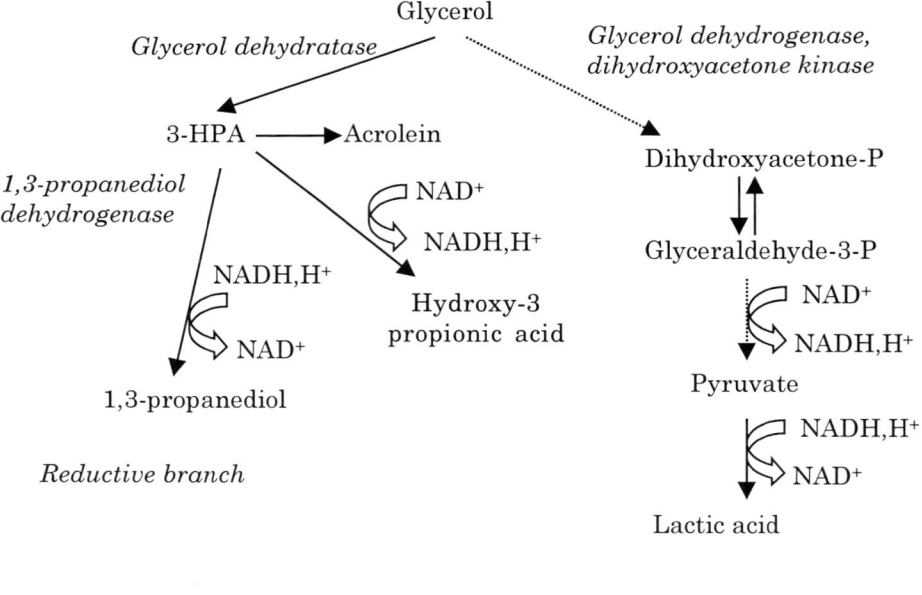

Figure 7. Metabolic pathways of glycerol in bacteria.

branch" [30]. Glycerol metabolism has been intensively studied in other bacteria (*Klebsiella aerogenes, Klebsiella pneumoniae, Enterobacter agglomerans, Citrobacter fruendii,* and *Clostridium butyricum),* which assimilate glycerol both by the "reductive" and the "oxidative" branches characterized by glycerol dehydrogenase (GDH) and glycerol dehydratase (GD) (Figure 7). When the four determining enzymes are active, the bacteria can grow anaerobically with glycerol as sole carbon source, since coenzymes are regenerated by one branch or the other. The genes encoding the four determining enzymes constitute the *"dha regulon".* The entire *"dha* regulon" of *Klebsiella pneumoniae* was expressed in *E. coli* [31].

When acrolein is produced, it means that some 3-HPA escapes the reduction to 1,3-propanediol. Accumulation of 3-HPA by *Enterobacter agglomerans* is intensively studied for industrial applications [32]. In some *Lactobacillus collinoides* strains isolated from bitter ciders, glycerol is metabolized only by glycerol dehydratase and gives 1,3-propanediol as the main product. Very small amounts of acrolein are determined corresponding to the escape of 3-HPA to reduction by 1,3-propanediol dehydrogenase. Anaerobiosis is necessary for this metabolism [33]. As a result of investigations on *L. collinoides* of ciders, a DNA probe and GD1/GD2 primers for PCR amplification have been isolated which allow the identification of acrolein-producing bacteria. An interesting result is that only DNA extracted from acrolein-producing strains hybridizes with the probes or the primers. There is a strict relationship between the phenotype and the result of detection (Figure 8).

Using DNA/DNA hybridization and PCR, glycerol-degrading bacteria were isolated from wines which contained acrolein. They were separated in two groups. One includes strains which used glycerol and produced acrolein, 1,3-propanediol, and 3-hydroxypropionic acid, which can result from oxidation (instead of reduction) of 3-HPA. In addition, these strains produced lactate,

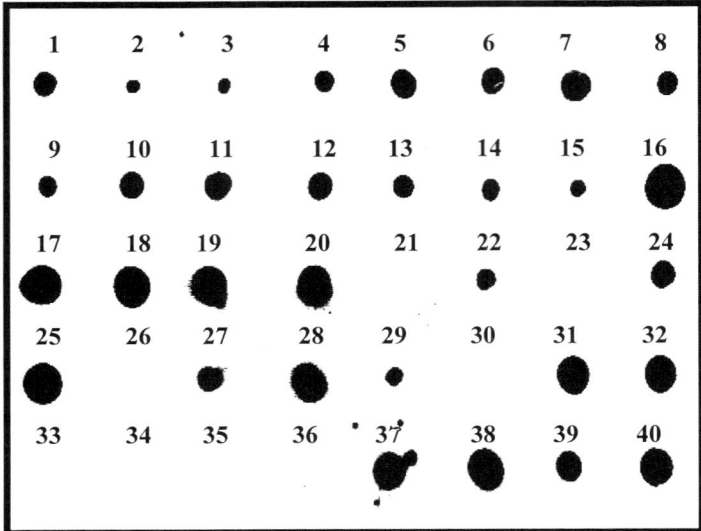

Figure 8. Detection of acrolein-producing strain by the glycerol dehydratase probe. DNA extracted from 40 strains isolated from cider was spotted. All are *L. collinoides* except 19 to 21 (*L. hilgardii*) and 30 (*L. mali*) . Strains 21, 23, 26 30, 33 to 36 were unable to degrade glycerol.

ethanol and acetate, supporting the existence of the GDH activity. They were identified as *L. hilgardii* and might contain the total *dha* regulon. The other group constitutes strains, which probably only have glycerol dehydratase activity, since only 1,3-propanediol, 3-hydroxypropionic acid, and acrolein were produced. These strains are heterofermentative lactobacilli close to *L. hilgardii* according to the 16S rDNA gene sequence but their DNA do not hybridize with a specific *L. hilgardii* DNA probe. Their metabolism differs from *L. collinoides* in the production of relatively high amounts of 3-hydroxypropionic acid. Therefore, 3-HPA-producing strains of wine belong to two different species distinct from *L. collinoides*.

In practice, bitterness is not very widespread in wine. It seems to be more frequent in cider. However, it can compromise the quality of wine due to the bad taste produced and the disappearance of glycerol, which is one of the principal components of wine and plays an important role on sensorial quality. For wines distillated to make Cognac or brandies, the issue is more important since acrolein is easily produced during the distillation process from 3-HPA. The product is depreciated not only because of the presence of undesirable carcinogenic acrolein but also because the taste is altered.

Production of ethylcarbamate (urethane) by lactic acid bacteria in wine

Urea produced by yeast from arginine during alcoholic fermentation and citrulline and carbamyl phosphate produced by lactic acid bacteria by the arginine deiminase (ADI) pathway, are the precursors of ethylcarbamate in wine (Figure 9).

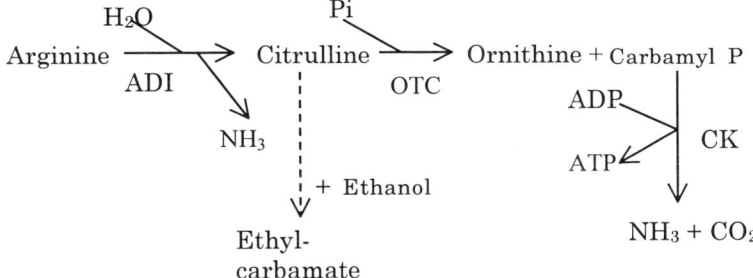

Figure 9. Reactions of the arginine deiminase pathway in lactic acid bacteria. ADI: arginine deiminase; OTC: ornithine transcarbamylase; CK: carbamate kinase.

Ethylcarbamate concentrations increase with time in the presence of ethanol, a process accelerated by high temperature. Its carcinogenic properties have been shown in laboratory experiments on animals [34]. It is naturally present in several kinds of food and beverages. In the quest for healthier and better food, this component might be considered in future as a marker of quality, and regulations governing its levels might be established. Winemakers are of course concerned, and recently the addition of urease to wine was permitted to minimize ethylcarbamate from yeast metabolism.

According to the previous classification of wine lactic acid bacteria [5], only *L. hilgardii* are able to use arginine by the ADI pathway. *O. oeni* was described as a non arginine-degrading species. In fact most, but not all *O. oeni* strains metabolize arginine by this pathway, leading to ethylcarbamate precursors [35]. For example, in a collection of 14 strains isolated from wine, 11 degraded arginine releasing ornithine, citrulline, and NH_3 in the medium with a substantial increase in pH. This metabolism provides ATP to bacteria and the comparison between arginine-degrading and non-degrading strains proved that the first could survive nutrient limitations better than the second and grew better in media arginine [36]. Arginine deiminase (ADI), ornithine transcarbamylase (OTC) and carbamate kinase (CK) are coded by three genes organized as an operon, as shown in Figure 10.

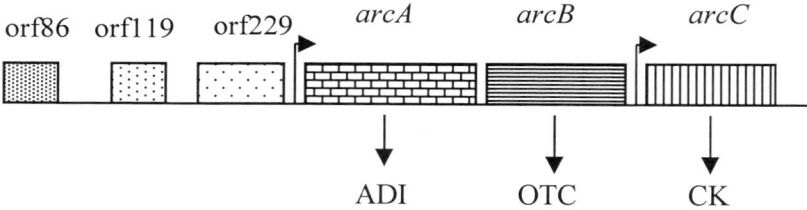

Figure 10. The *arcABC* gene cluster of the arginine deiminase pathway of *Oenococcus oeni* [37].

No relevant information is deduced from the sequence upstream *arcA*, except for orf229 which might be involved in transcription regulation. The gene encoding the arginine transporter described in other bacteria and notably in *Lactobacillus sakei* [38] has not yet been found for *O. oeni*.

Specific primers and probes have been obtained from DNA sequences of *arcA*, *arcB*, and *arcC*. They were used to detect the genes in a collection of *O. oeni* strains. All bacteria analyzed carried the three genes, including those which were unable to assimilate arginine. The phenotype is not dependent on the presence of these genes. The presence or absence of a gene transporter and/or regulatory genes might determine it. Therefore, when selecting *O. oeni* strains for preparation of starter cultures, a simple test, based on the detection of genetic determinants, is not yet available. The candidate strains must be assayed in optimum culture conditions for arginine degradation, and the substrate and the products of the metabolism must be monitored.

L. hilgardii is known for its capacity to use arginine according to the ADI pathway. It is one of the main criteria for the classification of these strains. In wine, this species usually disappears during alcoholic fermentation. However, high pH favours its survival after malolactic fermentation and it might be involved in the production of ethylcarbamate precursors. This is especially the case in spoiled fortified wines where *L. hilgardii* is the predominant species. In these wines the concentration of volatile acidity increases as a result of heterofermentation of hexoses as the concentration of citrulline does from arginine metabolism [39].

Production of an exopolysaccharide by bacteria: "ropy" wines

"Ropy" wines are excessively viscous and, therefore, cannot be marketed. Except for abnormal viscosity, they normally have no other defects such as volatile acidity or characterized *off-flavours*. Ropiness occurs by accumulation of an exopolysaccharide (EPS). With time the bacterial biomass sediments in a sticky deposit and the wine flows like oil. Even if it can occur in tanks, it is more often encountered in bottles. The bacterial biomass expands very slowly and the wine usually becomes viscous several months or years after bottling.

The lactic acid bacteria population of these wines is always very high 10^5-10^6 CFU.ml^{-1}, and so far only strains of *Pediococcus damnosus* have been isolated. Repetitive sub-cultures of these special strains in the laboratory lead to loss of the "ropy" phenotype, except if the medium contains 8-10% ethanol. During subcultures the wild strains are changed to variants, which are no longer ropy. The explanation for this is the loss of a plasmid, to which the phenotype is strictly correlated [40]. This occurs frequently when they are cultured without ethanol in the medium. Ethanol and, perhaps, other stress factors exert a selective pressure to maintain the plasmid-carrying strains. The polysaccharide produced is a glucan (homopolymer of glucose) composed of a β 1→3 chain with ramifications of one glucose branched in β 1→2 [41].

The plasmid sequence suggests the presence of a gene encoding a glucosyl-transferase which could be involved, at least in part, in EPS synthesis (unpublished results). Not all *P. damnosus* strains carry this plasmid, and common non-ropy *P. damnosus* strains are relatively widespread in the natural inoculum of grape must and in wines. The phenotype might be a means for bacteria to adapt better to wine conditions. EPS synthesis might be a response of the cell to the presence of inhibitors and to *nutritional deprivation*, as more EPS seems to be produced under nitrogen starvation. This explains why this sort of spoilage is of great concern when it occurs in a given cellar, since it is easily transmitted to non-spoiled wines. Moreover, compared to the banal *P. damnosus* strains, ropy ones are much less sensitive to acidity and to sulfur dioxide which is used in wine stabilization. Thus, once the problem is detected in a cellar, rigorous sanitation procedures must be undertaken.

Early detection of ropy strains is now rapidly and easily performed by PCR detection. From the sequence of the plasmid, several pairs of primers were assayed for specific amplification of DNA of ropy strains. Finally a pair was chosen which gives reliable and relevant results. Moreover, a fragment of the plasmid including the glucosyl transferase (GT) gene has been selected and labelled to be used as a DNA probe. Like the first random "ropy probe" isolated [40], this "GT probe" is used in colony hybridization (Figure 11).

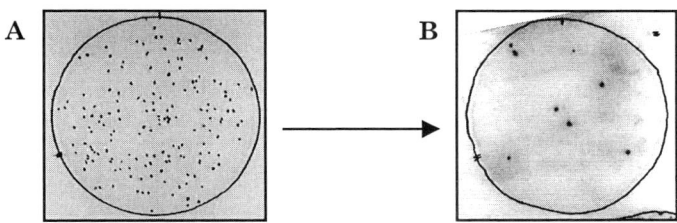

Figure 11. Detection of ropy *P. damnosus* strains in wine by colony hybridization [42]. A : *P. damnosus* specific DNA probe; B : Ropy *P. damnosus* DNA probe. Only ropy colonies hybridize.

Therefore, it is now possible to know whether ropy strains are present or not in a given wine. However, today it is almost impossible to precisely predict the fate of a wine which is contaminated with a low population (@ 10^3 CFU.ml^{-1}) of these *P. damnosus* strains. Years of observations have shown that in practice the disease, which appears after bottling, takes a very long time before making the wine undrinkable. The generation time of such bacteria in wine is very high. Therefore, when a PCR or a DNA probe detection test is positive, winemakers must treat the wine very carefully to eliminate even a low population. Filtration, which is normally performed just before the bottling, must be adjusted. However, since the cell size is often very small, another suitable means is heat-treatment.

Ropiness has become more and more frequent at least in some producing areas for recent vintages. There is no satisfactory explanation for this phenomenon. Perhaps the pH increase, which not only results from vine nutrition but also from over maturity of grapes at harvest, is one of the main reasons. Moreover, since EPS-producing strains seem to resist adverse conditions, a natural selection might occur in some cellars, where they are more and more represented.

FUTURE CONCERNS REGARDING WINE LACTIC ACID BACTERIA

Winemakers understandably consider lactic acid bacteria as essential in vinification. They are right as long as the population is composed exclusively of *O. oeni* strains. All the other species of *Lactobacillus, Leuconostoc,* and *Pediococcus* genera are not well-adapted to malolactic fermentation or are absolutely undesirable. However, before alcoholic fermentation occurs, a wide range of species and strains is present. The pH of the medium determines not only the concentration but also the nature of the bacteria. Today the general tendency is to harvest grapes at a more mature stage than in the past because the aroma precursors and red wine pigments are more abundant and extractable. Moreover, the health quality of grapes is now better controlled. This probably explains in part the general increase in grape must and wine pH. It corresponds to the preferences of consumers, but at the same time the medium contributes more to the growth of microorganisms.

As pH increases, bacteria and yeast grow more easily. Alcoholic fermentation is the natural mechanism of selection of lactic acid bacteria through the competition between yeasts and bacteria. It is essential to take advantage of this period to select *O. oeni* for malolactic fermentation The addition of sulfur dioxide on grapes or in must to prevent chemical and enzymatic oxidation lowers the lactic acid bacterial population in grape must and helps the selection of *O. oeni*. However, it is now established that several *O. oeni* strains grow and are responsible for malolactic fermentation. Yet some might be undesirable in view of the production of amines or ethylcarbamate precursors from amino acid metabolism. The reason for the presence and persistence of such strains in some cellars, but not in others located in the same area, is not yet explained. There are several hypotheses, but there is still insufficient accurate ecological data. This is probably one of the main research areas now concerning wine microbiologists.

The inoculation of wine after alcoholic fermentation by malolactic starters is now possible and is normally undertaken when malolactic fermentation does not start spontaneously. However, since efficient commercial cultures are now available, winemakers are more tempted to use them as soon as MLF is required. This has the advantage that in cellars, where the lactic microflora comprises undesirable *O. oeni* strains, the latter are eliminated by the massive addition of starters. In addition, sensorial analyses have shown that wines are different according to the starter used. Therefore, in the next near future starters manufacturers will select strains not only on basic enological criteria, such as adaptation to wine conditions and absence of undesirable phenotypes but also on their performance on taste and wine aromas. The wine industry has adopted the use of yeast (*"active dry yeasts"*) for controlling alcoholic fermentation and is now doing the same with *O. oeni* for malolactic fermentation.

The drawback is of course the possible lack of microbial diversity that this might induce. However, it is probably less risky than in other fermentation processes (as in dairy industry), since grapes carry yeasts and bacteria, which renew the microflora at each vinification. On the other hand, the *"standardization"* of wine by use of industrial starters is not a real risk. It is certainly detrimental for some wines produced in areas where yeasts mark the aroma too heavily. Concerning lactic acid bacteria, the impact of strains on sensorial quality is detectable but subtler than for yeast.

From good quality grapes, winemakers must optimize the fermentations. New starters can help them to do this. To preserve complexity and typicity, which are the bases of high quality wines, it is important to capitalize on the diversity of strains. We need to know more about the ecology of microorganisms living on grapes, how they survive on cellar equipment and how they grow in wines. Such studies were almost impossible in the past. Now the new molecular methods are opening new horizons for studying wine microorganisms, particularly the lactic acid bacteria.

REFERENCES

1   Seifert W. Z.Landwirstch.Versuchsu. Deut. Oest. 1901; 4: 980-992. 1901.

2   Lafon-Lafourcade S. In: Rehm HJ, Reed G, eds. Biotechnology. Verlag Chemie, 1983; 82-163.

3   Lafon-Lafourcade S, Carre E, Lonvaud-Funel A, Ribereau-Gayon P. Conn Vigne Vin 1983; 17: 55-71.

4   Nielsen JC, Prahl C, Lonvaud-Funel A. Amer J Enol Vitic 1996; 47:42-48.

5   Garvie EI. In: Sneath PHA, Mair NS, Sharpe ME, Holt JG, eds. Bergeys Manual of Systematic Bacteriology. Vol 2. Baltimore: Williams and Wilkins, 1986.

6   Garvie EI. J Gen Microbiol 1981; 127: 209-212.

7   Dicks LMT, Dellaglio F, Collins MD. Int J Syst Bacteriol 1995; 45: 395-397.

8   Lonvaud-Funel A, Joyeux A, Ledoux O. J Appl Bacteriol 1991; 71: 501-508.

9   Lonvaud-Funel A, Joyeux A, Desens C. J Food Sci Agric 1988; 44: 183-191.

10   Gindreau E, Joyeux A, de Revel G, Lonvaud-Funel A. J Int Vigne Vin, 1997; 31: 197-202.

11   Gindreau E, Joyeux A, Lonvaud-Funel A. In: Lonvaud-Funel A, ed. Oenologie 99. Paris: Technique et Documentation Lavoisier, 2000; 362-365.

12   Schutz M, Radler F. Arch Mikrobiol 1973; 91: 183-202.

13   Lonvaud M. Ph.D. Thesis, University of Bordeaux 2, 1975.

14   Denayrolles M, Aigle M, Lonvaud-Funel A. FEMS Microbiol Lett 1994; 116: 79-86.

15   Labarre C, Guzzo J, Cavin JF, Divies C. Appl Environ Microbiol 1996; 62: 1274-1282.

16   Groisillier A, Lonvaud-Funel A. Int J Syst Bacteriol 1999; 49: 1417-1428.

17   Poolman B. FEMS Microbiol Rev 1993; 12: 125-148.

18   Hugenholtz J. FEMS Microbiol Rev 1993; 12: 165-178.

19   Nielsen J, Prahl C. In: Lonvaud-Funel A, ed. Oenologie 99. Paris: Technique et Documentation Lavoisier, 2000; 317-320.

20   De Revel G, Martin N, Pripis-Nicolau L, Lonvaud-Funel A, Bertrand A. J Agric Food Chem 1999; 47: 4003-4008.

21   Henick-Kling T, Acree TE, Gavitt BK, Krieger SA, Laurent MH. In: Stockley CS, Lee TH, eds. Proceedings of the 8th Australian Wine Industry Conference, 1993.

22   Ribereau-Gayon P, Dubourdieu D, Doneche B, Lonvaud A. In: Ribereau-Gayon P, ed. Handbook of Enology. London: Wiley and Son Ltd, 1999; 129-148.

23   Zee JA, Simard RE, L'Heureux L, Tremblay J. Amer J Enol Vitic 1983; 34: 6-9.

24   Aerny J. Bull Off Intern Vin 1985; 657: 1016-1019.

25   Lonvaud-Funel A, Joyeux A. J Appl Bacteriol 1994; 77: 401-407.

26   Coton E, Rollan G, Bertrand A, Lonvaud-Funel A. Am J Enol Vitic 1998; 49: 199-203.

27   Moreno-Arribas V, Torlois S, Joyeux A, Bertrand A, Lonvaud-Funel A. J Appl Bacteriol 2000; 88: 584-593.

28   Voisenet E. CR Acad Sci 1910; 150: 1614-1616.

29   Sobolov M, Smiley KL. J Bacteriol 1960; 79: 261-266.

30   Schütz H, Radler F. System Appl Microbiol 1984; 5: 169-178.

31   Tong I.T, Liao HH, Cameron DC. Appl Environ Microbiol 1991; 57: 3541-3546.

32   Barbirato F, Soucaille P, Bories A. Appl Environ Microbiol 1996; 62: 4405-4409.

33   Claisse O, Lonvaud-Funel A. Food Microbiol (in press).

34   Mirvish SS. Adv Cancer Res 1968; 11: 1-42.

35   Mira de Orduna R, Liu S.Q, Patchett M.L., Pilone GJ. J Appl Microbiol 2000; 89: 547-552.

36   Tonon T., Lonvaud-Funel A. J Appl Microbiol 2000; 89: 526-531.

37   Tonon T, Lonvaud-Funel A. In: Lonvaud-Funel A, ed. Oenologie 99. Paris: Technique et Documentation Lavoisier, 2000; 238-241.

38   Zuniga M, Champonier-Verges M, Zagorec M, Perez-Martinez G. J Bacteriol 1998; 180: 4154-4159.

39   Hogg T, De Revel G, Couto J, Capela A, Pintado M. In: Lonvaud-Funel A, ed. Oenologie 95. Paris: Technique et Documentation Lavoisier, 1996.

40   Lonvaud-Funel A, Guilloux Y, Joyeux A. J Appl Bacteriol 1993; 74: 41-47.

41   Llauberes RM, Richard B, Lonvaud A, Dubourdieu D. D Carbohyd Res 1990; 203: 103-107.

42   Walling E, Gindreau E, Lonvaud-Funel A. Le Lait (in press).

Biotransformations: Bioremediation Technology for Health and Environmental Protection
V.P. Singh and R.D. Stapleton, Jr. (Editors)

Microbial transformation of aflatoxins

T. Shantha and M. Archana

Department of Food Microbiology, Central Food Technological Research Institute, Mysore - 570 013, India

INTRODUCTION

Microbes play an indispensable role in the life of all living systems, which include humans, animals, and plants. Centuries ago, great microbiologists like Pasteur, Jenner, and Fleming recognized their beneficial role in the field of medicine, pharmacology, fermentation, and agriculture. The pathogenic nature of certain species of fungi to plants, animals, and humans has also been recognized since the beginning of agriculture. These pathogens can produce metabolites that show toxic effects when they are ingested. Several examples in recent history exemplify this property. In 1960, 'Turkey-X' disease killed 1,00,000 turkeys, 14,000 ducklings and thousands of partridge and pheasant poults in England. In the middle 1930s and late 1970s, there were outbreaks of sickness in horses called *equineleukoencephalomalacia* in the United States; and alimentary toxic aleukia has been responsible for the distress and death of thousands of people, since it was first recorded in the 19th century in Russia. Although these syndromes are all very different, they have one thing in common i.e., they are all caused by mycotoxins.

Mycotoxins are secondary metabolites that are produced by fungi growing on cereals, nuts, soybeans and several other crops, including fruit. The outbreak of *Turkey-X* disease in England was traced to contaminated peanuts from Brazil and lead to the discovery of aflatoxins [1]. Mycotoxins are inherently a heterogenous group of metabolites that are formed along the polyketide chain (aflatoxins, patulin, penicillic acid, anthraquinones, and ergochromes) and the terpenes (trichothecenes) as well as from amino acids (ergot toxins and gliotoxins).

Among mycotoxins, aflatoxins have been recognized as the most potent ones. They are a group of heterocyclic, oxygen containing structures that possess bisfurano ring system. They are produced by certain strains of *Aspergillus flavus* and *A. parasiticus*. The ubiquity of toxin-producing fungi and potent biological activity of their mycotoxins at very low concentrations has stimulated a phenomenal amount of research in many different fields. They possess both mutagenic, carcinogenic and even teratogenic effects in a wide range of organisms. They have been implicated in primary liver cancer in humans and in Reye's syndrome [2].

The principal compounds found in plant products are aflatoxin B_1, B_2, G_1, and G_2 (Figure 1). Aflatoxin B_1 has been the subject of the most extensive study as its acute and chronic toxicity is substantially greater than that of other aflatoxins in most biological systems [3]. As a result, voluminous data are available on the natural occurrence of aflatoxins in cereals, influence of geographical and agricultural conditions, substrate, moisture and temperature, pH and competitive microorganisms on the production of aflatoxins. Interaction between *aflatoxigenic* molds and other microorganisms is a common phenomenon in nature. The interaction may result in a decrease or an enhancement of the production of aflatoxins. This communication presents the data available on the transformation, that may be brought about by the interacting organisms, during trophophase or in the idiophase.

In most of the cases, studies were restricted only to biosynthesis of aflatoxins by the toxigenic strains of *A. flavus* and *A. parasiticus* and the influence of other organisms when co-cultured. The influence exerted was not general.

Some organisms reduce synthesis of aflatoxin by toxigenic aspergilli, while some others enhance the production. A few organisms are reported to react with the preformed aflatoxin B_1, either by degrading it or by transforming it to less toxic compound.

MICROBES INTERACTING WITH A. *FLAVUS* AND AFLATOXIN

Table 1 lists the microorganisms which decrease or increase the synthesis of aflatoxin B_1 when co-cultured with the aflatoxin producing A. *flavus*.

Table 1
Microorganisms which influence aflatoxin production by aflatoxigenic aspergilli

Microbes	Influence on aflatoxin production
Rhizopus nigricans	Inhibition (near total)
Saccharomyces cerevisiae	Marginal (9%)
Aspergillus niger	Inhibition (> 90%)
Acetobacter aceti	Stimulation
Brevibacterium linens	Inhibition (marginal 12% & below)
Aspergillus candidus	Inhibition
Aspergillus chevalieri	Inhibition (near total)
Bacillus subtilis	Inhibition (near total)
Bacillus amyloliquefaciens	Stimulation
Trichoderma viride	Inhibition (near total)
Dactylium dendroides	Transformation
Absidia repens	Transformation
Mucor griseocyanus	Transformation
Lactobacillus casie pseudoplantarum	Inhibition
Phoma sp.	Inhibition
Penicillium raistrickii	Transformation
Pseudomonas sp.	Varied observation

Rhizopus nigricans, Trichoderma viride, A. niger, and *Phoma* sp. inhibit production of aflatoxin by *A. flavus,* when grown together in liquid culture medium without affecting the growth of *A. flavus.* Inhibition of aflatoxin synthesis by *A. chevalieri* and *A. candidus* was observed in rice medium [4,5].

TRANSFORMATION OF AFLATOXINS

Rhizopus

The *Rhizopus* species, studied by Cole and Kirskey [6], degraded aflatoxin G_1 (AFG$_1$) to aflatoxin B_3 (AFB$_3$) or aflatoxicol in wheat medium (Figure 1). The change was brought about by the culture when the vegetative growth had apparently slowed or ceased. This indicated that the transformation of AFG$_1$ occurs as a result of changes of the culture medium, or as a result of fungal growth or the action of enzymes produced during later periods of the growth cycle. The coversion of AFG$_1$ to AFB$_3$ was confirmed by using ^{14}C-AFG$_1$, wherein 45% of the radioactivity was recovered in the extracts of the culture containing B$_3$. The UV spectrum of AF-1 (converted AFG$_1$) has λmax in methanol at 330, 264 and 256 nm, which indicated that a part of UV chromophore of G$_1$ was altered. This UV spectrum was similar to that of tetrahydrodesoxo aflatoxin B$_1$, which has λmax in methanol at 332, 264 and 255 nm and indicates that the dilactone system of aflatoxin G$_1$ was absent in AF-1. This was supported by IR and NMR data, all of which coincided with the data obtained for standard aflatoxin B$_3$.

The same authors have reported conversion of aflatoxin B$_1$ [7] to hydroxy compounds by *Rhizopus arrhizus* isolates from Georgia peanut. The *Rhizopus* sp. was inoculated on to shredded wheat coated with purified aflatoxin B$_1$. The cultures were extracted with hot chloroform at 24 h intervals from 1 to 10 days. The crude extracts contained four blue flourescent compounds (as shown in Figure 1) on silica gel (TLC) with Rf values of 0.30 (I), 0.26 (II) while aflatoxin B_1 appeared at Rf 0.33. Compound (I) appeared on day 1, while (II) appeared only after 3 days. The third (III) and fourth (IV) spots were derived from (I) and (II), later identified as AF R_0 or aflatoxicol, as indicated by the concomitant decrease in the intensity of fluorescence of (I) and (II). The conversion of aflatoxin B_1 to these compounds were confirmed

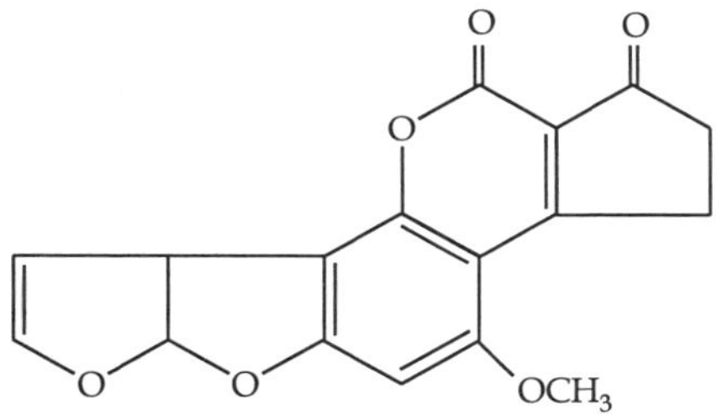

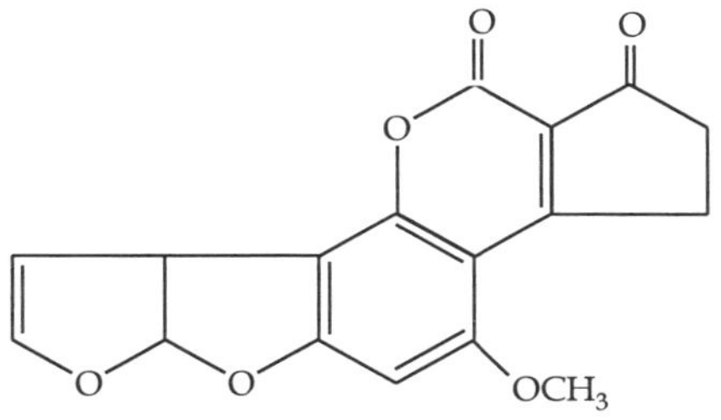

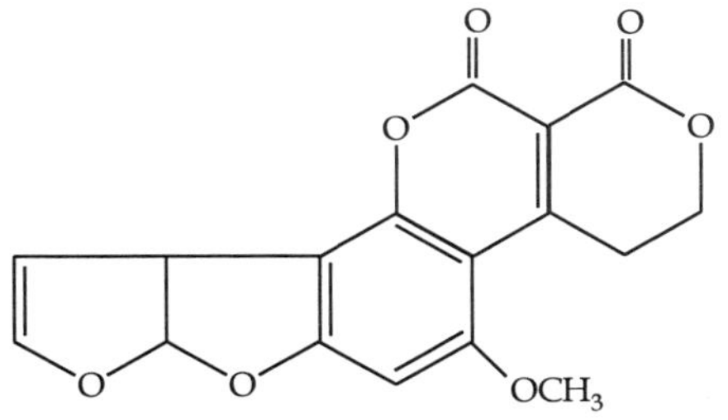

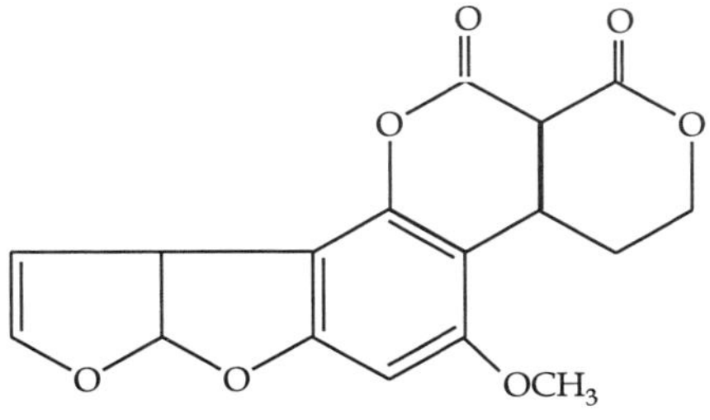

Figure 1. Chemical structures of aflatoxins: (I) aflatoxin B_1, (II) aflatoxin B_2, (III) aflatoxin G_1, (IV) aflatoxin G_2, (V) aflatoxin R_0.

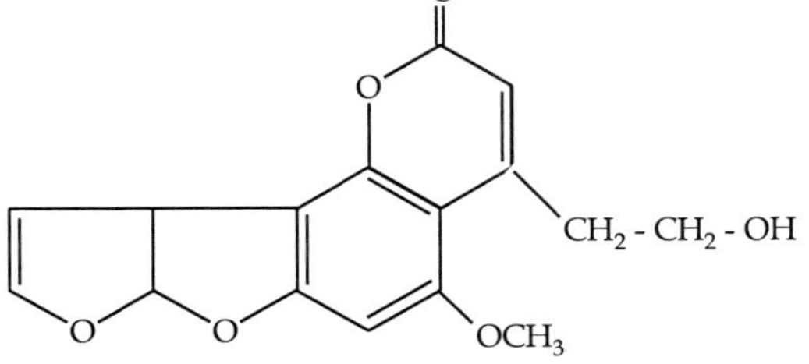

by supplying radioactive aflatoxin B_1 to the culture and obtaining the radioactivity in the derivatives mentioned. The four derivatives have been identified by UV, IR, NMR and MS. The UV and NMR spectral data were similar to those of the aflatoxin R_O reported earlier by the same authors and of tetrahydroxy aflatoxin B_1 (λmax in methanol at 332, 264, 265), which demonstrated the absence of ketone carbonyl in cyclopentane ring of aflatoxin B_1. This was also supported by the IR spectra of all the four compounds, which lacked absorption at 1760 and 1685 cm^{-1}, but exhibited absorption at 1735 cm^{-1}. The IR spectra of compounds I and II exhibited absorption at 3560 cm^{-1} suggesting a hydroxyl group.

The mass spectral analysis indicated that (III) and (IV) were derivatives of (l) and (II) formed spontaneously on silica column. Compounds (I) and (II) are only stereoisomers differing only in the conformation of hydroxyl group. Thus, the chemical alteration has occurred in the cyclopentane ring system and not in the highly reactive terminal furan ring system.

The biological alteration of aflatoxin B_1 differed from the biological alteration of aflatoxin G_1 by *Rhizopus* spp [6]. Metabolism of aflatoxin B_1 was incomplete (60%) and occurred within the first week of culture. No further metabolism occurred after the first week of fungal growth. The greatest breakdown of aflatoxin G_1 occurred between the second and third week, and degradation was more complete than that of aflatoxin B_1.

Tetrahymena pyriformis decreased 58% of AFB_1 in 24 h and 67% in 48 h [8]. An unknown blue fluorescent substance was produced with intensity about half of the unchanged B_1 with Rf 0.52 compared with 0.59 of B_1 and 0.55 for B_2 on TLC and with an UV spectrum showing maxima 253, 261 and 328 nm, the cells did not degrade AFG_1.

Rhizopus nigricans and Saccharomyces cerevisiae

The above two organisms effectively compete with A. parasiticus or A. flavus, when co-cultured and actually decrease growth and aflatoxin formation by the Aspergillus. S. cerevisiae retards the production of only aflatoxin G_1 without affecting aflatoxin B_1, perhaps by producing an end product, which could have served as a metabolic repressor for aflatoxin G_1 production [9].

Dactylium dendroides, Absidia repens and Mucor griseocyanus

Detroy and Hesseltine [10] report that several fungi like Dactylium dendroides, Absidia repens, and Mucor griseocyanus transform aflatoxin B_1 to compounds R_0 and R_1 with reduced biological activity, in shake culture flask. The conversion activity may be cell wall or membrane bound, since the cell free extracts were not able to convert aflatoxin B_1. They further postulate that these fungi are capable of oxidation and reduction of the steroid nucleus, a similar reaction with aflatoxin may be possible. The conversion is not affected by the age of the culture. Spectral data for R_0 and R_1 λ_{max} in methanol is 325, IR for R_0 3400 cm^{-1} (-OH), 1590 cm^{-1} (Phenyl), 1130 cm^{-1} (C-O-C) 2850 cm^{-1} (-O-CH$_3$); other bands at 1485, 1440, 1360 cm^{-1} like AFB$_1$, also bands at 1720 and 1610 cm^{-1}, i.e. a shift from 1760, 1685 cm^{-1} bands of AFB$_1$. R_1 is formed at a later stage of incubation. The conversion of B_1 to R_0 is maximum in 72 h and prolonged incubation to 90-100 h will yield another fluorescent compound R_1 with a changed Rf (0.5) as compared with an Rf of 0.69 and 0.57 for B_1 and R_0, which may represent a reduction products resulting from R_0. Same authors identified R_0 as aflatoxicol, which is 18 times less toxic than AFB$_1$.

Bacillus subtilis

Kimura and Hirano [11] screened about 1000 soil organisms and isolated a bacterium, which was identified as Bacillus subtilis. This strain inhibited the growth of toxigenic aspergilli even on peanuts or corn, probably by synthesizing an inhibitory compound for the growth and production of aflatoxin. The inhibitory activity was relatively thermostable and could survive *pasteurization*. The effect of proteolytic enzymes on the inhibitory activity suggested the proteinaceous nature of the inhibitory compound, and the loss of activity upon heating at 100°C was probably due to denaturation of the protein or cofactors. Similarly, the inhibitory principle from *Lactobacillus* culture was identified to be a small peptide, since it could diffuse through dialysis sack and lost the activity when treated with proteolytic enzymes [12].

Flavobacterium aurantiacum

Line et al. [13] studied the degradation pattern of aflatoxin B_1 by *Flavobacterium aurantiacum*. They incubated ^{14}C-aflatoxin B_1 with live and dead cells of the bacterium and analyzed for the ^{14}C content over time. It was found that the normally chloroform soluble aflatoxin B_1 was converted into water soluble degradation products. A portion of radioactivity was also obtained in carbon dioxide, evolved by the live cells. The dead cells could not carry out any degradation of aflatoxin B_1. These organisms were also tested for their ability to remove aflatoxin B_1 from partially defatted or non-defatted peanut milk, which was contaminated to a level of 1 μg/ml aflatoxin B_1. There was a 74% reduction of aflatoxin B_1 in partially defatted peanut milk treated with 48 h old *F. aurantiacum* cells and only 23% reduction in non-defatted peanut milk in 24 h [14]. Smiley and Draughon [15] have carried out the partial characterization of the factor which causes AFB_1 removal. They have proved that the factor is proteinaceous in nature by studying the loss in activity due to proteinase K treatment. They also carried out the ammonium sulphate precipitation of the factor. Fractionation by ultra-filtration demonstrated a protein fraction between 20,000 to 50,000 molecular weights, capable of AFB_1 removal. This, along with gel filtration and anion exchange chromatography, indicated the presence of *cytosolic* protein in *F. aurantiacum*, which is capable of AFB$_1$ removal. Bohra et al. [16] showed an increased expression of two proteins, 14.5 and 31 kD, respectively, on exposure to AFB$_1$ for 72 h. During this period, 78% AFB$_1$ was found to be degraded. D'Souza and Brackett [17] further proved the involvement of seryl and sulphydryl groups in the active site of the AFB$_1$-degrading enzyme of *F. aurantiacum*.

Aspergillus flavus and *A. parasiticus*

It has been observed by many investigators that the level of aflatoxin produced by the toxigenic fungi namely, *Aspergillus flavus* and *Aspergillus parasiticus* gradually increases starting from the second day of inoculation, reaches a maximum in 7-10 days and then decreases gradually [Shantha et al., unpublished] (Figure 2).

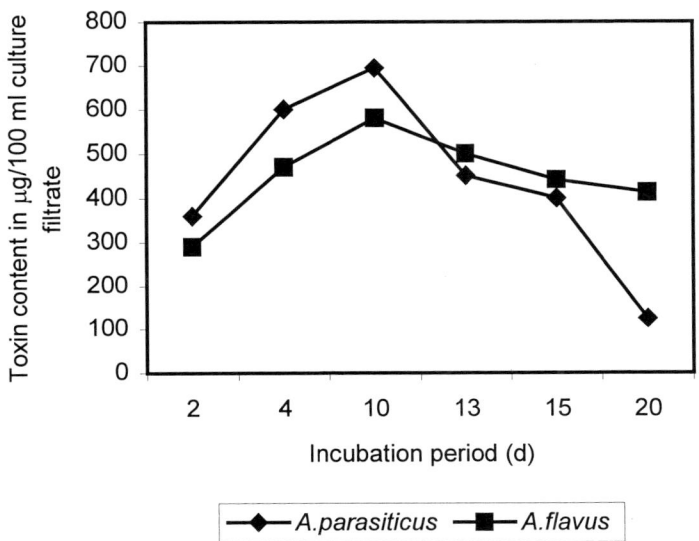

Figure 2. Aflatoxin synthesis in *Aspergillus flavus* and *A. parasiticus* cultures.

The decrease is not due to the binding of aflatoxin to the mycelial mat of the fungi (as the aflatoxin added could be recovered completely), but due to the degradation of aflatoxin B_1 by the aged mycelium. The synthesis as well as the degradation is faster in the case of fragmented mycelium than in the intact mycelium, the fragmented mycelium producing about 290 µg aflatoxin B_1 in 9 days as compared to 166 µg synthesized by the whole cells under the same conditions. There is a steep fall (96.6%) in the toxin level after 9 days. While the intact cells could reduce the toxin only by 20%. The factor responsible for degrading the toxin is intracellular because the culture filtrate of the fungus could bring down the toxin level only by 20% (Figure 3).

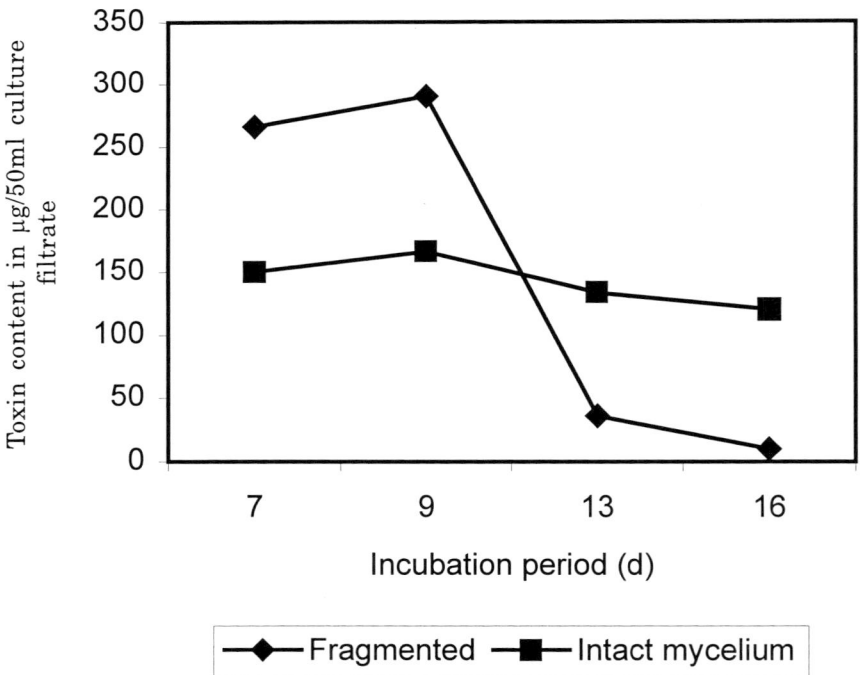

Figure 3. Aflatoxin synthesis in fragmented and intact mycelium of *A. parasiticus*.

The substrates, that support substantial growth of mycelia, yield mycelium having the greatest ability to degrade the toxin. Strains of *A. parasiticus* and *A. flavus*, that produced larger amounts of aflatoxin, generally degraded more aflatoxin. The rate of degradation increased as the amount of mycelia was increased. Also, the degradation activity increased as the amount of aflatoxin increased. Maximum activity occurred at 28°C and pH 5 to 6.5 [18]. Steaming mold mycelium reduced the degrading ability. These results indicate that the degrading factor is heat labile and intracellular. The enzymatic nature of this degrading factor was proved by adding classical enzyme inhibitors like cycloheximide, SKF 525-A or metyrapone to 6-8 day-old culture or the cell free extracts of *A. flavus* or *A. parasiticus* which resulted in almost complete inhibition of the *endogenous degradation* of aflatoxins [19]. Cytochrome c also lowered the activity of the cell free extracts, but only to some extent. KCN had no effect. Addition of NADPH, $NaIO_4$ or KIO_4 to the cell free extracts of *A. flavus* or *A. parasiticus* stimulated aflatoxin degradation. $NaIO_4$ increased the degradation of aflatoxin B_1 and G_1 from 19.02% and 22.05% to 45.88 and 47.85% [20]. Addition of KIO_4 to the cell free extracts of *A. parasiticus* stimulated the degradation from 16.6% to 33.0% [Shantha et al., unpublished]. The above factors strongly suggest the involvement of a group of enzymes belonging to the cytochrome P-450 monooxygenase system. Doyle et al. [18] suggested that peroxidase may be one of the enzymes, which catalyzes the *decomposition* of hydroperoxides, resulting in the generation of free radicals which react with aflatoxin. Some peroxidases produce hypochlorite and singlet oxygen in the presence of hydrogen peroxide and chloride ion. The singlet oxygen may react with the terminal furan ring.

Phoma sp. [20]

A number of fungi namely *Phoma* sp., *Mucor* sp., *Trichoderma harzianum*, *Trichoderma* sp., *Rhizopus* sp., *Alternaria* sp., and *Sporotrichum* spp. have been found not only to prevent aflatoxin B_1 synthesis by aflatoxigenic *A. flavus* in liquid medium but also are capable of degrading the preformed aflatoxin. *Phoma* sp. was found to be the most efficient organism (Figure 4). The cell-free extract of *Phoma* sp. was better than its own culture filtrate, destroying 99% aflatoxin B_1 added to a level of 500 ppb to a liquid medium in five days. The possibility of a heat stable enzymatic activity in the cell-free extract of *Phoma* sp. is proposed. It was observed that some of the fungi, which can suppress or degrade aflatoxin, enhance aflatoxin production if their addition to the toxigenic *A. flavus* culture was delayed by 72 h or more (e.g., *Mucor, Phoma, Trichoderma*) (Table 2). Wiseman and Marth [21] have also reported that under certain conditions, *S. lactis* stimulates growth and aflatoxin production by *A. parasiticus*.

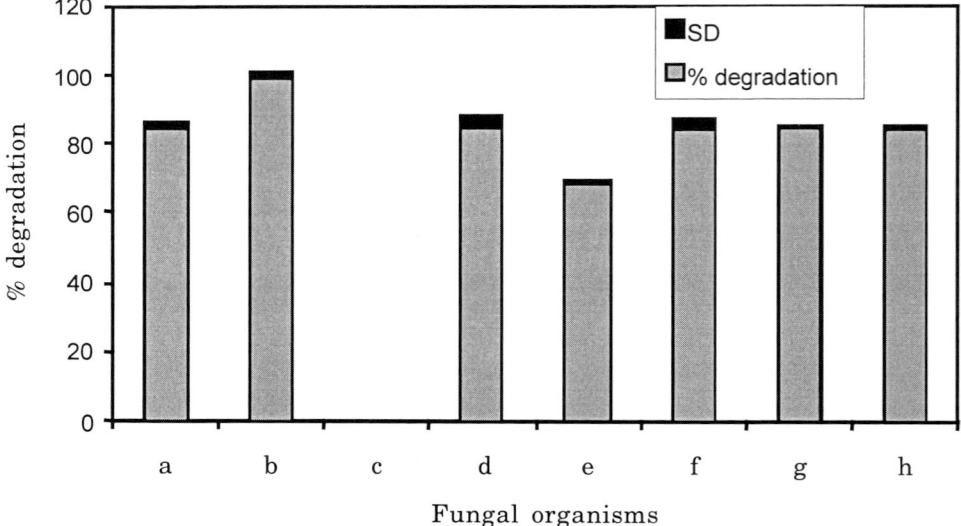

Figure 4. Degradation of added aflatoxin by different fungal organisms. a: *Trichoderma* sp. 639; b: *Phoma* sp.; c: *Rhizopus* sp. 663; d: *Rhizopus* sp. 668; e: *Rhizopus* sp. 720; f: *Sporotrichum* sp. ADA IV B14(a); g: *Sporotrichum* sp. SF VI BF(9); h: *Alternaria* sp.

Table 2
Aflatoxin produced by A. *flavus* in the presence of cultures of other organisms
added at different time period at 28 ± 2°C

Cultures	Aflatoxin B_1 produced (µg/100 ml medium) when other cultures inoculated at (h) ± s.d. after A. *flavus*				
	0	24	48	72	96
A. *flavus* (AF)	0.0	8.8±1.1	177.0±18.4	320.0±43.6	800.0±43.6
AF + *Phoma* sp.	53.0±2.64	44.0±6.0	44.0±6.2	1778.0±92.3	2667.0±94.2
AF + *Tr. harzianum*	53.0±2.6	1778.0±45.0	2666.0±93.2	3555.0±57.7	4266.0±342.2
AF + *Tr*. 639	30.3±2.0	44.0±2.00	53.0±0.5	66.0±8.7	89.0±18.5
AF + *Alternaria* sp.	17.7±1.7	35.5±1.9	66.6±2.9	66.0±3.6	457.0±35.2
AF + *Mucor* sp.	18.0±1.0	1333.0±103.1	1500.0±136.1	1860.0±99.9	3555.0±99.6

Tr = *Trichoderma*; s.d. = standard deviation
*The data from A. *flavus* show the time course of production of aflatoxin B_1.
Other cultures were added at different times after A. *flavus* and incubated
for a further 96 h. Hence, to establish the effect on aflatoxin B_1 level all data
points for the 5 cultures should be compared to the A. *flavus* result at 96 h
(i.e. 800 ± 43.6).

Pseudomonas sp.

Effect of several strains of *Pseudomonas* sp. have been studied by various
worker. Kimura and Hirano [11] reported that *Pseudomonas* did not have
the ability to degrade aflatoxin. Ciegler et al. [22] reported that a large
number of *Pseudomonas* species appear to degrade aflatoxin in shake culture
with TGY medium. But they attribute the degradation to the pH effect,
which rises to 8.5 in 24 h. Archana and Shantha [23] have observed that
Pseudomonas secretes an extracellular enzyme, which can degrade aflatoxins
very efficiently. The enzymatic reaction starts right from 5 min and completes
in 4 h. Some of the products are blue fluorescent compounds, one
predominantly appearing at an Rf of 0.2 as compared to that of B_1 at 0.6 Rf.
The other 3 products are some-what less prominent in intensity. In addition,
there may be other non-fluorescent water soluble compounds. The enzyme
can be used to detoxify milk, contaminated with aflatoxin B_1.

Penicillium raistrickii NRRL 2032

This organism partially converts aflatoxin B_1 to a compound similar to aflatoxin B_2 [22]. Spores of *A. terreus, A. flavus,* and *A. luchuensis* partially transformed AFB_1 to series of pale blue fluorescent compounds, having approximate Rf 0.87, 0.8, 0.3, 0.2, 0.08 and 0.03, showing spots at 0.87, 0.8 and 0.08 fluoresced intensely, while the rest were barely perceptible.

Armillariella tabescens

Liu et al. [24] reported that the multi-enzyme preparation from the mycelial pellet of *Armillariella tabescens* could effectively detoxify aflatoxin B_1. However, 200 μg protein could completely destroy 16 μM of aflatoxin B_1 in 30 min at 28°C. The efficiency of the enzyme decreased beyond this concentration of aflatoxin. The destruction was monitored by the loss in fluorescence. The loss in toxicity was verified by conducting experiments on animals, using rats. The rats were injected with enzyme-treated and untreated aflatoxin B_1 (@ 6 mg/kg body weight). Their body weights were determined after 72 h and sacrificed for histological and ultra-structural studies. It was observed that the animals maintained on multi-enzyme treated aflatoxin B_1 were almost similar to the controls (without aflatoxin B_1) in their body weight, structure of liver and biochemical composition of the blood serum, indicating that the *multi-enzyme system* efficiently detoxifies aflatoxin B_1. These authors proposed that the detoxification might have involved formation of AFB_1-epoxide, catalyzed by the fungal multi-enzyme, followed by hydrolysis of epoxide to give the dihydrodiol. The ring would open in subsequent step. The appearance of fluorescence products on the TLC plate at different Rf values with AFB_1 (0.376-treated AFB_1, 0.295-AFB_1) suggests the intactness of coumarin ring structure.

REFERENCES

1   Apsimon JW. Mycotoxins in Grain, USA, Eagen Press, 1994; 3.

2   Heathcote JG. Mycotoxins - Production, Isolation, Separation and Purification. New York, Basel, Honkong: Marcel Dekker Inc, 1984; 57.

3   Eaton DL, Ramsdell HS. Handbook of Applied Mycology, Vol. 5, 1992; 157.

4   Boller RA, Schroeder HW. Phytopathology 1973; 63: 1507-1510.

5   Boller RA, Schroeder HW. Phytopathology 1974; 64: 121-123.

6   Cole RJ, Kirskey JW. J Agric Food Chem 1971; 19: 222-223.

7   Cole RJ, Kirskey JW, Blankenship BR. J Agric Food Chem 1972; 20: 1100-1102.

8   Teunnisson DJ, Robertson JA. Appl Microbiol 1967; 15: 1099-1103.

9   Weckbach LS, Marth EH. Mycopathol 1977; 62: 39-45.

10   Detroy RW, Hesseltine CW. Can J Microbiol 1969; 15: 495-500.

11   Kimura N, Hirano S. Agric Biol Chem 1988; 52: 1173-1179.

12   Gourama H, Bullerman LB. Intl J Food Microbiol 1997; 34: 131-143.

13   Line JE, Brackett RE. Wilkinson RE. J Food Prot 1994; 57: 788-791.

14   Hao YY, Brackett RE. J Food Sci 1988; 53: 1384-1386.

15   Smiley RD, Draughon FA. Poster Abstr, 85th Ann Meeting IAMFES, Nashville, 1998; P. 34.

16   Bohra LK, Reuger SA, Phelous RK, Smith JS, Grieger D. Poster Abstr, 84th Ann Meeting IAMFES, Orlando, 1997; P. 28.

17   D'Souza DH, Brackett RE. Poster Abstr, 84th Ann Meeting IAMFES, Orlando, 1997; T. 32.

18   Doyle MP, Applelbaum RS, Brackett RE, Marth EH. J Food Prot 1982; 45: 964-971.

19   Hamid AB, Smith JE. J Gen Microbiol 1987; 113: 2023-2029.

20   Shantha T. Natural Toxins 2000 (in press).

21   Wiseman DW, Marth EH. Mycopathologia 1981; 73: 49-56.

22   Ciegler A, Lillehoj EB, Peterson RE, Hall HH. Appl Microbiol 1966; 14: 934-939.

23   Archana Mudbidri, Shantha, T. 38th Ann Meeting Association of Microbiologists of India, New Delhi, 1997; GMB-28.

24   Liu DA, Yao DS, Liang R, Ma L, Cheng WQ, Gu L-Q. Food Chem Toxicol 1998; 36: 563-574.

Biotransformations: Bioremediation Technology for Health and Environmental Protection
V.P. Singh and R.D. Stapleton, Jr. (Editors)
© 2002 Elsevier Science B.V. All rights reserved.

Biotransformations of tannery wastes

Ved Pal Singh

Department of Botany, University of Delhi, Delhi - 110 007, India

INTRODUCTION

Tanning industry is one of the oldest and important major industries in India where there are more than 2000 tanneries, and majority of them are situated in the states of Tamil Nadu, Uttar Pradesh, and West Bengal. Effluents from tanning industries are one of the complex types of pollutants released by them, which are toxic to plants, animals and soil microorganisms, thus posing potential threat to human health as well as the environment [1-3]. The properties and the chemical composition of tannery waste include the following: colour: yellow to brown; total solid: 479,000 ppm; suspended solid: 12,166 ppm; dissolved solid: 35,733.33 ppm; oil and grease: 5.15 ppm; pH: 8.00; COD: 6,113.3 ppm; BOD: 2,502 ppm; total nitrogen as N: 590.98 ppm; Cl^-: 11,258.7 ppm; phenol: 7.5 mg/litre; SO_4^{2-}: 1,499.2 ppm; NO_3^-: 5.34 ppm; F^-: 1.04 ppm; PO_4^{3-}: 0.77 ppm; and Cr(III): 0.6 ppm [4].

It has been observed that, tannery effluent contains *vegetable tannins*, in addition to solid organic matter, suspended solids, chromium, high chlorine and sulphide concentrations, and high pH. High chlorine content causes chlorosis in plants and high sodium content increases sodium absorption ratio and, thus adversely affecting the soil structure and plant growth [4].

As per estimate, tanneries release about 300 litres of waste water/kg of skin processed, and the annual discharge ammounts to 9,420 kl [5]. With such a massive pollution load and toxic nature of effluents, tanneries pose potential threat to our ecosystem and disrupt the food chain. Tannins appear to exhibit antimicrobial properties, as they inhibit growth of several bacteria, fungi, yeast and viruses [4]. However, certain bacteria and fungi are capable of growing on tannery wastes, and this property has been exploited by several workers to develop biodegradation models. Fungi have also been implicated in the utilization of tannery wastes for their growth; and the tannins, which are the important major constituents of tannery waste, have been used as carbon source in many cases [6]. Also, Makkar et al. [7] have reported biodegradation of tannins in oak (*Quercus incana*) leaves by *Sporotrichum pulverulentum*, decreasing the contents of total phenols and condensed tannins by 58 and 66%, respectively in 10 days. It has been found that simple phenols like catechol and pyrogallol, which are building blocks of tannins, could be effectively metabolized by bacteria like *Pseudomonas* and *Bacillus* [1,8].

TYPES OF TANNINS

By nature, tannins are water soluble polyphenols, which differ from most other natural polyphenolic compounds in their ability to precipitate proteins such as gelatin [9]. They are most abundant plant constituents, followed by cellulose, hemicellulose, and lignin [10]. In tribal pulse (*Bauhinia malabarica* Roxb.), tannins have been found to be one of the antinutritional factors, in addition to phenols, L-DOPA, and haemagglutinating activity [11]. However, cotton condensed tannin has been shown to interact with *Heliothis virescens* larvae and cry IA (C) *delta*-endotoxin of *Bacillus thuringiensis* [12]. They are added to the soil in the form of leaves, twigs, and fruits. Rice and Pancholy [13] have reported that 85 kg/ha of tannins are added every year to the forest soil. On the basis of their structures, tannins have been classified into two major groups:

(i)   Condensed tannins.

(ii) Hydrolyzable tannins.

(i)   Condensed tannins

They are polymers of catechin or similar flavans, that are connected by carbon linkages, and are more resistant to microbial attack as compared to hydrolyzable tannins. The non-hydrolyzable tannins are formed by flavan 3,4-diols (leucoanthocyanin and proanthocyanin). Catechin is a flavan-3-ol with two hydroxyl groups in the side ring. The catechins include gallic acid esters with the acid moieties attached to the hydroxyl groups. Flavan-3,4-diols are also termed as leucoanthocyanins (Figure 1) [1].

(ii)   Hydrolyzable tannins

Esters of sugars and phenolic acids or their derivatives are referred to as hydrolyzable tannins. They are composed of a molecule of carbohydrate, generally glucose, to which gallic acid or similar acids are attached by ester linkages (Figure 1) [14,15]. Hydrolyzable tannins are subdivided into gallotannins and ellagitannins. Gallotannins are esters of glucose or a polysaccharide and gallic acid or *m*-digallic acid. On hydrolysis, they yield gallic acid. The chief commercial tannin is tannic acid. Gallotannin contains 8 to 10 moles of gallic acid and its derivatives, especially ellagic acid, which is formed by lactonization of hexahydroxydiphenic acid [1].

MICROBIAL TRANSORMATIONS OF TANNINS

In general, tannins are *recalcitrant* molecules and are very much resistant to microbial attack because of their complexity in structure [16]. A holistic view of microbial degradation of tannins and the potential for manipulating the *detannification* of certain microbial strains for beneficial effects on food, beverages, feed, and fodder has been given by Bhat et al. [17]. Condensed tannins are more resistant than the hydrolyzable tannins, and are toxic to a variety of microorganisms [18].

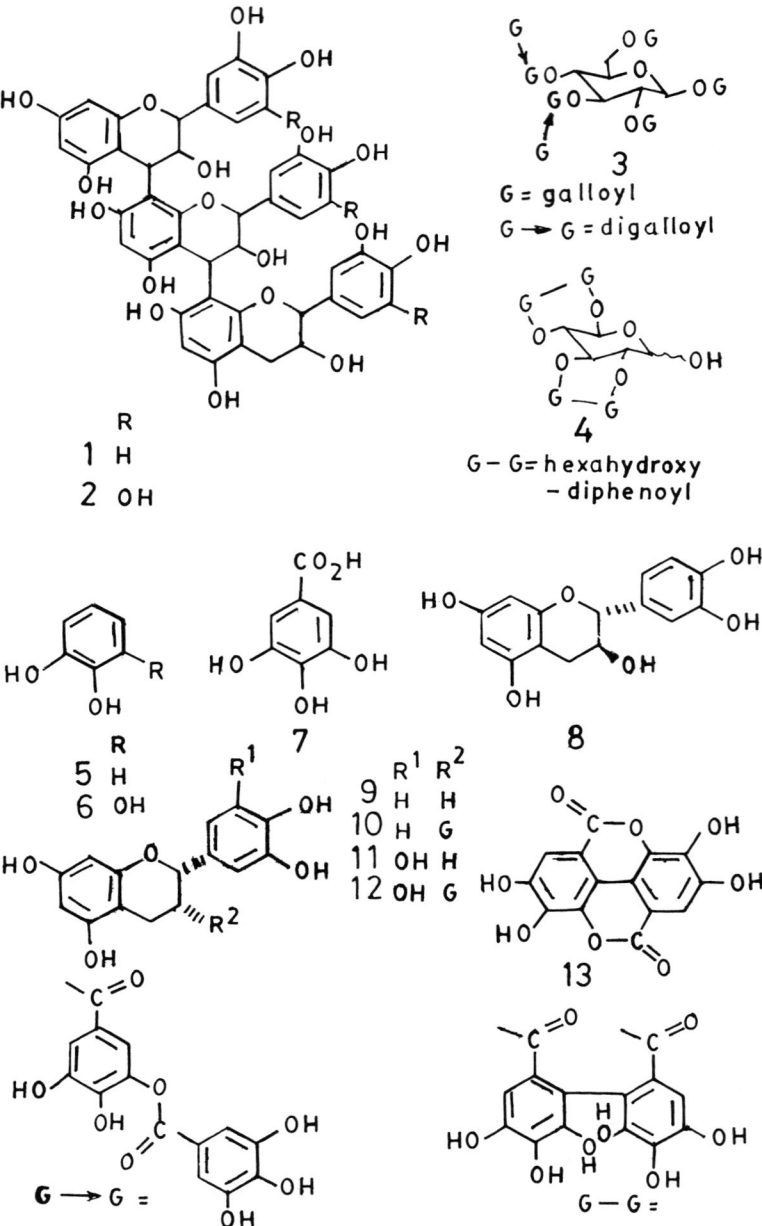

Figure 1. Chemical structures of proanthocyanicidins, hydrolyzable tannins and other structurally related phenols. 1, procyanidin trimer; 2, prodelphinidin trimer; 3, heptagalloylglucose; 4, pedunculagin; 5, catechol; 6, pyrogallol; 7, gallic acid; 8, (+)-catechin; 9, (-)-epicatechin; 10, (-)-epicatechin gallate; 11, (-)-epigallocatechin; 12, (-)-epigallocatechin gallate; 13, ellagic acid.

Tannin biotransformations by bacteria

There are only a few reports, which indicate the ability of bacteria to degrade tannins. However, it is well known that the tannic acid and tannins are *bacteriostatic* and are toxic compounds, which undergo non-reversible reactions with proteins. Deschamps et al. [19] reported the biodegradation of tannins by bacteria. Some bacteria, however, may also degrade many phenolic compounds, including natural ones like catechol, protocatechuic acid, etc. Fifteen bacterial strains, belonging to genera *Bacillus, Staphylococcus,* and *Klebsiella,* have been isolated by enrichment culture technique, using tannic acid as the sole source of carbon [19]. Nine of the isolated strains grew both on tannic acid as well as on gallic acid, which are obtained upon hydrolysis; but only four strains degraded catechol or catechin. Deschamps and Lebeault [20] isolated several bacterial strains capable of degrading hydrolyzable and condensed tannins, including chestnut, wattle, and Quebracho commercial tannin extracts by enrichment. The wattle tannin-degrading strains were identified as *Enterobacter aerogenes, Enterobacter agglomerans, Cellulomonas* sp., *Streptococcus* sp., *Corynebacterium* sp., and *Pseudomonas* sp. Phenolic acid was not detected in uninoculated and inoculated samples of wattle tannin; however, gallic acid was detected in samples inoculated with Quebracho tannin.

Achromobacter species degraded *hydrolyzable* tannins, but did not degrade catecin and wattle tannin; it effectively decomposed gallotannin at pH 6.0 [21]. Catechin was degraded by the cultures of *Pseudomonas* species; the order of decreasing degradability was: gallotannin, catechin, chestnut, and wattle tannins. The cell free extracts of *Pseudomonas putida*, grown on syringic acid as sole source of carbon, was quite effective in the oxidation of gallic acid, where 1 mole of oxygen was able to oxidize 1 mole of gallate with the liberation of 2 moles of pyruvic acid and 1 mole of carbon dioxide [22].

In the above process, oxaloacetate appeared as an intermediate (Figure 2) [1].

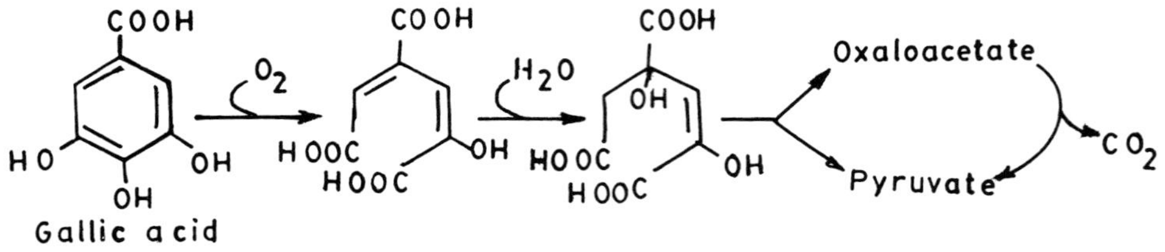

Figure 2. Degradation of gallic acid by *Pseudomonas putida*.

Myrobalan tannin was found to be utilized by *Pseudomonas solanacearum* with the uptake of oxygen, which was more than that observed when wattle tannin was utilized by this bacterium [23]. On the other hand, catechin was degraded by *Ps. solanacearum* [24] and *Rhizobium* species [25]. A first report, in which catechin was catabolized to form protocatechuic acid, came from the work of Muthukumar et al. [25], where a bacterium *Rhizobium japonicum* was found to degrade catechin. Along with *R. japonicum*, the other species and strains of *Rhizobium*, were also able to grow on catechin by utilizing/ catabolizing this compound [25].

Tannin biotransformations by fungi

Among fungi, filamentous fungi, especially the species of the genera *Aspergillus* and *Penicillium,* have been implicated in the biotransformations of tannins [26-28]. The role of fungi in tannin degradation dates back as early as 1900, when Fernbach [29] and Pottevin [30] independently reported the hydrolysis of tannins using cell free preparation of *Aspergillus niger.* Some microbes, including fungi, grow as pure culture on media containing tannins by decomposing and utilizing the same as sole source of carbon (Table 1) [21].

Table 1
Decomposition of tannins by microorganisms

Microorganisms	% Decomposition			
	Gallotannin	Chestnut tannin	Catechin	Wattle tannin
Achromobacter sp.	74	13	N	N
Streptomyces sp.	18	20	N	N
Aspergillus flavus	84	N	69	N
A. *fumigatus*	80	N	50	N
A. *niger*	100	10	100	N
Penicillium citrinum	90	43	93	N
P. *frequentans*	81	10	N	N
P. *janthinellum*	74	N	75	N
P. *purpurogenum*	83	23	N	N
P. *thomii*	74	30	50	N
Fusarium sp. 1	96	N	100	N
Fusarium sp. 2	85	N	77	N

Both condensed and hydrolyzable tannins were used as substrates, and several kinds of tannins were compared. *Aspergillus, Penicillium, Fomes, Polyporus, Poria,* and *Trametes* species were shown to grow better on gallotannin than on chestnut tannin (ellagitannin) or wattle tannin (condensed tannin). The other fungi, which were capable of degrading tannins or tannery waste constituents belonged to the genera *Chaetomium, Fusarium, Rhizoctonia, Cylindrocarpon, Trichoderma, Candida* [31], and *Psalliata* (particularly *P. campestris*) [1] as well as certain yeasts [32,33].

In a study carried out by Mahadevan and Sivaswamy [1], *P. campestris* can oxidize catechin and *A. niger* can degrade gallic acid; and the intermediates of this degradation pathway were *cis*-aconitic acid, α-ketoglutaric acid and citric acid. *A. fumigatus* decomposed gallotannin to gallic acid; and catechin, at 0.3% level, was degraded by this fungus without intermediate formation in 6-8 days.

ENZYMATIC TRANSFORMATIONS OF TANNINS

Initially, there were two independent studies, one by Fernbach [29] and the other by Pottevin [30], in which the enzymatic nature of tannin hydrolysis, using cell-free preparation of *A. niger* were first reported. The enzyme *tannase* (tannin acyl hydrolase), capable of cleaving tannin, has been isolated and characterized from various sources [1]. Production of tannase by *A. niger* with increasing concentration of tannin was reported as early as in 1913 [34]. This enzyme has also been isolated from *Penicillium* sp. [35,36], *A. niger* [37], and *Candida* sp. [38]. It was further proved that the tannase of *A. niger* was an inducible enzyme, which hydrolyzed the ester linkages of gallic acid [39]. However, the production of tannase by soil isolates of *Aspergillus* sp. and *Panicillium* sp. occurred even in the absence of inducer [40]. The rapid production of tannase has been found to be favoured by continuous supply of carbon source for growth [41]. Tannase from *P. chrysogenum* has been characterized by Suseela and Nandy [42,43]. According to them, tannase of this microbe had a molecular weight of 300,000 on Sephadex G-200, which is higher as compared to the tannases from other sources. Their studies, based on *denaturation analysis* revealed the molecular weight of this enzyme to be 158,000. This suggested that the *P. chrysogenum* tannase consists of two subunits of equal molecular weights. A detailed account of microbial role of tannase and its role in tannin degradation has been given by Bhat et al. [17].

BIOTECHNOLOGICAL CONSIDERATIONS

In recent years, microbial biotechnology has been of tremendous importance for the benefit of mankind, be it in the field of agriculture or be it in the field of health and environmental protection programmes. Since microorganisms can readily adapt to adverse and changing environmental conditions, scientists have explored the possibilities of exploiting their potential as tools to control pollution through biotransformations/degradation of xenobiotics of health and environmental concern. Investigations on plasmids coding for enzymes of catabolic pathways have led to the discovery of several degradative plasmids. Chakrabarty [44] demonstrated that enzymes of hydrocarbon degradation pathways are *plasmid-specific*, and he named the plasmids with genes specific for dissimilatory pathways as degradative plasmids [45].

A variety of *degradative plasmids*, such as camphor (CAM) plasmid, octane (OCT) plasmid, naphthalene (NAH) plasmid, toluene (TOL) plasmid, and salicylate (SAL) plasmid, have been identified by Wheelis [46]. Biotechnological aspects for microbial degradation of tannins, in the right perspective, including the emerging area of tannin degradation by gastrointestinal microbes of herbivores, have been reviewed by Bhat et al. [17], with particular emphasis on rumen microorganisms.

CONCLUSION

Tannins are one of the major constituents of effluents released by tanning industries and which are toxic to plants, animals and microorganisms, posing potential threat to both human health as well as the environment. However, certain microbes, such as bacteria and fungi including yeasts, have the potential to degrade tannins and utilize these toxic componds as sole source of carbon for their growth. There are reports on bacterial degradation of tannins; and among fungi, mostly the species of *Aspergillus* and *Penicillium* have been implicated in tannin biotransformation/degradation. Microbial enzyme (tannase)-dependent degradation of tannins and biotechnological approaches, with special reference to role of degradative plasmids in the *catabolism* of various xenobiotic compounds find a place of tremendous importance in health and environmental protection programmes.

REFERENCES

1   Mahadevan A, Sivaswamy SN. In: Mukerji KG, Singh VP, eds. Frontiers in Applied Microbiology. Lucknow: Print House India, 1985; 1: 327-347.

2   Saxena RK, Sharmila P, Singh VP. Progress in Industrial Microbiology 1995; 32: 259-270.

3   Singh VP, Bhatnagar AK. In: Rai B, Upadhyay RS, Dubey NK, eds. Trends in Microbial Exploitation. Varanasi: International Society for Conservation of Natural Resources, 1996; 219-223.

4   Nandan R, Raisuddin In: Mukerji KG, Singh VP, Dwivedi S, eds. Concepts in Applied Microbiology and Biotechnology. New Delhi: Aditya Books Pvt Ltd, 1996; 189-228.

5   Arora HC. In: Sundresan BB, ed. Proceedings of National Workshop on Microbial Degradation of Industrial Wastes. Nagpur: NEERI, 1982; 145-156.

6   Dalvesco G, Fiuello N, Vei Hiranus M. Allonis 1972; 17: 25-40.

7   Makkar HPS, Singh B, Kamra DN. Lett Appl Microbiol 1994; 18 (1): 39-41.

8   Chandra T, Madhavkrishna W, Neyudumma Y. Can J Microbiol 1969; 15 : 303-306.

9   Spencer CM, Cai Y, Martin R, Gaffney SM, Goulding PN, Magnolato D, Lilley TH, Haslam E. Phytochemistry 1988; 27: 2397.

10   White T. J Sci Food Agric 1957; 8: 377.

11   Vijaykumari K, Siddhuraju P, Janarshanan K. Plants for Human Nutrition (Dordrech) 1993; 44(3): 291-298.

12   Navon A, Hare JD, Federici BA. J Chem Ecol 1993; 19(11): 2485-2499.

13   Rice EL, Pancholy SK. Am J Bot 1973; 60(7): 691-702.

14   Hathway DE. In: Hills WE, ed. Wood Extractives. New York: Academic Press, 1962; 191-228.

15   Jurd L. In: Hills WE, ed. Wood Extractives. New York: Academic Press, 1962; 229-260.

16   Sivaswamy SN, Mahadevan A. J Indian Bot Soc 1986; 65: 95-100.

17   Bhat TK, Bhupinder Singh, Sharma OP, Singh B. Biodegradation 1998; 9(5): 343-357.

18   Haslam E. In: Priss J, ed. The Biochemistry of Plants. New York: Academic Press, 1981; 527-556.

19   Deschamps AM, Otuk G, Lebeault JM. J Ferment Technol 1983; 61: 55-59.

20   Deschamps AM, Lebeault JM. In: Moo-Young M, Robinson CW, eds. Advances in Biotechnology. New York: Pergamon Press, 1981; 639-643.

21   Lewis JA, Starkey RL. Soil Sci 1969; 107: 235-241.

22   Tack BF, Chapman PJ, Dagley S. J Biol Chem 1972; 247: 6438-6443.

23   Muthukumar G, Mahadevan A. Indian J Exp Biol 1981; 19: 1083-1085.

24   Orlitta A. Kozarstvi 1962; 12: 82-86.

25   Muthukumar G, Arunakumari A, Mahadevan A. Plant Soil 1982; 69: 163-169.

26   Freidrich M. Arch Microbiol 1956; 25: 297-306.

27   Cowley GT, Wittingham WF. Mycologia 1961; 53: 539-542.

28   Nishira H. Chem Abstr 1952-62; 54: 6864; 58: 1102; 60: 3298.

29   Fernbach MA. Compt Rend 1900; 131: 1214-1215.

30   Pottevin M. Compt Rend 1900; 131: 1215-1217.

31   Mahadevan A, Muthukumar G. Microbiologia 1980; 72: 73-79.

32   Jacob FH, Pingal MC. Mycopathologia 1975; 57: 139-148.

33   Otuk G, Deschamps AM. Mycopathologia 1983; 83: 107-111.

34   Knudson L. J Biol Chem 1913; 14: 159-206.

35   Nishira H. J Ferment Technol 1961; 39: 137-146.

36   Nishira H. Sci Rep Hyogo Univ Agric 1962; 5: 117-123.

37   Dhar SC, Bose SM. Leather Sci 1964; 11: 27-38.

38   Haslam E, Haworth RD, Jones K, Rogers MJ. J Chem Soc 1961; 8: 1829-1835.

39   Yamada H, Adachi O, Watanabe M, Gato O. Agric Biol Chem 1968; 32: 257-258.

40   Basarba J. Plant Soil 1964; 21: 8-16.

41   Ganga PS, Suseela RG, Nandy SC, Santappa M. Leather Sci 1977; 25: 203-209.

42   Suseela RG, Nandy SC. Leather Sci 1985; 32: 249-254.

43   Suseela RG, Nandy SC. Leather Sci 1985; 32: 278-280.

44   Chakrabarty AM. J Bacteriol 1972; 112: 815-823.

45   Chakrabarty AM. Proc Natl Acad Sci (USA) 1973; 70: 1641-1644.

46   Wheelis ML. Ann Rev Microbiol 1975; 29: 505-524.

Biotransformations: Bioremediation Technology for Health and Environmental Protection
V.P. Singh and R.D. Stapleton, Jr. (Editors)

Oxidation of organic and inorganic sulfur compounds by aerobic heterotrophic marine bacteria

Jose M. González[a], Ronald P. Kiene[b], Samantha B. Joye[a], Dimitry Yu. Sorokin[c] and Mary Ann Moran[a]

[a]Department of Marine Sciences, University of Georgia, Athens, GA 30602, USA.

[b]Department of Marine Sciences, University of South Alabama, Mobile, AL 36688, USA.

[c]Institute of Microbiology, Russian Academy of Sciences, pr. 60-letiya Oktyabrya 7, k. 2, Moscow 117811, Russia

INTRODUCTION

Microbial transformations of organic and inorganic sulfur compounds in marine environments are often the critical link in sea-atmosphere sulfur exchange. Since the oceans are the source of about half the sulfur compounds emitted to the atmosphere, the bacteria that mediate these transformations are of particular interest in understanding *sulfur cycling* on a global scale. While poorly understood, the oxidation of reduced sulfur compounds (organic and inorganic) could have important ecological consequences within marine food webs. Furthermore, the transformations of sulfur carried out by marine microorganisms may be of interest to biotechnologists involved in mitigating the adverse environmental effects of noxious reduced sulfur compounds.

Marine bacteria are responsible for degradation and transformation of volatile and *foul-smelling* organic sulfur compounds such as dimethylsulfide (DMS) and methanethiol. These compounds can originate from a variety of sources, but in aerobic marine habitats a major precursor is the plant osmolyte β-dimethylsulfonioprionate (DMSP). Studies over the last three decades have indicated that DMSP-derived DMS is the main pool for exchange of sulfur with the atmosphere [1,2], and that DMS plays an important roles in acidity of precipitation, cloud formation and climate regulation [3,4]. DMS is oxidized in the atmosphere to compounds that attract water droplets, increasing the likelihood of cloud formation [3,5]. DMS, its parent molecule DMSP, methanethiol and other organic sulfur compounds are typically associated with algae and vascular plants, because these organisms produce organic sulfur osmolytes. When marine plants decay or are grazed, DMSP is released and degraded to DMS or methanethiol. Much of this degradation occurs through the activities of aerobic bacteria. Concentrations of these organic sulfur compounds are highest in association with algal blooms [6,7,8] and coastal marshes [9,10].

Large quantities of reduced inorganic sulfur compounds in the form of sulfide are produced in salt marshes [10], generated from the reduction of sulfate during organic matter oxidation under anaerobic conditions. Over the past decades, large amounts of anthropogenic volatile sulfur compounds have also been discharged into the marine environment, potentially causing an imbalance in the natural processes of sulfur cycling. Reduced inorganic sulfur compounds are produced indirectly due to eutrophication processes that increase the rate of sulfate reduction. Organic and inorganic sulfides are also produced in large quantities by the paper industry, oil refineries, animal waste and wastewater treatment plants. The bacteria that degrade or transform this class of sulfur compounds are of biogeochemical and biotechnological interests.

Autotrophic sulfur oxidizers are of potential use in the transformation of inorganic and organic sulfur compounds of anthropogenic origin. Obligatey heterotrophic sulfur oxidizers, while less studied, are able to transform many compounds at rates comparable to those of autotrophic organisms [11-14]. Heterotrophic sulfur oxidizers are routinely isolated from the environment, and in some systems, such as the Black Sea, they can be retrieved in large numbers [13,14]. Fewer organisms have been isolated that oxidize the sulfur atom of organic compounds of environmental concern, but it appears that methylotrophic organisms may play a leading role in this type of transformation (see below).

The abundance of reduced sulfur compounds and the number of heterotrophic bacteria, that are isolated from the marine environment and which oxidize inorganic sulfur, indicate that it may be a widespread phenomenon in nature. This chapter focuses on the heterotrophic bacterial transformation of organic and inorganic molecules that are commonly found in the marine environment.

HETEROTROPHIC OXIDATION OF INORGANIC SULFUR COMPOUNDS

A variety of organisms oxidize sulfur as a source of energy, including chemosynthetic, photosynthetic and heterotrophic bacteria. Chemosynthetic sulfur oxidizers use sulfur compounds as a source of energy to fix CO_2. Phototrophic bacteria obtain reducing power from the oxidation of sulfur compounds. Heterotrophic bacteria oxidize inorganic sulfur compounds but relatively little is known about what energetic advantages they gain. Yet the widespread occurrence of this metabolism, as detailed below, argues for a need to review what is currently known of the process. These organisms are called chemolithoheterotrophic bacteria, living at the expense of an organic source of carbon (i.e. they are heterotrophic) and obtaining additional energy from inorganic sulfur oxidation [15].

Obligately heterotrophic inorganic sulfur oxidizers were isolated early in the last century from diverse environments, both marine and terrestrial [16]. Whenever characterized, they belonged to phylogenetically distant groups, including the genera *Pseudomonas* and *Xanthobacter*. Although not confirmed, certain species of *Mycobacterium, Arthrobacter, Flavobacterium,* and

Escherichia also appear to oxidize inorganic sulfur. Generally, these organisms oxidize one or more species of sulfur. Only in few cases, however, they have been shown to gain energy from such transformations. For example, facultative autotrophs (i.e. those that can grow both autotrophically and heterotrophically) have been shown to obtain energy from sulfur oxidation during heterotrophic growth, especially under limitation of organic substrate. Representatives of these organisms include some species of *Thiobacillus* and *Paracoccus* [17,18]. For strict heterotrophs, the rates of sulfur oxidation are surprisingly similar to the rates found in autotrophic organisms. The presence of reduced inorganic sulfur compounds increased the growth yield of several heterotrophic species, including unidentified marine pseudomonads [19], *Bosea* [20], *Sulfitobacter* [21,22], *Hyphomicrobium* [23], *Catenococcus* [24], and *"Thiobacillus"* strain Q [11]. Ruby et al. [25] isolated heterotrophic bacteria with a variety of phylogenetic affinities from a hydrothermal vent [26], whose growth was stimulated by the presence of reduced forms of inorganic sulfur compounds.

The ability to oxidize inorganic sulfur compounds appears to be widespread among lineages within the bacterial domain, and has been found in bacteria of widely different physiological types. For example, Sorokin et al. [27] detected oxidation of thiosulfate to tetrathionate both under aerobic and denitrifying conditions by the marine isolates belonging to different genomic groups of *Pseudomonas stutzeri*. Alkalophilic bacteria, belonging to the *Halomonas* cluster, were also found to oxidize sulfide, elemental sulfur, and thiosulfate to tetrathionate [28,29].

There are two pathways for the heterotrophic oxidation of inorganic sulfur compounds: one is leading to formation of tetrathionate and the other to sulfate (Figure 1). Although in autotrophic thiobacilli, polythionates (such as tetrathionate) are normal intermediates and can be further metabolized to

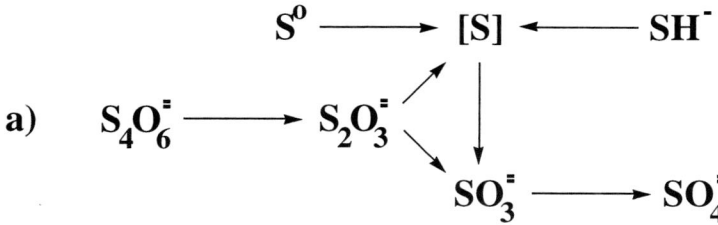

Figure 1. Pathways for the transformation of inorganic sulfur compounds in sulfate-forming (a) and tetrathionate-forming (b) heterotrophic bacteria. The initial transformation of tetrathionate ($S_4O_6^=$) is a reduction to thiosulfate ($S_2O_3^=$). [S] indicates sulfur covalently bound to a carrier, such as a protein carrier. In pure cultures of bacteria, as well as in the natural environment, chemical oxidations co-occur.

sulfate in obligate heterotrophs, the two pathways do not usually co-occur. Many heterotrophic bacteria oxidize thiosulfate, sulfide, and elemental sulfur to tetrathionate as the final product [12,19,30,31]. The oxidation of sulfur compounds may be more complex than a simple progression through the least to the most oxidized form, since compounds like elemental and sulfate are also formed during the oxidation of thiosulfate to tetrathionate by the Baltic Seas isolates [32].

Bacterial isolates from the Black Sea have served as model organisms for extensive studies of the oxidation of an array of sulfur compounds [13,31]. One of these versatile inorganic sulfur oxidizers is *Sulfitobacter pontiacus* [33]. This heterotrophic bacterium specializes in the oxidation of sulfite, which it is able to use as an additional energy source [22]. Other bacteria, that are closely related to *S. pontiacus* and carry out similar inorganic sulfur transformations, have been isolated from the marine environment [34]. In addition, two other species of *Sulfitobacter* have been recently described from the Mediterranean Sea (*S. mediterraneus* [35]) and Antarctica (*S. brevis* [36]).

Despite the potential energetic benefits offered by sulfur oxidation to heterotrophs and its widespread occurrence, the nature of this type of metabolism is not completely understood. For instance, the oxidation of inorganic sulfur compounds by heterotrophic bacteria could also be a defense mechanism against sulfur toxicity. The energy released during the oxidation of thiosulfate to tetrathionate is approximately eight times lower than that released by the oxidation of thiosulfate to sulfate. Yet, most heterotrophic sulfur oxidizers transform thiosulfate into tetrathionate, suggesting that there must be either an energetic or physiological advantage to this transformation. However, only a few studies have demonstrated that the addition of inorganic sulfur to the medium increases the efficiency of organic carbon utilization [11,19,24,37]. The formation of ATP during thiosulfate oxidation to tetrathionate has been demonstrated [38], pointing to the fact that organisms can conserve some energy in the oxidation of thiosulfate to tetrathionate. The fact that this metabolism is found in phylogenetically diverse bacteria suggests that it may be a widespread and successful mechanism to compete for limited amount of organic substrate (energy) [37], since growth yield on organic compounds can be higher when oxidized simultaneously with inorganic sulfur.

The model suggested by Sorokin et al. [24] for *Catenococcus thiocyclus* could explain the increase in growth yield for organisms that oxidize thiosulfate to tetrathionate. Tetrathionate (produced in the oxidation of thiosulfate) reacts chemically with sulfide to form elemental sulfur and regenerates thiosulfate. Thus thiosulfate is continuously recycled and only needs to be present at low concentrations to be used as a source of additional energy.

Such examples of energy gain from inorganic sulfur oxidation are not common; however, it is presently believed that most heterotrophic sulfur oxidizers do not obtain energy from this process. Yet, energy conservation coupled to the oxidation of reduced sulfur may be more common in heterotrophic organisms than currently recognized. Previous research on this aspect may not have accounted for the fact that enhancement of growth yield by sulfur oxidation may largely depend on experimental conditions [19]. For example, Sorokin and Lysenko [13] showed that an energetic advantage from the presence of reduced sulfur compounds could only be demonstrated in strain exhibiting high growth rates. Other bacteria may have the genetic capability to realize an energy gain from inorganic sulfur oxidation, yet do not express this ability at levels that can be detected under less-than-favorable growth conditions. In general, heterotrophs that are capable of complete oxidation of sulfur compounds to sulfate have more possibilities for gaining energy than tetrathionate-producing heterotrophs.

Heterotophic sulfur oxidizers may have an important ecological advantage in environments, where the flux of reduced sulfur compounds is insufficient for the proliferation of autotrophic species. Such environments would include the oxic/anoxi interface of the Black Sea, and the peripheral regions around hydrothermal vents and *volcanoes*. Studies have shown that heterotrophic bacteria, that oxidize thiosulfate to tetrathionate or sulfate, are abundant in the Black Sea oxic/anoxic interface, and appear to dominate over obligate chemolithotrophic organisms [13,14,39]. Studies with mixed natural populations of bacteria have shown chemolithoheterotrophic sulfur oxidation in other natural environments as well [32,40].

SOURCES OF INORGANIC SULFUR COMPOUNDS

Bacterial metabolism is the main source of reduced inorganic sulfur compounds in most marine environments. The formation of sulfide by sulfate reducing bacteria and Archaea is well studied, and the oxidative transformation by chemolithoautotrophic and phototrophic bacteria have also been well characterized. Reduced sulfur compounds of intermediate redox state are formed during the oxidative and reductive transformations carried out by sediment bacteria. For example, it is known that granules of elemental sulfur and polythionates are intermediates during growth of many chemolithoautotrophic thiobacilli [17]. Thiosulfate and polythionates (tetra- and trithionates) are found as products in the medium of pure cultures of sulfate reducers [41,42].

Besides the recognized role of sulfide, sulfite and elemental sulfur in the general sulfur cycle [43], Jorgenesen [44] and Jorgenesen and Bak [45] have pointed out that thiosulfate and tetrathionate are particularly important in the marine environment.

Podgorsem and Imhoff [32] proposed a tetrathionate cycle that could explain the formation of elemental sulfur in the Baltic Sea. Although tetrathionate is unstable and not expected to be present at high concentrations, up to 300 μM tetrathionate has been measured in coastal sediments [46]. Tetrathionate reacts chemically with sulfide to form elemental sulfur with the regeneration of thiosulfate. Such a cycle was previously proposed in the hydrothermal vent isolate *Catenococcus thiocyclus* (see above).

Other potential sources of inorganic sulfur compounds in the marine environment include reduced inorganic sulfur compounds, that are produced during the degradation of sulfur amino acids and organic sulfur compounds such as DMS, DMSP, methanethiol, and methanesulfonate. The parent organic compounds are present in significant concentrations in seawater. Another potentially important source of sulfide is the chemical hydrolysis of carbonyl sulfide (COS) in seawater, the latter is formed photochemically from DOM [47]. However, the source of most of the sulfide in open ocean seawater appears to be phytoplankton, particularly cyanobacteria [48], although the physiological reason for sulfide production is not clear [49]. Thus, because of its multiple sources and ubiquitous production, sulfide could be an important source of energy for heterotrophic sulfur oxidizers in the open ocean and other marine systems. Although not characterized by high standing stocks of sulfide, the pico- to nanomolar pools of sulfide in these systems may turn rapidly enough to support bacterial sulfide metabolism [48-50].

DEGRADATION OF ORGANIC SULFUR COMPOUNDS

Organic sulfur compounds are important substrates for microbial activity, since they are formed naturally in the environment and are also introduced via industrial processes. For example, the Kraft pulping process used to make paper releases significant amounts of DMS, methanethiol, DMDS, COS and CS_2 to the environment. Volatile sulfur compounds are also produced in wastewater treatment plants, brewing, and in compost piles. These volatiles give rise to offensive odors and contribute to acidification of atmospheric precipitation. Microbial desulfurization of organic sulfur compounds present in fossil fuels (i.e. thiophene compounds, benzothiophenes and dibezothiophenes) increases fuel quality and reduces emission of combustion products like sulfur dioxide. Moreover, a number of xenobiotics, such as pesticides and pharmaceuticals, contain a reduced sulfur atom and in many cases, the bioactive nature of the drug is due to sulfur moiety in the molecule.

In the aerobic marine environment, significant quantities of natural volatile organic sulfur compounds are formed due to the degradation of DMSP and other sulfur compounds. DMSP is produced by phytoplankton, macroalgae, and coastal vascular plants for which it serves as osmoprotectant. DMSP can be released into seawater by grazing of algal cells, cell lysis, and is naturally released by healthy plants to a limited degree. The concentration of DMSP in surface seawater generally ranges between 3 and 200 nM, but higher concentrations have been observed occasionally, and its turnover rate is high [51,52]. Until recently, the consensus was that most of the DMSP was degraded by bacteria to DMS (Figure 2). Recent studies have pointed out, however, that methanethiol is the main product of DMSP degradation in the surface ocean [7,53,54]. Since DMS is the primary compound involved in the exchange

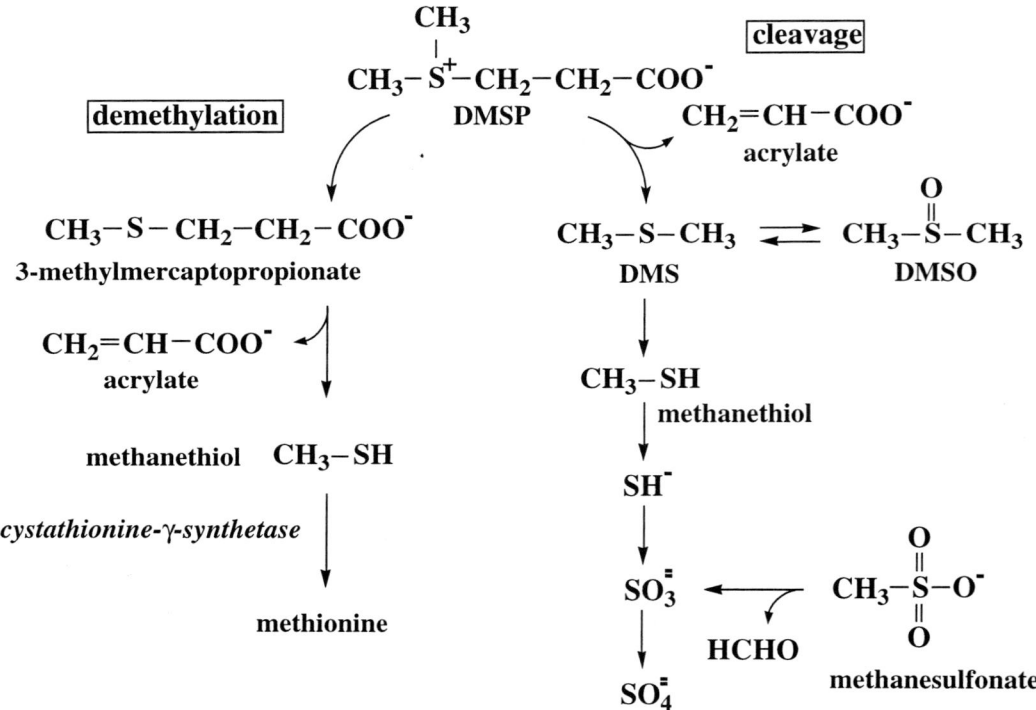

Figure 2. Pathways for the degradation of organic sulfur compounds in members of the *Roseobacter* group. At higher DMSP concentrations, most DMSP undergoes cleavage, whereas at lower concentrations, most DMSP follows the demethylation pathway with the production of methanethiol. Methanethiol is assimilated into cell material through cystathionine-γ-synthetase, a key enzyme in the assimilation of sulfur. The same mechanism operates in the marine environment.

of sulfur between the ocean and the atmosphere, understanding alternative pathway for DMSP degradation is clearly important.

In both aerobic and anaerobic bacteria, the enzyme DMSP lyase mediates the conversion of DMSP into DMS and acrylic acid, whereas a presently uncharacterized demethylae produces 3-methylmercaptopropionate (MMPA). Different phylogenetic groups of anaerobic bacteria demethylate DMSP [55] and, in a few cases, anaerobic isolates have been found to cleave DMSP into DMS and acrylic acid [56]. DMSP lyases have been purified and characterized [57,59], and the partial amino acid sequences of DMSP lyases were also reported from *Alcaligenes* sp. and *Pseudomonas doudoroffii* [58].

Aerobic and anaerobic bateria demethiolate MMPA, resulting in the formation of methanethiol and acrylic acid. Alternatively, MMPA is further demethylated to 3-mercaptopropionate (MPA). Methanethiol, that is formed from the demethiolation step, assimilated into bacterial cell proteins by means of the enzyme cystathionine-γ-synthetase, as documented with natural assemblages of marine bacteria [60]. MPA is degraded to acrylic acid and sulfide (Figure 2) [55]. Despite the fact that the DMSP demethylation is an important process in the natural environment, its biochemistry is for the most part unknown. In sulfate reducing bacteria, tetrahydrofolate is involved as methyl carrier for DMSP demethylation [61]. In other bacterial taxa, the methyl carrier has not been identified.

In the marine environment, DMSP supports up to 15% of the bacterial carbon demand in the surface ocean [62,63] and the sulfur moiety is efficiently assimilated into proteins [60]. The expression of the enzyme DMSP demethylase in the natural environment permits bacteria to regulate the amount of DMS released into the atmosphere by varying the ratio of DMSP, undergoing degradation through the lyase pathway versus the demethylase pathway. The ratio of DMSP degraded to methanethiol or DMS is not constant and varies seasonally [53], but the demethylation pathway can account for up to 80% of DMSP degradation in some oceanic surface waters [7,54,62,63]. DMSP demethylation may be regulated by bacterial demand for reduced forms of sulfur. Sulfate in seawater is at a very high concentration relative to DMSP, however, suggesting that this is an extreme adaptation by bacteria to use reduced forms of sulfur in preference to the abundant oxidized from. Reduced sulfur assimilation via methanethiol has also been documented in pure cultures grown in nutrient rich medium, where the assimilation of sulfate seawater should not have presented an energetic burden [34,61].

Bacteria, that use DMSP as carbon source, are routinely isolated from the marine environment [34,57,64-69]. The use of DMSP as a sulfur source is also likely to be a common characteristic of marine isolates [60]. Fewer bacteria have been isolated that degrade DMS or methanethiol [34,70-73]. Some chemolithotrophic bacteria can use oxidation of the reduced sulfur moiety of DMS and related compounds to support autotrophic growth [74,75]. Heterotrophic bacteria, such as members of the genus *Hyphomicrobium,* are also able to degrade DMS and related compounds while gaining energy from the oxidation of the sulfur moiety [23,71]. Methylotrophy is characteristic of the genus *Hyphomicrobium* [76] and possibly of other bacteria that degrade this type of C1 compound, such as *Roseobacter* group isolates [34]. Bacteria in the *Reseobacter* group also oxidize inorganic sulfur compounds, suggesting that they may obtain energy from the further oxidation of the sulfur moiety when degrading organic sulfur compounds. Since bacteria in the *Roseobacter* group often dominate *bacterioplankton* communities in coastal seawater and communities associated with algal blooms, the metabolism of these isolates may be representative of the transformations in natural marine environments [34,60,77,78]. The sulfur metabolic pathways that have been documented in members of the *Roseobacter* group are represented in Figure 2.

Additional sulfur transformations, carried out by members of the *Roseobacter* group, include the reduction of dimethylsulfoxide (DMSO) to DMS. For example, *Sagittula stellata* rapidly oxidizes DMS to DMSO, possibly to obtain energy or as a byproduct of the presence of oxidants, such as H_2O_2, that are produced during growth. DMS oxidation to DMSO was also found in *Pseudomonas acidovorans* isolated from a biofilter of a waste treatment plant [79]. In *Ps. acidovorans* as well as *S. stellata*, the reaction was reversible, since the two bacteria evolved DMS in the presence of excess DMSO. Phototrophic purple bacteria also use DMS as electron donor although their growth rate is generally much lower than that of heterotrophic bacteria.

In the marine environment, DMS degradation by bacteria is the main process that controls the concentration of DMS in surface waters [80]. DMS is also oxidized chemically and photochemically to DMSO, and the latter is found at relatively high concentrations in seawater, especially in association with phytoplankton [81]. DMS is also degraded in anaerobic environments by methanogens and sulfate reducers [82].

There are several known pathways for the degradation of DMS and DMSO; DMSO can be reduced to DMS by many types of bacteria, possibly cometabolically. In anaerobic environments DMSO is used as an electron acceptor with the regeneration of DMS [73]. During the degradation of DMS, the methyl groups are sequentially removed and oxidized to formaldehyde [23]. Visscher and Taylor [83] proposed a different mechanism for the degradation of DMS in *Thiobacillus* strain ASN-1, in which the methyl groups are not released as formaldehyde but rather a methyl carrier is involved, possibly a cobalamin carrier. Alternatively, DMSO can be oxidized to dimethyl sulfone, which is degraded to methanol, methane, and sufate [84].

Methanethiol, also ubiquitous in the marine environment, is derived from the degradation of animo acids and DMSP. The methylation of sulfide by bacteria is another source of methanethiol in aquatic environments. Methanethiol is generally very reactive and may interact with particles and metals in seawater before being subjected to biological transformation [6]. The isolation of methanethiol degraders, as well as DMS degraders, has been difficult since low concentration of these compounds are toxic to bacteria. However, methanethiol can be the sole carbon source for autotrophic isolates [74,75] and various facultative methylotrophs [23,70].

Another important organic compounds in the oceanic sulfur cycle is methanesulfonic acid. In the atmosphere, a portion of the DMS pool is oxidized to methanesulfonic acid, which returns to the Earth's surface via precipitation, where it is available for biological transformation. Methanesulfonic acid is degraded to sulfite and formaldehyde, and a few organisms capable of using it as a sole carbon source have been isolated [85-87].

Other sulfonates of biological origin can commonly be found in the marine environment, produced by *phytoplankton* and bacteria for various physiological function. Marine sediments contain substantial amounts of sulfonates of biological origin [88]. Household and industrial wastewater also contain sulfonates. For instance, linear alkylbenzene sulfonates (LAS) are major constituents of detergents that can be degraded in the environment [89,90]. The turnover rate of LAS in wastewater treatment plants, for example, is in the range of a few hours [89].

The gas produced during the burning of coal and Kraft pulping contains significant amounts of COS and CS_2. Other source of these gases are biomass burning and *fossil fuel* combustion. However, the major source of these two gases is biological production [91]. Autotrophic organisms can obtain energy from the breakdown of COS and CS_2 [75]. COS can also be cometabolized by CO-degrading bacteria [92].

Another important source of organic sulfur volatiles is the chemical integration between sulfides and organic acids and aldehydes in anaerobic environment [93]. Organic polysulfides result from chemical reactions of sulfide with simple organic molecules and may constitute a substantial pool of organic sulfur in the sediment [47,94].

CONCLUSION

Bacteria play important roles in both organic and inorganic sulfur transformations in the marine environment. Key component of the sulfur cycle is DMSP, perhaps the most important simple substrate that has been identified in seawater, since it contributes significantly to the bacterial carbon demand and is the main source of sulfur for bacteria in the marine environment. Thus, DMSP and its derivatives could be used for isolation of microorganisms that are either representatives of the bacteria in the natural environment or have potential utility for biotechnological applications. Heterotrophic sulfur oxidizers may be abundant in the marine environment, since there exists a variety of sources of reduced sulfur. Bacteria capable of heterotrophic sulfur oxidation may have an *energetic advantage* over those unable to obtain energy from the oxidation of inorganic sulfur.

REFERENCES

1   Andreae MO. The Role of Air-Sea Exchange in Geochange Cycling. New York: Reidel Publishing Co, 1986.

2   Kelly DP, Smith NA. Adv Microbiol Ecol 1990; 11: 345-385.

3   Charlson RJ, Lovelock JE, Andreae MO, Warren SG. Nature 1987; 326: 655-661.

4   Schwartz SE, Andreae MO. Science 1996; 272: 1121-1122.

5   Andreae MO. Mar Chem 1990; 30: 1-29.

6   Kiene RP, Linn LJ, Bruton JA. J Sea Res 2000 (in press).

7   Kiene RP. Mar Chem 1996; 54: 69-83.

8   Malin G, Kirst GO. J Phycol 1997; 33: 889-896.

9   Morrison MC, Hines ME. Atmos Environ 1990; 24: 1771-1779.

10 Steudler PA, Peterson BJ. Nature 1984; 311: 455-457.

11 Gommers PJF, Kuenen JG. Arch Microbiol 1988; 150: 117-125.

12 Mason J, Kelly DP. Microb Ecol 1988. 15: 123-134.

13 Sorokin Dyu, Lysenko AM. Microbiology (Engl Transl Mikrobiologiya) 1993; 62: 1018-1031.

14 Tuttle JH, Jannasch HW. Limnol Oceanogr 1972; 17: 535-543.

15 Kuenen JG, Beudecker RF. Phil Trans R Soc Lond B 1982; 298: 473-497.

16 Starkey RL. Soil Science 1935; 39: 197-219.

17 Kuenen JG, Robertson LA, Tuovinen OH. In: Balows A, Truper HG, Dworkin M, Harder W, Schleifer KH, eds. The Prokaryotes. Berlin: Springer-Verlag, 1992; 2638-2657.

18 Matin A. In: Strohl WR, Touvinen OH, eds. Microbial Chemoautotrophy, Colombus, Ohio: Ohio State University Press, 1984; 57-78.

19 Tuttle JH, Holmes PE, Jannasch HW. Arch Microbiol 1974; 99: 1-14.

20 Das SK, Mishra AK, Tindall BJ, Rainey FA, Stackebrandt E. Int J Syst Bacteriol 1996; 46: 981-987.

21 Sorokin DYu. Microbiology (Engl Transl Mikrobiologiya) 1994; 63: 255-259.

22 Sorokin DYu, Vedenina IYa, Grabovich MYu. Microbiology (Engl Transl Mikrobiologiya) 1999; 68: 14-20.

23 Suylen GMH, Stefess GC, Kuenen JG. Arch Microbiol 1986; 146: 192-198.

24 Sorokin Dyu, Robertson LA, Kuenen JG. FEMS Microbiol Lett 1996; 19: 117-125.

25 Ruby EG, Wirsen CO, Jannasch HW. Appl Environ Microbiol 1981; 42: 317-324.

26   Lane DJ, Harrison AP Jr, Stahl D, Pace B, Giovannoni SJ. Olsen GJ, Pace NR. J Bacteriol 1992; 174: 269-278.

27   Sorokin DYu, Teske A, Robertson LA, Kuenen JG. FEMS Microb Ecol 1999; 30: 113-123.

28   Sorokin DYu, Lysenko AM, Mityushina LL. Microbiology (Engl Transl Mikrobiologiya) 1996; 65: 326-338.

29   Sorokin DYu, Mityushina LL. Microbiology (Engl Transl Mikrobiologiya) 1998; 67: 78-85.

30   Durand P, Benyagoub A, Prieur D. Can J Microbiol 1994; 40: 690-697.

31   Sorokin Dyu. Microbiology (Engl Transl Mikrobiologiya) 1996; 65: 1-5.

32   Podgorsek L, Imhoff JF. Aquat Microb Ecol 1999; 17: 255-265.

33   Sorokin Dyu. Microbiology (Engl Transl Mikrobiologiya) 1995; 64: 354-365.

34   González JM, Kiene RP, Moran MA. Appl Environ Microbiol 1999; 65: 3810-3919.

35   Pukall R, Buntefuß D, Frühling A, Rohde M, Kroppenstedt RM, Brughardt J, Lebaron P, Bernard L, Stackebrandt E. Int J Syst Bacteriol 1999; 49: 513-519.

36   Labrenz M, Tindall BJ, Lawson PA, Collins MD, Schumann P, Hirsch P. Int J Syst Evol Microbiol 2000; 50: 303-313.

37   Tuttle JH. Appl Environ Microbiol 1980; 40: 516-521.

38   Vedenina DYa, Sorokin DYu. Microbiology (Engl Transl Mikrobiologiya) 1993; 61: 530-534.

39   Sorokin DYu. Izv Akad Nauk SSSR Ser Biol 1991; No. 4, 558-570.

40   Sorokin DYu. Microbiology (Engl Transl Mikrobiologiya) 1994; 63: 207-209.

41   Fitz RM, Cypionka H. Arch Microbiol 1990; 154: 400-406.

42   Sass H, Steuber J, Kroder M, Kroneck PMH, Cypionka H. Arch Microbiol 1992; 158: 418-421.

43   Kelly DP. Symp Soc Gen Microbiol 1988; 42: 65-98.

44   Jorgenesen BB. Symp Soc Gen Microbiol 1988; 43: 31-63.

45   Jorgenesen BB, Bak F. Appl Environ Microbiol 1991; 57: 847-856.

46   Luther GW, Church TM, Scudlark JR, Cosman M. Science 1986; 232: 746-749.

47   Radford-Knoery J, Cutter GA. Geochimica et Cosmochimica Acta 1994; 58: 5421-5431.

48   Walsh RS, Cutter GA, Dunstan WM, Radford-Knoery J, Elder JT. Limnol Oceanogr 1994; 39: 941-948.

49   Cutter GA, Krahforst CF. Geophys Res Lett 1988; 15: 1393-1396.

50   Luther GW, Tsamakis E. Marine Chemistry 1989; 27: 165-177.

51   Iverson RL, Nearhoof FL, Andreae MO. Limnol Oceanogr 1989; 34: 53-67.

52   Turner SM, Malin G, Liss PS, Harbour DS, Holligan PM. Limnol Oceanogr 1988; 33: 364-375.

53   Ledyard KM, Dacey JWH. Limnol Oceanogr 1996; 41: 33-40.

54   van Duyl FC, Gieskes WWC, Kop AJ, Lewis WE. J Sea Res 1998; 40: 221-231.

55   Taylor BF, Visscher PT. In: Kiene RP, Visscher PT, Keller MD, Kirst GO, eds. Biological and Environmental Chemistry of DMSP and Related Sulfonium Compounds. New York: Plenum Press, 1996; 265-276.

56   Wagner C, Stadtman ER. Arch Biochem Biophys 1962; 98: 331-336.

57   de Souza MP, Yoch DC. Appl Environ Microbiol 1995; 61: 21-26.

58   de Souza MP, Yoch DC. In: Kiene RP, Visscher PT, Keller MD, Kirst GO, eds. Biological and Environmental Chemistry of DMSP and Related Sulfonium Compounds. New York: Plenum Press, 1996; 293-304.

59   van der Maarel MJEC, Aukema W, Hansen TA. FEMS Microbiol Lett 1996; 143: 241-245.

60   Kiene RP, Linn LJ, González JM, Moran MA, Bruton JA. Appl Environ Microbiol 1999; 65: 4549-4558.

61   Jansen M, Hansen TA. Arch Microbiol 1998; 169: 84-87.

62   Kiene RP, Linn LJ. Geochim Cosmochim Acta 2000 (in press).

63   Kiene RP, Linn LJ. Limnol Oceanogr 2000 (in press).

64   Dacey JWH, Blough NV. Geophys Res Lett 1987; 14: 1246-1249.

65   de Zwart JMM, Kuenen JG. In: Kiene RP, Visscher PT, Keller MD, Kirst GO, eds. Biological and Environmental Chemistry of DMSP and Related Sulfonium Compounds. New York: Plenum Press, 1996; 413-426.

66   Kiene RP. Appl Environ Microbiol 1990; 56: 3292-3297.

67   Ledyard KM, DeLong EF, Dacey JWH. Arch Microbiol 1993; 160: 312-318.

68   Taylor BF, Gilchrist DC. Appl Environ Microbiol 1991; 57: 3581-3584.

69   Visscher PT, Taylor BF. Appl Environ Microbiol 1994; 60: 4617-4619.

70   de Bont JAM, van Dijken JP, Harder W. J. Gen Microbiol 1981; 127: 315-323.

71   Suyle GMH, Kenen JG. Antonie van Leeuwenhoek 1986; 52: 281-293.

72   Zhang L, Hirai M, Shoda M. J Ferment Bioeng 1991; 72: 392-396.

73   Zinder SH, Brock TD. Arch Microbiol 1978; 116: 35-40.

74   Kanagawa T, Kelly DP. Microb Ecol 1987; 13: 47-57.

75   Smith NA, Kelly DP. J Gen Microbiol 1988; 134: 3041-3048.

76   Pointexter JS. In: Balows A, Trüper HG, Dworkin M, Harder W, Schleifer K-H, eds. The Prokaryotes. Berlin: Springer-Verlag, 1992; 2176-2196.

77   González JM, Moran MA. Appl Environ Microbiol 1997; 63: 4237-4242.

78   González JM, Simo R, Massana R, Covert JS, Casamayor EO, Pedros-Alió C, Moran MA. Bacterial Community Structure Associated with a DMSP-Producing North Atlantic Algal Bloom, 2000 (submitted).

79   Zhang L, Kuniyoshi I, Hirai M, Shoda M. Biotechnol Lett 1991; 13: 223-228.

80   Kiene RP, Bates TS. Nature 1990; 345: 702-705.

81   Andreae MO. Limnol Oceanogr 1980; 25: 1054-1063.

82   Kiene RP, Visscher PT. Appl Environ Microbiol 1987; 53: 2426-2434.

83   Visscher PT, Taylor BF. Appl Environ Microbiol 1993; 59: 3784-3789.

84   Omori T, Saiki Y, Kasuga K, Kodama T. Biosci Biotech Biochem 1995; 59: 1195-1198.

85   Kelly DP, Murrell JC. Arch Microbiol 1999; 172: 341-348.

86   Kelly DP, Baker SC, Trickett J, Davey M, Murrel JC. Microbiology (Reading, England) 1994; 140: 1419-1426.

87   Thompson AS, Owens NJP, Murrell JC. Appl Environ Microbiol 1995; 61: 2388-2393.

88   Vairavamurthy A, Zhou W, Eglinton T, Manowitz B. Geochim Cosmochim Acta 1994; 58: 4681-4687.

89   Berna JL, Moreno A, Ferrer J. J Chem Biotechnol 1991; 50: 387-398.

90   Federle TW, Ventullo RM. Appl Environ Microbiol 1990; 56: 333-339.

91   Khalil MAK, Rasmussen RA. Atmos Environ 1984; 18: 1805-1813.

92   Smith KD, Klasson KT, Ackerson MD, Clausen EC, Gaddy JL. Appl Biochem Biotech 1991; 28-29: 787-796.

93   Mopper K, Taylor BF. In: Sohn ML, ed. Organic Marine Geochemistry. Washington D.C.: American Chemical Society 1986; 324-339.

94   LaLonde RT, Ferrara LM, Hayes MP. Org Geochem 1987; 11: 563-571.

Biotransformations: Bioremediation Technology for Health and Environmental Protection
V.P. Singh and R.D. Stapleton, Jr. (Editors)
© 2002 Elsevier Science B.V. All rights reserved.

Lignin degradation by bacteria

Archana P. Iyer and A. Mahadevan

Centre for Advanced Study in Botany, University of Madras, Chennai - 600 025, India

INTRODUCTION

Lignin is by far the most abundant aromatic substance present in the biosphere [1]. It is found in all higher plants [2]. Despite impressive data on the degradation of lignin by fungi [3], studies on the bacterial lignin degradation are woefully inadequate. Bacteria display versatile pathways to degrade aromatic substances, from simple phenols to highly complex lignin and related *xenobiotic* substances. The high rate of genetic adaptation exhibited by bacteria has contributed to enable them to develop this versatility through evolution.

In the cell wall, lignin is found intimately interspersed with hemicellulose, forming a matrix around the orderly cellulose microfibrils. Biosynthetically it arises from three substituted cinnamyl alcohols: *p*-coumaryl alcohol, coniferyl alcohol, and sinapyl alcohol. The relative contribution of these three alcohols to lignin biosynthesis in plants varies with phylogenetic origin. Lignins from gymnosperms are made up primarily of coniferyl alcohol with small amounts of *p*-coumaryl alcohol and traces of sinapyl alcohol. Angiosperm lignins are formed from approximately equal amounts of coniferyl and sinapyl alcohols and small amount of *p*-coumaryl alcohol. Lignins from graminaceous plants comprise of substantial amounts of *p*-coumaryl alcohol, coniferyl alcohol and large amount of sinapyl alcohol [3]. Oxidation of the precursor alcohols leads to the formation of free radical structures that polymerise to form the complex lignin polymer. It is a highly heterogeneous polymer, comprises of phenylpropanoid units linked through a variety of non-hydrolyzable C-C and C-O-C bonds [1]. Mounting environmental *pollution* has led to the growing interest in the potential use of microorganisms for the bioconversion of lignin-containing industrial waste, especially from paper mills.

COMPLEXITY OF LIGNIN DEGRADATION

Lignin breakdown involves multiple biochemical reactions, that have to take place more or less simultaneously: cleavage of intermonomeric linkages, demethylation, hydroxylation, side chain modifications, and aromatic ring fission, followed by disssimilation of the aliphatic metabolites produced. Insolubility of lignin and its lack of stereoregularity contribute to making it a substrate that is difficult for the microorganisms to degrade. Fungi overcome these constraints by virtue of a family of extracellular isoenzymes collectively called ligninases [4], which are peroxidases that act through a mechanism involving free radical formation [5]. Bacterial counterparts of these enzymes have not been found for a long time, although distinctive intracellular enzymes, catalyzing all the above reactions except cleavage of intermonomeric linkages have been described for low-molecular weight substrates [1].

Bacterial lignin peroxidases

An extracellular bacterial lignin peroxidase ALip-P3 was reported in *Streptomyces viridosporus* [6,7]. This enzyme catalyzed H_2O_2-requiring carbon-carbon bond cleavage in the side chains of phenolic and non-phenolic lignin substructure models. The reaction pattern resembled that of the reaction catalyzed by the fungal enzyme of *Phanerochaete chrysosporium* [8]. Microbial lignin peroxidase may play an important role in the initial biodegradation of lignin by oxidatively depolymerizing it within plant residues in soil.

Wang et al. [9] reported that *Streptomyces lividans* TK23-3651, a genetic variant having enhanced extracellular peroxidase and hydrogen peroxide production, significantly enhanced short term (30 days) organic carbon mineralization rates in soil. After its introduction into soil, the rate of organic carbon mineralization increased, particularly in non-sterile soil amended with lignocellulose. The enhancement was greater than that observed when wild-type isolate of *Streptomyces* was introduced into the soil. However, the significant enhancement was transient. Further studies showed that strain TK23-3651 was unstable, both on laboratory media and in soil. To obtain a stable *Streptomyces* strain, that produced enhanced levels of extracellular lignin peroxidase, Wang et al. [10] cloned the lignin peroxidase gene from the genome of *S. viridosporus* T7A into *S. lividans* TK64 in the high copy number plasmid pIJ702. Recombinant *S. lividans* strain TK64.1, expressing plasmid pIJ702.LP encoding the ALip-P3 lignin peroxidase gene, was the first lignin peroxidase clone isolated. It expressed high lignin peroxidase activity. ALip-P3 lignin peroxidase appears to affect the short term turnover rate of lignin-derived organic carbon in soil and normal low lignin peroxidase concentrations in soil may limit the initial turnover rate of lignified plant residues in soil.

Patterns of bacterial degradation of wood

An additional challenge for potential ligninolytic microorganisms is the need to gain access to the substrate by penetration of plant tissues. Fungi as well as lignocellulose-degrading actinomycetes accomplish this task by hyphal invasion of various cell wall layers [1]. Although physiological and taxonomic affiliations of wood-degrading bacteria are not well known, three main morphological forms of cell wall degradation have been recognized: (i) tunnelling, (ii) erosion, and (iii) cavitation [11]. *Tunnelling* bacteria migrate through the cell wall producing minute tunnels. *Erosion* bacteria attack the wall from the lumen into the secondary walls singly or in small groups.

Bacteria of the third category-*scavengers* are largely confined to the wood cell lumen, and apparently utilize products resulting from the activities of wood degraders [12].

Cometabolism

Despite impressive findings on the degradation of secondary substances, the factors that influence their metabolism have not gained much attention. Microbial degradation of lignin is a secondary process, and initially the energy requirement of the organism is met by easily metabolizable carbon sources such as cellulose or sugars [13]. Mahadevan et al. [14] reviewed the significance of *cometabolism* in the degradation of plant phenols. Utilization of guaiacol-glyceryl ether by *Bacillus subtilis* increased when 5 mM glucose was used as cometabolite.

LIGNIN-DEGRADING BACTERIA

Bacteria of several genera, including *Alcaligenes, Arthrobacter, Nocardia, Pseudomonas,* and *Streptomyces* readily degrade single ring aromatic substrates [2]. But most studies on lignin degradation by bacteria, carried out before 1977, led to unreliable results due to limitations in the experimental methodologies employed by various investigators. These limitations were overcome by the use of ^{14}C-labelled lignins, allowing researchers to measure the formation of $^{14}CO_2$, or to monitor the fate of the radioactive label.

Radiolabelling experiments have been carried out with various bacteria to determine their lignin degrading potential. A strain of *Nocardia* was reported to have released between 4% and 7% $^{14}CO_2$ from ^{14}C-ring labelled maize lignin or synthetic lignins (DHP : dehydrogenative polymerizate) in 15 days, and 6% to 15% of the label was present in the side chains and methoxyl groups [15]. In contrast, *Pseudomonas* exhibited only 1% $^{14}CO_2$ formation, irrespective of the location of the label [16]. *Bacillus megaterium* in turn, showed a distinct behaviour, mineralizing 12% of ^{14}C-side chain labelled spruce lignin, but only 0.3% of ^{14}C-ring DHPs in 20 days [17]. Janshekar and Fiechter [18] reported that *Nocardia, Pseudomonas,* and *Corynebacterium* strains degraded 1.0-10% of four different lignin preparations, as measured with a spectrophotometer. In this case, only unlabelled lignins were used, so these values may overestimate actual degradation. *Arthrobacter* sp. KB-1, growing on four lignin preparations from peanut hulls, mineralized 2.9% of the lignin component of *S. alternifolia* (^{14}C-lignin) lignocellulose in 10 days, although (^{14}C)kraft lignin from the same source was converted to $^{14}CO_2$ at only one-third of that rate [19]. Kern [20] reported *mineralization* of (^{14}C)DHP, ($O^{14}CH_3$)DHP, and (^{14}C-ring)DHP by resting cells of *Xanthomonas* sp. After 20 days, about 30% of the label of the DHP substrates had evolved as $^{14}CO_2$, the methoxyl groups being oxidized slightly faster than the rings and the side chains. Degradation of DHPs did not take place in the absence of oxygen and it occurred to the same extent upon incubation under air or 100% oxygen. The fate of the label was not only $^{14}CO_2$, about one-third of the fraction of DHP degraded by *Xanthomonas* sp. was incorporated into RNA, DNA, and protein, indicating that the bacterium utilized synthetic lignin as a carbon source.

METABOLISM OF LOW MOLECULAR WIEGHT LIGNIN MODEL COMPOUNDS

The most characteristic reaction taking place during biodegradation of a polymer is the enzyme catalyzed breakage of the intermonomeric linkages. Since there are so many different types of bonds in the lignin macromolecule, it is difficult to follow simultaneously the fate of all of them during the process. A widely accepted approach to overcoming this limitation consists of the use of lignin model substances, dimeric or tetrameric, possessing linkages that are typical of lignin. Besides simplifying the system, model substances are suitable for studying other reactions involved in lignin breakdown, such as side chain modification, ring fission, and demethoxylation.

The earliest studies employing model substances were done with α-conidendrin, a cyclolignan composed of two phenylpropane units. It is metabolized by *Flavobacterium* [21,22], *Achromobacter* [22], *Pseudomonas* [22,23], and *Agrobacterium* [24]. Substituted naphthalenes, isovanillic acid, vanillic acid, and guaiacol appeared as intermediates. *Pseudomonas putida* FK-2 [25], later found to consist of a consortium of bacteria [26] cleaved the carbon-carbon (C-C) bond of the coumarin ring of dehydrodiconiferyl alcohol, a β-5 type model, releasing ferulic acid and coniferyl alcohol as intermediates. *Pseudomonas putida* [27,28], *Beijerinckia* [29], *Ps. paucimobilis* [30], and *Ps. pseudoalcaligenes* [31] decompose biphenyl by introducing molecular oxygen into one of the rings to form a *cis*-hydrodiol, which is dehydrogenated to give 2,3-biphenyldiol. The modified ring is then cleaved between carbons 1 and 2 and further metabolized to produce benzoic acid from the intact ring.

A gene cluster encoding biphenyl metabolism in *Ps. pseudoalcaligenes*, has been cloned in *Ps. aeruginosa* [31]. The foreign DNA segment contains the first three enzymes of the *bph* operon, namely biphenyl dioxygenase, a dehydrogenase, and the ring-cleaving enzyme. The 5-5' biphenyl dehydrodivanillin is quantitatively consumed by *Ps. putida* and *Ps. fluorescens* via vanillic acid [32]. The utilization of various lignin and lignin-model compounds has been reported in *Acinetobacter* sp. from this laboratory [33]. It utilized ^{14}C-labelled teak wood lignin, veratrylglycerol-β-guaiacylether (Figure 1), 2-methoxy-4-formylphenoxyacetic acid (Figure 2), *p*-benzyloxyphenol, dehydrodivanillyl alcohol (Figure 3), α-conidendrin (Figure 4), dehydrodiisoeugenol (Figure 5), black liquor lignin, and indulin as sole carbon source [33].

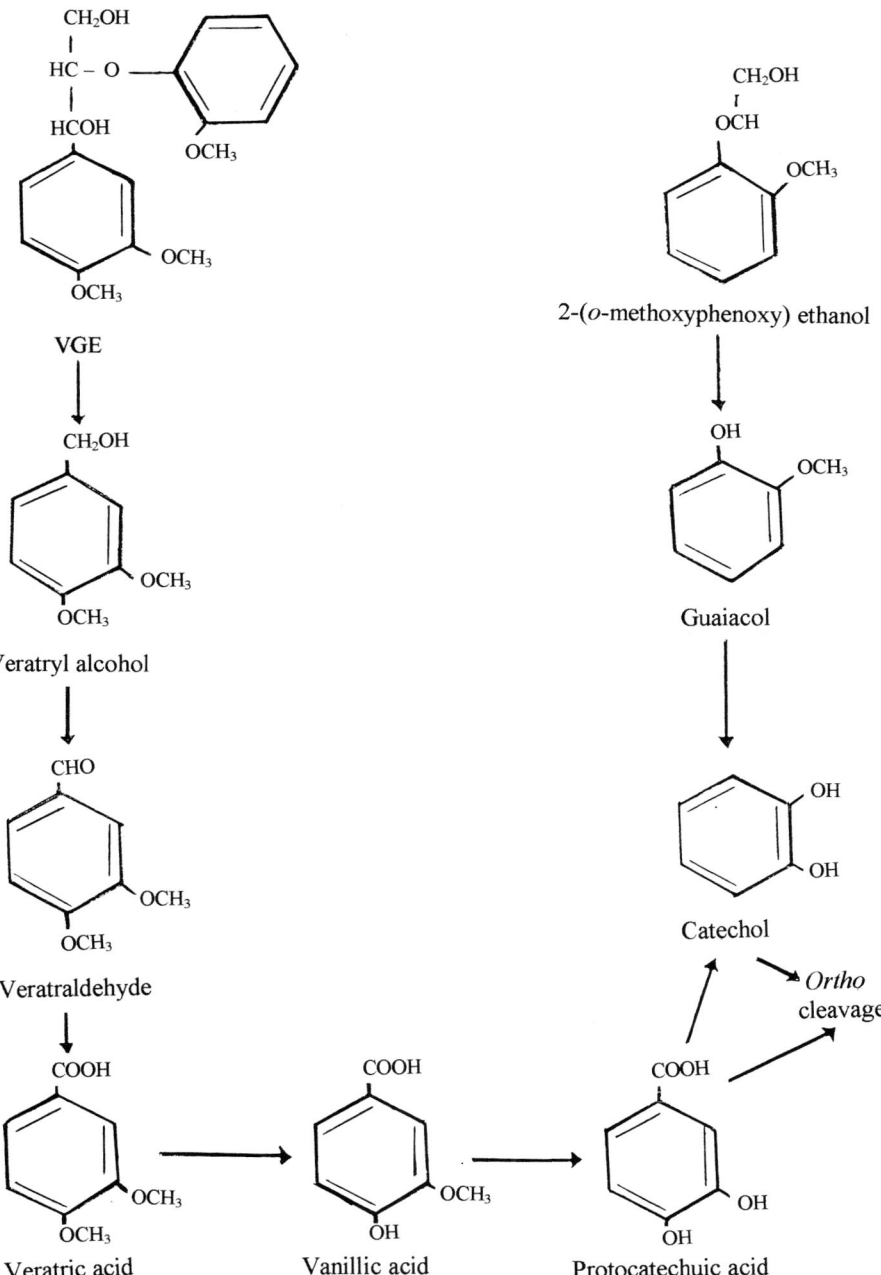

Figure 1. Proposed pathway for the degradation of veratrylglycerol-β-guaiacyl ether by *Acinetobacter* sp. [33].

318

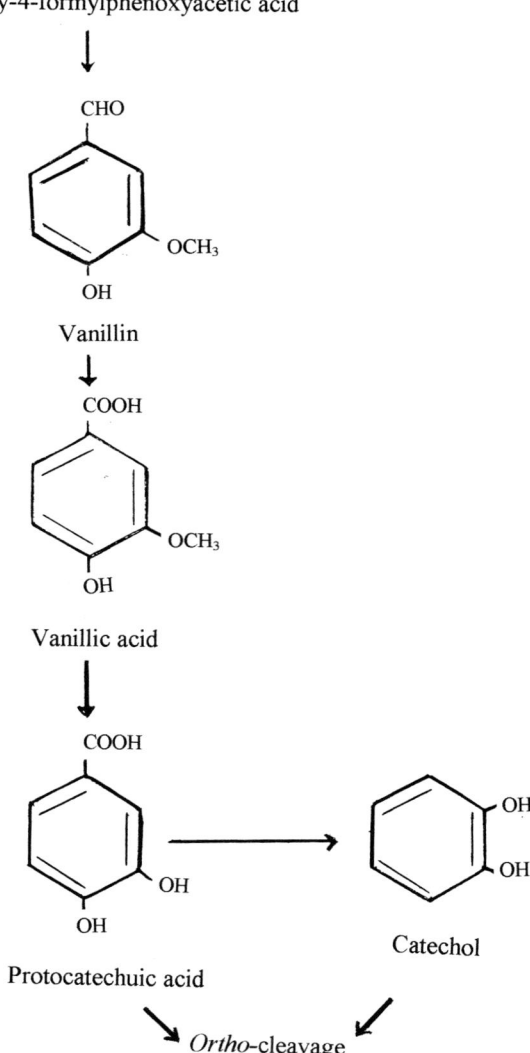

2-Methoxy-4-formylphenoxyacetic acid

Vanillin

Vanillic acid

Protocatechuic acid

Catechol

Ortho-cleavage

Figure 2. Pathway for the degradation of 2-methoxy-4-formylphenoxyacetic acid by *Acinetobacter* sp. [33].

The report by Muthukumar et al. [34], from this laboratory on the degradation of aromatic substances by rhizobia is a classic finding in soil

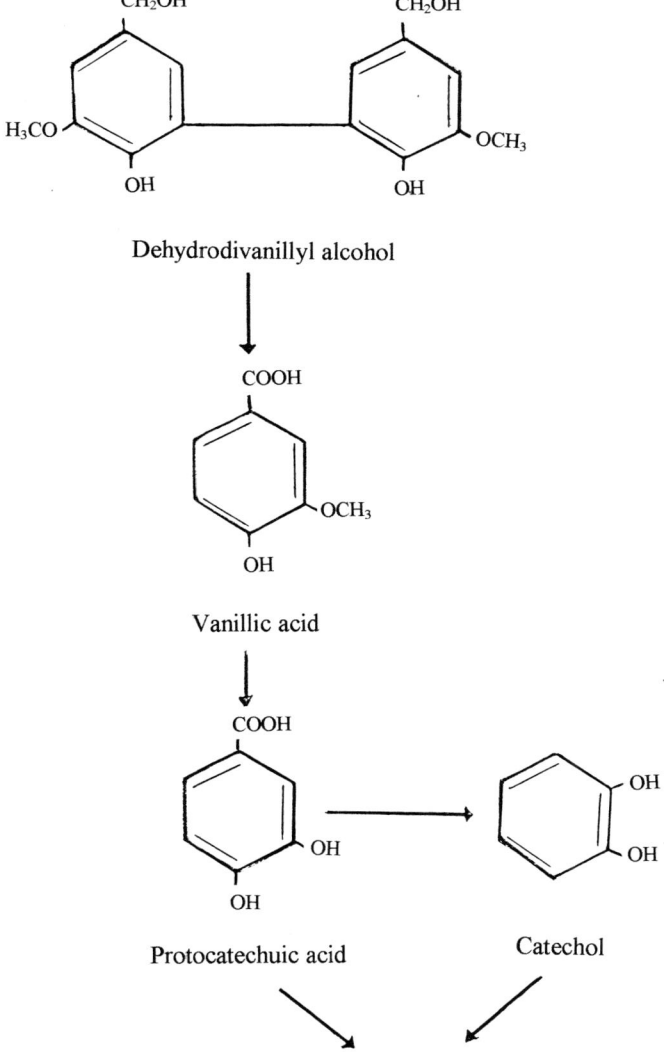

Figure 3. Pathway for the degradation of dehydrodovanillyl alcohol by *Acinetobacter* sp. [33].

microbiology. Subsequent studies in our laboratory [35,36] have clearly revealed that most rhizobia have the potential to degrade a wide spectrum of naturally occurring aromatic substances. Latha and Mahadevan [37] have reviewed the role of rhizobia in the degradation of aromatic substances. According to them, rhizobia have developed mechanisms to degrade and mineralize aromatic substances through the activity of chromosomally and plasmid-encoded oxygenases.

Plasmid-mediated dissimilation of ferulic acid has been reported in an isolate of *Bacillus subtilis* by Gurujeyalakshmi and Mahadevan [38]. It

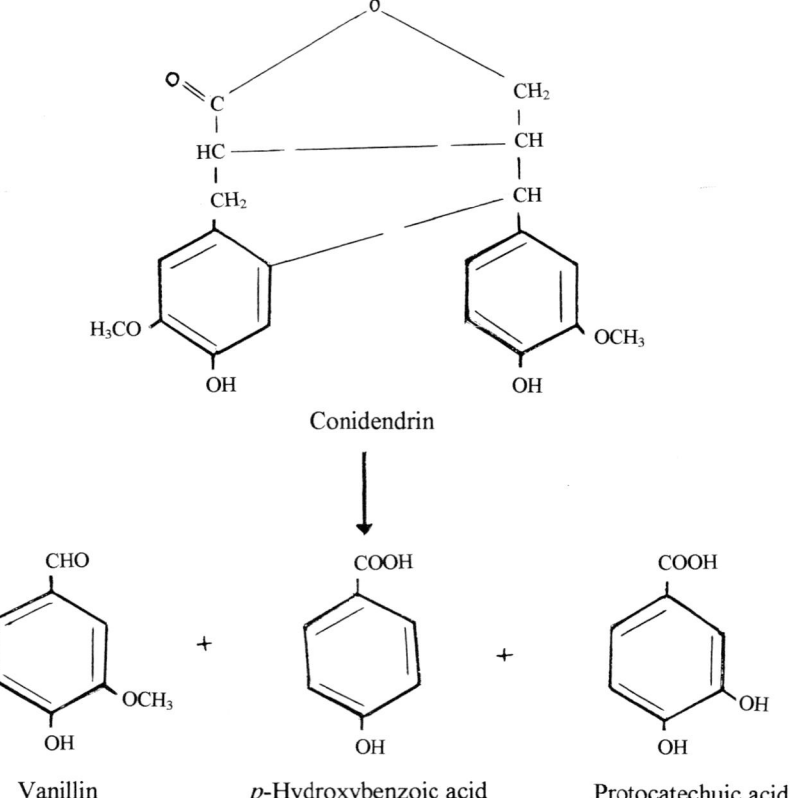

Figure 4. Degradation products of conidendrin degraded by *Acinetobacter* sp. [33].

produced vanillin, vanillic acid, and protocatechuic acid as intermediates (Figure 6). The enzymes of the ferulic acid degradative pathway, such as deacetylase, vanillin oxidase, vanillate-O-demethylase and protocatechuate 3,4-dioxygenase, were inducible in nature. This strain also utilized guaiacol glyceryl ether (GGE) as sole carbon source and catabolized it via catechol and guaiacol (Figure 7). Cell-free extracts of GGE grown cells contained high levels of catechol 1,2-dioxygenase and cleaved catechol via the *ortho* pathway

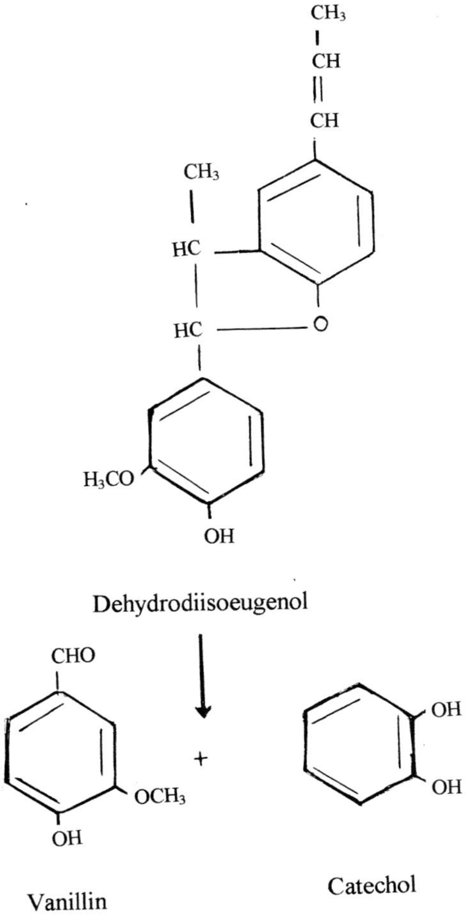

Dehydrodiisoeugenol

Vanillin

Catechol

Figure 5. Degradation products of dehydrodiisoeugenol degraded by *Acinetobacter* sp. [33].

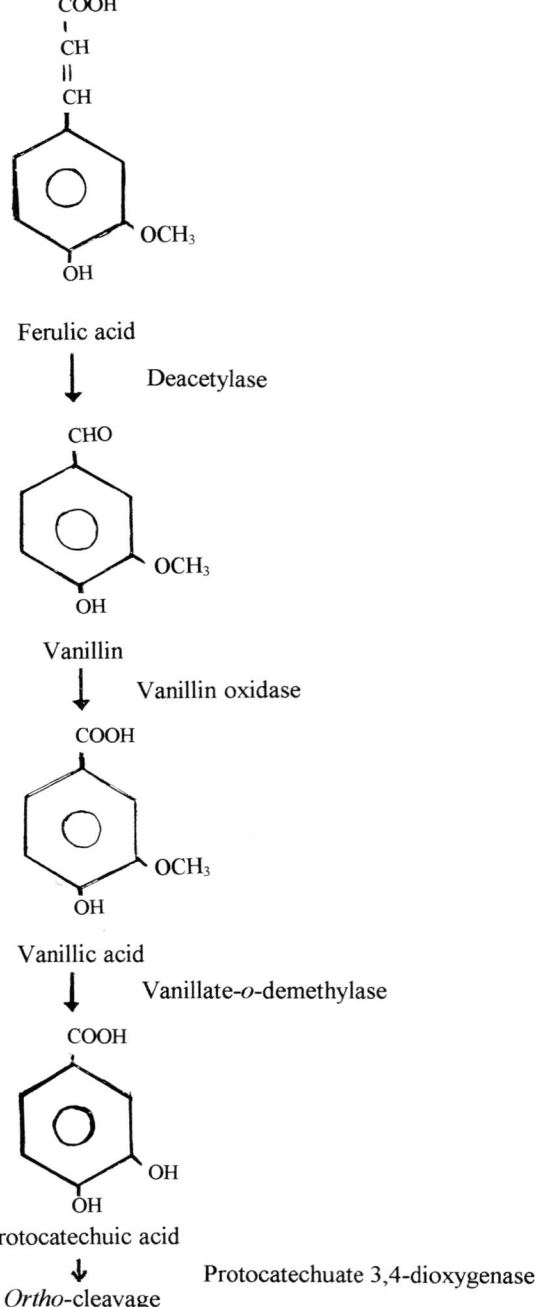

Ferulic acid

Deacetylase

Vanillin

Vanillin oxidase

Vanillic acid

Vanillate-*o*-demethylase

Protocatechuic acid

Ortho-cleavage

Protocatechuate 3,4-dioxygenase

Figure 6. Proposed pathway for the degradation of ferulic acid by *Bacillus subtilis* [38].

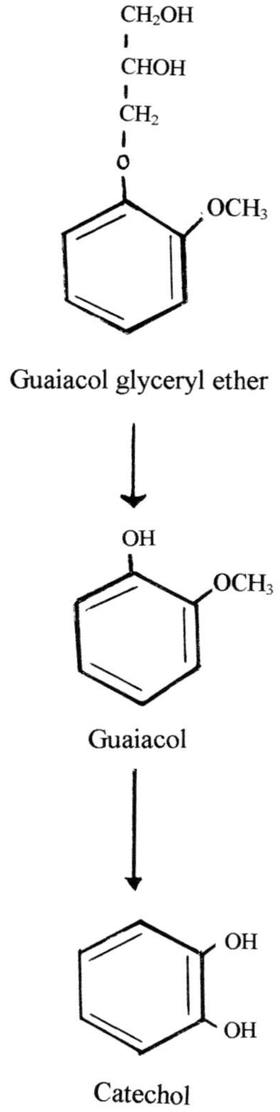

CH₂OH
|
CHOH
|
CH₂
|
O

OCH₃

Guaiacol glyceryl ether

OH

OCH₃

Guaiacol

OH

OH

Catechol

Figure 7. Pathway for the degradation of guaiacol glyceryl ether by *Bacillus subtilis* [38].

[38]. *Pseudomonas solanacearum* exhibited alternate pathways to dissimilate protocatechuic acid, which was concentration-dependent [39]. PCA at 2 mM was directly cleaved to β-carboxy *cis,cis*-muconic acid by protocatechuate

324

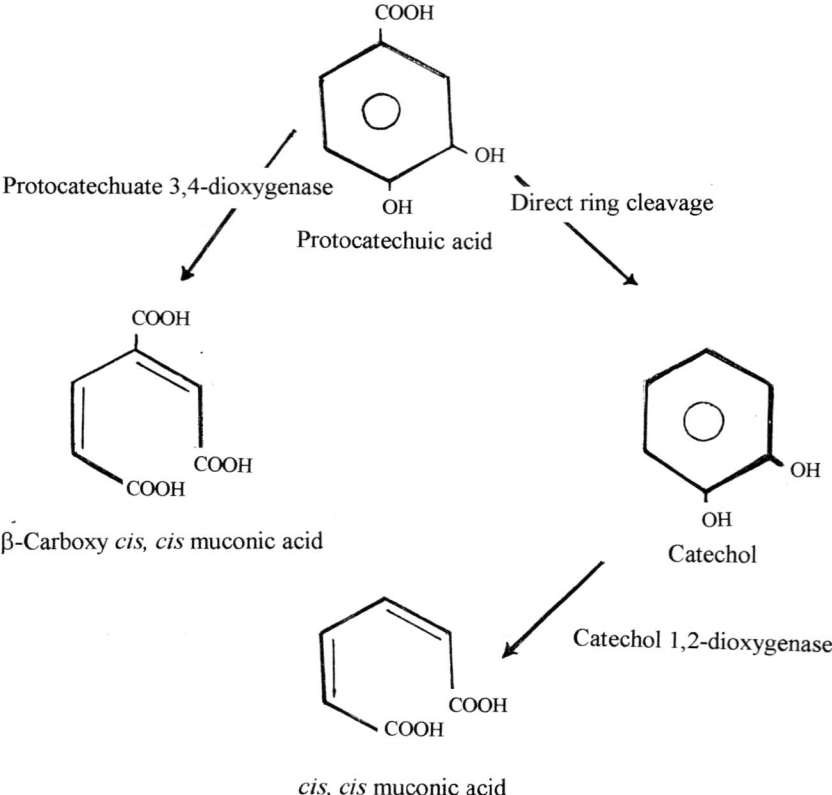

Figure 8. Cleavage of protocatechuic acid (PCA) by *Pseudomonas solanacearum* [39].

3,4-dioxygenase (Figure 8); but at 5 mM and 10 mM, it was decarboxylated to catechol.

Cloning to improve degradation efficiency

Cloning, sequencing and expression of genes encoding protocatechuate 3,4-dioxygenase from a marine isolate of *Bacillus subtilis* was achieved and assigned the accession no. AF057519 by GenBank,USA [40]. Growth of bacteria on model dimers implies that these bacteria produce enzymes that catalyze cleavage of *interunit* linkages. If so, why are bacteria so inefficient in degrading lignin? One explanation could be that bacterial enzymes are more specific than ligninase. If this is the case, complete depolymerization of lignin by a particular strain would require the simultaneous presence of several enzymes [1]. Comparative studies on the growth of natural bacterial isolates on different lignin model substances have shown that, in general, each species metabolizes only one type of dimeric model substance [41]. Enzyme specificity does not refer only to the nature of the atoms participating in each linkage, but also to their spatial configuration. Whereas lignin lacks stereoregularity as a result of its formation via free radical polymerization [42], most enzymes are highly stereospecific. An additional explanation for resistance of lignin to bacteria is that, although enzymes are produced, they are not secreted in the extracellular medium. In favour of this hypothesis is the evidence that ligninolytic strains exert very small changes in the average molecular weight of lignin substrates [43], suggesting degradation of only small fragments dissociated spontaneously from the polymer. Confirming this possibility, Kern and Kirk [44] observed an inverse correlation between lignin molecular weight and ability of *Xanthomonas* sp. to mineralize it, which led them to propose that only the fragments taken up by the cells undergo degradation.

GENETIC MANIPULATION

Application of *genetic engineering* techniques should provide valuable information on bacterial lignin degradation. Catabolic plasmids have been sought, because dissimilating pathways of several aromatic substances are coded for by extrachromosomal DNA [45]. In addition, plasmid DNA is easier to manipulate than chromosomal DNA. Salkinoja-Salonen and coworkers [46,47] proposed the involvement of plasmids in the degradation of low-molecular weight lignin-related compounds, based on metabolic instability of the strains, although direct evidence of their existence was not provided.

Gurujeyalakshmi and Mahadevan [48] provided evidence of plasmid-mediated ferulic acid dissimilation in *Bacillus subtilis*.

Another promising genetic approach is the construction of genomic libraries of ligninolytic strains. The aim of these experiments is to clone specific genes, hopefully those that code for enzymes catalyzing cleavage of interunit linkages. Selection for relevant transformants or conjugants can be performed on media containing dimers, that are metabolized by the strain whose DNA is being cloned, but not by the recipient. The main advantage of this strategy is that it allows cloning of genes, which code for enzymes that might not have been isolated [1]. Recently, in our laboratory, we have constructed the genomic library of *Bacillus* sp. isolated from decaying coir, capable of utilizing 0.25% indulin (kraft pine lignin) as sole carbon source, in the high copy number plasmid vector pUC19. Transformants were selected by α-complementation, and two clones MBA5 and MBA23 were selected, based on their ability to utilize indulin. The insert size was calculated as 4 kb.

A novel approach towards unravelling the mystery of bacterial lignin degradation is the study on the enzyme β-etherase. Arylglycerol-β-aryl ether is the most predominant linkage in most of the lignins. Therefore, cleavage of this linkage is the most important step in lignin degradation. Masai et al. [49] cloned and sequenced the genes encoding β-etherase enzyme from *Ps. paucimobilis* into *Escherichia coli*. Assay for β-etherase enzyme as well as hybridization studies using this fragment of known sequence can be used in large scale screening of potential *ligninolytic* bacteria.

PROSPECTS

Although studies on bacterial lignin degradation were initiated in 1940 [50], there are still several authentic gaps in this enigmatic puzzle. Perhaps the main reason for this is the less number of reports describing ligninolytic bacteria. Does this imply that very few bacteria degrade lignin or is it simply because only a few researchers are working on this aspect? However, the problem involves not only a qualitative aspect but also a quantitative one. The few bacterial strains described to date exhibit mineralizing activities significantly lower than those of ligninolytic fungi. In order to find out whether these strains are a representative sample of what can be found in nature, the search for new ones must continue. Further, the most recent advances in molecular biology and recombinant-DNA technology must be put to extensive use in not only large scale screening for potential ligninolytic bacteria but also in elucidating the molecular mechanisms underlying the phenomenon.

CONCLUSION

The complex aromatic polymer lignin comprises about 25% of the land-based biomass on earth, and its recycling is a vital component of the earth's carbon cycle. Its breakdown involves multiple biochemical reactions. Though the role of fungal lignin peroxidases has been well established only in recent years, bacterial lignin degradation has gained importance. Bacteria gain access to the polymer by mechanisms such as tunnelling, erosion, and scavenging. Bacteria of several genera such as *Alcaligenes, Acinetobacter, Pseudomonas, Rhizobium, Bradyrhizobium, Bacillus,* and *Streptomyces* readily degrade numerous lignin model compounds. *Acinetobacter* utilized veratrylglycerol-β-guaiacyl ether, indulin, 2-methoxy-4-formylphenoxyacetic acid, α-conidendrin and dehydrodiisoeugenol. Plasmid-mediated dissimilation of ferulic acid and guaiacolglyceryl ether has been established in *Bacillus subtilis.* Cloning, sequencing, and expression of the genes involved in the *dissimilation* of lignin substances have been done from *Bacillus subtilis* to improve the degradation efficiency. The most recent advances in genetic engineering are to create potential lignin-degrading bacteria and also to elucidate the molecular mechanisms involved.

ACKNOWLEDGEMENTS

Archana P. Iyer was financially supported by a Senior Research Fellowship from CSIR. A. Mahadevan's work was sponsored by Department of Environment, University Grants Commission and Indian Council for Agricultural Research, New Delhi.

REFERENCES

1   Vicuna R. Enzyme Microb Technol 1988; 10: 646-656.

2   Mahadevan A. Biochemical Aspects of Disease Resistance. Part II. Post Infectional Defence Mechanisms. New Delhi: Today and Tomorrow, 1991.

3   William F, Boominathan K, Vasudevan N, Gurujeyalakshmi G, Mahadevan A. J Sci Ind Res 1986; 45: 232-243.

4   Kirk TK et al. Enzym Microb Technol 1986; 8: 27-31.

5   Kersten PJ et al. J Biol Chem 1987; 262: 419-424.

6   Ramachandra M, Crawford DL, Pometto AL III. Appl Environ Microbiol 1987; 53: 2754-2760.

7   Ramachandra M, Crawford DL, Hertel G. Appl Environ Microbiol 1988; 54: 3057-3063.

8   Glen JK. et al. Biochem Biophys Res Commun 1983; 144: 1077-1083.

9   Wang Z, Crawford DL, Rafii F. Can J Microbiol 1989; 35: 535-543.

10   Wang Z, Bleakley BH, Crawford DL, Hertel G, Rafii F. J Biotechnol 1990; 13: 131-144.

11   Blanchette RA. Can J Bot 1995; 73: S999-S1010.

12   Singh AP, Nilsson J, Daniel GF. J Inst Wood Sci 1990; 12 (3): 143-157.

13   Gold MH, Mayfield MB, Cheng TM, Krisnaghura, Enoki A, Glen JK. Arch Microbiol 1982; 132: 115.

14   Mahadevan A, Sivasamy N, Sambandam T. In: P-Maal's Commemoration Volume, New Delhi: Today and Tomorrow, 1982; 42.

15   Trojanowski J, Haider K, Sundman V. Arch Microbiol 1977; 114: 149-153.

16   Kaplan DL, Hartenstein R. Soil Biol Biochem 1980; 12: 65-75.

17   Robinson LE, Crawford DL. FEMS Microbiol Lett 1978; 4: 301-302.

18   Janshekar H, Fiechter A. Eur J Appl Microbiol Technol 1982; 14: 1201-1206.

19   Kerr TJ, Kerr RD, Benner R. Appl Environ Microbiol 1983; 46: 1201-1206.

20   Kern HW. Arch Microbiol 1984; 138: 18-25.

21   Konetza NA, Pelczar MJ Jr, Gottlieb S. J Bacteriol 1952; 63: 771-778.

22   Tabak HH, Chambers CW, Kobler PW. J Bacteriol 1959; 78: 469-476.

23   Toms A, Wood JM. Biochemistry 1970; 9: 733-740.

24   Sundman V, Haro K. Finska Kemists Medd 1966; 75: 111-118.

25   Sundman V. Finska Kemists Medd 1962; 71: 26-35.

26   Samejima M et al. Mokuzai Gakkaishi 1985; 31: 956-958.

27   Catelani D et al. Biochem J 1973; 134: 1063-1066.

28   Catelani D, Colombi A. Biochem J 1974; 143: 431-434.

29   Gibson DT et al. Biochem Biophys Res Commun 1973; 50: 211-219.

30   Furukawa K, Simon JR, Chakrabarty AM. J Bacteriol 1983; 154: 1356-1362.

31   Furukawa K, Miyazaki T. J Bacteriol 1986; 166: 392-398.

32   Kawakami H. Mokuzai Gakkaishi 1976; 22: 537-538.

33   Vasudevan N, Mahadevan A. J Biotech 1989; 9: 107-116.

34   Muthukumar G, Arunakumari A, Mahadevan A. Plant Soil 1982; 69: 163-169.

35   Waheeta A, Mahadevan A. In: Bisen PS, Frontiers in Appl Microbiol. Delhi: CBS Publishers, 1994; 77-82.

36   Gajendiran A, Mahadevan A. Zentralbl Microbiol 1991; 146: 99-101.

37   Latha S, Mahadevan A. World J Microbiol Biotechnol 1997; 13: 601-607.

38   Gurujeyalakshmi G, Mahadevan A. Current Microbiol 1987; 16: 69-73.

39   Bhoominathan K, Mahadevan A. Zenralbl Microbiol 1992; 147: 483-487.

40   Gold DM, Mahadevan A. Ph.D Thesis, University of Madras,1998.

41   González B et al. Appl Environ Microbiol 1986; 52: 1428-1432.

42   Adler E. Wood Sci Technol 1977; 11: 169-218.

43   Kawakami H, Shumiya Y In: Higuchi T, Chang H, Kirk TK, eds. Recent Advances in Lignin Biodegradation Research. Chuo-Ku, Tokyo: Univ Publishers Co, 1983; 64-77.

44   Kern HW, Kirk TK. Appl Environ Microbiol 1987; 53: 2242-2246.

45   Farell R, Chakrabarty AM. In: Timmis KN, Puhler A, eds. Plasmids of Medical, Environmental and Commercial Importance. Amsterdam: Elsevier, North Holland Biochemical Press, 1979; 97-109.

46   Salkinoja-Salinen MS, Vaisanen E, Paterson A. In: Timmis KN, Puhler A, eds. Plasmids of Medical, Environmental and Commercial Importance. Amsterdam: North Holland Biochemical Press, 1979; 301-314.

47   Salkinoja-Salinen MS, Sundman V. In: Kirk TK, Higuchi T, Chang HM, eds. Lignin Biodegradation: Microbiology, Chemistry and Potential Applications. Florida: CRC Press, 1981; Vol I: 179-198.

48   Gurujeyalakshmi G, Mahadevan A. Appl Microbiol Biotechnol 1987; 26: 289-293.

49   Masai E, Katayama Y, Kubota S, Kawai S, Yamasaki M, Morohashi N. FEBS Lett 1993; 323: 1-2, 135-140.

50   Zobell CE, Stadler J. Arch Hydrobiol 1940; 37: 163-171.

Biotransformations: Bioremediation Technology for Health and Environmental Protection
V.P. Singh and R.D. Stapleton, Jr. (Editors)
© 2002 Elsevier Science B.V. All rights reserved.

Microbial bioremediation of textile effluents

R.S. Upadhyay

Department of Botany, Banaras Hindu University, Varanasi - 221 005, India

INTRODUCTION

Rapid industrialization is considered as a sign of development for developing and under-developed countries, but unfortunately most of the industries in these countries do not have proper waste treatment facilities. The discharged industrial effluents are mixed up with the natural water bodies either untreated or partially treated. Costly physico-chemical *decontamination* processes are often the only treatment alternatives available for such wastewaters. Untreated textile effluents are highly toxic, as they contain a large number of heavy metals [1,2]. Such toxic metal ions present in textile, kraft and paper mill effluents pose major environmental problem. Mutagenic, carcinogenic, and toxic potential of the heavy metals present in textile effluents have been extensively studied [3,4].

India, being a leading country in the world in exporting coloured textiles, carpets and crafts, is presently using millions of tons of dyes per year at small as well as at large scale for colouring cotton, silk, synthetic fibers, jute fibers, wool, etc. The colouring process discharges huge quantities of dye effluents, which pollute local terrestrial habitat, aquatic bodies as well as rivers. The major chemical groups of dyes are azo and reactive ones. Azo dyes have carcinogenic property and, therefore, people around water-bodies complain about skin diseases. It has also been observed that the effluents discharged in aquatic and terrestrial habitats persist for a longer period and create adverse effect on the vegetation.

There is no device that can be used at large scale for removing these dyes and making whole process *eco-friendly*. Recent studies have confirmed that strains/species of *Pseudomonas, Streptomyces, Phanerochaete,* and *Fusarium* have potential to colonize and decolourize *chromogenic* dye effluents. The cultural characteristics of strains, physiological experiments, effects of environmental factors in batch and continuous cultures, reductive pathways involved in reducing azo-group of dyes, and application of enzymes and microorganisms, both in free and in immobilized conditions, are reported in a few publications. In order to develop suitable technology to decolourize chromogenic dyes discharged in the effluent and to convert them into beneficial products simultaneously, a well-planned scientifically acceptable technology is needed.

The use of biological materials for heavy metal removal and recovery have gained important credibility during recent years because of good performance and low cost of treatment [5,6]. Microorganisms play a predominant role in solubilization, transport, and deposition of metals and minerals in the environment. The accumulation of metal ions from aqueous solution by a variety of microorganisms, including algae, bacteria, and fungi is well documented [7-9]. Walker et al. [10] demonstrated sorption of toxic metals to bacterial wall composites and Flaming et al. [11] extended this observation to desorption processes, that immobilized adsorbed metals from such biomineralized complexes. The high metal uptake capacity of many bacteria has also encouraged the application of microbial biomass for detoxification of effluents and also metal recovery [12-14].

Azo dyes, one of the largest classes of dyes used in textile industry, are released into the aquatic and terrestrial environments through the effluents emerging from textile and dye-stuff industries and are normally not removed by conventional wastewater treatment system [15]. The characteristic chemical structure of azo dyes makes them recalcitrant to biological break down.

Phanerochaete chrysosporium, a lignolytic white-rot fungus, degrades a wide variety of structurally diverse organopollutants [16]. The bacteria, *Streptomyces* and *Pseudomonas,* are also reported by few workers to degrade the colour as well as other pollutants of the effluent [17,18]. Fungal biosorption has been studied more extensively because of the availability of large amount of waste fungal biomass from fermentation industry and the amenability of the microorganisms to genetic and morphological manipulations [19]. Paszczynski et al. [17] have examined a new approaches to increase the susceptibility of azo dyes to degradation by aerobic microorganisms, especially by *Streptomyces* spp. and *P. chrysosporium.*

THE EFFLUENTS

Textile industry generates a large volume of effluents that are extremely variable in composition and their pollution load. At some point in the manufacture of most textile goods, chemical wet processing operations are necessary to properly prepare, purify, colour or finish the product. This results in the production of wastewater, the *pollution load* of which arises from the discharge of impurities as well as residual chemical reagents used for processing [20]. The wastewater compositions vary widely due to the variety of recipes, techniques, machinery, raw materials, and fabrics. In sizing, starch, polyvinyl alcohol, polyvinyl acetate, carboxymethyl cellulose, and gums are applied to increase tensile strength and smoothness. Wastewater also results from the cleaning of sizing boxes, size mixer, and eventually from the drainage of sizing solution [21]. The pollution load of designing effluent results from additives used in the size recipes such as surfactants, enzymes, acids, or alkali [22]. The generated wastewaters can be the largest contributors to the biochemical oxygen demand (BOD) and total suspended solids (TSS) in *textile effluents* [23]. Caustic soda is used to remove the natural waxes, pectins, oil and other non-cellulosic components. Pentachlorophenol is used to control mildew during storage and transportation [24].

Raw wool scouring is the highest pollution operation within the textile industry. In *bleaching* processes, hydrogen peroxide and sodium hypochlorite are used. Auxiliary chemicals such as sulfuric acid, hydrochloric acid, caustic soda, surfactant and chelating agents are also used [21,23]. There are hundreds of dyes reported in literature [25], which are classified by their chemical structure or in terms of their application to the fiber type. Heavy metals can arise from either the metallic part of the dye molecule, as is the chromium in acid dyes or copper in direct dye or from other materials used in the dyeing process [20,26,27].

BIOREMEDIATION OF THE TEXTILE EFFLUENTS

The biodegradation of dyes, using biomaterials as adsorbents is approximately a decade old. Literature reveals removal of astrazone blue by sawdust [28], lanosan black and sandocryl orange by peat and rice hulls [29], methylene blue by coconut husk [30], reactive dyes by bamboo pulp and jute fibers [31], congo red, acid violet and rhodamine-B by banana pith [32-34], and astrazone blue, maxilon red and telon blue by bagasse pith and maize cob [35], and basic dyes by sawdust [36]. Cationic dyes are reported to be removed by mosses [37]. Yoshida et al. [38] reported chitosan as an economical bioadsorbent for removing brilliant yellow and acid dyes. Chitin, fungal cell wall material, has been used as a bioadsorbent for removal of textile effluent [39].

The understanding on microbial degradation and decolouration of azo and reactive dyes is limited and has been studied by only a few groups of workers. However, results indicate that maximum dyes adopt reductive process of degradation. In this endeavour, Kulla [40] employed strains of *Pseudomonas* in chemostat culture for removal of dyes. Some anaerobic bacteria [41,42] and *Streptomyces* [17] have been characterized for decolouration of chromogenic dyes. A basidiomycetous fungus *Ganoderma lucidum* has been found as a suitable organism for removal of rhodamine-B and sandolan rhodine [43]. The *decolouration* of dyes, containing toxic chlorinated phenols used in Kraft bleach dyeing, has been observed by several workers [44-48]. Bennette et al. [49] reported that brownish colour of the effluent in the textile industry is due to chlorolignin. The ability of microfungi to degrade this component of the effluent, which is an enzymatic process, has been studied by Cammarota and Santa-Anna Jr. [48].

A basidiomycetous fungus *P. chrysosporium* possesses great potential for its commercial use in bioremediation of dyes and lignocellulosic materials present in textile effluents. This fungus has now become a model example for its biotechnological use in bioremediation. Glen and Gold [50] reported decolourization of three polymeric dyes namely Poly B-411, Poly R-481, and Poly Y-606 by *P. chrysosporium*. They found that these dyes serve as substrates for lignin degradation system of the fungus. The lignolytic degradation activity of the fungus was responsible for decolouration of the dyes. *P. chrysosporium* degrades a wide variety of diverse organopollutants [16] through its non-specific H_2O_2-dependent extracellular lignin-degrading enzyme, which is produced during secondary metabolism by the fungus in nitrogen limited medium. This fungus has been shown to degrade azo and heterocyclic dyes [51] and crystal violet [52,53]. Capalash and Sharma [54] reported degradation of eighteen different commercially used textile dyes by this fungus. The degradation varied from 40 to 70% based on the colour removal. Recently, Yesilada [53] reported decolouration of crystal violet by *Coriolus versicolor* and *Funalia trogii*. Bakshi et al. [55] found enhanced biodecolouration of synthetic commercial textile dyes by *P. chrysosporium* by improving Kirk's medium in respect to buffer, C : N ratio, Mg^{2+} and Zn^{2+}, temperature shifts, agitation, and sunflower oil addition. Bioremediation of dyes, containing toxic chlorinated phenols used in *Kraft bleach dyeing*, has been observed by several workers [44-48]. In a controlled experiment Ferrell et al. [56] compared the colour removal efficiency of lyophilized fungal culture with those of lignin peroxidase and horse-radish peroxidase. Both lyophilized culture and enzyme decoloured Kraft bleach dye.

Use of biological methods for detoxification of hazardous waste is an expanding technology with great potential as an effective and inexpensive alternative to past methods for clean-up of polluted environments. Various microorganisms have been used for the uptake of heavy metals and degradation of colour from aqueous system. Nakajima and Sakaguchi [57] studied metal binding capacity of 83 microorganisms and concluded that actinomycetes show higher capacities to bind metal ions as compared to fungi and bacteria. Fungi, in common with other microbial groups such as bacteria, cyanobacteria, and algae, are capable of accumulating heavy metals and radionuclioids even from dilute external solution [58-63]. Fungi possess some unique attributes that, in many ways, reflect their morphological and physiological diversity. There may be marked differences in uptake mechanisms between living and dead cells and, therefore, they have potential applications in various industries. Living cell can accumulate metals intracellularly often to higher levels than dead cells, and may also precipitate metals in and around cell wall, or in the external medium by various products of metabolism. Zhou and Kift [64] demonstrated uptake of copper from aqueous solution by immobilized fungal biomass. Biosorption columns of immobilized *Rhizopus arrhinus* have been found to be effective in removing low concentration of Cu^{2+} ions from aqueous solution.

BIOREMEDIATION OF TOXIC METALS FROM THE EFFLUENTS

A variety of mechanisms occur in fungal cells for the removal of heavy metals from the solution, which depend on intracellular metal accumulation or extracellular precipitation of metals. Metal uptake by living fungi can be divided into two main phases: (i) uptake by dead cells, and (ii) energy-dependent intracellular metal uptake across the cell membrane. In growing cells either or both of the phases of uptake may be observed by additional aspects of metabolism such as extracellular products, which may complex or precipitate metals outside the cells.

Three main types of adsorption involve electrical attraction, Vander-wall attraction, and chemical attraction of the solute to the adsorbent. First type is related to ion-exchange and is called *exchange adsorption* [65], occurring widely in fungal biomass, and adsorption has been defined as the attraction of positively charged ions to negatively charged ligands in cell materials [66]. A variety of ligands are involved in metal binding, including carboxylamine, hydroxyl phosphate, and sulfhydril groups. Biosorption can, therefore, be affected by the composition of the biomass and by other factors. In *R. arrhinus* adsorption appeared to be related to the ionic radius of La^{3+}, Mn^{2+}, Cu^{2+}, Zn^{2+}, Cd^{2+}, Ba^{2+}, Hg^{2+}, and Pb^{2+} , but not for Cr^{3+} or the alkali metal cations Na^+, and K^+ [67]. The biomass concentration has a significant effect on adsorption. Yeast cells adsorb more metal ions at low cell densities than at high cell densities. The binding of Hg^{2+}, Ag^{2+}, Cd^{2+}, Al^{3+}, Ni^{2+}, Cu^{2+}, and Pb^{2+} shows strong dependence in cell concentration.

Various treatments may be employed to increase the capacity of biomass for metal adsorption. Certain killing treatments may increase the adsorption capacity, and powdering of dried biomass exposes additional binding sites [67]. Compounds derived from fungal biomass may act as efficient biosorption agents. A wide variety of these materials are found in fungal cell wall, which are mannans, glucans, phosphomannans, melanins, chitin, and chitosan [68]. Chitin phosphate and chitosan phosphate have been found to adsorb greater amounts of V than Cu, Cd, Mn, Al , Co, Mg, and Ca. Fungal phenolic polymers and melanins contain, among other components, phenolic units, peptides, carbohydrates, aliphatic hydrocarbons, and fatty acid, the major oxygen containing groups being carboxyl, phenolic, alcoholic, hydroxyl, carbonyl, and methoxyl groups. Such groups are involved in transition metal binding to form metal-organic complexes [69]. Dilute (0.1 M) mineral acids (HCl, HNO$_3$ and H$_2$SO$_4$) have been found effective in removing Cu^{2+} from *Candida utilis* [70], *R. arrhinus,* and *Penicillium italicum* [9].

Metal ions are transported into cells across the cell membrane, which is a slower process than adsorption, and is inhibited by low temperature. In yeast, much greater amount of metal is accumulated by such a process than by metabolism-independent accumulation [71,72]. Yeast as well as other fungi may be able to precipitate metals around the cells as a result of metabolic processes, and are also capable of synthesizing intracellular metal-binding protein. Energy-dependent uptake of several metals by fungi has been demonstrated [62,73]. External factors like pH, presence of the anion and cations and organic materials can affect intracellular uptake.

Many extracellular fungal products can complex or precipitate heavy metals. Citric acid can be an efficient metal ion chelator, and oxalic acid can interact with metal ions to form insoluble oxalate crystals around cell walls in the external medium [74,75]. Iron is of fundamental importance to living cells, and many fungi release high-affinity Fe-binding molecules called siderophores. The externally formed Fe^{3+} chelates are subsequently taken up into the cell [76,77].

Bacterial sorption of metal ions has received attention mainly due to its environmental and biotechnological implications. The interaction between metal ions and bacteria attached to soil particles or sediment may influence the transport and fate of heavy metals in nature. Kauri et al. [78] observed that some bacterial strains, isolated from soil near Nagasaki, Japan, were able to bind Pu^{2+} during growth in the presence of low concentration of this ion. Walker et al. [10] demonstrated the sorption of toxic metals to bacterial wall-clay composites. Flemming et al. [11] extended these observations to desorption processes, that remobilize adsorbed metal from such biomineralized complexes. The high metal uptake capacity of many bacteria has also encouraged the application of microbial biomass to detoxification of effluents and also metal recovery [12-14]. The passive sorption of metal ion on to bacterial cells is related to the chemistry and structure of bacterial surface [79]. Some capsulated bacteria possess large amounts of anionic extracellular polysaccharides surrounding their cell walls. The important role of these *biopolymers* in the metal sorption has been studied in *Klebsiella aerogenes, Pseudomonas putida* [80], and *Arthrobacter viscous* [81]. Physical entrapment of precipitated metals in the polymer matrix and the complexation of soluble species by charged constituents of polymers have been suggested to be important in metal removal. Although microbial polymers consist mainly of neutral polysaccharides, they also contain compounds such as uronic acid, hexa-amines, and organically bound phosphate, which complex soluble ions by charged constituents of polymers. There are some reports indicating that the active synthesis of these polymers is induced in the presence of toxic metals. Lawson et al. [82] and Cheng et al. [83] isolated polymers from *activated-sludge* processes and studied the complexation capacity of these for metals such as copper, cadmium, and nickel. Acidic polysaccharides and proteins, which are present in the sheath of the *Sphaerotilus leptothrix* group, have been suggested to serve as the matrix for ferromanganese-oxide deposition [84].

MECHANISM OF DYE DEGRADATION

Secretion of azoreductases, which catalyse azo-linkage, has been discovered in *Pseudomonas* strains. The degradation mechanism described for *Pseudomonas* involves an oxygen insensitive azoreductase, which catalyzes the reductive cleavage of the azo group using NAD(P)H as electron donor [85]. Various anaerobic bacteria have been reported to degrade azo dyes. However, under aerobic condition these dyes have been considered to be essentially non-biodegradable. Cripps et al. [51], however, reported that *P. chrysosporium* degrades polycyclic hydrocarbons, containing azo and sulphur groups. Paszczynski et al. [17] adopted a new approach to increase susceptibility of azo dyes to degradation by aerobic microorganisms, especially by *Streptomyces* spp. and *P. chrysosporium*. They made azo dyes more degradable by introduction of a metabolizable substituent into the dye's chemical structure. It was found that lignase and manganese peroxidase preparations from *P. chrysosporium* were involved in dye degradation. Ferrell et al. [56] reported that *P. chrysosporium* produces multiple lignin peroxidase isoenzymes. They purified and characterized these heme-containing peroxidases. Leisola et al. [86] used analytical isoenzyme focussing to distinguish lignin peroxidase (LiP) enzymes in the extracelluar growth medium. The enzyme secretion by this fungus has been reported to increase by immobilizing the fungus on polyurethane foam (PUF) [87]. Buswell et al. [88] have demonstrated that *P. chrysosporium* INA-12 produces high LiP activity under nitrogen sufficiency, when glycerol is used as a carbon source. Supplementing the medium with oleic acid and soybean phospholipids [89] as well as temperature shifts [90] can also enhance LiP production. In a controlled experiment Ferrell et al. [56] compared the colour removal efficiency of lyophilized fungal culture with those of lignin peroxidase and horse-radish peroxidase. Both lyophilized culture and enzyme decoloured Kraft bleach dye. A well-defined lignin-degrading enzyme system, consisting of lignin peroxidase [91], Mn (II)-dependent peroxidase (MnP) [92-94] and glyoxal oxidase [95], was found to be involved in the dye degradation. The LiP and MnP are glycosylated heme-proteins containing photoporphyrin IX. Enhanced secretion of LiP and MnP in immobilized cell cultures of *P. chrysosporium* has been reported by Bonnarme et al. [87]. These authors discovered that soybean phospholipid fractions in the culture of *P. chrysosporium* increase succinate dehydrogenase, a mitochondrial marker and cytochrome-oxidoreductase, an endoplasmic marker. Reinekeke and Knackmuss [96] constructed bacteria, degrading haloaromatic compounds from a wild strain. The mechanism of degradation of chlorophenols and related compounds are also worked out [97]. Harayama and Don [98] stated the roles of *catabolic plasmids* in bacterial cells during utilization of halophenolics.

THE TECHNOLOGY

Several bioreactor types have been evaluated from Kraft bleach dye removal, using *P. chrysosporium*. Eaton et al. [45] used immobilized cells on rotating biological reactors but it failed due to fungal half-life, high oxygen requirement by the system and biomass losses [99]. Packed-bed or fixed-bed reactors have also been standardized, and have been found satisfactory [30]. The colour removal efficiency in Kraft bleach dye effluent, however, was increased by controlling hydraulic retention time in the packed-bed reactor [48].

Production of LiP and MnP in a low shear environment has been successfully achieved in airlift bioreactors [100]. Immobilization of *P. chrysosporium* on various supports, such as polyurethane foam or nylon web, has been done with success [101,102]. A thermodynamic model has been developed by these workers to predict the adhesion of conidiophores or mycelium to various solid carriers. Polyurethane foam was found to be the best amongst all the carriers. Zhen and Yu [103] studied the effect of hydraulic shear pressure and oxygen starvation stress on immobilized *P. chysosporium* in submerged culture. The immobilized fungal hyphae formed an even mycelial mat on a rotating cylindrical stainless steel net showing little sensitivity to the shear stress. However, at low aeration rate the lignase activity was considerably dropped.

Both living and dead fungi can accumulate heavy metals and radionuclioids by a variety of mechanisms. The current perspective appears to be that dead biomass has more advantages than living material in that it may be obtained inexpensively as waste. For use in packed-bed or fluidized-bed reactors, immobilized or pelleted biomass is of greater potential. Immobilized biomass has advantage in the sense that high flow rates can be achieved and clogging can be minimized, particle size can be controlled, and high biomass loadings are possible [104-106].

CONCLUSION

The search for efficient microorganisms and using them for large-scale use in dye degradation in textile effluents is in focus that will enable textile industries eco-friendly. When there is lack of proper technology for dye removal, more information gathered on microbial bioremediation through scientific research will provide useful technology for colour removal from dye effluents. Microbial biomass and other byproducts from *fermentation* industries will yield useful substances for bioremediation. It is clear from this review that bioremediation of toxic industrial effluents by microorganisms can be an effective method for removal of textile effluents, and can substitute the conventional remediation processes. Broad screening of microbial biomass type should be undertaken for developing new technologies. The current consensus is that, for improved commercial use of microorganisms, immobilization or pellet preparations in suitable carriers will prove economical for their use. In relation to other parameters of industrial relevance, microorganisms can be highly efficient in removing the toxic components from the effluents. The nonspecific lignin-degrading system of *P. chrysosporium* can be significantly utilized for the aerobic degradation of heterogeneous mixture of *recalcitrant* dyes in wastewaters. Growing evidence indicates that lignase (lignin peroxidase) is the key lignin-degrading enzyme of this white-rot fungus, which has potential for its commercial application.

REFERENCES

1   Walsh GE, Bahner LH, Horming WB. Environ Pollut Series A 1980; 21: 169.

2   Crine M. Tribune de l'eav 1993; 561: 3-19.

3   Delclos KB, Tarpley WG, Miller, EC, Miller JA. Cancer Research 1984; 44: 2540-2550.

4   Joachim F, Burrell A, Anderson J. Mutation Research 1985; 156: 131-138.

5   Volesky B. TIBTECH 1987; 5: 96-101.

6   Gadd GM. In: Rehm HJ, Reed G, eds. Biotechnology 1988; 6b: 401-430.

7   Sakaguchi T, Nakajma A, Horikosh T. Eur J Appl Microbiol Biotechnol 1981; 12: 84-89.

8   Charley RC, Bull AT. Arch Microbiol 1979; 123: 239-244.

9   de Rome L, Gadd, GM. Appl Microbiol Biotechnol 1987; 26: 84-90.

10   Walker SG, Flemming CA, Ferris, FG, Beveridge TJ, Bailey GW. Appl Environ Microbiol 1989; 55: 2976-2984.

11   Flamming CA, Ferris FG, Beveridge TJ, Bailey GW. Appl Environ Microbiol 1990; 56: 3191-3209.

12   Hutchins SR, Davidson MS, Brierley JA, Brierley CL. Ann Rev Microbiol 1986; 40: 311-336.

13   Gadd GM. Experientia 1990; 46: 834-840.

14   Volesky B. In: Volesky B, ed. Biosorption of Heavy Metals. Boca Raton: CRC Press, 1990; 7-43.

15   Pagga U, Brown D. Chemosphere 1986; 15: 479-491.

16   Bumpus JA, Aust SD. Bioassay 1986; 6: 166-170.

17   Paszczynski S, Pasti MB, Goszczynski S, Crawford DL, Crawford RL. Enzyme Microbiol Technol 1991; 13: 378-384.

18   Srivastava SK, Srivastava AK, Jain N. Indian J Exp Biol 1995; 33: 962-966.

19   White C, Gadd GM. J Chem Tech Biotechnol 1990; 19: 331-343.

20   ADMI. Dyes and the Environment: Report on Selected Dyes and Their Effects. American Dye Manufacturers Institute USA, 1973.

21   Cooper SG. The Textile Industry: Environmental Control and Energy Conservation. Noyes Data Co, Park Ridge, New Jersey, 1978.

22   Smith B. Textile Chem Color 1992; 24(6): 30.

23   Nolan WF. Analysis of Water Pollution Abatement in the Textile Industry. M.Sc. Thesis Chemeson University, Chemson USA, 1972.

24   Lockerbie M. In: Proceedings of the Symposium on Textile Industry Trade Effluents. Institute for Water and Environmental Management, Rochdale, UK, 1994.

25   Colour Index. The Society of Dyers and Colourists. American Association of Textile Chemist and Colourist, USA, 1987.

26   Horning RH. Textile Dyeing Waste Water Characterization and Treatment, US. Department of Commerce. National Technical Information Service, USA, 1978.

27   Smith B. American Dye Stuff Reporter 1989; 78 (4): 26.

28   Asfour HM, Nasar MM, Fadali OA, El-Geundi, . J Chem Technol 1985; 35: 28-35.

29   Nawar SS, Doma HS. The Sci Total Environ 1989; 79:271-277.

30   Low KS, Lee CK. Pertanika 1990; 13: 221-228.

31   Shukla SR, Sahardane, VD. J Appl Polymer Sci 1990; 41: 2655-2663.

32   Kanchana, M. M. Phil. Dissertation, Bhartiar University, India 1991.

33   Namasivayam C, Kanchana N. Chemosphere 1992; 25: 1691-1706.

34   Namasivayam C, Kanchana N, Yamuna RT. Water Management 1993; 13: 24.

35   Nassar MM, El-Geundi MS. Chem Technol Biotechnol 1991; 50: 257-264.

36   Khattri S D, Singh MK. Indian J Chem Technol 1999; 5: 230-234.

37   Lee CK, Low KS. Pertanika 1987; 10: 335-339.

38   Yoshida H, Fukuda S, Okamoto A, Kataoka J. J Water Sci Technol 1991; 23: 1667-1676.

39   Mac Kay HG, Otterburn MS, Sweeney AG. Water Res 1982; 14: 15-20.

40   Kulla, HG. In: Leisinger T, Cook AM, Nuesch J, Hutter T, eds. Microbial Degradation of Xenobiotics and Recalcitrant Compounds. London: Academic Press, 1981; 387-399.

41   Wuhrmann K, Mechsner K, Kappler T. Eur J Appl Microbiol Biotechnol 1980; 9: 325-338.

42   Meyer V. In: Leisinger T, Cook AM, Nuesch J, Hutter T, eds. Microbial Degradation of Xenobiotic and Recalcitrant Compounds. London: Academic Press, 1981; 371-385.

43   Mittal AK, Venkobacher C. Indian J Environ Health 1989; 31: 105-111.

44   Marton J, Stern AM, Marton T. Tappi J 1969; 52 (10): 1975-1981.

45   Eaton DC, Chang DC, Joyce TW, Jeffries TW, Kirk TK. Tappi J 1982; 5(6): 89-92.

46   Renganathan V, Mikik K, Gold MH. Biochemistry 1987; 26: 5127-5132.

47   Pellinen J, Yinc F, Joyce TW, Chang H. J Biotechnol 1988; 8: 67-76.

48   Cammarota MC, Santa-Anna Jr GL. Environ Technol 1992; 13: 65-71.

49   Bennett DJ. Tappi J 1971; 554(12): 2019-2026.

50   Glen JK, Gold MH. Appl Environ Microbiol 1983; 45: 1741-1747.

51   Cripps C, Bumpus JA, Aust SD. Appl Environ Microbiol 1990; 56: 114-1118.

52   Bumpus JA, Brock BJ. Appl Environ Microbiol 1988; 54: 1143-1150.

53   Yesilada O, World J Microbiol Biotechnol 1995; 11: 501-502.

54   Capalash N, Sharma P. World J Microbiol Biotechnol 1992; 8: 309-312.

55   Bakshi DK, Gupta KG, Sharma P. World J Microbiol Biotechnol 1999; 15: 507-509.

56   Ferrell I, Dezotti M, Duran N. Biotechnology Lett 1991; 13 (8): 577-82.

57   Nakajima A, Sakaguchi T. Appl Microbiol Biotechnol 1986; 24(1): 59-64.

58   Zajic JE, Chiu YS. Dev Ind Microbiol 1972; 13: 91-100.

59   Tuovineno H, Kelly DP. Intl Metall Rev 1974; 19: 21-31.

60   Gadd GM, Griffiths AJ. Microbios Lett 1978; 6: 117-124.

61   Shumate, SE, Strandberg GW. Comprehensive Biotechnology 1985; 4: 235-240.

62   Gadd GM. In: Herbert RA, Codd GA, eds. Microbes in Extreme Environments. London: Academic Press, 1986; 135-147.

63   Trevorse JT, Stratton GW, Gadd GM. Can J Microbiol 1986; 32: 447-464.

64   Zhou J, Kift RJ. J Chem Tech Biotechnol 1991; 52: 317-330.

65   Weber WJ. In: Weber WJ, ed. Physico-chemical Processes for Water Quality Control. New York: Wiley, 1972; 199-259.

66   Brierley JA, Brierley CL. In: Westbroke P, de Jong EW, eds. Biomineralization and Biological Metal Accumulation. Dordrecht: Reidel, 1983; 499-509.

67   Tobin JM, Copper DG, Neufeld RJ. Appl Environ Microbiol 1984; 46: 821-824.

68   Muzzarelli RAA, Tanfani F. In: Mirano S, Tokura S, eds. Chitin and Chitosan. Japan: The Japanese Society of Chitin and Chitosan to Hori, 1982; 183-186.

69   Saizjimenez C, Shafizadeh F. Curr Microbiol 1984; 10: 281-286.

70   Khovrychee, MP. Microbiology 1973; 42: 745-749.

71   Norris PR, Kelly DP. J Gen Microbiol 1977; 99: 317-324.

72   Norris PR, Kelly DP. Dev Indust Microbiol 1979; 20: 299-308.

73   Borst-Pauwels, GWFH. Biochem Biophys Acta 1981; 650: 88-127.

74   Murphy RJ, Levy JE. Trans Brit Mycol Soc 1983; 81: 165-168.

75   Shutter HP, Jones EBG, Walchi, O. Fr Mater Org 1983; 18: 243-263.

76   Raymond KN, Muller G, Matzanke BF. Topics Curr Chem 1984; 123: 49-102.

77   Adjimani JP, Emery T. J Bacteriol 1987; 169: 3664-3668.

78   Kauri T, Santry DC, Kudo A, Kushner D. Environmental Toxicology and Water Quality 1991; 6: 109-112.

79   Beveridge, TJ. Ann Rev Microbiol 1989; 43: 147-171.

80   Brown MJ, Lester JN. Water Res 1982; 16: 1539-1548.

81   Scott JA, Sang GK, Palmer SJ, Powell DS. Biotechnol Lett 1986; 8: 711-714.

82   Lawson PS, Steritt RM, Lester JN. J Chem Tech Biotechnol 1984; 34B: 253.

83   Cheng MG, Patterson JW, Minear RA. J Water Pollut Control Fed 1975; 47: 362.

84   Ghirorse WC. In: Ehrlich HL, Holmes DS, eds. Biotechnology for Mining, Metal-refining and Fossil Fuel Processing Industries. New York: Wiley, 1986; 141-148.

85   Zimmerman T, Kulla HG, Leisinger T. European J Biochem 1982; 129: 197-203.

86   Leisola MS, A Kozulic B, Meusdoerffer F, Fiechter A. J Biol Chem 1987; 262: 419-424.

87   Bonnarme P, Delatter M, Corriev G, Asthar M. Enzyme Microbiol Technol 1991; 13: 727-733.

88   Buswell JA, Mollet B, Odier E. FEMS Microbiol Lett 1984; 2: 295-299.

89   Asther M, Lesage L, Drapon R, Corrieu G, Odier E. Appl Microbiol Biotechnol 1988; 27: 393-398.

90   Asther M, Capdevila C, Corrieu G. Appl Environ Microbiol 1988; 54:3194-3196.

91   Kirk TK, Croans TM, Murtagah KE, Farrell RL. Enzyme Microbial Technol 1986; 8: 27-32.

92   Anderson LA, Renganathan V, Chiuaa L, Gold MH. J Biol Chem 1985; 260: 6080-6087.

93   Huynh VB, Crawford RL. FEMS Microbiol Lett 1985; 28: 119-123.

94   Asada Y, Myabe M, Kikkawa M, Kuwaham K. Fermento Technol 1987; 65: 483-487.

95   Kersten PJ, Kirk TK. J Bactriol 1986; 8: 2195-2201.

96   Reinkeke W, Knackmuss HJ. Nature (Lond) 1979; 277: 385-386.

97   Chapman PJ. In: Omenu GS, ed. Environmental Biotechnology. NewYork: Plenum Press, 1988; 81-95.

98   Harayama S, Don RH. In: Sethow JK, Hollander A, eds. Genetic Engineering: Principles and Methods, Vol. 7 New York: Plenum Corporation, 1985; 283-307.

99   Campbell AG. Ph. D. Dissertation, North Carolina State University USA. 1983.

100   Bonnarme P, Jeffries TW. J Ferment Bioeng 1990; 70: 158-163.

101   Kirpatrick N, Palmer JM. Appl Microbiol Biotechnol 1987; 27: 129-133.

102   Linko S. Enzyme Microbiol Technol 1988; 10: 410-417.

103   Zhen Z, Yu J. The Canadian J Chem Eng 1998; 76: 784-789.

104   Shumate SE, Strandberg GW, Whirter DA, Parrott JR, Bogaiki, GM, Locke, BR. Biotechnol Bioeng Symp 1980; 10: 27-34.

105   Tsezos, M. In: Eccelas H, Haut S, eds. Immobilization of Ions by Biosorption. Chichestor, UK: Ellis, Horwood, 1986; 201-218.

106   Yakubu NA, Dudency AWL. In: Eccles H, Hunt S, eds. Immobilization of Ions by Biosorption. Chichester, UK: Ellis Harwood, 1986; 183-200.

Biotransformations: Bioremediation Technology for Health and Environmental Protection
V.P. Singh and R.D. Stapleton, Jr. (Editors)
© 2002 Elsevier Science B.V. All rights reserved.

Biodegradation of diaryl esters: Bacterial and fungal catabolism of phenylbenzoate and some of its derivatives

Stefan Schmidt

Abteilung Mikrobiologie, Institut für Allgemeine Botanik der Universität Hamburg, Ohnhorststraße 18, 22609 Hamburg, Germany

INTRODUCTION

The ability of microorganisms to tackle organic compounds, that do or might pose a threat to the environment, is of tremendous importance. Using appropriate organisms is a potential means to eliminate or at least detoxify unwanted contaminants. However, although many bacterial and fungal strains have been isolated and subsequently characterized for their ability to degrade selected man-made pollutants, the number of new and existing chemicals that might enter the environment, being it via waste disposal, industrial activities, or *accidental spillages*, does still outnumber the microbial specialists so far available. Therefore, as long as we lack sufficient information on the biodegradability of not as yet evaluated chemicals, screening for microorganisms with a hitherto not reported desire for a specific contaminant is still an important and rewarding task for the environmental microbiologist. The information concerning the biodegradability is important for an ecotoxicological assessment in order to enable a more realistic estimate of the exposure concentration of a particular pollutant. This review aims to provide a brief overview of the biodegradability of the diaryl ester phenylbenzoate and some of its derivatives under aerobic conditions by comparing bacterial and fungal strategies. In addition, some areas for future research will be indicated.

MAN-MADE DIARYL ESTERS

Ester bonds are a notable structural feature present in a substantial number of industrially manufactured organic compounds of environmental concern such as pharmaceutical products, pesticides, and polymers [1-3]. Indeed, a large number of microorganisms with the ability to hydrolyze ester bonds present in environmentally relevant compounds has been reported [4-10]. The diaryl ester phenylbenzoate (Figure 1) is the basic structure present in a plethora of man-made compounds such as pharmaceuticals [1,11], *plasticisers* [1,12], mesogenic structures used for the production of liquid crystalline polymers employed in electro-optical devices such as LCD-displays [13-18], or cross-linking agents exploited to generate epoxy networks with

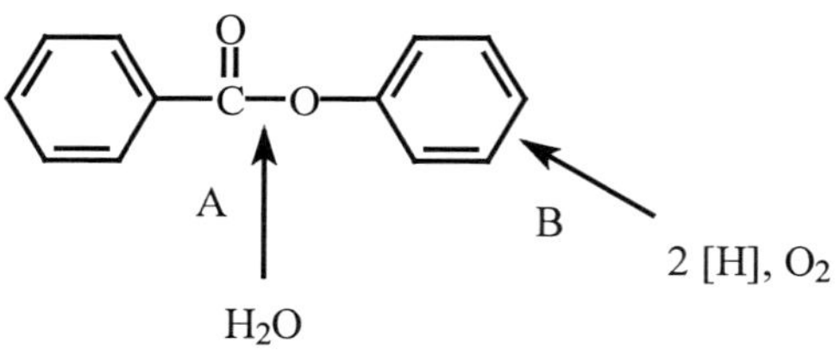

Figure 1. Phenylbenzoate. The arrows indicate potential sites for an initial hydrolytic (A) or ring hydroxylating (B) microbial attack.

ordered structures [19]. Some examples of man-made compounds, featuring the phenylbenzoate moiety, are shown in Figure 2. Due to its use in the industrial production of plastics and perfumes [20], phenylbenzoate has been frequently detected in industrial effluents [21]. This compound was in fact reported in the EPA TSCA inventory and obtained a hazard rating of 2 [22]. Some substituted diaryl esters, such as mixed diesters of *ortho*-phthalic acid, were found to persist for long periods in different soils and in sewage sludge [12] and compounds such as DPHP (diphenylphthalate) are regarded as toxicologically relevant chemicals [23]. To enable a more reliable prediction

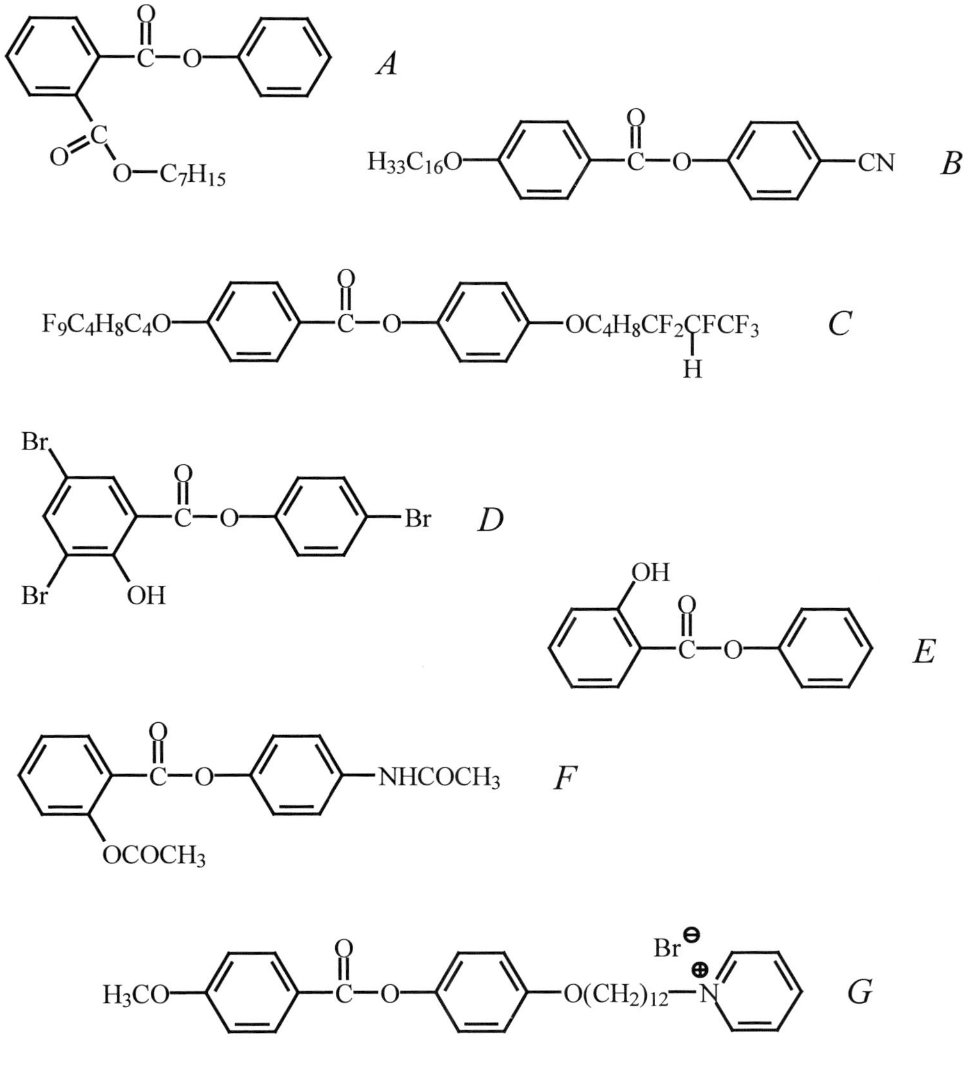

Figure 2. Synthetic compounds containing the phenylbenzoate moiety. *A*. A mixed phenyl alkyl ester of *ortho*-phthalate, a plasticiser [12], *B*. A mesogenic structure [17], *C*. A novel phenylbenzoate liquid crystal [15], *D*. Cordol, a bactericidal compound [1], *E*. Salol, a plasticiser [1], *F*. Benorylate®, an analgesic [11], *G*. An LMW mesogen [18].

of the fate and biodegradability of these compounds in the environment, the catabolic sequences involved in their biotransformation or biodegradation need to be elucidated.

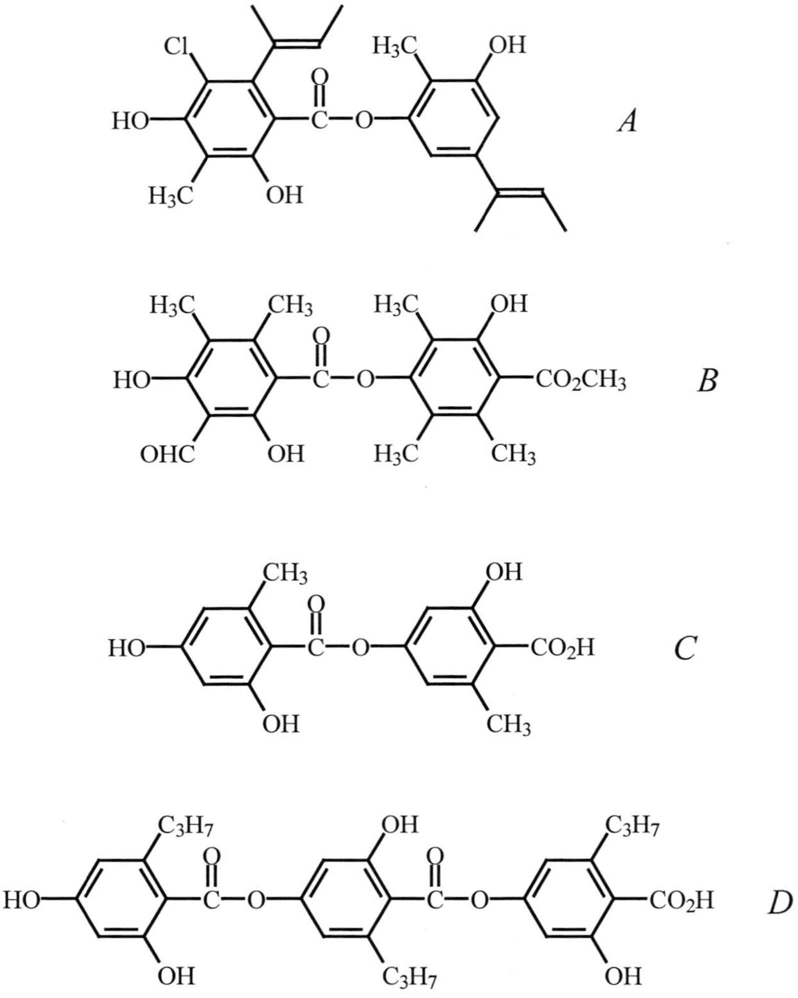

Figure 3. Diaryl esters of biological origin. A. Guisinol (*Emericella unguis*) [31], B. Pseudocyphellarin A (*Pseudocyphellaria endochrysea*) [32], C. Lecanoric acid (*Stereocaulon curtatum*) [33], D. Trivaric acid (*Ramalina americana*) [34].

DIARYL ESTERS ARE NOT ALIEN TO NATURE

Environmental microbiologists involved in estimating the biodegradability of known or potential pollutants are well aware, that the man-made compounds that will be most readily and effectively catabolized are those that have chemical structures and motifs in common with compounds occurring naturally. Microorganisms present in soil, water, and sediments will most certainly have evolved enzymes to degrade such pollutants, if they have been exposed to identical, or at least similar structures previously. In the case of the diaryl ester phenylbenzoate and its derivatives, there is evidence indicating the presence of diaryl esters of biological origin in the environment. First of all, several of the so-called gallotannins such as those isolated from *Rhus javanica* [24], *Haematoxylon campechianum* [25], a *Spirogyra* species [26] or *Ceratonia siliqua* [27] are known to contain diaryl ester structures. In addition, several diaryl esters containing the phenylbenzoate moiety are synthesized by fungi and lichens (Figure 3). It has been suggested that these compounds might serve in lichens as a chemical defence against microbial attack [28], as herbivore deterrents [29], as compounds with allelopathic effect, as light absorbing agents protecting the algal symbionts from damage by UV-radiation, or as chelating agents for the acquisition of micronutrients [30]. Although microorganisms have thus most probably been exposed to diaryl esters of biological origin, commercial compounds might be less degradable due to the presence of individual substituents or combinations thereof, that have not been encountered by these microorganisms previously.

HOW TO THRIVE ON PHENYLBENZOATE — THE BACTERIAL WAY

A prerequisite for the successful degradation of the diaryl ester phenylbenzoate, a compound featuring an ester bond connecting two aromatic rings, is either an initial attack that cleaves the bond linking the two aromatic nuclei or an initial reaction transforming one ring to enable further catabolic reactions (Figure 1). However, the ester linkage connecting the two aromatic nuclei in phenylbenzoate does appear to be the structure most vulnerable to an initial attack. In the case of the structurally related aryl ester benzylbenzoate, an *Acinetobacter* species able to use benzylacetate as sole source of carbon and energy, was shown to attack the ester linkage present to yield benzylalcohol and benzoic acid [35]. However, the cleavage products benzylalcohol and benzoic acid were not utilized for growth by this organism. Strains of *Bacillus pumilis*, *Klebsiella pneumoniae,* and *Corynebacterium* spp. were found to produce a tannase with depsidase activity, thus able to hydrolyze diaryl ester bonds present in tannic acid [36,37].

A *Pseudomonas* species, enriched from soil samples, used phenylbenzoate and two of its substituted derivatives as sole sources of carbon and energy [38]. The doubling times reported, ranging from 3 to 9 hours, showed that the simplest diaryl ester phenylbenzoate was a better growth substrate than the other two derivatives (*p*-tolylbenzoate and 4-chlorophenylbenzoate) tested. The biodegradation of phenylbenzoate was initiated in this organism by an inducible esterase, which catalyzed the hydrolysis of the diaryl esters tested, thus producing stoichiometric amounts of two monoaromatic metabolites. These metabolites were subsequently identified by the authors as benzoate and phenol in the case of phenylbenzoate, benzoate and 4-methylphenol in the case of *p*-tolylbenzoate, and benzoate and 4-chlorophenol in the case of 4-chlorophenylbenzoate, respectively. Catabolic enzymes required for the degradation of the metabolites phenol and benzoate, which were evidently induced after growth with phenylbenzoate, indicated that these monoaromatic intermediates were utilized via known catabolic sequences. Phenylbenzoate-grown cells exhibited benzoate 1,2-dioxygenase, phenolhydroxylase and high catechol 1,2-dioxygenase activity. Thus, upon attack of an inducible hydrolase, the ester bridge linking the two aromatic nuclei in phenylbenzoate was cleaved, yielding phenol and benzoate. These two monoaromatic structures

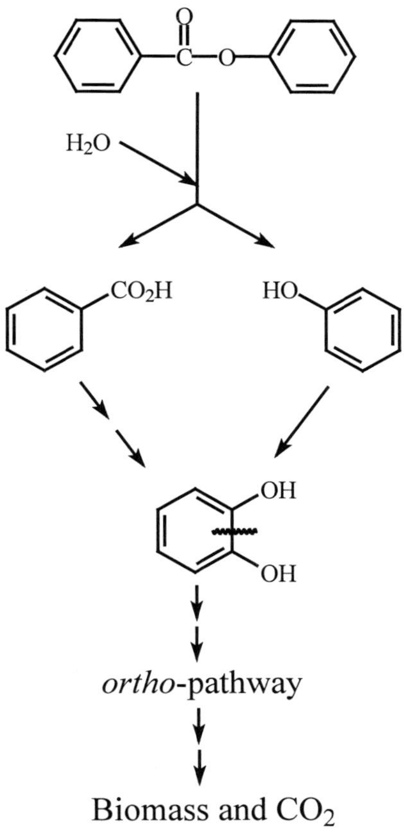

Figure 4. Biodegradation of the diaryl ester phenylbenzoate by *Pseudomonas* sp. strain TR3. Adapted from FEMS Microbiology Letters 1999; 176: 477-482, with permission.

were transformed by the benzoate 1,2-dioxygenase and the phenolhydroxylase respectively, subsequently entering the *ortho*-pathway for complete degradation via catechol and 3-oxoadipate.

A pathway for the *productive degradation* of phenylbenzoate by *Pseudomonas* sp. strain TR3 was proposed and is shown in Figure 4. The two other derivatives of phenylbenzoate (*p*-tolylbenzoate and 4-

chlorophenylbenzoate) that were utilized by the strain for growth, were catabolized analogously. However, in addition to catechol (produced from benzoate as in the case of phenylbenzoate), either the substituted derivative 4-methyl- or 4-chlorocatechol was obtained due to the formation and subsequent hydroxylation of the corresponding 4-methyl- (in the case of *p*-tolylbenzoate) and 4-chlorophenol (in the case of 4-chlorophenylbenzoate). This strain employed a similar pathway to exploit benzylbenzoate, which was enzymatically hydrolyzed to produce benzoate and benzylalcohol. Benzylalcohol was oxidized to benzoate, and the benzoate obtained from both the initial hydrolysis of benzylbenzoate and the oxidation of benzylalcohol was channelled via catechol into the *ortho*-pathway [Reich, Schmidt and Fortnagel; unpublished results].

HOW TO THRIVE ON PHENYLBENZOATE — A FUNGAL STRATEGY

As mentioned above for the bacterial strains, able to catabolize tannic acid, a tannase with typical depsidase activity was extracted from a strain of *Aspergillus niger* grown with Chinese gallotannin or similar compounds as carbon source [37,39,40]. This enzyme was able to cleave the diaryl ester *m*-digallic acid hydrolytically. However, this tannase did only attack those structures effectively, which carried at least two phenolic hydroxyl groups in the acid component and, in addition, it was found that the esterified carboxyl group must not be *ortho* to one of the hydroxyl groups [39]. Another tannase isolated from a *Penicillium* species was shown to hydrolyze several gallotannins effectively and was, therefore, employed as a tool for structure elucidation [27]. While evaluating the catabolic versatility of *Scedosporium apiospermum*, a saprophytic hyphomycete able to productively utilize the pollutants, phenol and *p*-cresol, as sole sources of carbon and energy [41], it was discovered that this fungus can grow with phenylbenzoate. Growth experiments employing this compound demonstrated, that this diaryl ester was an excellent carbon source for *Scedosporium apiospermum*, giving rise to molar growth yields significantly higher than those obtained for the monoaromatic compounds phenol, p-cresol, benzoate, or the nonaromatic carbon source, glucose. A possible reaction mechanism, leading to the production of phenol and benzoate from phenylbenzoate by involving a cysteine and a histidine residue as typical for cysteine-dependent hydrolases [42], is outlined in Figure 5.

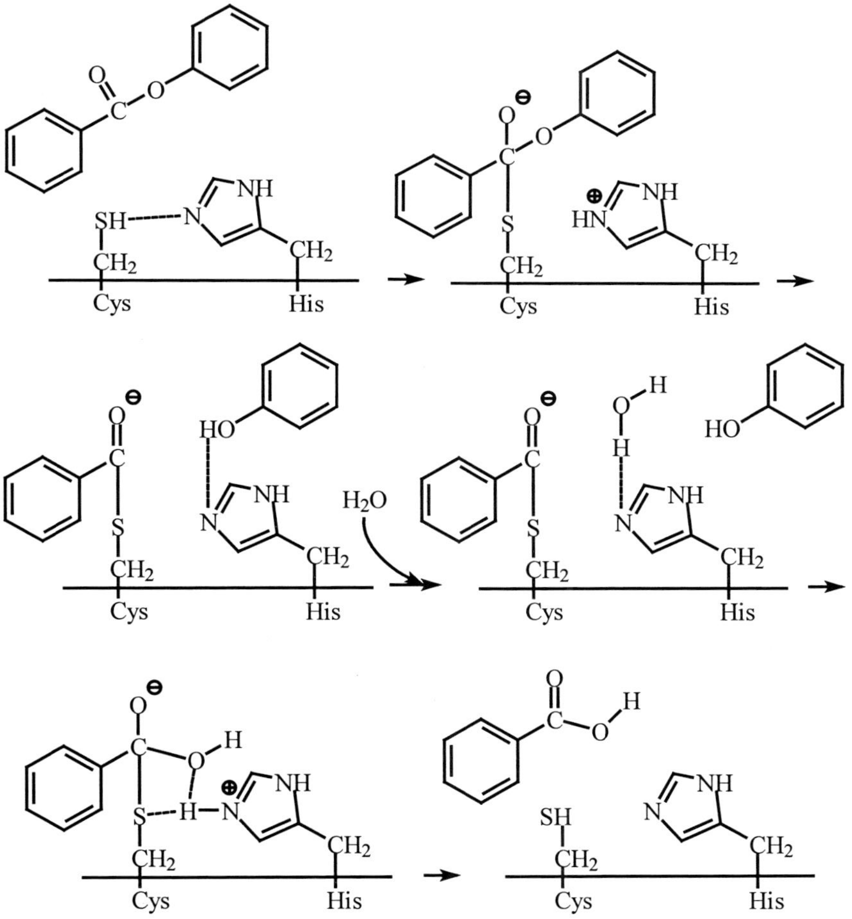

Figure 5. A possible mechanism for the hydrolysis of phenylbenzoate by the esterase of Scedosporium apiospermum.

The biological degradation of phenylbenzoate was initiated by an inducible esterase that catalyzed the hydrolysis of the ester bond present in this diaryl ester to produce stoichiometric amounts of phenol and benzoate. Results of inhibition experiments, using crude preparations and purified enzyme, clearly indicated that this esterase contains a cysteine residue essential for its catalytic activity [Schilbach, Fortnagel and Schmidt; unpublished observation]. Besides phenylbenzoate the esterase hydrolyzed, albeit to a lesser extent, *p*-tolylbenzoate and 4-chlorophenylbenzoate. Analogous to the hydrolysis of phenylbenzoate, the reaction led to the production of stoichiometric amounts

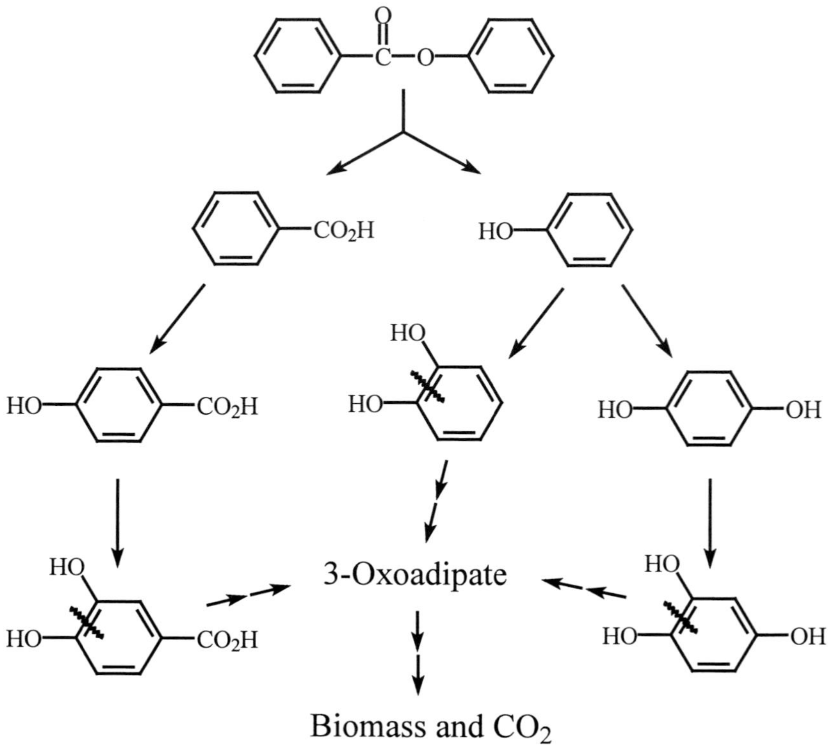

Figure 6. Productive degradation of phenylbenzoate by *Scedosporium apiospermum*. Modified from Research in Microbiology 1999; 150: 413-420, with permission.

of the corresponding catabolites *p*-cresol and benzoate, and 4-chlorophenol and benzoate, respectively.

The catabolism of the resulting cleavage products of phenylbenzoate was catalyzed in *Scedosporium apiospermum* by different sets of enzymes, that were induced after growth with phenylbenzoate. Thus the benzoate was hydroxylated to 4-hydroxybenzoate and further degraded by the well-known protocatechuate branch of the fungal *ortho*-pathway [44,45]. However, this fungus catabolized phenol both via catechol and *cis,cis*-muconate as well as via hydroquinone, hydroxyhydroquinone and maleylacetate, hence using a branched pathway to furnish the central *ortho*-pathway intermediate 3-oxoadipate [41].

Figure 6 summarises the pathway proposed for the productive degradation of phenylbenzoate by this fungal strain. The catabolic sequence employed for the degradation of *p*-tolylbenzoate was simpler, as *p*-cresol was catabolized by *Scedosporium apiospermum* in a rather straightforward manner via the protocatechuate pathway [41].

Hence, both *p*-cresol and benzoate, the products obtained from the hydrolysis of *p*-tolylbenzoate, were channelled via 4-hydroxybenzoate into the protocatechuate branch of the *ortho*-pathway (Figure 7).

Although the growth of *Scedosporium apiospermum* with 4-chlorophenylbenzoate led to the production of fungal biomass, it was a rather poor substrate as compared with phenylbenzoate and *p*-tolylbenzoate [42]. This was due to the fact that this fungal strain transformed the 4-chlorophenol obtained upon the hydrolysis of 4-chlorophenylbenzoate into 4-chlorocatechol and the corresponding chlorinated muconic acid [Deppe and Schmidt; unpublished results], which was not utilized further (Figure 8).

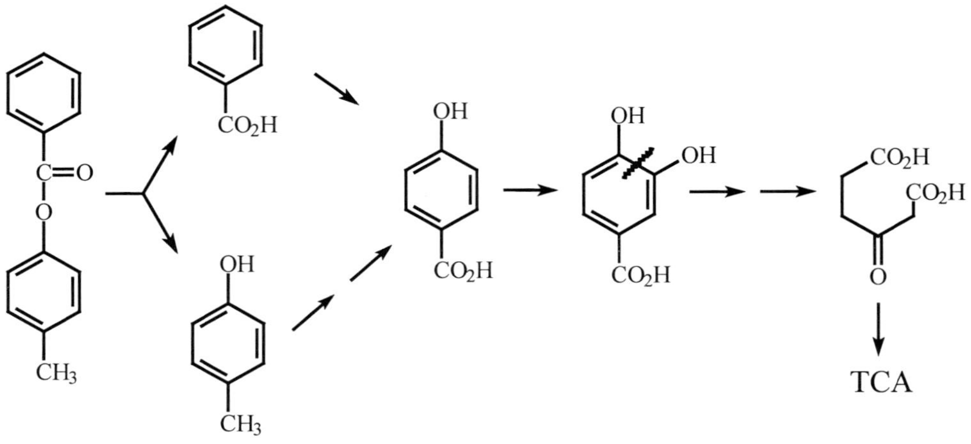

Figure 7. Productive degradation of *p*-tolylbenzoate by *Scedosporium apiospermum*. Modified from Research in Microbiology 1999; 150: 413-420, with permission.

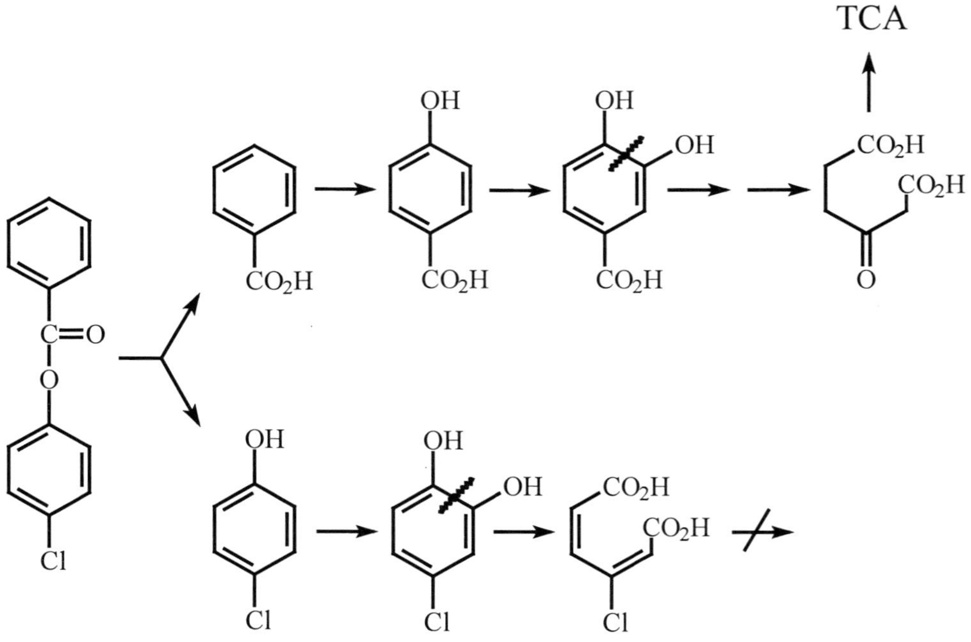

Figure 8. Catabolism of 4-chlorophenylbenzoate by *Scedosporium apiospermum*.

CONCLUSION

The ability of microorganisms to degrade the simplest diaryl ester phenylbenzoate effectively, does suggest that this structure, and most probably simple derivatives of it, can be eliminated by using biological means. However, it remains to be seen if some of the more complex compounds shown in Figure 2, or similar structures, are biodegradable. Even if the basic diaryl ester structure can be hydrolyzed by microorganisms, such as those described in this short review, the impact of substituents present on the resulting cleavage products might interfere with degradation. Therefore, more research is required to elucidate the environmental fate of these structures and, consequently, the current pure culture studies should be expanded not only to involve microbial consortia but as well extended into anaerobic systems. This will not only help to evaluate the potential environmental fate of such compounds, but furthermore such research might prove to be a rather useful tool for the development of novel structures that are biodegradable.

ACKNOWLEDGEMENTS

I would like to thank my colleagues M. Claußen, U. Deppe, T. Reich, and J. Schilbach for their enthusiastic participation in, and their contributions to this work, and P. Fortnagel for his generous support and continuous encouragement.

REFERENCES

1   Budavari S (ed). The Merck Index (11th edn) Rahway, N.J.: Merck & Co., Inc., 1989.

2   Matolcsy G, Nadasy M, Andriska V. Pesticide Chemistry. New York: Elsevier, 1988.

3   Scott G. Polymers and the environment. Cambridge: Royal Society of Chemistry, 1999.

4   Golovleva LA, Aharonson N, Greenhalgh R, Sethunathan N, et al. Pure Appl Chem 1990; 62: 351-364.

5   Maloney SE, Maule A, Smith ARW. Appl Environ Microbiol 1988; 54: 2874-2876.

6   Crabbe JR, Campbell JR, Thompson L, Walz SL, et al. Int Biodet Biodeg 1994; 33: 103-113.

7   Suyama T, Hosoya H, Tokiwa Y. FEMS Microbiol Lett 1998; 161: 255-261.

8   Howard GT, Blake RC. Int Biodeg Biodet 1998; 42: 213-220.

9   Nakajima-Kambe T, Shigeno-Akutsu, Nomura N, Onuma F, et al. Appl Microbiol Biotechnol 1999; 51: 134-140.

10   Kavelman R, Kendrick B. Mycologia 1978; 70: 87-103.

11   Raab WP. Arzneim Forsch 1971; 21: 1662-1664.

12   Muszkat L, Bir L, Raucher D. Bull Environ Contam Toxicol 1997; 58: 348-355.

13   Krücke B, Zaschke H, Kostromin SG, Shibaen VP. Acta Polymerica 1985; 36: 639-643.

14   Chang S, Han CD. Macromolecules 1996; 29: 2103-2111.

15   Liu H, Nohira H. Ferroelectrics 1998; 207: 541-553.

16   Shiota A, Körner H, Ober CK. Liquid Crystals 1998; 25: 199-206.

17   Naoum MM, Seliger H, Happ E. Liquid Crystals 1997; 23: 247-253.

18   Masson P, Gramain P, Guillon D. Macromol Chem Phys 1999; 200: 616-620.

19   Strehmel V. J Polym Sci A: Polym Chem 1997; 35: 2653-2688.

20   Verschueren K. Handbook of environmental data on organic chemicals. New York: van Nostrand Reinhold, 1977.

21   EPA (Environmental Protection Agency). (1975), Identification of organic compounds in effluents from industrial sources. EPA-560/3-75-002.

22   Sax NI, Lewis RJ. Dangerous properties of industrial materials. New York: van Nostrand Reinhold, 1989.

23   Parkerton TF, Konkel WJ. Ecotox Environ Safety 2000; 45: 61-78.

24   Taniguchi S, Takeda S, Yabu-Uchi R, Yoshida T, et al. Phytochemistry 1997; 46: 279-282.

25   Kandil FE, El-Sayed NH, Micheal HN, Ishak MS, et al. Phytochemistry 1996; 42: 1243-1245.

26   Nishizara M, Yamagichi T, Nonaka GI, Nishioka I, et al. Phytochemistry 1985; 24: 2411-2413.

27   Nishira H, Joslyn MA. Phytochemistry 1968; 7: 2147-2156.

28   Huneck S. Naturwissenschaften 1999; 86: 559-570.

29   Lawrey JD. Bryologist 1989; 92: 326-328.

30   Lawrey JD. Bryologist 1986; 89: 111-122.

31   Nielsen J, Nielsen PH, Frisvad JC. Phytochemistry 1999; 50: 263-265.

32   Huneck S. Phytochemistry 1984; 23: 431-434.

33   Hamada N, Ueno T. Phytochemistry 1990; 29: 678-679.

34   Culberson CF, LaGreca S, Johnson A, Culberson WA. Bryologist 1999; 102: 595-601.

35   Jones RM, Collier LS, Neidle EL, Williams PA. J Bacteriol 1999; 181: 4568-4575.

36   Deschamps AM, Otuk G, Lebeault JM. J Ferment Technol 1983; 61: 55-59.

37   Lekha PK, Lonsane BK. Adv Appl Microbiol 1997; 44: 215-260.

38   Reich T, Schmidt S, Fortnagel P. FEMS Microbiol Lett 1999; 176: 477-482.

39   Dyckerhoff H, Armbruster R. Hoppe-Seylers Zeitschrift Physiol Chem 1933; 219: 38-56.

40   Haslam E, Stangroom JE. Biochem J 1966; 99: 28-31.

41   Claußen M, Schmidt S. Res Microbiol 1998; 149: 399-406.

42   Claußen M, Schmidt S. Res Microbiol 1999; 150: 413-420.

43   Cain RB, Bilton RF, Darrah JA. 1968. Biochem J 1968; 108: 797-828.

44   Jamaluddin M, Subba Rao PV, Vaidyanathan CS. J Bacteriol 1970; 101: 786-793.

Biotransformations: Bioremediation Technology for Health and Environmental Protection
V.P. Singh and R.D. Stapleton, Jr. (Editors)
© 2002 Elsevier Science B.V. All rights reserved.

Degradation of natural rubber products by *Nocardia* species

A. Tsuchii and Y. Tokiwa

National Institute of Bioscience and Human-Technology, Tsukuba, Ibaragi 305, Japan

INTRODUCTION

Rubber baloons and plastic bags are now considered to be a danger to wild animals when disposed off in the natural environment. It is important to know the potential for microbial degradation of rubber products from the point of view of prolongation of usage and of waste disposal problems. One particular concern is the disposal of used tire, because a large number of tires are produced and discarded annually [1].

Although many reports have been submitted on the microbial breakdown of natural rubber (NR), only a few reports are available on the modes of microbial action and the mechanisms involved [2]. One reason may be that it is quite resistant in comparison with other natural polymers, and the actions of microorganisms are very slow under normal ecological or laboratory conditions. The colonization of *Nocardia asteriodes* on the surface of a hard type NR vulcanizate was monitored by scanning electron microscopy (SEM). When the colonies were removed, the rubber beneath was found to be pitted and local penetration to depth of 220 μm was recorded [3]. *Colonization* and penetration of latex gloves by strains of actinomycetes were also demonstrated by SEM [4]. The rate and the extent of degradation reported by these investigators, however, were very slow and restricted. The degradation of the hard type vulcanizate amounted only 3% after two years [3]; whereas, the weight reduction of rubber pieces from the latex glove was 8 to 18% after 6 weeks [4].

NR consists of hydrocarbon molecules (*cis*-1,4-polyisoprene) with molecular weight higher than 10^5, and the vulcanizate has a cross-linked network structure insoluble in any organic solvent. It is readily expected that the growth of microorganisms on the pieces of rubber products is very slow in contrast to growth in common media, where carbon substrate is dissolved. Furthermore, the effect of culture conditions upon microbial growth on an insoluble solid substrate, like rubber, has not been well characterized. The adherence of microorganisms to an insoluble solid substrate, like cellulose or pyrite, is considered to be a key factor in the further utilization of the insoluble substrate [5,6]. It was reported that spherical holes were formed on a PHB film as a result of colonization [7]. Among the various kinds of rubber-degrading organisms in our culture collection, *Nocardia* sp. 835A, the strongest decomposer of solid rubber, was selected for further study. After a year of trial-and-error experiments, we found that long strips of NR vulcanizates in stirred flasks experienced the highest rate of rubber degradation by the organism [8]. These experimental conditions gave a result with sufficient reproducibility in a relatively short time period, and the method enabled us to examine various factors affecting microbial rubber degradation. We found that the colonization of rubber pieces plays an important role in microbial decomposition of NR, and we would like to describe the unique characteristic feature of growth of the rubber-degrading organism on rubber pieces in this report.

CHAIN SCISSION REACTIONS

The strain 835A of *Nocardia* sp. grows well on unvulcanized NR and synthetic isoprene rubber. NR and synthetic isoprene rubber A (content of *cis*-1,4-structure, 97 to 98%) were completely degraded after an 8-weak cultivation period, while only half of isoprene rubber B (*cis*-1,4-content, 78%) was degraded. The results strongly suggest that the high molecular weight hydrocarbon polymers of *cis*-1,4-polyisoprene were utilized as the sole source of carbon and energy by the organism. Other types of synthetic rubber, like butadience rubber and SBR (styrene butadiene rubber), were hardly attacked by the organism.

Not only unvulcanized but vulcanizate NR products also were more or less utilized by the organism as the growth substrate. During microbial growth on the films from a latex glove, isoprene oligomers with molecular weight from 10^3 to 10^4 were produced. ^{1}H- and ^{13}C-NMR spectra of the oligomer were those expected of *cis*-1,4-polyisoprene with an acetoaldehyde group at one terminus, and a methyl ketone group at the other. The proposed chemical structure of low molecular weight fragments of rubber suggests that the organism cleaved NR at the double bond shown by a wavy line in the formula in Figure 1 [9].

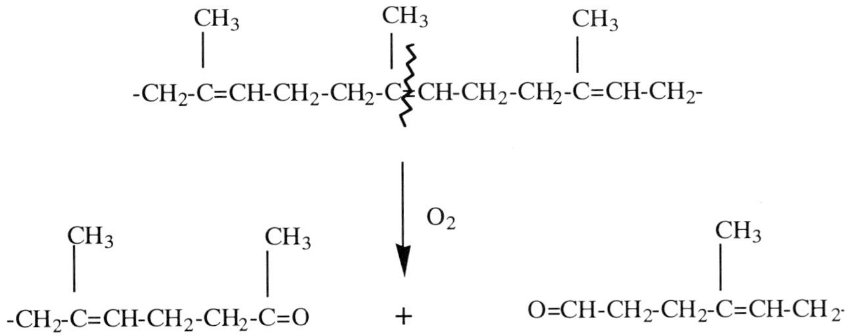

CH$_3$ CH$_3$ CH$_3$

-CH$_2$-C=CH-CH$_2$-CH$_2$-C$\lessgtr$CH-CH$_2$-CH$_2$-C=CH-CH$_2$-

$\downarrow$ O$_2$

CH$_3$ CH$_3$ CH$_3$

-CH$_2$-C=CH-CH$_2$-CH$_2$-C=O   +   O=CH-CH$_2$-CH$_2$-C=CH-CH$_2$-

Figure 1. Schematic diagram of chain scission reactions of NR [9].

These results suggest that the network of polymeric chains in the vulcanizate was attacked by the biological action, and that the oligomers produced by the scission of longer polymers were consumed by the organism. This special ability to cause scissions of polymeric chains makes rubber degradation possible [10].

COLONIZATION OF RUBBER PIECES

The *Nocardia* sp. strain 835A grows well on the thin film of a latex glove. The growth occurs essentially on the insoluble rubber substrate and the cells are tightly bound to the rubber pieces in the initial stage of growth. The colony diameter reached 30 to 100 μm on day-7. Although the rubber surface was mostly covered with dense growth of the organism, there were few or no suspended cells in the culture medium in spite of vigorous stirring.

The actively growing colonies of stain 835A on the rubber surface were visualized clearly by staining with Schiff reagent. The purple color produced by the reagent was evidence that isoprene oligomers containing aldehyde group were produced and accumulated during the microbial degradation of rubber. Only colonies were stained with the reagent and no color was observed

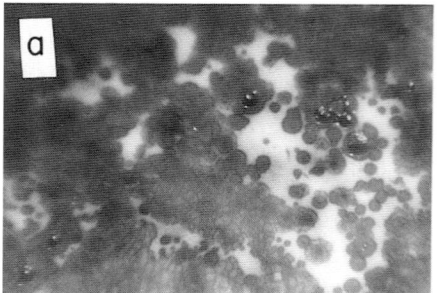 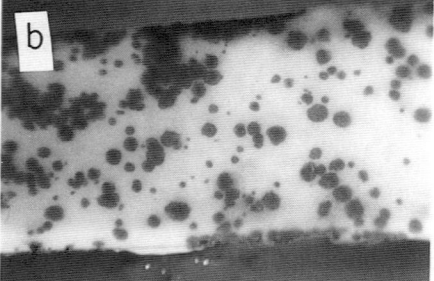

Figure 2. The effect of width upon colonization. Colonization of rubber pieces after 14 days was followed by staining with Schiff reagent [8]. (a), a wide strip from a latex glove (6 mm in width); (b), a short and narrow strip (3 mm in width).

apart from the colonies. These facts led us to conclude that rubber-degrading reactions occurred only at points of direct contact with the colonies (Figure 2) [8].

The rate and the degree of degradation and colonization of the rubber pieces were affected by various factors and experimental conditions. In most cases, we found that the rate of degradation was in good correlation with the number of colonies per unit area of the rubber surface. Under laboratory conditions, long strips of NR *vulcanizates* in stirred flasks were most rapidly colonized and degraded by strain 835A. With this method, more than 90% loss in weight of a thin film (0.1 mm in thickness) from latex glove after 2 weeks was obtained. On the other hand, the colonization and degradation of short and narrow pieces, which had essentially identical surface area with long strips, were substantially slower and less extensive. The movement of rubber pieces in the culture medium differed, depending on the shape of the pieces and the method of shaking the flasks. As a result, rubber pieces of different shapes have different flow field around the rubber pieces, and hence different efficiency of colonization is observed at the rubber surface.

Jar-fermentor experiments for microbial degradation of a commercial surgical glove also demonstrated that rapid and uniform colonization of the surface of the rubber pieces is the most important rate-determining step [11]. The rubber pieces (0.2 mm in thickness) added into the jar were completely degraded within 2 or 3 weeks.

THE EFFECT OF CROSS-LINK DENSITY

Various kinds of vulcanized NR products were utilized by strain 835A, to some extent, as the growth substrate. It grew well on soft type products, such as rubber bands and tubing. Growth on hard type products, like tire tread or rubber stopper, was poor. We have prepared a series of vulcanizates and examined their relative susceptibilities against microbial attack [12]. The resistance of the vulcanizates was in good correlation with the cross-link density (Figure 3), and addition of carbon black as a filler made the vulcanizate apparently more resistant to biological actions.

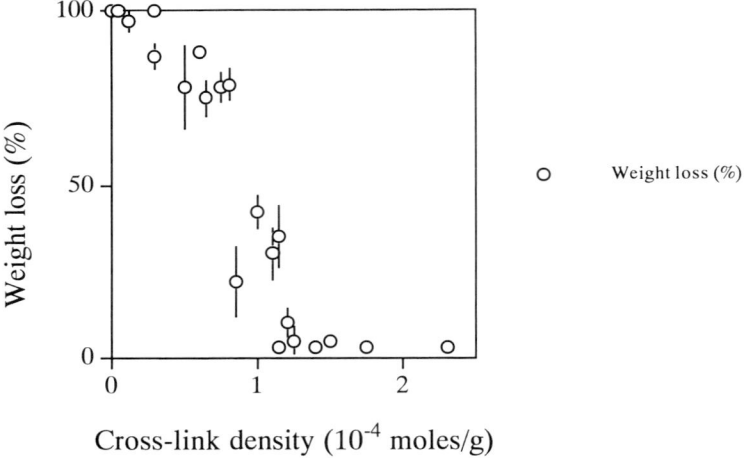

Figure 3. The effect of cross-link density on the microbial degradation of NR vulcanizates [12].

DEGRADATION OF THE RUBBER IN TRUCK TIRES

While truck tires are made principally from NR, the rubber in the tires is highly resistant to microbial attack. Although a strip from a truck tire tread was degraded only slightly when it was used as the sole growth substrate for stain 835A, its degradation was markedly enhanced by addition of a strip of glove rubber, which the organism readily utilized as a growth substrate [13]. It was found that about 28% of the tread strip was microbiologically converted to very small particles (mostly less than 30 μm in diameter) and that 23% was truly degraded and mineralized.

Colonization of the tire strip and microbial growth at the expense of rubber was observed with SEM. The growth on tire strip was very slow and colonies with diameters of 10 to 20 μm were observed on day-7. The degradation of tire strip proceeded in accordance with the progress of colonization and with the development of the colony size. Although the rubber beneath the colonies was severely disintegrated and pitted, the plain smooth surface areas remained between the pits (Figure 4).

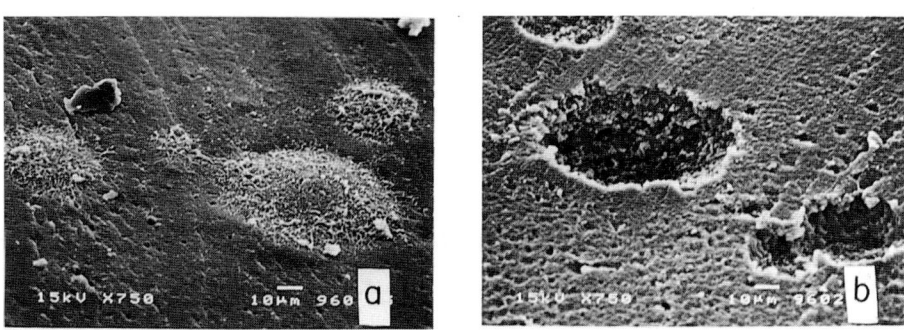

Figure 4. SEM of the surface of a tire strip [13]. (a) 14 days micro-colonies before washing; (b) cavities appeared after washing.

These observations strongly suggest that microbial disintegration occurred only at the points of direct contact with the colonies. They also suggest that microbial cells suspended in the medium may not directly contribute to the tire rubber degradation, but may contribute only through stimulation of the colonization.

Quite recently, we found that a mutant strain Rc of *Nocardia* sp. 835A disintegrates tire rubber about 6 times more efficiently than the parental stain, even without the addition of glove rubber [14]. The weight losses and the degrees of mineralization of tread compounds from four kinds of truck tires commercially produced by different tire-making companies are shown in Figure 5. It is clear that the extent of microbial attack was consistently influenced by the NR content.

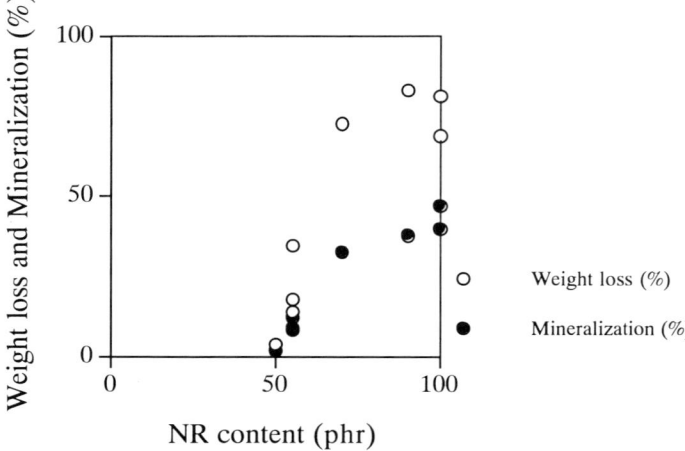

Figure 5. Effect of NR content on microbial decomposition of tread rubber [14].

DISCUSSION

NR is biosynthesized as a finely dispersed emulsion (latex), and the products made of NR are bulky solids like films, bands and so on. One would hardly expect that solid and compact mass of the products in its coagulated state would be subjected to biological action to the same extent as the same material in the finely dispersed condition, in which it exists as latex in plants.

Strain 835A of *Nocardia* sp. was a very strong decomposer of vulcanized NR products and was able to cause scissions of polymeric chains in the network structures. This suggests that NR products are principally biodegradable. The rate and the extent of degradation, however, is affected by various factors like cross-link density, NR content, etc. We can expect that soft type products, like bands and gloves, are practically biodegraded completely. But in the case of hard type products, like car tires, only a part of the rubber is mineralized and a part is disintegrated to very small particles.

We have found that strain 835A has a typical attack mode for rubber degradation (i.e. colonization of rubber pieces). The strain has a high inoculum potential at the surface of rubber pieces [15], and the rubber-degrading reaction occurred only at the points of direct contact with the colonies. The degradation of rubber pieces proceeded in accordance with the progress of colonization and with the development of colony size [8,13]. This unique attack mode of the organism makes it possible to efficiently decompose rubber in the state of coagulated mass [16].

Natural rubber is synthesized by over 1800 plant species. Consequently, rubber-degrading organisms can be expected to be widely distributed in the natural environment. *Streptomyces* and *Nocardia* have been reported to play a major role in degradation of NR [2]. Recently, microorganisms belonging to some other genera of bacteria (*Gordona* [17] and *Xanthomonas* [18]) and actinomycetes (*Amycolatopsis* and *Micromonospora*) [19] were also reported to decompose rubber. Strain 835A has a high ability to colonize rubber pieces, but no extracellular enzyme was detected in the culture medium. On the contrary, *Xanthomonas* sp. 35Y secreted rubber-degrading enzyme into the extracellular culture medium [18]. The ability of the strain to colonize and to decompose solid rubber, however, was poor. Not only taxonomically but also physiologically, different characteristics could be seen among various types of rubber-degrading organisms. Better understanding of the diversity of the microorganisms will help us to make better use of their versatility.

CONCLUSION

Nocardia sp. strain 835A was shown to possess an ability to cause scissions of polymeric chains of natural rubber (NR), and the organism grew well on unvulcanized natural and synthetic isoprene rubber. Not only unvulcanized but also vulcanized NR products were more or less utilized by the organism as the growth substrate and the resistance of the vulcanizates was in good correlation with the *cross-link density*. Strain 835A had a high ability to colonize rubber pieces and microbial attack occurred only at points of direct contact with the colonies. A mutant strain Rc of *Nocardia* sp. 835A with a very high ability of disintegrating tire rubber was isolated. Disintegration of tread compounds from truck tires by strain Rc was found to be consistently influenced by the NR content.

REFERENCES

1   Bressi G. In: International Directory of Solid Waste Management 1993/ 1994. Copenhagen: The ISWA Year Book 1994; 131-138

2   Tsuchii A. In: Singh VP, ed. Progress Indust Microbiol 1995; 32: 177- 187.

3   Hanstveit AO, Gerritse GA, Scheffers WA. Na Rubber Technol 1988; 19: 50-58.

4   Heisey RM, Papadatos S. Appl Environ Microbiol 1995; 61: 3092-3097.

5   Lamed R, Bayer EA. Adv Appl Microbiol 1988; 33: 1-46.

6   Kumar SR, Gandhi KS, Natarajan KA. Appl Microbiol Biotechnol 1991; 36: 278-282.

7   Nishida H, Tokiwa Y. J Appl Polym Sci 1992; 46: 1467-1476.

8   Tsuchii A, Takeda K, Suzuki T, Tokiwa Y. Biodegradation 1996; 7: 41- 48.

9   Tsuchii A, Suzuki T, Takeda K. Appl Environ Microbiol 1985; 50: 965- 970.

10   Tsuchii A. 1993; Thesis in the University of Hokkaido, under the supervision of Prof. Dr. F. Tomita.

11   Kajikawa S, Tsuchii A, Takeda K. Nippon Nogeikagaku kaishi 1991; 65: 981-986.

12   Tsuchii A, Hayashi K, Hironiwa T, Matsunaka H, Takeda K. J Appl Polym Sci 1990; 41: 1181-1187.

13   Tsuchii A, Takeda K, Tokiwa Y. Biodeterioration 1997; 7: 405-413.

14   Tsuchii A, Tokiwa Y. J Biosci Bioeng 1999; 87: 542-544.

15   Dwivedi RS, Garrett SD. Trans Br Mycol Soc 1968; 51: 95-101.

16   Tsuchii A, Tokiwa Y. In: Steinbuechel A, ed. Proceedings of the International Symposium, Muenster: Wiley–VCH, 1998; 258-264.

17   Linos A, Steinbuechel A, Kautsch Gummi Kunst 1998; 51: 496-499.

18   Tsuchii A, Takeda K. Appl Environ Microbiol 1990; 56: 269-274.

19   Jendrossek D, Tomasi G, Kroppenstedt RM. FEMS Microbiol Lett 1997; 50: 179-188.

Biotransformations: Bioremediation Technology for Health and Environmental Protection
V.P. Singh and R.D. Stapleton, Jr. (Editors)
377

Sewage treatment systems: Microbiological aspects

Ved Pal Singh and Kartiki Bhatnagar

Department of Botany, University of Delhi, Delhi - 110 007, India

INTRODUCTION

The utility of microbiological techniques in the amelioration of pollution is illustrated by municipal wastewater treatment which is based on the consumption by bacteria of human waste and carbonaceous substances in environmental *runoff* wastewater. Sewage treatment involves a combination of engineered processes, that typically include at least one biological process because the bulk of the carbon entering waste treatment system is largely in the form of biodegradable compounds, the bacteria convert it into carbon dioxide and biomass [1].

In describing the treatment of sewage to reduce the microbial pathogenesis and biochemical oxygen demand, it is conventional to refer to the process applied in terms of the type of the treatment involved as the sewage passes through the three basic processes of sewage treatment, which includes the preliminary treatment for the removal of gross solids and grit, secondary treatment for the biooxidation and the tertiary treatment for the removal of fine solids from the effluents [2].

This chapter is concerned with determining the functioning of the microorganisms in the overall efficiency and effectiveness of the treatment systems along with the analysis of the biotransformations as the basic unit of the waste treatment scenario.

BIOLOGICAL TREATMENT REACTIONS

Efficacy of the biological treatment of the pollutants depends on the microorganisms living in the system, these multispecies conglomerates are able to interact with each other and the physical environment through biologically catalyzed reactions, which decompose the various organic compounds in the waste streams. Several examples of these type of reactions are provided in Table 1 [3].

BIOTRANSFORMATIONS IN THE CONTAMINATED ENVIRONMENT

Biotransformation is the chemical alteration of the molecular structure of organic and inorganic compounds, resulting in different complexity or the loss of some characteristic property with no loss of molecular complexity [4]. Transformation occurs due to microorganism-induced oxidation of chemicals with molecular oxygen and hydroxylations, mediated by hydroxylases at specific durations [5]. Microorganism-induced *epoxidation* is also one of the methods to transform the insecticides such as aldrin and heptachlor. During epoxidation, incorporation of an oxygen atom to a double bond occurs [6].

Cometabolism, a process in which microbial populations growing on one compound may fortuitously transform a contaminating chemical, is the important mode of biotransformation occurring in the nature [7]. Initially, biotransformation was most commonly thought to be associated with bacteria but fungi, algae, rotifers, crustaceans, and protozoans transform the compounds in their own way.

Fungi-based transformation systems

Fungi capable of degrading a wide range of environmental pollutants like polyaromatic hydrocarbons, DDT, TNT, and pentachlorophenol are now

Table 1
Biological treatment reactions

___

I. **Anaerobic nonphotosynthetic reactions (molecular oxygen absent)**

Nitrate reduction (denitrification)

$5CH_3COOH + 8NO_3^- \rightarrow 10CO_2 + 4N_2 + 6H_2O + 8OH^-$

$5S + 6NO_3^- + 2H_2O \rightarrow 5SO_4^= + 3N_2 + 4H^+$

Sulfate reduction

$2CH_3CHOHCOOH + SO_4^= \rightarrow 2CH_3COOH + H_2S + 2OH^-$

$4H_2 + SO_4^= \rightarrow 2H_2O + H_2S + 2OH^-$

Organic carbon reduction (fermentation)

$CH_3COOH \rightarrow CH_4 + CO_2$

$4CH_3OH \rightarrow 3CH_4 + CO_2 + 2H_2O$

$C_6H_{12}O_6 \xrightarrow{\text{bacteria}} 3CH_3COOH$

$C_6H_{12}O_6 \xrightarrow{\text{yeast}} 2CH_3CH_2OH + 2CO_2$

Carbon dioxide reduction

$2CH_3CH_2OH + CO_2 \rightarrow 2CH_3COOH + CH_4$

$4H_2 + CO_2 \rightarrow CH_4 + 2H_2O$

$4H_2 + 2CO_2 \rightarrow CH_3COOH + 2H_2O$

II. **Aerobic nonphotosynthetic bacterial reactions**

Oxygen limited systems

$CH_3CH_2OH + O_2 \rightarrow CH_3COOH + H_2O$

$2CH_3CHO + O_2 \rightarrow 2CH_3COOH$

$2CH_3CHOHCH_3O_2 \rightarrow 2CH_3COCH_3 + 2H_2O$

Table 1 continued

Complete oxidation

$CH_3COOH + 2O_2 \rightarrow 2CO_2 + 2H_2O$

$2H_2 + O_2 \rightarrow 2H_2O$

$CH_4 + 2O_2 \rightarrow CO_2 + 2H_2O$

Nitrification

$2NH_3 + 3O_2 \rightarrow 2NO_2^- + 2H^+ + 2H_2O$

$2NO_2^- + O_2 \rightarrow 2NO_3^-$

Sulfur oxidation

$2H_2S + O_2 \rightarrow 2S + 2H_2O$

$2S + 2H_2O + 3O_2 \rightarrow 2SO_4^= + 4H^+$

$S_2O_3^= + H_2O + 2O_2 \rightarrow 2SO_4^= + 2H^+$

Nitrogen fixation

$N_2 \rightarrow$ Nitrogenous organics

III. **Photosynthetic reactions**

$$CO_2 + 2H_2S \xrightarrow{\text{light}} (CH_2O) + H_2O + 2S$$

$$3CO_2 + 2S + 5H_2O \xrightarrow{\text{light}} 3(CH_2O) + 4H^+ + 2SO_4^=$$

$$CO_2 + 2H_2O \xrightarrow{\text{light, algae}} (CH_2O) + H_2O + O_2$$

$$9CH_3COOH \xrightarrow{\text{light}} 2CO_2 + 4(C_4H_6O_2) + 6H_2O$$

$$CO_2 + 2H_2 \xrightarrow{\text{light}} (CH_2O) + H_2O$$

$$2CH_3COOH + H_2 \xrightarrow{\text{light}} (C_4H_6O_2) + 2H_2O$$

being isolated [8]. *Phanerochaete chrysosporium,* cultivated in a biofilm reactor system, fiber reactor, and silicone membrane reactor, produced lignin peroxidase, which is a potential degrading enzyme [9]. *P. chrysosporium* was used to study the potential for degradation of trinitrotoluene (TNT) [10], in the wastewater from a munition plant. TNT has also been found to be successfully degraded by *P. chrysosporium* immobilized on a rotating biological contactors [11].

Various fungal species in the bioremediation of the wastewater are *Coriolus versicolor* and *P. chrysosporium,* when the substrate is pulp and paper mill waste [12,13]. *Candida utilis* degrades distillery wastes by bioconversion [14] and effluents containing Kraft bleaching are degraded by *P. chrysosporium* [15]. *Aspergillus,* and *Penicillium* degrade tannery effluents [16]; sulphite waste is found to be degraded by *Candida utilis* and *Paecilomyces varioti,* and brewery waste and molasses are degraded by *Saccharomyces cerevisiae* and *S. carlbergensis* [17]. Various aliphatic compounds are degraded by *Cladosporium* species, which are commonly isolated from fuel tanks [18-20]. Filamentous fungus, *P. chrysosporium* has also been found to degrade xenobiotics like benzo(a)pyrene, DDT, alkyl halide insecticide — phanthracene, azo dyes, and atrazine in the environment [21-26].

The attributes, that filamentous fungi show in the field of biotransformations, determine them to be a suitable candidate for *mycoremediation* research. Among filamentous fungi, *P. chrysosporium* is emerging as the model system for studying biotransformations. Aquatic fungi, mycorrhizal fungi, anaerobic fungal degraders in association with pollutant tolerant higher plants provide an opportunity for developing new bioremediation prospects.

Bacteria-based transformation systems

Numerous bacteria have been studied to transform aliphatic components of the crude oil; these are *Achromobacter, Acinetobacter, Arthrobacter, Bacillus, Flavobacterium, Pseudomonas,* and *Vibrio. Mycobacterium* has been shown to mineralize/degrade pyrene, naphthalene, and chlorinated hydrocarbons. Aerobic *mineralization* and *photodegradation* may be carried out by *Pseudomonas* and *Rhodopseudomonas, Pseudomonas* and *Moraxella,* respectively. Pentachlorophenol is found to be degraded by *Flavobacterium* and *Rhodococcus* [19]. *Clostridium,* isolated from the sludge, has been shown to degrade 3-hydroxy propanoate, lactate, and pyruvate to acetate and propionate [27].

Transformation of nitrilotriacetic acid by *Chelatobacter heintzii* and *Chelatococcus asaccharovorans* is also known [28]. Reduction of oxidized nitrogen to elementary nitrogen by denitrification is another example of biotransformation. The denitrifiers are the bacteria responsible for returning fixed nitrogen to the atmosphere. They are, therefore, very valuable in reducing nitrogen pollution. Denitrifiers, such as *Paracoccus denitrificans,* may accumulate nitrite or N_2O after exposure to anoxic periods [29]. Activated sludge is rich in the ammonia-oxidizing activity, as illustrated by enumeration of various *Nitrosomonas* species [30].

Pseudomonas transforms various chemical compounds, which are found as contaminants like the polycyclic compounds, including benzo(a)pyrene and bicyclic halogenated compounds like PCB and chlorobenzoate [31-33]. Alkylated compounds, like dibutyl phthalate, are degraded by *Brevibacterium* [34]. Polychlorinated biphenyls (PCBs) are introduced in the environment from industrial effluents and are found to be recalcitrant in nature, but they can be aerobically degraded by the bacteria utilizing biphenyl pathway [35]. Plasmid-based degradations of PCB, by incorporating genes involved in biphenyl- and chlorobenzoate-degrading pathways, were combined to propose a bacteria-based transformation of PCB [36].

Sander et al. [37] isolated a *Pseudomonas* sp. strain PS14, which degraded monochlorobenzene, all three isomers of dichlorobenzene, 1,2,4-

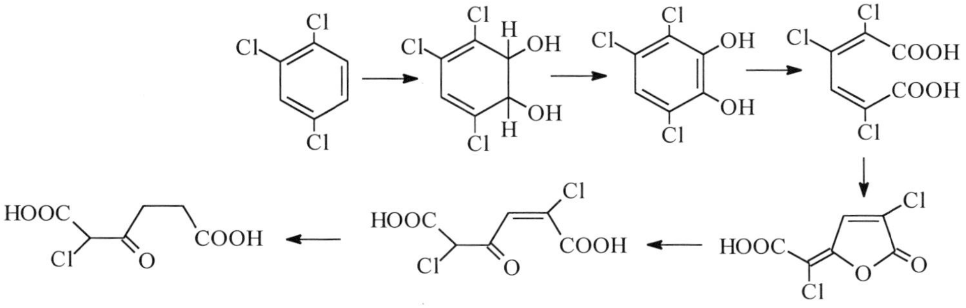

Figure 1. The degradation of 1,2,4-trichlorobenzene by *Pseudomonas* sp. strain PS14 [37].

trichlorobenzene, and 1,2,4,5-tetrachlorobenzene; the degradation pathway for 1,2,4-trichlorobenzene is shown in Figure 1. The genes responsible for the degradation of chlorobenzene from *Pseudomonas* sp. strain P51 have been cloned and characterized [38].

Microbial metal transformations

This section aims to provide an understanding of the metal bioremediation of the effluents from the industries, transuranic waste from nuclear reactors, mining, and mineral leaching. Various microbial species degrade toxic constituents used in the industries and immobilize soluble heavy metals by metal biosorption. Bacteria like *Sulfolobus*, *Metallosphaera*, *Sulfobacillus*, and *Sulfurococcus* oxidize iron and sulfur compounds under acidic conditions at temperature ranging from 55 to 85°C [39,40].

Effluents containing cyanide are very strong contaminants in the environment; cyanide oxidation by microbes was detailed elsewhere [41,42], Lien and Altringer [43] isolated *Pseudomonas pseudoalcaligenes,* which precipitates metals directly on the cell surface and is extensively used for *immobilization* of heavy metals and radionuclides. Example of this kind is represented by *Citrobacter* sp. [44]. Microbes can transform heavy metals by oxidation, reduction, methylation, and demethylation [45]. Arsenic-loaded sewage, treated with arsenite oxidase producing bacteria, can remove arsenic from the industrial effluents [46]. Ferrous and radioactive elements can also be transformed by *Alteromonas* through reduction. Chromate (CrO_4^{2-}) reducing bacterium *Enterobacter cloacae* is resistant to chromate and can anaerobically reduce CrO_4^{2-} to Cr (III), which gets precipitated and eliminated from the effluents [47]. Metal absorption by the microbes may be due to the presence of metallothioneins and other metal-binding proteins, which chelate the metals and help in the metal biosorption and transformation [48].

Novel approach in the field of biotransformation

One dimensional fourier-transform proton nuclear magnetic resonance (^{1}H-NMR) spectroscopy can be used to study biotransformations *in situ. In vivo* and *in aqua* the absolute stereochemistry of some reactions can also be determined and assessments of *metabolic fluxes* can be made, the technique can be used in many fields of organic chemistry, biotechnology and microbiology for analytical approaches in research [49]. HPLC is another technique to observe biotransformation of xenobiotics under anaerobic conditions [50].

BIOREMEDIATION TECHNOLOGIES FOR SEWAGE TREATMENT

Several well recognized technologies have been developed and new technologies are being developed aggressively, the following are the descriptions of existing technologies:

PRELIMINARY TREATMENT

In this first stage of water treatment, gross solids are removed from the flow of sewage by different methods. The most commonly used method is the removal of the solids by the use of screens, which are either manually raked or automatically raked, with the debris being removed for disposal. Macerator can also be used for preliminary treatment, which shreds the solids; this type of treatment removes the grit and is a physical process [51].

SECONDARY TREATMENT

Bio-oxidation of the sewage by the microbial consortia present in the system is being exploited here. In this treatment variations are common, each having its own advantages and disadvantages.

Following are the descriptions of existing technologies for the secondary treatment of the effluents:

AEROBIC TREATMENT

In the aerobic treatment system, oxygen is the ultimate hydrogen acceptor with large release of energy. These treatments are as follows:

Trickling filter or percolating filter

Trickling filter (biological bed) processes rely on the activity of microbes. Liquid waste streams are sprayed over the surface of the filter with the help of jets; this liquid waste trickles down through a tower, which is randomly packed with a solid medium, either mineral or synthetic plastic. The aerobic process is the fundamental activity of the microbes in the trickling filter, which results in the production of enzymes for the bio-oxidation of the organic compounds. On the other hand, the anaerobic decomposition of organic compounds to CO_2 and methane occurs in the areas adjacent to the media. Advantages of trickling filter include attached biomass with good oxygenation and the toxicity limits in the tolerable range. The disadvantages include the odour, which anaerobic biofilm creates and the failure of mechanical parts [1]. Trickling filters have been used in treating oil refinery waste. The microbial consortia in the trickling filter biofilm is identified *in situ* by 16S rRNA-targeted oligonucleotide probe (NEU), specific for lithoautotrophic ammonia-oxidizing bacteria. Based on the comparative sequence analysis, *in situ* hybridization of the biofilms with probe NEU allowed the detection of dense cell clusters of ammonia-oxidizing bacteria [52].

Sulfate-reducing bacteria have been detected in the photosynthetic biofilms of the trickling filter. The vertical distribution of these bacteria was investigated by fluorescent dye-conjugated oligonucleotides as the *phylogenetic probes* binding to 16S rRNA. A negative correlation between the vertical distribution of bacteria and the O_2 profiles was found [53].

Rotating biological contactors (RBCs)

Another treatment with lower energy cost and maintenance than trickling filter is the rotating biological contactors in which rotating disks support the growth of biofilms; this disk is rotated with 40% of the surface submerged in the water. RBCs have been used in the treatment of petroleum and textile wastes, phenol, sulfides and ammonia compounds. They can withstand shock, since the biomass is fixed with no sludge to recycle, and there is no loss of culture.

Activated-sludge

Activated-sludge is a mixed culture of the microorganisms cultivated in bubler tanks on organic substances present in the wastewater. This mixed culture suspension forms activated-sludge due to wastewater aeration by metabolic reactions of the microbes. After a particular incubation period the sludge is sent to a clarifier to remove the biomass [54].

The sludge from the settlement tank is dewatered and is disposed off. Advantages of activated-sludge include high oxidation rates, recycling of the biomass, short residence time, and good containment. Disadvantages include toxicity beyond the tolerable limit of the microbes [1].

Activated-sludge is composed of the flocs, these flocs are composed of discrete clumps of microorganisms like heterotrophic bacteria, rotifers, and protozoans, which hydrolyze and oxidize the carbon compounds. These flocs are bound together with the organic and inorganic substances such as lipids, proteins, and polysaccharides along with slime forming bacterium, *Zoogloea ramigera* [55].

Microbial composition of the activated-sludge is found to be very vast. The microbes, most frequently occurring in activated-sludge, are *Achromobacter, Flavobacterium, Pseudomonas, Chromobacterium, Acinetobacter, Alcaligenes, Arthrobacter, Nocardia,* and *Lophomonas* [56,57]. Along with the bacteria and fungi, yeasts are also found in the activated-sludge [56]. Various filamentous organisms like *Sphaerotilus, Beggiatoa,* and *Leptomitus,* along with the glycocalyx (polysaccharides) form the flocs in the activated-sludge [58]. Cilliated protozoans like *Vorticella* and *Opercularia* also constitute the microbial consortia of the activated-sludge [59]. Activated-sludge is also probed with modern analytical techniques like *immunofluorescence test*, polyamine pattern analysis, fatty acid and protein patterns to analyze the diversity of the microbes and their interactions.

Polyphosphate accumulating *Acinetobacter* strains were isolated from the activated-sludge by analyzing the polyamine pattern, fatty acid and protein patterns [60,61]. Activated-sludge was probed with fluorescently labelled 16S and 23S ribosomal RNA-targeted oligonucleotide probes, specific for defined phylogenetic groups of bacteria; *in situ* hybridization was used to detect flamentous bacteria in the samples [62]. *Polyclonal antibodies*, raised against Gram-negative nitrilo triacetic acid utilizing species of *Chelatobacter heintzii* and *Chelatococcus asaccharovorans,* enabled their determination by an indirect immunofluorescence test in the activated-sludges [28]. Two more kinds of activated sludge are recognized. The floating activated-sludge is produced when the particles of the activated-sludge have a density less than that of water; sludge rising and floating problems are due to excessive growth of *Nocardia amarae* ST6. Toxicity test showed that chlorination was done to control *Nocardia* foaming in an activated-sludge process [63].

Another cause of floating of the activated-sludge is the formation of nitrogen bubbles from the reduction of nitrites and nitrates; nitrogen adheres to sludge flocs and raises them to the surface of the activated-sludge. Bulking activated-sludge is another kind of sludge, which forms when the sludge has decreased settling velocity and expressed as zone settling velocity (ZSV) or sludge volume index (SVI) [64]. Bulking occurs by filamentous organisms like *Thiothrix, Beggiatoa,* and *Nocardia* [65], and it can be suppressed by removal of sulphide in the sludge. Non-filamentous organisms also cause the bulking of the activated-sludge, the non-filamentous bulking occurs by *Zoogloea ramigera* [66].

Aerated lagoons

Aerated lagooning is a technique based on the activated-sludge system. They are mechanically agitated by means of aerators installed in a stabilization ponds. Recycling of the biomass is not accomplished in the aerated lagoons. Aerated lagoons are inexpensive with higher oxidation rates, but disadvantageous in the sense that sludge removal is essential after a particular period of incubations [1]. Aerated lagoons have been utilized in the treatment of *industrial effluents* containing petrochemicals, textile, and paper mill wastes.

Waste stabilization ponds

A modified version of aerated lagoons consists of shallow open air ponds in which the waste is incubated without agitation. Both aerobic and anaerobic degradation pathways operate in these ponds. Algae like *Chlorella* and *Oscillatoria,* and diatoms, *Pandorina, Chlorogonium,* and *Scenedesmus* degrade the carbonaceous compounds present in the waste [3].

Fixed film system

In fixed film system, microbial biofilm grows on a fixed bed; the waste stream flows down over the microbes and the microbes degrade the carbonaceous product in the sewage. Advantages of a fixed film system include low power requirements and simple to use. Disadvantage includes the need for large space requirement for the system to operate [67].

ANAEROBIC TREATMENT

In anaerobic treatment systems, the ultimate hydrogen acceptor may be nitrate or sulfate instead of oxygen with low release of energy as compared to aerobic systems.

Descriptions of these treatments are given below:

Anaerobic digesters

An anaerobic digester is a closed reactor with no mechanical agitation. This process employs separation of solids and their recycling, which increases the organism concentration and the retention time of the biomass. Separation of solids is accomplished by gravitational forces [3]. Anaerobes degrade the sewage matter into organic acids, alcohols, H_2, and CO_2, which can be converted to methane by methanogenic bacteria.

Anaerobic filters

A packed bed of support material, on which the wastewater is sprayed from the top works like trickling filter; the microbial biomass degrades the waste in a symbiotic relationship between acid-forming and the methane-forming bacteria to constitute anaerobic system [68].

New approaches for anaerobic bioremediation

Effluents from the industries can contaminate the aquifers by seeping into the ground and is the source of groundwater pollution. Anaerobic bioremediation is the technique for treating the effluents in the regions where anaerobic conditions are prevailing. Recent advances have been made in this field by exploiting the anaerobic microbial metabolism [69]. The addition of alternative soluble electron acceptors, like nitrate or sulfate to the subsurface, can result in the removal of contaminants by stimulating the growth of the microbes, utilizing the anaerobic pathway. The degradation of petroleum has been attempted in this way. Anaerobic benzene degradation is reported in petroleum contaminated aquifers; degradation of benzene is found to be coupled with the iron-reducing conditions because *Geobacter* species, found in these aquifers, can couple the oxidation of aromatic compounds to the reduction of Fe (III) [70-72]. Degradation of hydrocarbons involves *in situ* biological production of molecular oxygen by perchlorate reducers to aerobic organisms, which degrade organic compounds in the presence of chlorite. Thus, coupling of the degradation of the organic compounds and the reduction of perchlorate occurs [73-75]. Predominant perchlorate-reducing bacteria in the environment appear to belong to *Dechloromonas*. Another example of anaerobic degradation is the transformation of the nitroaromatics by ferrous-reducing bacteria. *Geobacter*, a member of the family Geobacteraceae and sulfate-reducing bacteria are also found to immobilize heavy metals like technetium and uranium from the waste streams. The examples cited above are the indication of the potential of anaerobic metabolism as a key remediative technology for the industrial effluents percolating in the groundwater.

TERTIARY TREATMENT

Tertiary treatment is the removal of residual solids from the settled bio-oxidized effluents.

Gravel bed hydroponics

Gravel bed hydroponics is a constructed wetland system for sewage treatment, which has proved effective for the tertiary treatment. Improvement in the effluent quality was observed in the beds, planted with the *Phragmites* (common reed), resulting in the large reduction in BOD, suspended solids, and ammoniacal nitrogen [76]. As the effluents percolate through the gravel beds, treatment is effected by chemical, physical, and microbial processes. In the roots of the *Phragmites*, the microbial biofilm exists, which changes with the bed distance [77]. The biofilms attached to the roots of *Phragmites australis* supported higher potential rates of nitrogen transformations in the gravel bed [78,79].

Sand filtration

Shallow tanks, containing a layer of special sand about 10 cm deep overlaying a similar layer of graded gravel resting on the underdrains constitute the process of sand filteration. The tank is flooded with the effluents from secondary tanks and pass down through filters, solids are strained out from the upper layer from time to time [2].

Microstrainers

Microstrainers consist of large drum, closed at one side and covered with a finely woven steel fabric and rotating on a horizontal axis. The drum is submerged in a tank and divided into compartments [2]. Microstrainers are used for the straining purposes of the effluents.

RECENT TECHNIQUES FOR EFFLUENT TREATMENT

Application of a reactor with the polyurethane foam cubes as attached medium was used for the BOD removal and nitrification. Nitrifiers grew preferentially on the polyurethane foam cubes, and its nitrifying activity reached 0.33 for cubes of 8 cm^3. Removal of BOD was completed in 4 hours, and nitrification completed in 10 hours [80].

Immobilization of bacteria in the polymer beads to observe the inhibition of microflora

Nitrifying bacteria are immobilized in polymer beads (alignate beads), which are then placed in a biological reactor. Wastewater is pumped into the reactor, together with ammonium. The nitrate formed in the reactor is analyzed and is used to monitor the nitrifying activity on the reactor. Thus a drop in the nitrate production indicates that an inhibition of the microflora has taken place by the inhibiting compounds [81].

In situ analysis of nitrifying bacteria in the sewage

In situ analysis of the nitrifying bacteria in the sewage treatment eliminates the limitations of traditional microbiological methods in the identification of ammonia- and nitrite-oxidizing bacteria, which was based on comparative sequence analysis. A collection of 16S rRNA-targeted oligonucleotide probes was designed for all validly described members of the genus *Nitrobacter*. Whole cell hybridizations of target and reference cells with fluorescent probe derivatives were used to determine the optimal hybridization stringency for each of the probes. These probes were applied together with a recently developed probe for important members of the genus *Nitrosomonas* for simultaneous identification of ammonia- and nitrite-oxidizing bacteria in natural and *engineered* systems [82].

Multispecies plant systems for wastewater quality improvement and habitat enhancement

The use of macrophytic plants for the treatment of municipal wastewater is growing rapidly; aquatic plants, such as *Eichhornia crassipes, Lemna* spp., *Arundo donax, Scirpus olneyi, Salix nigra,* and *Populus fremontii* are used. Data from the water quality studies indicated that these constructed ecosystems reduce BOD_5 consistently to below the 10mg/litre BOD_5 tertiary standard. Beside this, these ecosystems also remove nitrogen and pathogens [83].

ROLE OF GENETICALLY ENGINEERED ORGANISMS IN THE BIOREMEDIATION PROCESS

Genetically engineered organisms may solve the environmental pollution problems. Specific genetic traits may be placed together such that one trait brings about a detectable signal, that acts as an indicator of the presence or absence of the activity of the gene. These "*reporter genes*" are exemplified by the *lux* gene sequence, isolated from the fluorescent bacterium, which causes the bacterium to give off visible light, when the gene sequence is active [1]. *Pseudomonas* is a favoured organism for genetic manipulations, including the development of recombinants with biodegradative biotransformative or biosorptive pathways [18].

Mutant bacteria in the wastewater treatment

Mutant bacteria, developed by *in situ* genetic engineering techniques, are being manufactured. Stock culture of these bacteria is accomplished through cultivation of pure organism culture on a sterilized substrate.

Bacterial *mutation technology* has removed the industrial effluents like phenols, ammonia, sulfides, alcohols, and oil. Mutant bacteria additives are now available for *bioaugmentation* in the activated sludge, trickling filters, and lagoons. These bacterial cultures are sold as stock cultures in the products [84].

EVALUATION OF THE EFFICIENCY OF THE SEWAGE TREATMENT SYSTEMS

Evaluation of sewage treatment requires measurement of several parameters, which provide an idea of the biological and chemical constituents of the water and help to analyze the condition of the effluents. Efficiency of the sewage treatment system can be enhanced by developing a framework based on the factors like toxicity, oxygen requirement, microbial constitution, and pathogenicity.

Prediction of the rate of dissolved oxygen by the microbes

Idea of the biochemical oxygen demand (BOD) is absolutely necessary for the proper deoxygenation of the wastewater. Oxygen demand can be measured by BOD_5 test, which is based on the measurement of dissolved oxygen consumed by the microbes in the 5-day period. BOD_5 of the effluents provides an idea of what the oxygen demand would be if treatments were not provided, and the difference between that value and the BOD_5 of the effluents is a measure of the reduction in the deoxygenation of the receiving water.

Chemical oxygen demand (COD) test

To find out the oxygen equivalent of the organic content in the effluents, along with the microbial biomass, the COD test is used by standardizing the dichromate solution in terms of molecular oxygen.

Total organic carbon (TOC) test

TOC test is used to measure the organic matter in the wastewater. TOC test gives the concentration of hazardous wastes in water by oxidizing the carbon to carbon dioxide, heat, and oxygen.

Algal growth test

Algal assays are used to determine nitrogen and phosphorous in the effluents and can be used to analyze what reductions in the nitrogen and phosphorus content of the effluents would be required to improve the quality of the wastewater. An alga *Selenastrum capricornutum* is used in the algal growth test [85].

Microtox test and the WET bioassay

For testing the toxicity of the effluents, the whole effluent toxicity (WET) test is performed by using algae, invertebrates, and fishe. WET bioassay, which is done by using the microbes, is the Microtox tests. The Microtox test utilizes a luminescent bacterium *Photobacterium phosphoreum* as a Microtox reagent. This bacterium in the reagent is exposed to the wastewater, and the decrease of light emission, as a result of toxicity, is taken as an indicator of pollution [86,87].

CONCLUSIONS

There is now great awareness of the potential dangers of the water pollution. The removal of these pollutants from the contaminated environment by living or non-living biomass and its products can provide an economically feasible and efficient means for element recovery and environmental protection. Both organic and inorganic materials can be transformed in the absence of molecular oxygen, suggesting that anaerobic bioremediation is of tremendous use. Experimentation on an integrated aerobic and anaerobic system for the treatment of sewage to achieve a greater COD/BOD removal is the present-day research need in the field of effluent treatment. Parallel advances are being made in *genetic engineering* for raising the microorganisms, capable of degrading the constituents in physiologically stressed environment, which can be used as immensely treasured biological resources along with the integrated anaerobic and aerobic system for the sewage treatment purposes.

REFERENCES

1   Dobbins DC. In: Nierenberg WA, ed. Encyclopedia of Environmental Biology Vol. 1, London: Academic Press Inc, 1995; 169-182.

2   Herschy RW. In: Herschy R W, Fairbridge RW, eds. Encyclopedia of Hydrology and Water Resources. Dordrecht: Kluwer Academic Publishers, 1998; 608-617.

3   Matlosz MM, Trattner RB. In: Cheremisinoff, PN, ed. Encyclopedia of Environmental Control Technology, Vol 4, Texas: Gulf Publishing Company, 1990; 411-445.

4   Hoeppel RE, Hinchee RE. Environ Sci Pollut Control Ser 1994; 6: 311-431.

5   Dagley S, American Scientist 1975; 63: 681-689.

6   Matsumura F, ed. Survival in Toxic Environments: Microbial Degradation of Pesticides. New York: Academic Press, 1974; 29-54.

7   Reineke W. In: Gibson DT, ed. Microbial Degradation of Organic Compounds. New York: Marcel Dekker Inc, 1984; 319-353.

8   Bumpus JA In: Bollag JM, Stotzky G, eds. Soil Biochemistry, Vol 8, New York: Marcel Dekker Inc, 1993; 65-100.

9   Venkatadri R, Irvine RL, Water Res 1993; 27: 591-596.

10   Bumpus JA, Tatarko M. Curr Microbiol 1994; 28: 185-190.

11   Sublette KL, Ganapathy EV, Schwartz S. Appl Biochem Biotechnol 1992; 34: 709-723.

12   Kirk TK In: Smith JE, Berry DR, Kristiansen B, eds. The Filamentous Fungi, Vol 4, Fungal Technology. London: Edward Arnold, 1983; 266-295.

13   Kirk TK. In: Gibson DT, ed. Microbial Degradation of Organic Compounds. New York: Marcel Dekker Inc, 1984; 399-437.

14   Friedrich J, Cimerman A, Perdih A , In: Arora DK, Elander RP, Mukerji KG, eds. Hand Book of Applied Mycology, Vol 4, Fungal Biotechnology. New York: Marcel Dekker Inc, 1992; 963-992.

15   Eriksson KE, Kirk TK. In: Robinson CW, Howell JA, eds. Comprehensive Biotechnology, Vol 4, The Practice of Biotechnology Speciality Products and Service Activities. Oxford: Pergamon Press, 1985; 271-294.

16   Berka RM, Dunn-Coleman N, Ward M. In: Bennette JW, Klich MA, eds. *Aspergillus* Biology and Industrial Applications. Boston: Butterworth, 1992; 155-214.

17   Litchfield JH. In: Porubcan RS, Sellars RL, eds. Microbial Technology, Vol 1, London: Academic Press, 1979; 93-155.

18   Faison BD. In: Lederberg J, ed. Encyclopedia of Microbiology, Vol 2, New York: Academic Press Inc, 1992; 335-349.

19   Lindley ND. In: Arora DK, Elander RP, Mukerji KG, eds. Handbook of Applied Mycology, Vol 4, Fungal Biotechnology, New York: Marcel Dekker Inc, 1992; 905-929.

20   Cerniglia CE, Sutherland JB, Crow SA. In: Winkelmann G, ed. Microbial Degradation of Natural Products, Germany: VCH Press, 1992; 193-217.

21   Haemmerli SD, Leisola MSA, Sanglard D, Feichter A. J Biol Chem 1986; 261: 6900-6903.

22   Kohler A, Jager A, Willershausen H, Graf H. Appl Microbiol Biotechnol 1988; 29: 618-620.

23   Kennedy DW, Aust SD, Bumpus JA. Appl Environ Microbiol 1990; 56: 2347-2352.

24   Morgan P, Lewis ST, Watkinson RJ. Appl Microbiol Biotechnol 1991; 34: 693-696.

25   Paszczynski A, Pasti-Grigsby MB, Goszczynski S, Crawford RL, Crawford DL. Appl Environ Microbiol 1992; 58: 3598-3604.

26   Mougin C, Laugero C, Asther M, Dubroca J, Frasse P, Asther M. Appl Environ Microbiol 1994; 60: 705-708.

27   Tanaka K. J Ferment Bioeng 1995; 79: 503-505.

28   Wilberg E, El-Banna T, Auling G, Egli-T. System Appl Microbiol 1993; 16: 147-152.

29   Robertson LA, Kuenen JG. In: Fry JC, Gadd GM, Herbert RA, Jones CW, Watson-Craik IA, eds. Microbial Control of Pollution. New York: Cambridge University Press, 1992; 227-267.

30   Suwa Y, Imamura Y, Suzuki T, Tashrio T, Urushigawa Y. Water Res 1994; 28: 1523-1532.

31   Barnsley EA. Can J Microbiol 1975; 213: 1004.

32   Sayler GS, Colwell RR. Environ Sci Technol 1976; 10: 1142.

33   Focht DD, Shelton D. Appl Environ Microbiol 1987; 53: 1846-1849.

34   Engelhardt G, Wallnoffer PR, Hutzinger O. Bull Environ Contam Toxicol 1975; 13: 342.

35   Lunt D, Evans WC. Biochem J 1970; 118: 54.

36   Furukawa K, Chakrabarty AM. Appl Environ Microbiol 1982; 44: 619-626.

37   Sander P, Wittich RM, Fortnagel P, Wilkes H, Francke W. Appl Environ Microbiol 1991; 57: 1430-1440.

38   van der Meer JR, van Neerven ARW, de Vries EJ, de Vos WM, Zehnder AJB. J Bacteriol 1991; 173: 6-15.

39   Kelley BC, Tuovinen OH. In: Salomons W, Forsnter U, eds. Chemistry and Biology of Solid Waste: Dredged Material and Mine Tailings. Berlin: Spinger Verlag, 1988; 33-53.

40   Tuovinen OH, Kelley BC, Groudev SN. In: Zeikus JG, Johnson EA, eds. Mixed Cultures in Biotechnology. New York: McGraw Hill, 1991; 373-427.

41   Howe RHL. Int J Air Water Pollut 1965; 9: 463-478.

42   Knowles CJ. Bacteriol Rev 1976; 40: 652-680.

43   Lien RH, Altringer PB. In: Torma AE, Apel ML, Brierley CL, eds. Biohydrometallurgy Technologies. Warrendale: The Minerals, Metals and Materials Society, 1993; 219-227.

44   Tolley MR, Strachan LR, Macaskie LE. J Ind Microbiol 1995; 14: 271-280.

45   Lovley DR, Phillips EJP, Gorby YA, Landa ER. Nature 1991; 350: 413-416.

46   Williams JW, Silver S. Enzyme Microbial Technol 1984; 6: 530-537.

47   Fujii E, Toda K, Ohtake H. J Ferment Bioeng 1990; 69: 365-367.

48   Rayner MH, Sadler PJ. In: Poole RK, Gadd GM, eds. Metal-Microbe Interactions. Oxford: IRL Press, 1989; 39-47.

49   Brecker L, Ribbons DW. Trends Biotechnol 2000; 18: 197-202.

50   Regan KM, Crawford RL. Biotechnol Lett 1994; 16: 1081-1086.

51   Ireland D. In: Herschy RW, Fairbridge RW, eds. Encyclopedia of Hydrology and Water Resources. Boston: Kluwer Academic Publishers, 1998; 607-608.

52   Wagner M, Rath G, Amann R, Koops HP, Schleifer KH. System Appl Microbiol 1995; 18: 251-264.

53   Birger-Ramsing N, Kuhl M, Barker Jorgensen B. Appl Environ Microbiol 1993; 59: 3840-3849.

54   Ardern E, Lockett WT. J Soc Chem Ind 1914; 33: 523.

55   Gray NF. In: Nierenberg WA, ed. Encyclopedia of Environmental Biology, Vol 3, London: Academic Press Inc, 1995; 463-478.

56   Benedict RG, Carlson DA. Water Res 1971; 5: 1023.

57   Pike EB, Carrington EG. Water Pollut Control 1972; 71: 2.

58   Costerton JW, Irvin RT. Ann Rev Microbiol 1981; 32: 299.

59   Curds CR, Fey GJ. Water Res 1969; 3: 853.

60   Duncan A, Vasiliadis GE, Bayly RC, May JW. Biotechnol Lett 1988; 10: 831-836.

61   Kampfer P, Bark K, Busse HJ, Auling G, Dott W. System Appl Microbiol 1992; 15: 409-419.

62   Wagner M, Amann R, Lemmer H, Manz W, Schleifer KH, Pujol R. Wat Sci Tech 1994; 29: 15-23.

63   Wong PK, Chung WK. J Environ Sci Health 1993; 28: 1615-1628.

64   Mohlman FW. Sew Wks J 1933; 5: 74.

65   Blackbeard JR, Ekama GA, Marais GVR. Water Pollut Control 1986; 85: 90.

66   Pipes WO. J Water Pollut Control Fed 1969; 41: 714.

67   James A. J Bacteriol 1964; 27: 197.

68   Mosey FE. Water Pollut Control 1978; 77: 370-378.

69   Anderson RT, Lovley DR. Adv Microb Ecol 1997; 15: 289-350.

70   Weiner J, Lovley DR. Appl Environ Microbiol 1998; 64: 775-778.

71   Anderson RT, Lovley DR. Bioremed J 1999; 3: 121-135.

72   Anderson RT, Lovley DR. Environ Sci Technol 2000; 34: 2261-2266.

73   Herman DC, Frankenberger WT Jr. J Environ Qual 1998; 27: 750-754.

74   Urbansky ET. Bioremed J 1998; 2: 81-95.

75   Herman DC, Frankenberger WT Jr. J Environ Qual 1999; 28: 1018-1024.

76   Williams J, Bahgat M, May E, Ford M, Butler J, Kadlec RH, Brix H. Wat Sci Tech 1995; 32: 49-58.

77   Loveridge RF, Williams JB, Butler JE, Bahgat M, El-Shatoury S. In: Proceedings of the 9th International Congress on Soilless Culture. St Helier, Jersey, Channel Islands, UK, 1996: 273-286.

78   Williams JB, May E, Ford MG, Butler JE, Bavor HJ, Mitchell DS. Wat Sci Tech 1994; 29: 29-36.

79   Butler JE, Loveridge RF, Awad A. In: Proceedings of the 8th International Congress on Soilless Culture, Hunters Rest, South Africa, 1993; 97-113.

80   Tsuno H, Somiya I, Matsumoto N, Sasai S. Wat Sci Tech 1992; 26: 2035-2038.

81   Persson A, Andersson C, Welander T, Gunnarsson L. Vatten 1993; 49: 41-48.

82   Wagner M, Rath G, Koops HP, Flood J, Amann R, Bally D, Asano T, Bhamidimarri R, Chin KK, Grabow WOK, Hall ER, Ohgaki S, Orhon D, Milburn A, Purdon CD, Nagle PT. Wat Sci Tech 1996; 34: 237-244.

83   Karpiscak MM, Gerba CP, Watt PM, Foster KE, Falabi JA, Angelakis A, Asano T, Diamadopoulos E, Tchobanoglous G. Wat Sci Tech 1996; 33: 231-236.

84   Deutsch DJ. Chemical Engineering 1979; 86: 100-102.

85   USEPA. US Envronmental Protection Agency, Corvallis, Oreg: EPA-600/9-78-018, 1978.

86   Kaiser KLE, Palabrica VS. Water Pollut Res J Can 1991; 26: 361-431.

87   Asami M, Suzuki N, Nakanishi J, Nyholm N, Jacobsen BN. Wat Sci Tech 1996; 33: 121-128.

Biotransformations: Bioremediation Technology for Health and Environmental Protection
V.P. Singh and R.D. Stapleton, Jr. (Editors)
© 2002 Elsevier Science B.V. All rights reserved.

Electro-physical properties of microbial cells during the aerobic metabolism of toxic compounds

O.V. Ignatov[a,b], S.Yu. Shchyogolev[a], V.D. Bunin[c] and V.V. Ignatov[a]

[a]Institute of Biochemistry and Physiology of Plants and Microorganisms, Russian Academy of Sciences, 13 Pr. Entuziastov, Saratov 410015, Russia

[b]Saratov State University, 83 Ul. Astrakhanskaya, Saratov 410005, Russia

[c]Institute of Applied Microbiology, Obolensk 142289, Russia

INTRODUCTION

Methods for analyzing the electro-physical properties of particles of diverse colloidal systems (microbial cells, subcellular fractions, viruses, etc.) have been in an intensive development over the past two decades. This is attested by the improvements made to existing approaches and by the emergence of new ones [1-9]. Among the techniques, most commonly referred to in the literature, are dielectric spectroscopy, dielectrophoresis, electrorotation, and electric-field orientation.

One of the most challenging fields of application of these techniques is studying the fine structural organization of cells and the changes occurring in it under the influence of physico-chemical and biological factors. Yet, in the vast majority of publications the emphasis has so far been placed on methodological aspects, with relatively few concrete biological problems being dealt with. This is possibly because investigators who develop the physico-chemical fundamentals of these methods and realize analytical equipment are first concerned with the validity of their mathematical models of cell behaviour in an electric field and with interpreting the results of electro-physical experiments. In so doing, they tend to neglect to examine the peculiarities of the physiological and biochemical processes taking place in a living cell. On the other hand, to secure financial backing for research, they often try to demonstrate the purely applied aspects of their methods in the context of routine problems in industrial microbiology, which are often not meant for discussion in the literature. However, of undeniable interest to investigators in various fields of microbiology, cell biophysics, and biotechnology may, in particular, be the possibility of establishing the interrelationship between cellular electro-physical properties and cell metabolic reactions.

In this review, we are concerned primarily with the results of our experimental studies of the effect of the aerobic metabolism of toxic compounds on the electro-physical properties of microbial cells. These studies were by use of the method of electro-optical analysis of cell suspensions [1-3,10-12]. No other reports investigating the electro-physical properties of microbial cells during the metabolism of toxic (or non-toxic) low-molecular-weight compounds have so far come to our notice. Works closest in theme to ours are those using electro-physical analysis to assess the damaging effect produced on microbial cells by various physico-chemical factors (including toxic compounds). They have as their main object the devising of methods for the rapid analysis of the viability of microbial cells, for the evaluation of the degree of cell injury and of the heterogeneity of cell cultures, etc. [10,13-16]. The fundamental difference between our studies and others' is that we used microbial strains with enzyme systems of the initial metabolism of certain toxic compounds. During the electro-optical investigations, independent monitoring of cellular metabolic activity towards the corresponding substrates was also performed by a number of traditional techniques.

PHYSICO-CHEMICAL FUNDAMENTALS OF THE ELECTRO-OPTICAL ANALYSIS OF CELL SUSPENSIONS

The orientational ordering of disperse systems, established and used in electro-optical analysis, gives rise to the effect of *optical anisotropy*. As this takes place, the substance itself of the particles of a disperse phase may be optically isotropic with the proviso that their shape is other than spherical (anisotropy of the form [17]). As a result, the dependence arises of the optical properties of suspensions on the degree of the orientation of anisometric particles ensured by application of a DC or an AC electric field.

Among these properties are birefringence, orientational dichroism and the *orientational turbidimetric effect* [12,18,19]. The last-named property is manifested as the dependence of the characteristics of attenuation of light by a suspension, which results from its scattering by particles, on the orienting field strength, frequency and direction with respect to the direction of the light beam. To record the orientational turbidimetric effect, one can use the transmission coefficient or optical density values, $D = \tau l/2.3$ (where τ is turbidity and l is the light beam path in the disperse system) in both polarized and unpolarized lights. This phenomenon was employed as the major electro-optical effect in the experimental investigations discussed in the next section of this review.

We note that combining the electro-optical [1-3,10-12,18-20] and spectroturbidimetric [21-23] analyses of disperse systems permits the use of a broad spectrum of the probing electromagnetic radiation (from the radio-frequency range to the optical range) by accounting for the stationary and relaxation characteristics of the systems. In so doing, it becomes possible to determine a large set of various parameters of the suspensions under study: morphometric (particle size, shape, concentration and refractive index), electro-physical (functions of the polarizability tensor components), aggregative (change in aggregate number and volume, fractal parameters), adsorptive (dependence of the amplitude and phase of the electro-optical signal on the adsorbate concentration), etc. Given sufficiently prompt operation of the appropriate measuring devices [10,11,24], these methods may be exploited to analyze the kinetics of rapid processes, which is of particular value in the study of surface phenomena (e.g., structural transformations of the cell membranes), particle aggregation, adsorption of low- and high-molecular-weight compounds on particles, etc.

The major metrological problem, which is solved by combining the above methods [11,12,20], is that of determination of the turbidity spectra of suspensions $\tau(\lambda)$ (where λ is the light wavelength *in vacuo*), having different degrees of orientational ordering with concurrent estimation of the relaxation and electro-physical properties of particles. The electro-physical properties are manifest by the force (dependence on the orienting field strength) [18,19] and dynamic (dependence on the orienting field frequency ω) [10,24] characteristics of the electro-optical effect.

The action of the electric field gives rise to induced charges on/in the suspended particles. The charge distribution and magnitude depend on the polarizability mechanism operating under given experimental conditions [25,26]. Characteristic of a bulk mechanism is the rise of induced charges at the interfaces between adjacent media with different complex dielectric permeability values. If the particles under study are cells, the interfaces are represented by contact areas: double electric layer - cell wall, cell wall - cytoplasmic membrane, cytoplasmic membrane - cytoplasm, etc.

The magnitude of the charges induced at the interfaces is proportional to the electric field strength, E, and depends on the relationship between the dielectric permeability values for the constituent structures of the media. The phenomenological parameter, describing this effect, is the particle polarizability tensor. For axisymmetric non-spherical particles, which are good models for many cell types, the polarizability tensor has at least two distinct components [26]: a longitudinal one, γ_a (corresponding to the direction of the long axis of the particle) and a transverse one, γ_b (corresponding to the orthogonal direction). The induced charges of opposite sign, that are distributed by particle volume, form an effective dipole, the interaction of which with the external electric field leads to particle orientation.

The preferable direction (along or across the strength vector direction of the orienting electric field) and the degree of particle orientation (characterized by the width of the function of particle distribution by orientation angles) are dependent on the sign of the *polarizability tensor anisotropy*, $\Delta\gamma = \gamma_a - \gamma_b$, and on the value of the parameter:

$$q = \Delta\gamma E^2 / (2kT), \tag{1}$$

where k is the Boltzmann constant and T is the absolute temperature. This value is determined by the relationship between the energy of the orienting electric field and the energy of Brownian motion, interfering with particle orientation. To a relatively weak degree of orientation, there corresponds the range of values $q \ll 1$.

As noted above, the complete or partial orientation of particles leads to substantial changes in the optical properties of suspensions, due to the effect of optical anisotropy and to the orientational turbidimetric effect [11,12,18-20]. To illustrate, Figure 1a portrays oscillograph traces (*oscillograms*) [11], reflecting the temporal change in the transparency of cell suspensions that occurs under the influence of the orienting electric field (trains of successive rectangular pulses of variable frequencies ω, but of equal duration (about 1 s). A beam of unpolarized light at $\lambda = 810$ nm, directed perpendicular to the field direction with the strength amplitude $E = 100$ V.cm^{-1}, was used (Figure 1b). An electro-optical signal consists of three segments corresponding to [12,27]: (i) relatively fast clarification of the suspension; (ii) quasistationary state with a given degree of particle orientation (with weak dynamic modulation at twice the frequency of the orienting field [27]; and (iii)

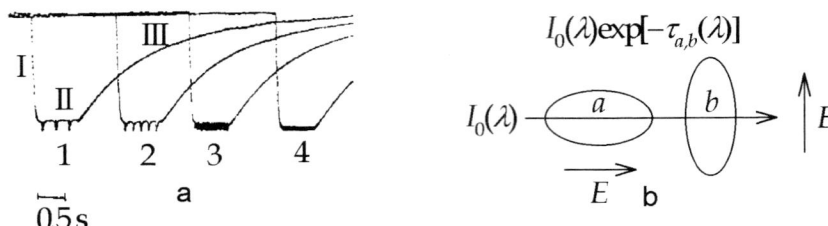

Figure 1. a: Oscillograms of the electro-optical signals observed for *Escherichia coli* cell suspensions at ω values of 2.5 (1), 7 (2), 14 (3) and 25 (4) Hz. b: Schematic representation for determination of integral parameters [11]. I_0 is the intensity of the incident light beam.

relaxation transition of the suspension to the state of random particle orientation, with a relaxation time t_R = 820 ms.

The turbidity spectra $\tau_{a,b}(\lambda)$, corresponding to the stationary segments, are used to determine the particle shape parameter (the axial ratio a/b) and the phase shift of the light wave, which occurs when the wave travels along the diameter of equivalent volume sphere. For example, for large enough, prolate spheroids with a refractive index close to that of a dispersion medium (good approximation for many bacteria), $a/b = (\tau_a/\tau_b)1/(n-1)$, where $n = -\delta \ln\tau/\delta \ln\lambda$ and τ is the turbidity value at random particle orientation. The phase shift depends on both the size and refractive index of particles. For this reason, for simultaneous estimation of these most important cell structural parameters, the mean cell size is determined independently by the value of the rotary diffusion coefficient, $\theta = 1/6t_R$, found from relaxation curves [11,20]. The next steps are to determine [21-23] (i) numerical cell concentration; (ii) mass-volume cell concentration (biomass concentration); and (iii) effective total area of the surface of particles contained per unit volume of the suspension, which is used in adsorption studies [28]. Finally, by determining the E dependence of the turbidity changes, $\Delta\tau_{a,b}/\tau$ (due to cell orientation at a given E), with account taken of the particle parameters mentioned above, the cell polarizability $\Delta\gamma$ can be found [18,19]. Thus, by combining the stationary and relaxation spectral characteristics of the electro-optical effect, determined in a single optical experiment, a complete set of seven above-mentioned major integral parameters of cell suspensions are obtained.

The essential restriction on the use of the above algorithm to solve the complete inverse problem of measuring the structural parameters of cell suspensions is the requirement for full cell orientation. This is achievable at comparatively high values of the field strength E, at which the electrolytic properties of a given dispersion medium may, in particular, prove most essential. In such cases, a need arises for its substitution using special sample preparation units [10,24], which may, in principle, have unwanted effects from the standpoint of nativity of the objects under study.

To an extent, this restriction can be overcome by using what are known as *orientational spectra* (OS) [10,24]. They are of particular interest in the context of this review. At relatively low degrees of particle orientation (q<<1, see above), these spectra may be represented by the dependences

$$\delta\Delta(\omega) = (\Delta_a - \Delta_b)/\Delta = \Delta\gamma(\omega)E^2 \ F \qquad (2)$$

on the frequency ω of the orienting electric field. In formula (2), D_a and D_b are the optical density values of the suspension, measured during the propagation of a light beam along (a) and across (b) the orienting field direction, D is the optical density at random particle orientation, and F is the coefficient, including what is called an optical factor, which depends among other things, on the cell size and refractive index [18,19]. All the quantities, associated with equation (2) are measured at a fixed value of the light wavelength λ.

Thus, at relatively low degrees of orientation, the orienting field frequency dependence of δD coincides with the frequency dispersion of the anisotropy of the particle polarizability tensor $\Delta\gamma(\omega)$ with an accuracy of the constant. However, there is reason to believe that at relatively high degrees of particle orientation as well, the OS of the type (2) reflect (at least qualitatively) the peculiarities of the frequency dispersion $\Delta\gamma(\omega)$, associated with the dependence of cell electro-physical properties on the structural transformations inside the cell or on its surface. We note that increasing the degree of orientation ensures better accuracy of δD measurements. Additionally, at low and moderate degrees of orientation, the informativity of the relaxation curves (Figure 1a) as a means of estimating mean particle size (and size distributions) is retained [1-3,10-12,18-20,24]. In such cases, however (unlike the above scenario of complete orientational ordering of disperse systems), the axial ratio and refractive index of cells can, strictly speaking, be determined only from an independent experiment.

The informativity of the OS as an attribute of the cell structure and of the subtle changes therein that occur under the influence of various extra- and intracellular factors is ensured by the high sensitiveness of the cellular electro-physical properties to such changes [29,30]. The *frequency dispersion* $\Delta\gamma(\omega)$ is known to mirror the effects of different cell structural elements depending on the orienting field frequency. They are: (i) cell surface biopolymers and the high- and low-molecular-weight compounds (coming from the environment), that are associated with them and form a double electric layer directly at the cell/environment interface (ω of the order of unities and tens of Hz); (ii) the components of the cell wall and cytoplasmic membrane (ω of the order of tens and hundreds of kHz); and (iii) the elements of the cell inner structure (cell organelles) (ω of the order of unities and tens of MHz). This makes it possible to obtain information on the various physical-chemical and physiological-biochemical processes, occurring on the surface of and inside the cell. The applications include analysis of biospecific interactions, determination of the proportion of viable cells in a population [10,24], and the investigation on the bacterial metabolism of industrially important compounds (see next section).

We now terminate this section by noting that, in our experiments, we used the electro-optical analyzer ELBIC (Figure 2), which had been configured, designed and implemented by the State Science Research Institute of Applied Microbiology, Obolensk, Russia. This apparatus has the following attributes :

(i)   Automatic preparation of samples of the analyzed cell suspensions and, when needed, the optimization of the dispersion medium composition, in particular with regard to the medium conductivity and to the cell osmotic parameters.

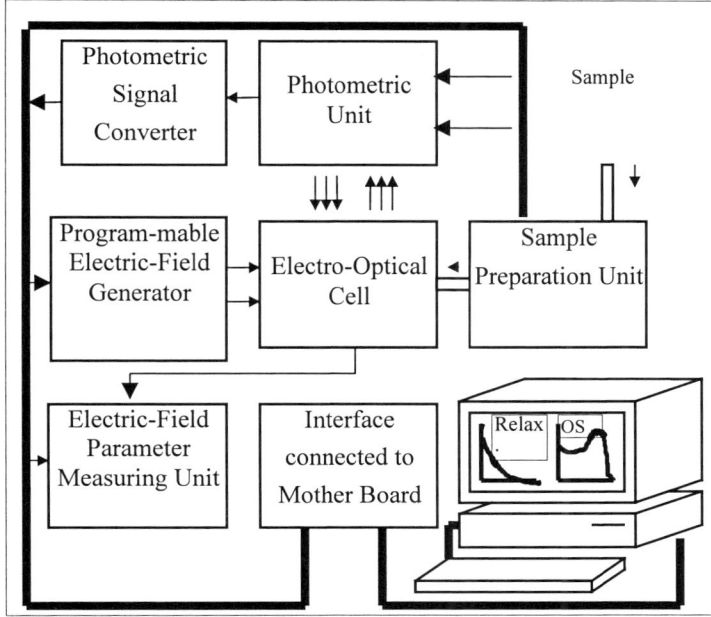

Figure 2. Schematic diagram for the ELBIC electro-optical analyzer.

(ii)   Generation in the electro-optical cell of electric fields of different forms and directions and with desired parameters over the ω range of 10 Hz ÷ 10 MHz and the E range of $10^2 ÷ 10^4$ V.m^{-1}.

(iii)   Recording of OS and orientational relaxation curves for the suspensions at a fixed λ value in the visible range of the light spectrum and with a relative error of measurement of the light beam intensity of the order 10^{-5}. Determination of the specific conductivity of the cell suspensions.

(iv)   Processing of the data obtained and their computer output via a standardized interface as the size distributions of the cells, numerical cell concentrations, estimates of culture heterogeneity by electro-physical or physiological parameters, and as values of the electro-physical parameters of individual cellular structures.

EXPERIMENTAL RESULTS

Materials and methods

Microorganisms: The microorganisms used were the bacteria *Acinetobacter calcoaceticum* A-122, which can use *p*-nitrophenol (PNP) as its sole carbon source [31], and *Brevibacterium* sp. 13PA, which contains the inducible enzyme, amidase, and can grow with acrylamide and/or acrylic acid as its sole carbon sources [32]. The strains were kindly provided by the 'Biocatalysis' Science Research Institute's Laboratory of Microbiological Transformation (Saratov, Russia).

OS measurements: The OS of the cell suspensions for unpolarized light were measured with the electro-optical analyzer ELBIC at a light wavelength of 670 nm. Analysis conditions were as follows: volume of the measuring cell,

1 ml; cell concentration, in units of optical density, $D_{670} = 0.45 \div 0.50$. Before analysis, the cells were washed three times by centrifugation at $5000 \times g$ for 10 min and resuspended in a little deionized water. To remove cellular aggregates, the cell suspension was repeatedly centrifuged at $1000 \times g$ for 1 min. Further work was carried out using the suspension which remained in the supernatant liquid. A discrete set of the orienting electric field frequencies (10,52,104,502,1000,5020, and 10000 kHz) were exploited. According to the discussion of equation (1), the general view of the OS is essentially determined by the frequency dependence of the anisotropy of the cell polarizability $\Delta\gamma(\omega)$. The design of the ELBIC apparatus provides for the representation of the OS as the frequency dependence of the ratio

$$\delta D_{ru} = \delta D / (E^2 FB) \approx A\Delta\gamma \tag{3}$$

which was used in our experiments. In this equation, the abbreviation ru stands for relative units, $B = I/I_0$ (where I is the intensity of transmitted light) is transmittance, and the constant A is the scale multiplier which makes for easy reading of the results of determination of δD_{ru}, the magnitude of which in this case had values of the order $10^2 \div 10^3$. At sufficiently low degrees of cell orientation (q<<1, see equation (1), such normalization ensures the independence of δD_{ru} from the cell concentration, the orienting field strength E, and the attenuation of the light beam during its passage through the scattering medium.

Determination of specific respiratory activity: The respiratory activity of the cells was determined with a Clark-type oxygen electrode and an OH-105 polarograph (Radelkis, Hungary). The assay conditions were: volume of the measuring cell, 1 ml; temperature T=35°C. To assay respiratory activity, the cells were sedimented by centrifugation at $5000 \times g$ for 10 min, resuspended in 0.01 M phosphate buffer of pH 6.8 (*A. calcoaceticum* A-122) or pH 7.6 (*Brevibacterium* sp. 13PA), and resedimented as above. The pellet was resuspended in a small amount of the same buffer, and the cell concentration present in the suspension was determined.

The oxygen electrode employed to follow current intensity was placed in the measuring cell filled with a buffer solution. The microbial cell suspension was injected into the cell with a dosing syringe, and endogenous respiratory activity measured by the change in current intensity over a given time period ($\Delta I_1 / \Delta t_1$). A sample of an aqueous solution of the substrate under study was then injected into the cell, and respiratory activity measured by following the change in current intensity over a given time period ($\Delta I_2 / \Delta t_2$). The difference between the values for substrate-dependent respiratory activity (i.e. in the presence and absence of the substrate) under study, was designated SRA and expressed in units of $\mu A.min^{-1}.mg.(dry\ cell\ wt)^{-1}$

$$SRA = (\Delta I_2 / \Delta t_2 - \Delta I_1 / \Delta t_1)/m, \tag{4}$$

where m is cell mass (dry cell wt).

Orientational spectra of *Brevibacterium* sp. 13PA and *A. calcoaceticum* A-122 during the metabolism of acrylamide and PNP

The respiratory activity of these strains, which contain enzyme systems of the initial metabolism of certain toxic low-molecular-weight compounds, was previously investigated in [33]. The results showed that it was possible to use SRA for determining the level of the corresponding substrates.

One can speculate that the microbial metabolism of toxic compounds is associated with structural alteration of the cells. This alteration results in redistribution of the cell areas with different dielectric permeabilities and, consequently, of their contribution to the frequency dependence of polarizability $\Delta\gamma(\omega)$. In such cases, we had a reason to anticipate characteristic OS changes, which could be recorded by the method of electro-optical analysis.

Brevibacterium sp.13PA and *A. calcoaceticum* A-122, both having an induced enzyme system of the metabolism of acrylamide and PNP, respectively, were incubated with the substrates, and the OS of the suspensions were recorded. To evaluate possible OS changes due to the non-specific (not associated with metabolism) action of the substrate on the cell, controls were performed which used cells of the same strains void of the induced systems of metabolism of acrylamide and PNP. Figure 3 depicts the results of experiments with *Brevibacterium* sp.13PA. The most dramatic OS changes occurred typically at the first five frequencies exploited (over the ω range of ≈10-1000 kHz) [34].

Similar experiments were conducted with *A. calcoaceticum* A-122. The cells were incubated with various PNP concentrations. A direct relationship was established between the characteristic parameter of the electro-optical

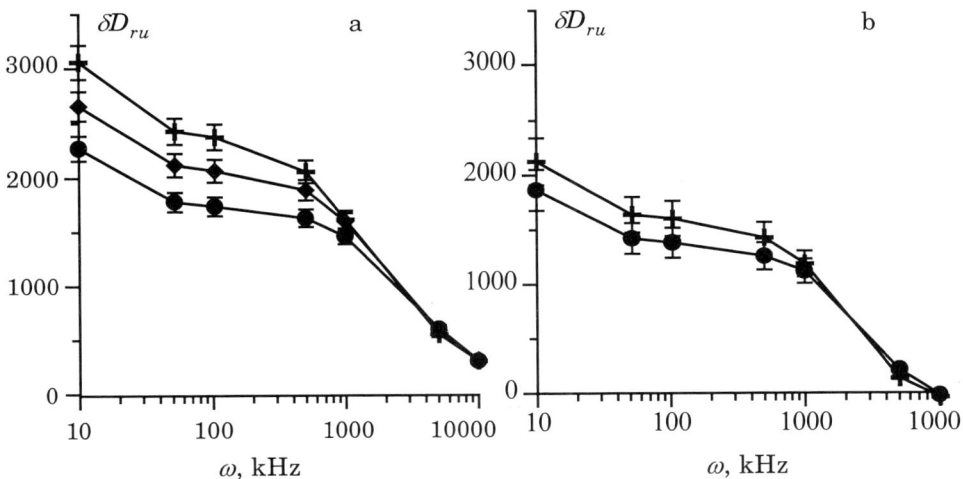

Figure 3. a: OS of *Brevibacterium* sp. cells upon acrylamide incubation (0.1 g/l; 0.5 g/l). +, cells incubated in deionized water; ◆, acrylamide-incubated cells (0.1 g/l); ●, acrylamide-incubated cells (0.5 g/l). b: OS of *Brevibacterium* sp. cells without amidase activity upon acrylamide incubation (0.5 g/l). +, cells incubated in deionized water; ●, acrylamide-incubated cells.

effect observed in the cell suspension and the PNP level in the incubating medium (Figure 4a) [35]. That was, therefore, the first time that the phenomenon of change in electro-physical properties during the metabolism of toxic low-molecular-weight compounds was recorded.

Comparative analysis of the electro-physical characteristics and specific respiratory activity of *A. calcoaceticum* A-122

In view of the high potential sensitiveness of electric-field orientation to diverse (including non-specific) actions on the cells, the above results needed additional experimental verification by independent methods. We believe that

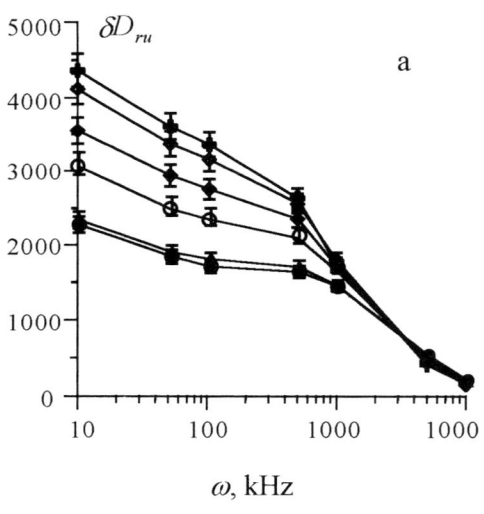

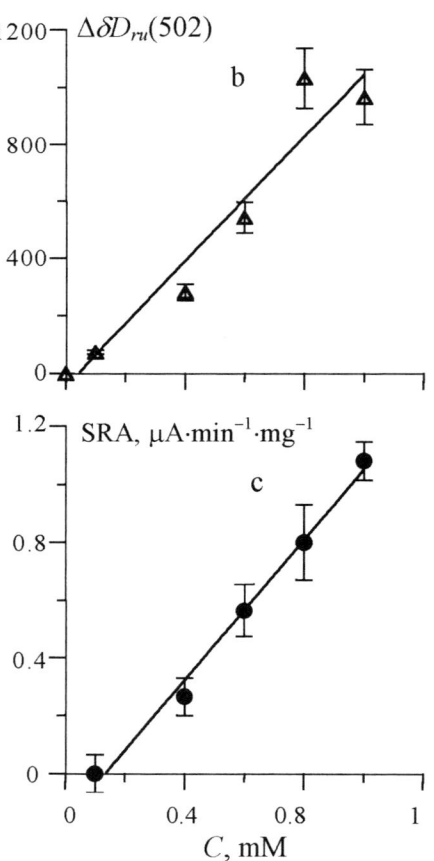

Figure 4. a: OS of the suspension of PNP-incubated *A. calcoaceticum* A-122 cells. +, control; ◇, 0.1; ◆, 0.4; □, 0.6; ●, 0.8; and Δ, 1.0 (mM PNP). b: The dependence of the difference $\Delta\delta D_{ru}$ between the ΔDru values (control - experiment) at a 502 kHz frequency on the PNP concentration C. c: SRA of the cells towards PNP as a function of C.

one of such tests during aerobic metabolism is evaluating the metabolic activity towards this or other substrates as the result of SRA measurements. Upon optimization of conditions for the determination of SRA of A-122 towards PNP, the SRA was plotted as a function of the PNP concentration in the oxygen cell. A comparison of the data obtained with the results of electro-optical experiments (Figure 4b, c) showed a good correlation between the SRA of cells and their electro-physical characteristics [36].

Changes in electro-physical properties as the result of inhibition of microbial metabolism

This sub-section considers another approach that aims to verify experimentally the dependence of the electro-physical properties of cell suspensions on the specific enzymatic processes associated with the metabolism of toxic compounds. To this end, the inhibition of the enzyme system of initial metabolism in the cells used in the electro-optical experiments was studied.

In principle, various natural and synthetic inhibitors of cellular enzyme activity would do in this case. However, their non-specific effect on cellular electro-optical characteristics cannot be excluded, which calls for additional control experiments. So, to inhibit cellular enzyme activity, we followed methods that do not require addition of chemical ingredients into the suspension medium.

The enzyme activity was changed by limiting the oxygen content in the medium and by changing the medium temperature and pH. In the first case, we used the effect of depletion of the oxygen reserve in the suspension as the result of the endogenous respiratory activity of the cells during their incubation without any atmospheric oxygen supply. It was found for both the strains used that the enzyme systems of initial metabolism were directly associated with the reactions of substrate oxidation. It could be anticipated, therefore, that the shortage of oxygen in the medium would lead to an inhibition of the enzyme activity of the cell systems and thus to a change in the electro-optical characteristics of the suspensions.

Cells of *A. calcoaceticum* A-122 were incubated in deionized water, without any atmospheric oxygen supply, for several hours at 35°C. The oxygen concentration in the vessels was monitored with a Clark oxygen electrode. Next, PNP was added to the cell suspension (final concentration 1.0 mM) and incubation was done for 30 min at 35°C. The substrate was added so that the supply of oxygen in air was minimal. The suspensions were then subjected to OS measurements. No changes in electro-physical properties were recorded in the presence of the substrate. Following these experiments, the cell suspensions were saturated with oxygen by barbotage and the substrate was added at 1.0 mM. This was followed by a measurement of the OS. The suspensions exhibited changes typical for the experiments described in the preceding sub-sections. A similar experiment was run with cells of *Brevibacterium* sp.13PA, yielding identical results [37]. Thus, the changes in the cellular electro-physical properties were found to be associated with the reaction of substrate oxidation, with this process being of a reversible character.

Figure 5 depicts the results of comparative estimation of the SRA of A-122 towards PNP and the changes in the electro-optical parameters of a suspension of the same strain that occurred when cellular enzyme activity was inhibited by changing the temperature of the incubating medium. The data show that in this case too, the results of the electro-optical experiments were correlated with the experiments to determine cellular SRA. The same correlation was observed when the cells were incubated at different pH of the medium. Similar results were obtained for *Brevibacterium* sp.13PA.

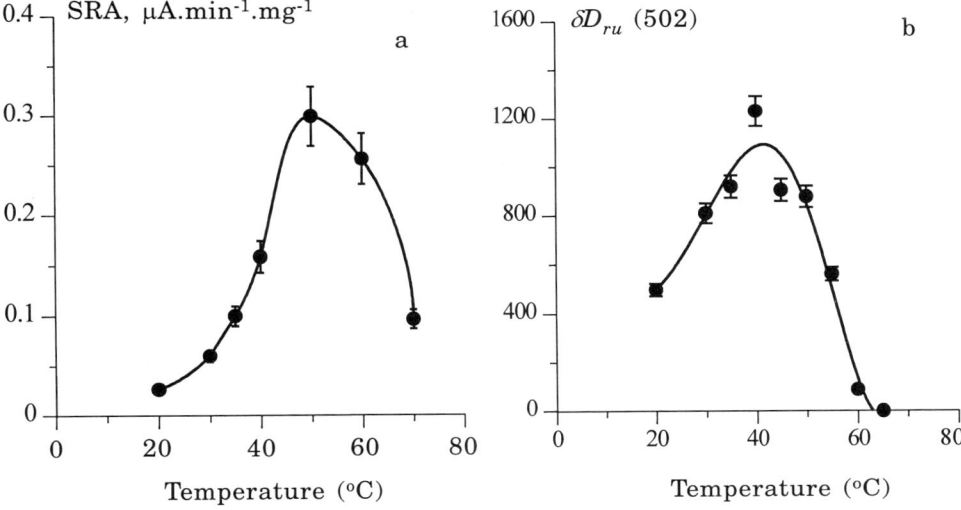

Figure 5. Temperature dependence of the SRA of *A. calcoaceticum* A-122 for PNP (a) and of the δD_{ru} at a 502 kHz frequency (b).

Scenarios for the practical exploitation of measurements of cellular electro-optical characteristics during the metabolism of toxic compounds

An important aspect of our research is estimating different scenarios for the practical exploitation of the results obtained. We thought they might be of importance primarily in solving the problems of measurement of the enzyme activity of microbial cells using their electro-optical characteristics. To this end, experiments were set up on batch cultivated *Brevibacterium* sp. cells with amidase activity. The cells were grown in a mineral medium with acrylamide as the sole source of carbon. Samples were periodically withdrawn for the acrylamide, acrylic acid and ammonium ion concentration assays and for the measurement of cellular amidase activity. The amidase activity during acrylamide metabolism was determined by the change in $\delta D_{ru}(502)$ and also in the usual way by the change in substrate concentration that occurs during the *microbial hydrolysis* of acrylamide to acrylic acid. A decline in amidase activity during growth was recorded in either case [38].

Noteworthy is the fact that these data check nicely with the results of studies of the specific respiratory activity of *Brevibacterium* sp. 13PA towards acrylamide [33]. Similar results were obtained when this strain was cultivated in a medium containing acrylic acid as a carbon source. The same experiment was run with cells of *A. calcoaceticum* A-122. The activity of A-122 increased as more PNP was degraded and reached the maximum at 6 to 8 h after the start of the experiment (when there was essentially no substrate in the growth medium). An explanation for this might be that during cell growth in mineral medium with PNP as the sole carbon source, the microorganisms synthesize, and maintain in an active state a special enzyme system of the initial metabolism of this substrate. The further incubation of the cells resulted in a decrease in their activity towards PNP, as determined by the electro-optical results. The data obtained are in accord with the views that cell metabolism proceeds in a very economical way. It is known that the enzymes involved in the utilization of substrate and in the inclusion of its decomposition products in intermediate metabolism are synthesized only when the substrate is present in the nutrient medium. As more substrate is utilized, accumulation of intermediate products occurs, which, in turn, exercise a regulating influence of enzyme biosynthesis and activity on the principle of *catabolite repression*.

Another field of practical exploitation of these results might be the determination of substrate concentration by the change in cellular electro-optical characteristics during metabolism. We studied the substrate concentration dependence of changes in the δD_{ru} of *A. calcoaceticum* A-122 suspensions (Figure 4). A comparison of the data obtained with the measured results for the SRA of this strain towards PNP shows a good correlation between the independent determinations (Figure 4b, c).

Let us consider the prospects for the use of electro-physical analysis from the standpoint of *biosensor technology*. Most microbial sensor systems use cellular enzyme activity to produce this or other effect, which is then recorded with a certain physico-chemical instrument. In essence, the change in the cellular electro-physical properties, that occurs due to the redistribution of charges in the cells during the metabolism of definite chemical compounds, is virtually a complete analogue to a microbial biosensor system based, e.g., on cellular respiratory activity. In either case, the changes are due to the operation of a specific enzyme system effecting the metabolism of a substrate. We used an electro-optical analyzer to follow the changes in the cellular electro-physical characteristics, but in principle, other devices measuring cellular electro-physical properties may also be employed. Compared with traditional *microbial sensor systems*, this method has advantages in that in this case, signal recording is not associated with the use of this or other (oxygen, ammonium, pH, etc.) electrode.

CONCLUSION

The study of the electro-optical characteristics of microbial cells permits the estimation of parameters, whose change may serve as a quantitative measure of the enzymatic processes taking place in the cells. Specifically, it may determine the concentration of substrate used in enzymatic reactions. Despite the long-standing history of the corresponding investigative techniques [1-3], this aspect of electro-physical analysis of cell systems is still in its initial stage. The prospects for the use of electro-physical analysis are associated by us primarily with progress in microelectronics and computer technologies. This will allow the realization of detector cells comparable to microbial cells in size and highly effective processing of the data obtained. It may be noted that, in principle, an experimental basis has already been created for the development of biosensors, using cellular electro-physical characteristics. However, it should be adapted to the solution of specific problems.

We suggest that the development of the electro-physical analysis of microbial cells will take several directions.

(i)   Further improvement of existing instruments.

(ii)   Further improvement of the methods of cell electrorotation [5-8], dielectric spectroscopy and their analogues [4,9], as well as the electric-field orientation method [1-3], to effectively analyze and use the electro-physical properties of microbial cells.

(iii) Use of the electro-physical analysis for the metabolic control analysis (MCA) of cells, which may serve as the basis for the development of research in one of the new directions in the area of electro-physical analysis of the cellular metabolism of toxic low-molecular-weight compounds.

ACKNOWLEDGEMENT

We thank D.N. Tychinin (IBPPM RAS, Saratov, Russia) for preparing the English version of the manuscript.

REFERENCES

1   Stoylov SP. Colloid Electro-Optics; Theory, Techniques and Applications. London: Academic Press, 1991.

2   Jennings BR, Stoylov SP, eds. Colloid and Molecular Electro-Optics 1991. Bristol and Philadelphia: IOP Publishing, 1992.

3   Tolstoy NA, Spartakov AA. Electro-Optics and Magneto-Optics of Disperse Systems. St. Petersburg: St. Petersburg University Publishing House, 1996 (in Russian).

4   Gimsa J, Glaser R, Fuhr G. In: Schutt W, Klinkmann H, Lamprecht I, Wilson T, eds. Physical Characterization of Biological Cells. Berlin: Verlag Gesundheit GmbH, 1991; 295-323.

5   Holzel R, Lamprecht I. Biochim Biophys Acta 1992; 1104: 195-200.

6   Burt JPH, Chan KL, Dawson D, Parton A, et al. Ann Biol Clin 1996; 54: 253-257.

7   Pruger B, Eppmann P, Donath E, Gimsa J. Biophys J 1997; 72: 1414-1424.

8   Hughes MP. Phys Med Biol 1998; 43: 3639-3648.

9   Gimsa J. In: Riu PJ, Rosell J, Bragos R, Casas O, eds. Electrical Bioimpedance Methods. Applications to Medicine and Biotechnology. New York: Ann New York Acad Sci, 1999; 287-298.

10 Bunin VD, Voloshin AG. J Colloid Interface Sci 1996; 180: 122-126.

11 Shchyogolev SYu, Khlebtsov NG, Bunin VD, Sirota AI, et al. In: Chance B, ed. Quantification and Localization Using Diffused Photon in a Highly Scattering Media. Bellingham: SPIE, 1994; 167-176.

12 Khlentsov NG, Melnikov AG, Bogatyrev VA, Sirota AI. J Quant Spectr Radiat Transf 1999; 63: 469-478.

13 Zhou X-F, Burt JPH, Pethig R. Phys Med Biol 1998; 43: 1075-1090.

14 Hodgson CE, Pethig R. Clin Chem 1998; 44: 2049-2051.

15 Holzel R. Biochim Biophys Acta 1998; 1425: 311-318.

16 Huang Y, Holzel R, Pethig R, Wang X-B. Phys Med Biol 1992; 37: 1499-1517.

17 Tanford C. Physical Chemistry of Macromolecules. New York: Acad Press, 1961.

18 Khlebtsov NG, Melnikov AG, Bogatyrev VA. J Colloid Interface Sci 1991; 146: 463-478.

19 Khlebtsov NG, Melnikov AG, Bogatyrev VA, Sirota AI. In: Jennings BR, Stoylov SP, eds. Colloid and Molecular Electro-Optics 1991. Bristol and Philadelphia: IOP Publishing, 1992; 13-20.

424

20   Khlebtsov NG, Melnikov AG, Bogatyrev VA. Colloids Surfaces A 1999; 148: 17-28.

21   Klenin VJ. Thermodynamics of Systems Containing Flexible Chain Polymers. Amsterdam: Elsevier, 1999.

22   Shchyogolev SYu, Khlebtsov NG. In: Jennings BR, Stoylov SP, eds. Colloid and Molecular Electro-Optics 1991. Bristol and Philadelphia: IOP Publishing, 1992; 141-146.

23   Shchyogolev SYu. J Biomed Opt 1999; 4: 490-503.

24   Bunin VD, Brezgunov VN, Voloshin AG, Svetogorov DE. In: Jennings BR, Stoylov SP, eds. Colloid and Molecular Electro-Optics 1991. Bristol and Philadelphia: IOP Publishing, 1992; 207-211.

25   Landau LD, Lifshits EM. Electrodynamics of Solid Media. Moscow: Gostekhizdat Publishing House, 1957 (in Russian).

26   Bottcher CF. Theory of Electrical Polarisability. New York: Academic Press, 1978.

27   Khlebtsov NG, Bogatyrev VA, Sirota AI, Melnikov AG. Biofizika 1990; 35: 173 (in Russian).

28   Khlebtsov NG, Fomina VI, Sirota AI. In: Jennings BR, Stoylov SP, eds. Colloid and Molecular Electro-Optics 1991. Bristol and Philadelphia: IOP Publishing, 1992; 177-180.

29   Bathlor A. Theory of Polarisability. Amsterdam: Elsevier, 1982.

30   Sherbet GV. The Biophysical Characterization of the Cell Surface. New York: Academic Press, 1978.

31   Singirtsev IN. Microbial Degradation of Nitrophenols. Cand Sc Thesis. Saratov, 1996 (in Russian).

32   Moiseyeva, TN, Kozulin SV, Kulikova LK, Voronin SP. Biotekhnologiya 1991; 6: 79-83 (in Russian).

33   Ignatov OV, Rogatcheva SM, Khorkina NA, Kozulin SV. Biosensors Bioelectron 1997; 12: 105-111.

34   Ignatov OV, Khorkina NA, Shchyogolev SYu, Khlebtsov NG, et al. Anal Chim Acta 1997; 347: 241-247.

35   Ignatov OV, Khorkina NA, Singirtsev IN, Bunin VD, et al. FEMS Microbiol Lett 1998; 165: 301-304.

36   Ignatov OV, Khorkina NA, Shchyogolev SYu, Singirtsev IN, et al. FEMS Microbiol Lett 1999; 173: 453-457.

37   Ignatov OV, Khorkina NA, Shchyogolev SYu, Singirtsev IN, et al. FEMS Microbiol Lett 1999; 178: 259-264.

38   Ignatov OV, Khorkina NA, Shchyogolev SYu, Bunin VD, et al. FEMS Microbiol Lett 1998; 162: 105-110.

Biotransformations: Bioremediation Technology for Health and Environmental Protection
V.P. Singh and R.D. Stapleton, Jr. (Editors)
© 2002 Elsevier Science B.V. All rights reserved.

Microbial degradation of sulfur compounds present in coal and petroleum

B.K. Gogoi and R.L. Bezbaruah

Biochemistry Division, Regional Research Laboratory, Jorhat - 785 006, India

INTRODUCTION

All fossil fuels contain sulfur compounds. The presence of sulfur in fossil fuels causes corrosion of pipelines, pumping and refining equipment, and premature breakdown of combustion engines. Sulfur also contaminates or poisons catalysts used in refining and combustion of *fossil fuels*. Moreover, the burning of fossil fuels releases large quantities of sulfur dioxide into the environment, contributing significantly to air pollution and being the principal cause of acid rain. Regulations, such as Clean Air Act of 1990 [1] require the removal of sulfur, either pre- or post-combustion, from all fossil fuels, and it becomes increasingly difficult to conform with such regulations.

There are several physicochemical methods for depleting the sulfur content of fossil fuels prior to combustion. For pre-combustion cleaning of coal, both physical [2] and chemical [3] separation methods can be used for removing inorganic forms of sulfur; however, the separation of organic sulfur from the complex coal matrix presents a major problem in coal desulfurization. For example, *hydrodesulfurization* (HDS) is not particularly effective for the desulfurization of coal. HDS is currently used in the oil refineries for the removal of organic sulfur from petroleum. In HDS, the fossil fuel is contacted with hydrogen gas at high temperature (>300°C) and high pressure (>100 atm) in the presence of a metal catalyst. In this process organic sulfur is removed by the reductive conversion of sulfur bond of carbonaceous molecules to H_2S, a corrosive gaseous product which is separated from the treated fuel by stripping. Although HDS can easily remove simple organic sulfur compounds such as mercaptans, thioethers, and disulfides but is not effective for removing complicated polycyclic sulfur compounds present in fossil fuels. Recently, with the depletion of low-sulfur containing petroleum reserves and increasingly stringent regulations, adequate desulfurization by conventional HDS alone is becoming progressively more difficult to achieve. The petroleum refining companies have recognized the fact that hydrodesulfurization units for the high extent of desulfurization are extremely expensive to build and operate. Therefore, as an alternating promising method to HDS, *microbial desulfurization* (MDS), which harnesses the metabolic processes of suitable bacteria to the desulfurization of fossil fuels, has attracted attention of many researchers. Consequently, in recent years there has been considerable effort to develop bioprocesses for fossil fuel desulfurization.

NATURE OF SULFUR COMPOUNDS IN FOSSIL FUELS

Sulfur in coals

Total sulfur content in coals varies from 0.5% to 11%, depending on the geographical location of the coal source [4]. Sulfur exists in coal in two basic forms: organic and inorganic. The relative amount of either of these two forms make up as much as 20-80% of the total sulfur in individual coals [5]. The predominant inorganic form of sulfur found in coal is iron pyrite (FeS_2), which may be dispersed throughout the coal matrix but are not bound to it and, therefore, can be removed to a significant extent by various commercial processes. Generally, pyritic sulfur equals or exceeds organic sulfur in coal. Minor amounts of sulfate may also be present in coal (mainly as gypsum, $CaSO_4 . 2H_2O$).

Organically bound sulfur is present as an integral part of the coal matrix. Organic sulfur in bituminous coal appears to consist predominantly of dibenzothiophene (DBT) and benzothiophene with lower amounts of disulfide joining cyclic structures, sulfide linked to alkyl groups and thio groups attached either to an aromatic ring or to an alkyl group [6-9].

The organic sulfur content of the coal is generally estimated by substracting the pyritic sulfur from the total sulfur [10]. The errors possible by such an indirect analytical method contribute to the difficulties of interpreting organic sulfur removal from coal [6-11]. Other methods such as electron microbeam technique [12] and extended X-ray absorption fine-structure (EXAFS) technique [13] can also be used to examine specifically the organic sulfur found in coal, but they are too specialized and are not readily adaptable for wide-spread use.

Sulfur in crude oils

Crude oils can typically range from 5% down to about 0.025% organic sulfur [14,15]. Crude oils obtained from the Persian Gulf and Venezuela can be particularly high in sulfur content [16,17]. Sulfur can also be present as elemental sulfur, H_2S and pyrite, dissolved or suspended in the oil. The sulfur content in crude oil fractions generally increases in the sequence: saturates<aromatics<resins<asphaltenes.

Organic sulfur compounds in petroleum crude include sulfides, disulfides, thiols, thiophenes, substituted benzo- and dibenzothiophenes and many other more complex materials [18,19]. The distribution and character of these different compounds reflect the source, maturity and alteration processes in the development of a particular crude.

Thiols in crude oils are generally of low molecular weight (less than 8 carbon atoms)[20]. They are present at petroleum fraction, boiling below 200°C and are readily removed during refinery processing. Aliphatic sulfides (cyclic or acyclic) are major components of the sulfur containing fraction of petroleum products boiling above 200°C, e.g., diesel fuels and heating oils [20]. Low molecular weight sulfides in crude oil are easily removed during the production and refining processes but the bulkier alkyl- and aryl-, mono- and disulfides are retained. Aromatic sulfides are of lower concentration in the heavier cuts [20]. Thiophene derivatives such as benzo-, dibenzo- or naphthobenzothiophenes are important constituents of high-sulfur oils and are the mostabundant sulfur compounds in distillates and residues, such as heavy fuel oils and bitumens [20]. Up to 70% of the sulfur in some Taxa crude oil has been reported as dibenzothiophene (DBT) and in several Middle East crude oils, substituted benzo- and dibenzothiophenes constitute more than 40% of the organic sulfur [21]. Dibenzothiophene and substituted dibenzothiophenes are some of the most recalcitrant organic sulfur compounds relative to hydrodesulfurization of crude oils and refined transportation fuels [22]. The ubiquity of aromatic derivatives in virtually all crude oils has led to the use of DBT as model compound in investigations of crude oil desulfurization.

MICROBIAL ATTACK ON HETEROCYCLIC SULFUR-CONTAINING HYDROCARBONS

Dibenzothiophene

Degradation/desulfurization pathways

DBT is generally regarded as a model compound representative of the forms of organic sulfur found in fossil fuels. It is the substrate used in most enrichment culture experiments designed to find organisms that can remove organic sulfur from fossil fuels. Another characteristic of DBT and derivatives thereof is that, following a release of fossil fuel into the environment, these sulfur-bearing heterocycles persist for long periods of time without significant biodegradation [23]. Thus, most prevalent naturally occurring microorganisms do not effectively metabolize and break down sulfur-bearing heterocycles.

Three major pathways of DBT degradation by aerobic microorganisms have been reported. The first is the ring-destructive oxidative pathway, in which DBT is partially oxidized to water soluble intermediates, but the sulfur of DBT remains intact [24-27]. In this pathway, originally proposed by Kodama et al. [24], DBT is degraded by hydroxylating one of the benzoid rings to DBT-dihydrodiol and DBT-diol, cleaving the ring between the diol substituents, and then producing, in sequence, 4-2-(3-hydroxy)-thionaphthenyl -2-oxo-3-butenoic acid and 3-hydroxy-2-formylbenzothiophene (HFBT), as shown in Figure 1. The occurrence of such ring-destructive pathway in DBT degradation appears to be widespread among soil bacteria, being reported in strains of *Pseudomonas* [17,28-31], *Micrococcus* [31], *Beijerinckia* [25] and a mixture of *Acinetobacter* and *Rhizobium* [32] and *Rhizobium meliloti* [33], and in unidentified soil isolates [34]. Degradation of DBT by all these microorganisms involved cometabolism in which DBT was degraded when the organism was grown in an alternate carbon source, such as yeast extract-peptone medium or glucose.

Although HFBT biodegradation was observed in other *Pseudomonas* strains, no sulfur release was detected with those microorganisms [35]. Microbial degradation of organic sulfur-containing carbonaceous materials by ring-destructive oxidative pathway results in net carbon loss and reduction of calorific value of the carbonaceous fuel. In addition, it has been observed that the pathway for naphthalene metabolism closely resembles the ring-destructive pathway, raising the possibility that further undesired metabolism of structurally related non-sulfur-containing fuel components may occur. Moreover, some pigments were formed in the bacterial cultures during DBT degradation, which will lead to trouble in the final oil product. It is, therefore, desirable to follow a *microbial degradation* route which removes sulfur from

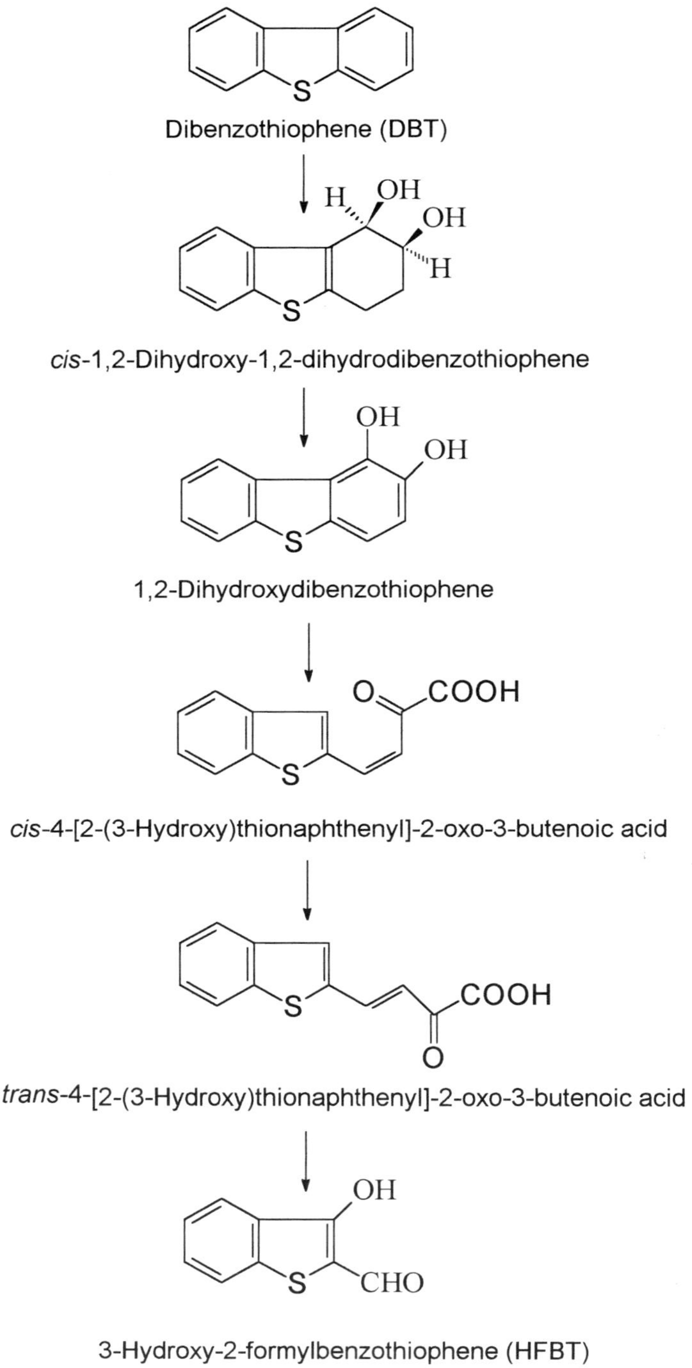

Dibenzothiophene (DBT)

cis-1,2-Dihydroxy-1,2-dihydrodibenzothiophene

1,2-Dihydroxydibenzothiophene

cis-4-[2-(3-Hydroxy)thionaphthenyl]-2-oxo-3-butenoic acid

trans-4-[2-(3-Hydroxy)thionaphthenyl]-2-oxo-3-butenoic acid

3-Hydroxy-2-formylbenzothiophene (HFBT)

Figure 1. Microbial degradation pathway of DBT.

the molecule without removing carbon from the molecule, thereby retaining calorific value of the fuel to a greater degree than is possible by *ring-destructive* pathways.

The second pathway of DBT degradation is sulfur-specific pathway in which DBT is desulfurized by the selective cleavage of the carbon-sulfur bond, resulting in the accumulation of 2-hydroxybiphenyl (2-HBP) as the end product. The specific cleavage of the carbon-sulfur bond is preferred for microbial desulfurization technology so that sulfur is removed, but the carbon and calorific values remain intact. Several effective screening techniques, such as spread-plate UV-fluorescence assay [36] and sulfur *bioavailability* assay [37,38] have been developed for the detection and characterization of microbial cultures that can specifically remove sulfur from organic substrates. Based on the sulfur bioavailability assay, using a continuous culture bioreactor selection and mutagenesis procedure with DBT as the sole source of sulfur, a strain of *Rhodococcus rhodochrous*, IGTS8, able to cleave the carbon-sulfur bond of DBT was isolated [37,39]. DBT was metabolized by this organism to 2-HBP and sulfate, the amount of DBT degradation being directly proportional to the concentration of bacterial cells present and corresponding to the minimum amount of sulfur required for bacterial growth (0.1 mM sulfur allows growth of 10^9 cells.cm^{-3}). The sulfur hetero atom was either released as sulfate or it was assumed to have been incorporated into biomass. The DBT desulfurization pathway in IGTS8 was proposed.

According to the sulfur-specific pathway, as shown in Figure 2, DBT is metabolized to 2-hydroxybiphenyl (2-HBP) via DBT 5'-sulfoxide (DBTO), DBT 5'-sulfone (DBTO$_2$), and 2'-hydroxybiphenyl 2-sulfinic acid (HBPS). The specific conversion of DBT to 2-HBP by the strain IGTS8 had been confirmed by analysis of reaction products using gas chromatography-mass spectroscopy (GC-MS) [37]. Many research groups have since then studied the desulfurization of DBT by the sulfur-specific pathway, demonstrating the

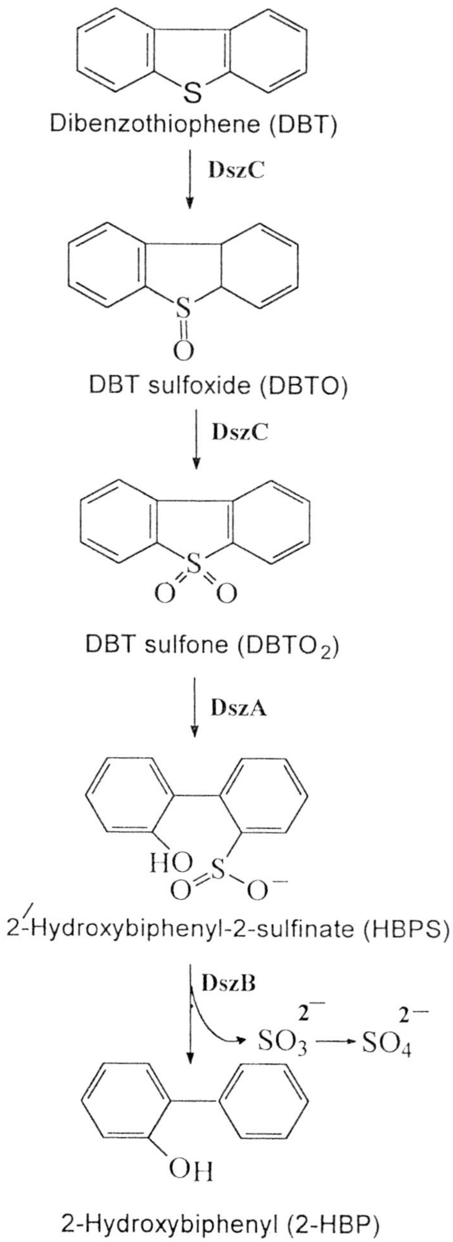

Dibenzothiophene (DBT)

DszC

DBT sulfoxide (DBTO)

DszC

DBT sulfone (DBTO$_2$)

DszA

2-Hydroxybiphenyl-2-sulfinate (HBPS)

DszB

$SO_3^{2-} \longrightarrow SO_4^{2-}$

2-Hydroxybiphenyl (2-HBP)

Figure 2. Sulfur-specific pathway for microbial DBT desulfurization without a cleavage of carbon-carbon bonds.

capacity to generate 2-HBP from DBT in *Rhodococcus* strains UM3 and UM9 [40], *Rhodococcus erythropolis* D-1 [41], *Rhodococcus erythropolis* N1-36 [42], *Corynebacterium* species SY-1 (43), two other *Corynebacterium* isolates [44], *Agrobacterium* MC501 [45], and *Mycobacterium* species G3 [46]. All these strains were Gram-positive bacteria, and no Gram-negative bacteria with a sulfur-specific pathway was discovered.

The third pathway of DBT metabolism is a completely destructive pathway (Figure 3), in which DBT is mineralized to CO_2, sulfite, and water. In this metabolism, DBT serves as the sole source of carbon, sulfur, and energy. *Brevibacterium* species mineralized DBT, and CO_2, sulfite, and water were formed as the final products [47]. *Arthrobacter* DBTS2 was reported to oxidize DBT sulfoxide, but not DBT, with formation of sulfate and benzoate [48]. Only these two microorganisms were known to have such a metabolic pathway, as shown in Figure 3. They desulfurize DBT, and also break the carbon skeleton of the hydrocarbon.

Direct oxidation at the sulfur hetero atom

Aerobic metabolism of DBT and similar aromatic heterocycles by fungal culture such as *Cunninghamella elegans* [49,50], *Rhizopus arrhizus* [49], and *Mortierella isabellina* [49] yielded the corresponding sulfoxide and sulfones. Kodama et al. [28] reported that *Pseudomonas* converted DBT to DBTO by an alternate pathway. A strain of *Pseudomonas putida* could transform DBT to $DBTO_2$ via DBTO [51]. Both these *Pseudomonas* cultures also degraded DBT to HFBT via the previously described ring-destructive pathway.

Anaerobic biodesulfurization

Anaerobic biodesulfurization of sulfur-containing hydrocarbons was first presented by Kurita et al. [52]. A Gram-negative anaerobe desulfurized thiophene under N_2 or H_2 in medium containing lactate and sulfate. Desulfurization was monitored by measuring H_2S; however, there was no documentation of a desulfurized end product. Later, Kohler et al. [53] reported that mixed cultures containing sulfate-reducing *Desulfovibrio* strains, supplied

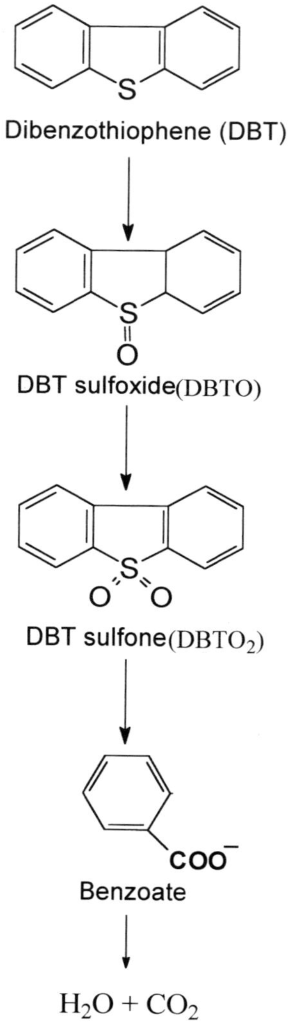

Dibenzothiophene (DBT)

DBT sulfoxide(DBTO)

DBT sulfone(DBTO₂)

Benzoate

$H_2O + CO_2$

Figure 3. Metabolic pathway for the degradation of DBT by *Brevibacterium* sp.

with hydrogen and lactate desulfurized a variety of model compounds, including DBT, benzothiophene, dibenzylsulfide (DBS), and dibenzyldisulfide (DBDS). DBS was converted to H_2S, benzyl mercaptans and toluene. DBT was stated to be degraded at a similar rate to DBS, but no analyses for reaction products were reported [53]. *Desulfovibrio desulfuricans* M6, selected for high hydrogenase activity and grown in sulfate free medium was able to degrade benzothiophene, DBT, phenyl sulfide, benzyl sulfide, benzyldisulfide, ethanethiol, and butanethiol, when supplied with electrochemically generated reducing equivalent [54,55]. The degradation ratio of DBT was 42% [54]. DBDS was reductively degraded by a methanogenic mixed culture derived from a sewage digester [56]. More recently, various sulfate-reducing bacteria such as *Desulfotomaculum orientalis* and *Desulfovibrio desulfuricans* (grown at 30°C) and *Thermodesulfobacterium commune* (grown at 60°C) were shown to use DBT as their sole source of sulfur and electron acceptor with the release of H_2S [57]. Other sulfate-reducing bacteria [58] isolated from oil-fields also had the ability to convert DBT to biphenyl anaerobically, but the conversion ratio was very low (0.22 to 1.14%).

Thiophenes

In petroleum, organic sulfur occurs mainly in the form of dibenzothiophene [17], but in some high-sulfur coals, 70% of organic sulfur can comprise simple mononuclear thiophene derivatives [59]. For this reason, the microbial degradation of thiophenes is of considerable current interest.

Studies made so far have shown that unsubstituted thiophene is not metabolized by naturally occurring aerobic microorganisms [60-63]. However, microorganisms capable of degrading 2-substituted sulfur heterocycles have been isolated by enrichment on thiophene-2-carboxylic acid (T2C). *Flavobacterium* species [61], *Rhodococcus* species [63], *Vibrio* species [64], and a yellow Gram-negative rod [62] were able to use T2C for growth, releasing the sulfur hetero atom as sulfate. *Pseudomonas* species released the sulfur of T2C as sulfide [65]. Other thiophenes substituted in the 2 and 5 positions with methyl, carboxyl or acetyl groups, were also attached with the release of the sulfur hetero atom [62-65]. Conversely, substitution at the 3 position prevented attack by bacteria [61-63]. Abdulrashid and Clark [59] mapped three genes involved in the *genetic adaptation* of *Escherichia coli* to grow on T2C, but otherwise there has been no attempt on the genetic analysis of thiophene degradation. Alkane-degrading bacterial cultures [66] were shown to degrade aerobically *n*-alkylthiophenes viz. 2-hexadecyl-5-methylthiophene, 2-methyl-5-tridecylthiophene, and 2-butyl-5-tridecylthiophene by attacking the long alkyl chains of these thiophenes, yielding 5-methyl-2-thoiophene acetic acid, 5-methyl-2-thiophene carboxylic acid, and 5-butyl-2-thiophene carboxylic acid, respectively.

Benzothiophenes

Sandhya et al. [67] reported that some *Nocardioides* strains can use benzothiophene as their sole source of carbon and energy. However, most other reports indicate that benzothiophene and methylbenzothiophenes cannot serve as sole source of carbon and energy for microorganisms, but these condensed thiophenes can be cometabolized [68-71]. For example, Bohonos et al. [68] used naphthalene as a growth substrate for mixed cultures, and tentatively identified 2,3-dihydrobenzothiophene-2,3-diol, benzothiophene-2,3-dione and benzothiophene sulfoxide as metabolites of benzothiophene. The 2,3-dione was identified in the extracts of cultures of *Pseudomonas* strain BT1 grown on 1-methylnaphthalene, glucose or peptone [69]. Eaton and Nitterauer [72] grew *Pseudomonas putida* RE204, an isopropylbenzene-degrading bacterium on succinate and yeast extract and identified trans-4-[3-hydroxy-2-thienyl] -2-oxobut-3-enoate as a product of cleavage of the homocyclic ring, and 2-mercaptophenylglyoxalate as a product of cleavage of the heterocyclic ring of benzothiophene.

Sulfoxides, sulfones, and 2,3-diones were commonly found as metabolites of cultures [69,71,73]. Some of these cultures were also capable of oxidizing the methyl groups of the methylbenzothiophenes, yielding benzothiophene-methanols, and benzothiophene-carboxylic acids [71]. Saftic et al. [73] demonstrated that 2,3-dimethylbenzothiophene was oxidized to its sulfone and sulfoxide by *Pseudomonas* strain BT1. Kropp et al. [74] used six of the 15 possible isomers of dimethylbenzothiophene for biotransformation studies with three *Pseudomonas* isolates that oxidize a variety of condensed thiophene, including methylbenzothiophenes and methyldibenzothiophenes. Each isolate was grown on 1-methylnaphthalene or glucose in the presence of one of the dimethylbenzothiophenes. Sulfoxide and sulfones were commonly found metabolites in the culture extracts from the 2,3-2,7-, and 3,7-isomers, whereas 2,3-diones, 3 (2H)-ones and 2(3H)-ones were formed from 4,6-and 4,7-isomers.

High molecular weight products, some of which were tentatively identified as tetramethylbenzo[b]naphtho[1,2-d]thiophenes, were detected in the extracts of cultures incubated with 4,6- or 4,7-dimethylbenzothiophene. The methyl groups of all the isomers, except 4,6-, were oxidized to give hydroxymethyl-methylbenzothiophenes and methylbenzothiophene-carboxylic acids, and these were the only products detected from the oxidation of 3,5-dimethylbenzothiophene. Bressler et al. [75] showed that a *Pseudomonas* strain DB1 could grow, using benzothiophene sulfone, 3-methylbenzothiophene sulfone and 5-methylbenzothiophene sulfone as its sole source of carbon, sulfur, and energy. Sulfate and sulfide were detected in the culture medium when the sulfones were consumed, providing evidence of thiophene ring cleavage.

MICROBIAL ATTACK ON NON-HETEROCYCLIC SULFUR-CONTAINING HYDROCARBONS

Alkyl and aryl sulfides

Metabolism of phenylbenzylsulfide with *Aspergillus niger* NRRL337 produced phenylbenzylsulfoxide, showing 23% yield along with a small quantity (9%) of the corresponding sulfone [76]. A similar metabolism of methyl β-naphthyl sulfide with *A. niger* produced methyl β-naphthyl sulfoxide and methyl β-naphthyl sulfone [76]. *Paecilomyces* sp. TLi, a coal solubilizing fungus, was shown to degrade organic sulfur-containing coal substructure compounds by oxidative attack localized at the sulfur atom [77]. Ethylphenyl sulfide and diphenyl sulfide were degraded to the corresponding sulfones. A variety of products were formed from dibenzyl sulfide, presumably via free radical intermediates [77]. Diphenyl disulfide and dibenzyldisulfide were cleaved to the corresponding thiols and other single-ring products [77]. A *Pseudomonas* isolate metabolized benzylmethyl sulfide as a sole source of sulfur via a *sulfur-specific pathway*, yielding benzoate and sulfate [78]. Several polysulfidic model compounds with bulky alkyl and aryl groups connected by short chains of sulfur (S_3-S_5) and polymeric polysulfides were reduced by an extreme thermophile *Pyrococcus furiosus* to H_2S [79]. However, polysulfidic sulfur is not believed to represent a significant fraction of the sulfur in coals.

Non-aromatic cyclic sulfur-containing hydrocarbons

n-Alkyl-substituted tetrahydrothiophenes (C_{10}-C_{30}) are found in petroleum, which have not undergone biodegradation in their reservoirs [80]. Biodegradability of 2-*n*-dodecyltetrahydrothiophene (DTHT) and 2-*n*-undecyltetrahydrothiophene were tested by using five Gram-positive, *n*-alkane-degrading bacterial isolates. The alkyl side chains of these compounds were oxidized, and the major intermediates found in 2-*n*-undecyltetrahydrothiophene and DTHT-metabolizing cultures were

2-tetrahydrothiophenecarboxylic acid (THTC) and 2-tetrahydrothiophene-acetic acid (THTA), respectively. Four *n*-alkane-degrading fungi were also shown to degrade DTHT, yielding both THTC and THTA. These transformations were consistent with the degradation of the alkyl side chains by the removal of acetate units through β-oxidation pathway. Quantitation of tetrahydrothiophene ring-containing products in 28-day-old bacterial and fungal cultures suggested that THTC and THTA were metabolized further to unidentified products [80].

IDENTIFICATION OF GENES FOR DBT DEGRADATION/DESULFURIZATION

Gene for DBT degradation

Microbial degradation of DBT by ring-destructive pathway to characteristic water soluble products in *Pseudomonas* sp. strains DBT2 and DBT4 is shown to be plasmid-mediated [81]. The two isolates harboured a 55-megadalton plasmid; growth in the presence of novobiocin resulted in both loss of the plasmid and loss of the ability to oxidize DBT. The DBT-*degradative genes* from *Pseudomonas alcaligenes* DBT2 were subsequently cloned into a cosmid vector to create plasmid pC1 [82]. The cloned DBT-degradative genes from *Pseudomonas alcaligenes* DBT2 (on plasmid pC1) were introduced into a spontaneous DBT mutant of *Pseudomonas* HL7b. Acquisition of plasmid pC1 simultaneously restored oxidation of DBT and naphthalene to the transconjugant, although the primary DBT metabolite produced by transconjugant HL7bR (pC1) corresponded to that produced by wild-type strain DBT2, rather than that from wild-type strain HL7b. It was shown that transconjugant HL7bR (pC1) is a mosaic of the parental types regarding DBT metabolite production, regulation, and use of carbon sources. Denome et al. [83] isolated a 9.8 kb DNA fragment from a soil isolate C18 of a *Pseudomonas* sp., that conferred the DBT-metabolizing phenotype on

Pseudomonas putida and *Escherichia coli*. This DNA was sequenced and found to contain ten open reading frames (ORFs) greater than 100 amino acids in length. These ORFs were designated as *dox* genes, since they were cloned by their ability to degrade DBT. The nucleotide sequence revealed that several of *dox* ORFs have sequence identities with genes for enzymes known to degrade other aromatic compounds. These include naphthalene, biphenyl, ben·zene, and toluene dioxygenases of *Ps. putida*.

Genes for DBT desulfurization

The genes for DBT desulfurization of *Rhodococcus* sp. strain IGTS8 were cloned by complementing *dsz* deletion mutants (84,85). DNA sequence and molecular subclone analyses revealed that a single operon containing three frames are involved in the conversion of DBT to 2-HBP. The three genes, designated as *dszA, dszB,* and *dszC,* encode enzymes DszA, DszB, and DszC, respectively, which are sufficient for the conversion of DBT to 2-HBP (84,85) (Figure 2). The three genes are clustered on a 120-kb linear plasmid and are transcribed in the same direction. The termination codon for *dszA* and the initiation codon for *dszB* overlap, and there is a 13-bp gap between *dszB* and *dszC*. Subclone analyses revealed that the DszC converts DBT to $DBTO_2$ and that DszA and DszB act in concert to convert $DBTO_2$ to 2-HBP.

ENZYMES FOR DBT DESULFURIZATION

The oxidation of DBT to 2-HBP by the sulfur-specific route in *Rhodococcus rhodochrous* strain IGTS8 has been linked to the enzymes DszA,B,C encoded by the *dszA,B,C* genes as shown in Figure 2. Among the three Dsz enzymes, DszC was first purified to homogeneity and characterized from *Rhodococcus erythropolis* D-1 [86] and *dszC* overexpressing *Escherichia coli* strain [87], respectively. DszC is a monooxygenase and is involved in two consecutive steps of oxidation of DBT to $DBTO_2$. DszC was shown to require a flavin reductase for its catalytic activity [87,88]. It was shown that DszC enzyme of *R. rhodochrous* IGTS8 binds one flavin mononucleotide (K_d, 7 μM) or reduced flavin mononucleotide ($FMNH_2$) ($K_d < 10^{-8}$ M) per 90200-Da homodimer, and $FMNH_2$ is an essential co-substrate for its activity [87]. On the basis of isotope labelling patterns with $H_2{}^{18}O$ and ${}^{18}O_2$, DBTO and $DBTO_2$ obtained their oxygen atom(s) from molecular oxygen rather than water in their formation from DBT [87]. DszA, also a monooxygenase, was shown to be involved in the hydroxylation of $DBTO_2$ to form HBPS without any evidence of a stable, detectable intermediate. Like DszC, DszA also required a flavin reductase for the catalytic activity [88]. DszB is a sulfinase that catalyzes the desulfination of HBPS to form 2-HBP. The sulfur product was identified as sulfite by a coupled enzyme assay in which the reaction from HBPS to 2-HBP was done in the presence of sulfite oxidase and cytochrome c. All the four enzymes involved in DBT desulfurization, DszA,B,C, and the flavin reductase were purified from *R. rhodochrous* IGTS8 [89]. The native molecular masses were 180 kDa for DszC, a tetramer, 100 kDa for DszA, a dimer, and 40 kDa for DszB, a monomer. The flavin reductase had an apparent molecular mass of about 25 kDa and is specific for NADH and FMN and, like all Dsz enzymes, it also did not contain tight association with flavin cofactor [89].

Very recently, DszA enzyme of *R. erythropolis* D-1 was purified to homogeneity and characterized [90]. DszA was found to have a molecular weight of 97 kDa, consisting of two subunits with identical molecular weight of 50 kDa each. The N-terminal amino acid sequence of the purified DszA completely coincident with the previously deduced amino acid sequence for gene *dszA* of *R. rhodochrous* IGTS8, except for Met residue at the latter N-terminus. The optimal temperature and pH for DszA activity were 35°C and ~ 7.5, respectively. DszA was inhibited by Mn^{2+}, Ni^{2+}, 2,2'-bipyridine and 8-quinolinol, suggesting that a metal might be involved in its activity.

MICROBIAL DESULFURIZATION OF COAL

The research on microbial desulfurization of coal has concentrated on the removal of pyritic sulfur. Microbial metabolism of pyritic sulfur by its oxidation, using mesophilic bacteria such as *Thiobacillus ferrooxidans* and *Thiobacillus thiooxidans* [91-96], and a thermophilic bacterium *Sulfolobus acidocaldarius* [97-99], is known. These chemolithotrophic organisms can utilize inorganic pyritic sulfur compounds as energy sources and are capable of removing 90% or more of the inorganic pyritic sulfur from coal within a few days. Patents by Detz and Barvinchak [100] and Madgavkar [101] described the use of *Thiobacillus ferrooxidans* and *Thiobacillus* species for the removal of pyritic sulfur from coal and coal-derived liquid, respectively. A patent by Kopacz [102] described the use of *Bacillus sulfasportare* ATCC 39909 for sulfur removal from coal without differentiation between pyritic and organic sulfur. A patent by Stevens and Burgess [103] described the use of an unidentified mixed culture of seven Gram-negative rods (ATCC 39327) prepared by *in situ* growth enriched with sulfur compounds and subsequently grown in the presence of coal, which could reduce the sulfur content from coal by about 20% per day with a substantial portion being reduced to organic sulfur. Although several processes on microbial pyritic sulfur removal have advanced to the pilot plant scale, there are no commercial processes based on this approach, because there are faster and less expensive physical and chemical methods for removing pyrites.

The major problem in microbial desulfurization of coal is the removal of organic sulfur, which is interwoven into the molecular matrix of coal. The microbial removal of organic sulfur from coal requires microorganisms capable of cleaving carbon-sulfur bonds and the accessibility of these bonds to microorganisms. It is imperative that to achieve a 90% reduction of organic sulfur, the coal must be ground to a size fine enough to expose 90% of the coal mass as surface area. Chandra et al. [104], used a mixed mesophilic heterotrophic culture consisting of strains of *Bacillus, Micrococcus,* and *Pseudomonas* to remove organic sulfur from pulverized high-organic sulfur (83%) Assam coals (6.64 wt% total sulfur). A peptone-beef extract broth containing 10 wt% powdered coal lost 38-45% of its total sulfur content after 10 days, irrespective of the bacterial strain used. The media pH fell from an initial pH 7 to pH 2.5.

Gokcay and Yurteri [105] used a thermophilic [50°C] *Thiobacillus* type organism for the removal of organic sulfur from Turkish lignite. Almost 50% of organic sulfur removal was reported in 25 days of incubation. Kargi and Robinson [106] used *thermophilic* sulfur-oxidizing organism, *Sulfolobus acidocaldarious* to remove organic sulfur from inorganic sulfur free coal samples. Nearly 44% of the initial organic sulfur was removed from 10% coal slurries at 70°C in about 4 weeks. *Pseudomonas putide* growing in commercial nutrient broth medium was claimed to reduce organic sulfur of lignites by up to 37% in 5-7 days [107]. A patent issued to Stevens et al. [108] claimed the reduction of organic sulfur in coals treated with yeast, *Hansenula cifferri*. Up to 46% reduction in total sulfur was claimed (of which the 'most part' was organic sulfur). A patent by Kibane II [109] claimed that continuous growth of the organic sulfur-specific *mutant microorganism, Rhodococcus rhodochrous* (ATCC No 53968), in the presence of sulfur-containing coal at 28-30°C, results in the removal of more than 90% of the organically bound sulfur. The thermophilic bacterium, *Sulfolobus brierley,* can be adapted to metabolize preferentially either organic or inorganic sulfur [110]. As much as 45% of organic sulfur has been removed in coal that contains equal amounts of organic and inorganic sulfur, while the inorganic sulfur contents remain essentially the same. The same bacterial culture removes 90% of inorganic sulfur from coal. However, this adaptation phenomenon is not currently understood.

MICROBIAL DESULFURIZATION OF PETROLEUM

Key advantages of using petroleum in biodesulfurization experiments are that the access of microorganisms to organic sulfur compounds in petroleum is far easier; the separation of microorganisms after biotreatment is easier; and additionally, more conclusive analytical methods such as gas-chromatography-sulfur chemiluminescence detection (GC-SCD) and gas-chromatography-atomic emission spectrometry (GC-AES) can be applied to the analysis of the reaction products.

The application of bacteria that metabolize sulfur heterocycles to oil desulfurization has been considered for many years. Early patents by Strawinski [111,112], Zobell [113], and Kirshenbaum [114] described microbial systems for solubilization of sulfur from oil. Hartdegen et al. [115] proposed a system for the biodesulfurization of petroleum, using a genetically altered derivative of *Pseudomonas alcaligenes* that oxidizes DBT and related molecules by ring-destructive pathway to water soluble products. Lack of specificity of these microbial catalysts for sulfur-containing hydrocarbons present in petroleum causes loss of valuable carbons (and consequently their fuel value). Walker et al. [116] observed a loss of 40% of the benzothiophenes and dibenzothiophenes and a loss of 50% of the naphthobenzothiophenes in the oil reserved from microbial cultures and concluded that sulfur-containing aromatics were roughly twice as refractory as non-sulfur analogues.

Atlas et al. [117] reported that the C_3-dibenzothiophenes were very persistent in residual oil from the Amoco Cadiz spill, and they used them as markers to follow the loss of other compounds, including hydrocarbons, dibenzothiophene, and C_1- and C_2-dibenzothiophenes. In their studies on the formation of heavy oils, in addition to the loss of n-alkanes, many of the lower molecular weight aromatic hydrocarbons, benzothiophenes, and dibenzothiophenes were lost. Fedorak and Westlake [78] reported the removal of alkylbenzo[b]thiophenes, dibenzothiophene, and C_1- and C_2-dibenzothiophenes from Purdhoe Bay crude oil by microbes of marine origin. Many of the sulfur heterocycles were metabolized without nutrient supplementation, although the number and the extent of the compounds degraded increased with nutrient addition. The order of susceptibility of the sulfur heterocycles in homologous series was found to be as follows: C_2-benzo[b]thiophenes $>$ C_3-benzothiophenes, dibenzothiophene $>$ C_1-dibenzothiophenes $>$ C_2-dibenzothiophenes. *Pseudomonas* HL7b [30] can degrade not only the alkyl-benzothiophenes, DBT, and alkyl-dibenzothiophenes but also aromatic compounds, such as alkyl-naphthalenes, biphenyl, methyl-biphenyls, phenanthrene, and alkyl-phenanthrenes of crude oil within 7 days of incubation. This organism was incapable of altering the saturated aliphatic fraction of crude oil, even after 21 days of incubation.

Microorganisms, which can degrade DBT by sulfur-specific pathway to 2-HBP, will have to be potential candidates for practical microbial desulfurization of petroleum. *Rhodococcus* sp. strain IGTS8 is shown to be able to remove organic sulfur from petroleum and water soluble coal-derived materials without decreasing the calorific value of these substrates [118]. Biodesulfurization of two different crude oils in the 22-31° API specific gravity (1-2% total sulfur content) is demonstrated in 1-dm³ batch stirred reactors using *Rhodococcus* sp. IGTS8 [119]. While analysis of the crudes, before and after biodesulfurization, did not reveal a decrease in total sulfur, GC-MS did reveal significant (43-99%) desulfurization of dibenzothiophene and substituted dibenzothiophenes. Fractionation of the whole crude, followed by analysis in using GC-SCD of the aromatic fraction of the Van Texas crude oil, demonstrated a reduction of sulfur in this fraction from 3.8% to 3.2%. This research indicates that IGTS8 may be capable of desulfurization of refined products such as gasoline and diesel, whose predominant sulfur species are dibenzothiophenes [119].

Recently, the possibilities of *biocatalytic desulfurization* were demonstrated in diesel oils [42,120,121]. The total sulfur in diesel oil, middle distillate unit feed, decreased from 0.15% to 0.06% in 12 h by the resting cells of *Gordona* strain CYKS1 [121]. The ratio of diesel oil was 10% in the reaction mixture. Thermophilic *Paenibacillus* species strains A11-1 and A11-2 [120] grew in the presence of the light gas oil (20%), and the content of sulfur in the oil phase decreased from 800 ppm to 720 ppm after the cultivation. By resting cells of *Rhodococcus erythropolis* KA2-5-1, the total sulfur in the light gas oil, which had been hydrodesulfurized, decreased from 800 ppm to 310 ppm [122]. The ratio of the light gas oil was 50% in the reaction mixture. In this treatment, the sulfur content significantly decreased, but the carbon content was the same as compared with that in the oil before the treatment. *Rhodococcus erythropolis* I-19, containing multiple copies of key desulfurization genes (*dsz*), was used to desulfurize alkylated dibenzothiophenes (C_x-DBTs) found in a hydrodesulfurized middle-distillate petroleum (MD1850) [123]. Initial desulfurization rates of DBT and MD1850 by strain I-19 were 5.0 and 2.5 µmole/g dry weight/min, which were 25-times higher than that for wild-type bacteria. According to Sulfur K-edge X-ray absorption near-edge structure (XANES) analysis, thiophene compounds accounted for >95% of the total sulfur found in MD1850, predominantly C_x-DBTs and alkylated benzothiophenes. Extensive biodesulfurization resulted in a 67% reduction of total sulfur from 1850 ppm to 615 ppm sulfur. XANES analysis of the 615 ppm material gave a sulfur distribution of 75% thiophenes, 11% sulfide, 2% sulfoxides, and 12% sulfones. The strain I-19 preferentially desulfurized DBT and C_1-DBTs, followed by the more highly alkaylated C_x-DBTs.

The feasibility of using sulfate-reducing bacteria to desulfurize high sulfur fossil fuels as a *pre-combustion technology* has gained considerable attention. It has been demonstrated that *Desulfovibrio desulfuricans* M6 reduced organic sulfur in petroleum to sulfide under anaerobic conditions, using electro-chemically supplied electrons [55,124]. An anaerobic process for sulfur removal would be attractive because it does not liberate sulfate as a byproduct that must be disposed off by some appropriate treatment. However, anaerobic microorganisms, effective enough for practical petroleum desulfurization, have not been found yet.

CONCLUSION

Many microbial desulfurization processes have been developed to desulfurize fossil fuels. However, the reaction rates and specific activities (sulfur removed/hour/g of biocatalyst), that have been reported in the literature, are much lower than those necessary for *commercialization* technology. Nevertheless, as the genes involved in the sulfur-specific metabolic pathway are identified and cloned, amplification of genes or enzymes responsible for desulfurization may lead to microbial cultures that remove organic sulfur efficiently from fossil fuels in near future. The *biocatalytic desulfurization* is a better option because of its low capital and operating expenses.

REFERENCES

1   Ember L. Chemical Engng News 1990; 68: 4-5.

2   Harrison AP. Annu Rev Microbiol 1984; 38: 265-292.

3   Meyer RA. Coal Desulfurization. New York: Marcel Dekker, 1977.

4   Chakraborti JN. Analytical Methods for Coal and Coal Products. New York: Academic Press, 1978; 1: 279-322.

5   Hessley RK, Reasoner JW, Riley JT. Coal Science. New York: John Wiley & Sons, 1986; 183-220.

6   Calkins WH. Fuel 1994; 73: 475-484.

7   Dugan PR. Biotechnol Bioeng Symp No. 16, 1986: 185-203.

8   Couch GR. Biotechnology and Coal. Publication No. ICTIS/TR38, IEA
        Coal Research, London, 1987.

9   Isbister JD, Kobaylinski EE. Coal Sci Technol 1985; 9: 627-641.

10   Annual Book of ASTM Standards. Method No. D2492, American Society
        of Testing Materials, 1984.

11   Olson GJ. Fuel Process Technol 1994; 40: 103-114.

12   Annual Book of ASTM Standard. Method No. D2493, American Society
        of Testing Materials, 1984.

13   Huffman GP, Huggins FE, Shah N et al. In: Chugh YP and Caudle
        RD, eds. Processing and Utilization of High Sulfur Coals ll. London,
        New York: Elsevier, 1987; 3.

14   Malik KA. Process Biochem 1978; 13: 10-12,35.

15   Speight JG. Chemistry and Technology of Petroleum. New York: Marcel
        Dekker Inc, 1980.

16   Monticello DJ, Kilbane JJ. Practical Considerations in Bio-
        desulfurization of Petroleum, IGT's 3rd Intl Symp on Gas, Oil, Coal
        and Environmental Biotechnol, New Orleans, 1990.

17   Monticello DJ, Finnarty WR. Ann Rev Microbiol 1985; 39: 371-389.

18   Coleman HJ, Hopkins RL, Thompson CJ. Int J Sulfur Chem, B 1971;
        (Quart Rept): 41-61.

19   Rall HT, Thompson CJ, Coleman HJ, Hopkins RL. Sulfur Compounds
        in Crude Oil, U.S. Department of the Interior Bureau of Mines, Bulletin
        659, 1972.

20   Shennan JL. J Chem Tech Biotechnol. 1996; 67: 109-123.

21   Takaoka S. Process Economics Program Report 47, Mento Park, CA:
        Stanford Res Inst, 1969.

22   Monticello DJ. Hydrocarbon Processing 1994; February: 39-45.

23   Gundlach ER, Boehm PM, Marchand M, Altas RM, Ward DM, Wolfe
        DA. Science 1983; 221: 112-129.

24   Kodama K, Nakatani S, Umehara K, Shimizu K, Monoda Y, Yamada K. Agric Biol Chem 1970; 34: 1320-1324.

25   Laborde AL, Gibson DT. Appl Environ Microbiol 1977; 34: 783-790.

26   Yamada K, Minoda Y, Kodama K, Nakatani S, Akasaki T. Agric Biol Chem 1968; 32: 840-845.

27   Kodama K. Agric Biol Chem 1977; 41: 1305-1306.

28   Kodama K, Umehara K, Shimizu K, Nakatani S, Minoda Y, Yamada K. Agric Biol Chem 1973; 37: 45-50.

29   Hou CT, Laskin Al. Dev lnd Microbiol 1976; 17: 351-362.

30   Foght JM, Westlake DWS. Can J Microbiol 1988; 34: 1135-1141.

31   Bezbaruah RL, Gogoi BK, Pillai KR. Indian J Microbiol 1994; 34: 49-53.

32   Malik KA, Claus D. Proc Fifth International Fermentation Symposium, Berlin, Abstract 23.03, 1976; 421.

33   Frassinetti S, Setti L, Corti A, Farrinelli P, Montevecchi P, Vallini G. Can J Microbiol 1998; 44: 289-297.

34   Ochman M, Klubek B, Boydstun J, Clark D, Nabe S. Microbios 1990; 63: 79-91.

35   Mormille MR, Atlas RM. Appl Environ Microbiol 1988; 54: 3183-3184.

36   Dutt D, Krawiec S. Abstract of the Annual Meeting of the American Society for Microbiology K 112, ASM, 1988.

37   Kilbane JJ. Proceedings of the 7th Annual International Pittsbugh Coal Conference 1990; September 10-14: 373-381.

38   Krawiec S. Dev Ind Microbiol 1990; 31: 103-114.

39   Kilbane JJ, Bielaga BA. Chemtech 1990; 20: 747-751.

40   Purdy RF, Lepo JE, Ward B. Curr Microbiol 1993; 27: 219-222.

41   Izumi Y, Ohshiro T, Ogino H, Hine Y, Shimao M. Appl Environ Microbiol 1994; 60: 223-226.

42   Wang P, Krawiec S. Arch Microbiol 1994; 161: 266-271.

43   Omori T, Monna L, Saiki Y, Kodama T. Appl Environ Microbiol 1992;
       58: 911-915.

44   Constanti M, Giralt J, Bordons A. World J Biotechnol 1994; 10: 510-
       516.

45   Constanti M, Giralt J, Bordons A. Enzyme Microbiol Technol 1996;
       19: 214-219.

46   Nekozuka S, Nakajima-Kambe T, Nomura N, Lu J, Nakahara T.
       Biocatal Biotransform 1997; 15: 17-21.

47   van Afferden M, Schacht S, Klein J, Truper HG. Arch Microbiol 1990;
       153: 324-328.

48   Sato H, Clark DP. Microbios 1995; 83: 145-159.

49   Holland HL, Khan SH, Richards D, Riemland E. Xenobiotica 1986; 16:
       733-741.

50   Crawford DL, Gupta RK. Curr Microbiol 1990; 21: 229-231.

51   Mormille MR, Atlas RM. Can J Microbiol 1989; 35: 603-605.

52   Kurita S, Endo T, Nakamura H, Yagi T, Tamiya N. J Gen Appl Microbiol
       1971; 17: 185-198.

53   Kohler M, Genz IL, Schicht B, Echart V. Zentralbl Mickrobiol 1984;
       139: 239-247.

54   Kim HY, Kim TS, Kim BH. Biotechnol Lett 1990; 12: 761-764.

55   Kim TS, Kim HY, Kim BH. Biotechnol Lett 1990; 12: 757-760.

56   Miller KW. Appl Environ Microbiol 1992; 58: 2176-2179.

57   Lizama HM, Wilkins LA, Scott TC. Biotechnol Lett 1995; 17: 113-116.

58   Armstrong SM, Sankey BM, Voordouw G. Biotechnol Lett 1995; 17:
       1133-1136.

59   Abdulrashid N, Clark DP. J Bacteriol 1987; 169: 1267-1271.

60   Knecht AT. Microbial Oxidation of Dibenzothiophene and its Possible
       Application in the Desulfurization of Coal and Petroleum. Ph.D Thesis,
       Louisiana State University, 1961.

61   Amphlett MJ, Callely AG. Biochem J 1969; 112: 12.

62   Cripps RE. Biochem J 1973; 353-366.

63   Kanagawa T, Kelly DP. Microb Ecol 1987; 13: 47-57.

64   Evans JS, Venables WA. Appl Microbiol Biotechnol 1990; 32: 715-750.

65   Klubek B, Clark D. Microbial Removal of Organic Sulfur from Coal. In: Degradation of Sulfur-containing Heterocyclic Compounds. Report to US Department of Energy, Washington, 1985; Contact No. DE-FC01-83FE 60339.

66   Fedorak PM, Peakman TM. In: Akin C, Markuszewski R, Smith JRW, eds. Microbial Degradation of Some n-alkylthiophene. IGT, Chicago Gas, Oil, Environ Biotechnol IV, (Pap Int Symp) 4th 1991 (Pub. 1992); 307-323.

67   Sandhya S, Prabu SK, Sundari RBT. J Environ Sc Health; 1995; 30: 2006.

68   Bohonos N, Chou TW, Spanggord RJ. Jpn J Antibiotic 1977; 30 (suppl): 275-285.

69   Fedorak PM, Gribic-Galic D. Appl Environ Microbiol 1991; 57: 932-940.

70   Sagardia F, Rigau JJ, Martinez-Lahoz A, Fuentes F. Lopez C, Flores W. Appl Microbiol 1975; 20: 722-725.

71   Kropp KG, Goncalves JA, Andersson JT, Fedorak PM. Environ Sci Technol 1994; 28: 1348-1356.

72   Eaton RW, Nitterauer JD. J Bacterol 1994; 176: 3992-4002.

73   Saftic S, Fedorak PM, Andersson JT. Environ Sci Technol 1992; 26: 1759-1764.

74   Kropp KG, Saftic S, Andersson JT, Fedorak PM. Biodegradation 1996; 7: 203-221.

75   Bressler DC, Norman JA, Fedorak PM. Biodegradation 1998; 8: 297-311.

76   Dodson RM, Newman N, Tsuchiya HMJ. Org Chem 1962; 27: 2707-2708.

454

77   Faison BD, Clark TM, Lewis SN, Ma CY, Sharkey DM, Woodward CA. Appl Biochem Biotechnol 1991; 28/29: 237-251.

78   Fedorak PM, Westlake DWS. Can J Microbiol 1983; 29: 291-296.

79   Tilstra L, Eng G, Olson GJ, Wang FW. Fuel 1992; 71: 779-783.

80   Fedorak PM, Payzant JD, Montgomery DS, Westlake DWS. Appl Environ Microbiol 1988; 54: 1243-1248.

81   Monticello DJ, Bakker D, Finnerty WR. Appl Environ Microbiol 1985; 49: 756-760.

82   Foght JM, Westlake DWS. Can J Microbiol 1990; 36: 718-724.

83   Denome SA, Stanley DC, Olson ES, Young KD. J Bacterial 1993; 175: 6890-6901.

84   Denome SA, Oldfield C, Nash LJ, Young KD. J Bacteriol 1994; 176: 6707-6716.

85   Piddington CS, Kovavevich BR, Rambosek J. Appl Environ Microbiol 1995; 61: 468-475.

86   Ohshiro T, Suzuki K, Izumi Y. J Ferment Bioeng 1997; 83: 233-237.

87   Lei BF, Tu SC. J Bacteriol 1996; 178: 5699-5705.

88   Xi L, Squires CM, Monticello DJ, Child JD. Biochem Biophys Res Commun 1997; 230: 73-75.

89   Gray KA, Pogrebinsky OS, Mrachko T, Xi L, Monticello DJ, Squires CH. Nature Biotechnol 1996; 14: 1705-1709.

90   Ohshiro T, Kojima T, Torii K, Kawasoe H, Izumi Y. J Biosci Bioeng 1990; 88: 610-616.

91   Silverman MP. J Bacteriol 1967; 94: 1046-1051.

92   Silverman MP, Rogoff MH, Wender I. Appl Microbiol 1961; 9: 491-496.

93   Dugan PR, Apel WA. In: Murr LE, Torma AE, Brierley JA, eds. Metallurgical Applications of Bacterial Leaching and Related Microbiological Phenomena. New York: Academic Press, 1978; 223-250.

94   Detz CM, Barvinchak G. Mining Congress J 1979; 65: 75-86.

95   Hoffman MR, Faust BC. Panda FA, Kook HH, Tsuchiya HM. Appl Environ Microbiol 1981; 42: 259-271.

96   Kargi F, Biotechnol Bioeng 1982; 24: 749-752.

97   Kargi F, Robinson JM, Biotechnol Bioeng 1982; 24: 2115-2121.

98   Kargi F, Robinson JM. Appl Environ Microbiol 1982; 44: 878-881.

99   Eligwe CA. Fuel 1988; 67: 451-458.

100   Detz CM, Barvinchak G. US Patent No 4, 1980; 206-288.

101   Madgavkar AM. US patent No 4, 1989; 861-723.

102   Kopacz EP. US Patent No 4, 1986; 632-606.

103   Stevens Jr. SE, Burgess WD. US Patent No 4, 1987; 659-670.

104   Chandra D, Mishra AK. In: DL. Wise, ed. Bioprocessing and Biotreatment of Coals, New York: Marcel Dekker, 1990; 631-652.

105   Gokcay CF, Yurteri RN. Fuel 1983; 62: 1223-1224.

106   Kargi F, Robinson JM. Fuel 1986; 65: 397-399.

107   Rai C, Reyniers JP. Biotechnol Prog 1988; 4: 225-230.

108   Stevens Jr. SE, Burgess WD. Patent No 4, 1989; 851-350.

109   Kilbane JJ II. US Patent No 5104801, 1992.

110   Bhattacharyya D, Kermode Rl, Hsieh M, Aleem HMI, Kahlid A. Liquefaction Sci Update 1988; 4: 1-4.

111   Strawinski RJ. US Patent No 2, 1950; 521,761.

112   Strawinski RJ. US Patent No 2, 1951; 574-070.

113   Zobell CE. US Patent No 2, 1953; 641-564.

114   Kirshenbaum I. US Patent No 2, 1961; 975, 103.

115   Hartdegen FG, Coburn JM, Roberts RL. Chem Eng Prog 1984; 80: 63-67.

116   Walker JD, Colwell RR, Petrakis L. Can J Microbiol 1975; 21: 1760-1767.

117   Atlas RM, Boehm PD, Calder JA. Estuarine, Coastal Shelf Sci 1981; 12: 589-608.

118   Kilbane JJ II, Jackowski K. Biotechnol Bioeng 1992; 40: 1107-1114.

456

119 Kaufman EN, Borole AP, Shong R, Sides JL, Juengst CJ. Chem Technol Biotechnol 1999; 74: 1000-1004.

120 Konishi J, Ishii Y, Onaka T, Okumura K, Suzuki M. Appl Environ Microbiol 1997; 63: 3164-3169.

121 Rhee SK, Chang JG, Chang YK, Chang HN. Appl Environ Microbiol 1998; 64: 2327-2331.

122 Ishii Y, Kobayashi M, Konishi J, Onaka T, Okumura K, Suzuki M. Nippon Kagaku Kaishi 1998; June: 373-381.

123 DiGrazia PM, Werner J, Palmer S. Appl Environ Microbiol 1999; 65: 4967-4972.

124 Kim BY, Kim HY, Kim TS, Park DH. Fuel Process Technol 1995; 43: 87-94.

Biotransformations: Bioremediation Technology for Health and Environmental Protection
V.P. Singh and R.D. Stapleton, Jr. (Editors)
© 2002 Elsevier Science B.V. All rights reserved.

Algae-dependent bioremediation of hazardous wastes

Inderdeep Kaur and A.K. Bhatnagar

Environmental Biology Laboratory, Department of Botany, University of Delhi, Delhi - 110 007, India

INTRODUCTION

Algae, with 200,000 species, belonging to class Thallophyta of the plant kingdom, are a group of unicellular, multicellular and macrophytic (*seaweeds*)[*] organisms, which occur in aquatic ecosystems all over the world, including the Arctic zone. Although these grow in a wide range of habitats, the greatest diversity is seen on rocky seashores and coral reefs. Most of the algae are photosynthetic. In fact, the amount of sun's energy trapped by algae is believed to be 10 times the amount trapped by all terrestrial plants. They are the major primary producers of organic compounds and play a key role in food chains. However, there are a few parasitic forms, which are not able to photosynthesize but are classified as algae because of their close resemblance to photosynthetic forms. The reproductive structures of algae lack sterile cells and do not form embryos.

---

[*]The term seaweed is a misnomer, perhaps derived from the fact that several macromarine algae, such as species of *Fucus, Sargassum, Gracilaria, Polysiphonia,* and *Ulva* caused obstruction to navigation in the sea and at the ports. Considering their valuable ecological role and economic potential, it would be more appropriate to call these plants sea plants, sea algae, sea flora, or marine algae. This will also recognize the fact that the marine plants, like their terrestrial counterparts, are benign and need to be investigated for their economic value and conserved for posterity. Considering that the sea is the ultimate sink of all pollutants, perhaps the greatest utility of the marine algae will be as bioremediators. They seem to be the most appropriate organisms to breakdown the xenobiotic substances brought to the sea – the last refuge of all non-biodegradable pollutants.

A few algae are used as foods, especially in Asian countries. In many coastal environments, algae are a source of diatomaceous earth and hydrocolloids, such as alginic acid, carrageenan, and agar.

Algae are extremely important not only ecologically but also phylogenetically. It is thought that all the major groups of animals and plants originated in the sea and that, even today, this is where one can find representatives of many ancient evolutionary lineages. Thus, if we are to understand the diversity and the phylogeny of the plant world, it is of fundamental importance, indeed essential, to investigate algae [1].

Lately, a great deal of interest has been centered around algae as potential candidates for bioremediation of polluted water bodies [2,3]. The groups of algae with such potential are Cyanobacteria (blue green algae), microalgae (generally green) and macroalgae. The term Cyanobacteria is often preferred since the members show their affinity with bacteria (cell wall composition and prokaryotic nature) and identify themselves as microbes. These algae have great capacity to accumulate dissolved metals and hold an important position in 'green clean' technology.

The technology deals with metal ion uptake that has been suggested as taking place in two stages. The first stage is rapid and reversible reaction called physical adsorption or ion-exchange, which occurs at the cell surface. The subsequent stage is slower and called chemisorption. Dead cells accumulate heavy metal ions to the same or greater extent than the living cells. This kind of adsorption is called biosorption [4].

Despite the algal diversity and relatively inexpensive algal biomass, there has been little commercial exploitation of these plants for metal removal, recovery, and prospecting. This chapter reviews algal diversity and economic importance with special reference to the processes of algal biosorption and bioaccumulation. While listing out the limitations of algal-based systems, the chapter also identifies areas where research should concentrate in order to give to the world an *algal-based bioremediation* industry for cleaning the water bodies.

ALGAL SYSTEMATICS AND CLASSIFICATION

As early as 1866, Haeckel suggested the term Protista for unicellular microorganisms - bacteria, algae, fungi, and protozoa. Bacteria were referred to as lower protists and the rest were the higher protists. The researches on algae revealed great diversity in pigments, food reserves, number of flagella, and algal cell wall composition. These characteristics have since then formed the basis for algal classification. Fritsch [5] gave one of the earliest and widely accepted classifications, where eleven classes were identified, which are given as follows: (i) Cyanophyceae, (ii) Chlorophyceae, (iii) Xanthophyceae, (iv) Chrysophyceae, (v) Bacillariophyceae, (vi) Cryptophyceae, (vii) Dinophyceae, (viii) Chloromonadineae, (ix) Euglenineae, (x) Phaeophyceae, and (xi) Rhodophyceae. For long, the Cyanophyceae were the only prokaryotic algae; but later in early 1980s, a green prokaryotic alga - *Prochloron* - was discovered. The classification needed a revision.

In the light of Whittaker's [6] work on classification of living organisms, it was suggested that the blue green algae be placed under Kingdom Monera with archaebacteria and eubacteria as two groups. Eubacteria included two phyla or divisions - Cyanophyta (= Cyanobacteria) and Prochlorophyta (= Chloroxybacteria). The latter was constituted to include *Prochloron*.

In 1977, Stainer [7] proposed a new name for blue green algae (BGA). This new name - Cyanobacteria - emphasizes two aspects of BGA better than the traditional name Cyanophyta, namely the prokaryotic nature of blue green algal cell and the fairly close relationship between BGA and Eubacteria.

On the other hand, the traditional name reflects the algal characteristics of BGA, such as their ability to perform photosynthesis and the similarity in structure. The phycologists, continue to retain Cyanophyta (or blue green algae) among algae. The relatively recent classification of Parker [8] is an example, which is outlined as follows:

At present, both names are being used alongside each other in literature. This practice is also followed in this chapter. However, for convenience, the

Kingdom	Division	Class	
Plantae	Prokaryota	Cyanophycota	
		Prochlorophycota	
	Eukaryota	Rhodophycota	Rhodophyceae
		Chromophycota	Chrysophyceae
			Prymnesiophyceae
			Xanthophyceae
			Eustigmatophyceae
			Bacillariophyceae
			Dinophyceae
			Phaeophyceae
			Raphidophyceae
			Cryptophyceae
		Euglenophycota	Euglenophyceae
		Chlorophycota	Chlorophyceae
			Charophyceae
			Prasinophyceae

term Cyanophyta or BGA has been used for the sections dealing with diversity and utility of algae. The term Cyanobacteria is more appropriately used under 'Bioremediation', which is also an emerging field in microbiology. The microbiologists place Cyanobacteria along with bacteria. However, no such role is known yet for the Prochlorophyta or Chloroxybacteria.

DISTRIBUTION AND DIVERSITY

Algae occur in stagnant ponds and reservoirs, flowing rivers as well as in the deap oceans. Ecologically, algae are worldwide in distribution, occurring on the land surface, on all types of soil excepting sandy deserts, on permanent ice, and snow fields. However, the major centres of distribution are in the waters, which cover 70% of the earth's surface. These are the primary and major organic producers. Besides occurring in the form of microscopic planktons, these occur as microscopic and macroscopic forms along the sea shores. In the ocean, the littoral habitat is occupied by few plants other than algae. Hence, these are known to occur practically in every habitable environment on earth. On the land, these are important constituents of the flora of soils, moist rocks, and stone surfaces.

Algae exhibit a remarkable diversity of form and size, ranging from flagellated swimming green cells as minute as 1 micron in diameter to brown kelps up to 70 m in length. These have a long fossil history, some possibly extending back to the time of origin of photosynthetic cellular plants. Algae are generally considered to be the group from which all the more complex cryptogams and, ultimately, the seed-bearing plants arose. These appear as rose-red feathery fronds in deep clear marine coastal waters (Rhodophyceae), green scum on a quiet pond (Chlorophyceae), Sargasso weed in the massive Atlantic gyre (Phaeophyceae) and as organisms responsible for colouration on mountain snow.

Algae also occur on shores and coasts, attached to the bottom (benthic species) or remain suspended in the water itself (planktonic species). Benthic marine algae show a greater range in size than members of fresh water benthos. The larger marine algae are readily visible and include representatives of the red, brown, and green algae. Microscopic algae are also common in coastal environments and include representatives of red, brown, and green algae.

Algal types, found on a reef, include endosymbiotic algae, turf algae, crustose algae, and calcareous algae. *Endosymbionts* are associated with corals and other invertebrates, whereas turf algae and encrusting algae aid in binding the reefs together. Herbivorous fish and invertebrates are common on reefs. Some algae are restricted to parts of the reef where herbivore activity is low or compensate for losses to grazers by rapid growth. Other algae possess hard coverings or chemical deterrents against herbivores. It is, therefore, evident from the foregoing description that these, the so called 'simple' plants, are not simple at all.

INDUSTRIAL, HEALTH AND ENVIRONMENTAL SIGNIFICANCE

Algae, being cheap and readily accessible, play a small but important role in the direct economy of many countries. Four major categories of industrial products agar, algin, carrageenans and diatomite are derived from algae. Algae are directly used as food, fodder, and fertilizer. A number of industrial products, antibiotics and medicines are also produced from some algal species. Besides beneficial aspects, the harmful effects of algae cannot be ignored. Their role as water blooms, fouling and corrosion agents also needs to be mentioned. The economic value and location-specific ecological significance of algae have to be considered alongside, while utilizing them as *bioremediators*.

Manurial value

Marine algae contain growth hormone-like substances. Hence, marine macroalgae have been used in agriculture and horticulture for many years.

There is evidence for abscisic acid in *Ascophyllum nodosum* [9] and cytokinins in *Ecklonia maxima* [10]. Wide range of beneficial effects, such as increased crop yield, nutrient uptake, and resistance to frost and stress conditions, have been observed with the application of such algae to farm soil. It is also known that algin of marine algae consolidates and binds sandy soils, and their repeated application can help in improving the soil texture. The ability of microalgae, especially Cyanophyta, to produce copious amounts of polysaccharides has made these also act as soil conditioning agents and *biofertilizers* [11]. In some cases, the thick mats of algae provide efficient substratum for the successful germination of tiny seeds of grasses by providing suitable moisture conditions.

This group has an inherent capacity to fix atmospheric nitrogen. For this simple reason, since long, species of *Oscillatoria, Anabaena, Spirulina,* and *Nostoc* are widely used to increase soil fertility, reducing the consumption of nitrogen fertilizers. The most significant role of BGA (blue green algae) is in reclamation of alkaline 'USAR' soils in India [12].

Medicinal value

Algae have also been used against a wide range of ailments. The role of algae and algal products in medicine is an area of knowledge restricted to coastal communities. Curative powers from selected algae, particularly the tropical and subtropical marine forms, are reported since ancient times. Algae may be used as vermifuges, anaesthetics, antipyretics, cough remedies, wound healing compounds, thirst quenching remedies, and for treatment of gout, gall stones, goitre, hypertension, diarrhoea, constipation, dysentry, burns, ulcers, skin diseases, lung diseases, and semen discharge [13,14]. Thus, the use of algae for cures or preventives against diverse medical problems is widespread in certain regions.

The kelps (Laminariales) and sargassums (Fucales) are rich source of iodine, and are used in treatment of goitre. The algal product, agar, is also used against prolapsed stomach. A series of Cyanophyta were screened at National Cancer Institute, USA, for anti-human immunodeficiency virus (HIV-1) properties. Cellular extracts from cultured *Lyngbya lagerheimia* and *Phormidium tenue* have been found to have the ability to protect the human lymphoblastoid T cells from the cytopathic effect of HIV-1 infection [15]. Macroalgae are also known to have good proportions of provitamin A, ascorbic acid, cobalamine, thiamine, riboflavin, nicotinic acid, pantothenic acid, lipoic acid, and choline. Two green algae - *Scenedesmus* and *Chlorella* - are known to be the richest sources of vitamins [16].

Animal feed

A few marine algae have also been employed in animal feed in various countries. *Laminaria saccharina* is relished by horses. *Laminaria* spp., *Alaria* spp., *Chorda filum,* and *Pelvetia canaliculata* are included in cattle feed and *Ascophyllum* spp. are popular pig feed.

Food for man

Algae are used as food delicacies or supplements in some maritime countries but have probably never become a staple item [17]. Atleast 170 species are used as food [18]. In Japan and China, Wales, Iceland, and the Canadian maritimes, selected macroscopic marine algae are served as delicacies. These are a source of vitamins and minerals [13]. Algae, both macroscopic and microscopic, are considered as potential inexpensive and effective dietary supplements for the malnourished [19]. The salt tolerant green alga *Dunaliella* is known to produce large amount of β-carotene [20].

The common red algae, consumed by people the world over, are *Gymnogongrus vermicularis, Asparagoposis sanfordiana, Grateloupia filicina,*

Gigartina stellata, Chondrus crispus, Laurencia pinnatifida, Iridaea edulis, Eucheuma isiforme, Gracilaria, Rhodymenia palmata, Porphyra perforata, Suhria vittata, Gracilaria compressa, Griffithsia sp., *Gigartina teedii, Acanthopeltis japonica,* and *Grateloupia divaricata.* The brown algae included in human diet are *Laminaria* spp., *Ecklonia stolonifera, Heterochordaria abietina, Eisenia bicyclis, Undaria,* and *Padina australis.* Among the green macroalgae, *Ulva lactuca, Enteromorpha* spp., *Monostroma* spp., and *Codium* spp. are consumed in various countries.

Microalgae are also used as food. *Chlorella* is a green microalga that has been produced and consumed in substantial quantities since 1960s primarily in the Asian countries like China and Japan. It is promoted and sold primarily as a health food product in the form of tablets or powder. *Chlorella pyrenoidosa* is considered to be an important food source for space flights. It multiplies rapidly and is known to synthesize a rich harvest of food utilizing CO_2 and liberating sufficient O_2 as a byproduct for use. It also assists in the decomposition of human wastes.

In China, not only the marine algae but also the blue green algae, such as *Nostoc commune, N. ellipsosporum,* and *Nematonostoc flagelliforme* form important food items. *Spirulina,* a common BGA, is now a popular health food, and is known to be an efficacious single cell protein (SCP).

Algal products

Major commercial algal products at present are based on only 10-20 algal species, almost all of them macroalgae. Polysaccharide products derived from algae, referred to as hydrocolloids or phycocolloids are available in all continents. At present their combined market value is well over US$ 500 million. Carrageenans and agars are obtained from different species of red algae. Alginates are obtained from species of brown algae.

Diatomite, is a rock-like deposit on the floor of the sea formed from indestructible siliceous frustules of the past diatoms over many millions of years. These deposits are mined in several parts of the world to obtain the diatomaceous earth, which is put to several commercial uses. It is highly porous, insoluble with abrasive qualities and is chemically inert. It is fire-proof and highly absorbent. It is used for clearing solvents and as filters for oils and other solutions.

Funori, a sizing agent and glue, is made in Japan from the marine alga *Gloiopeltis furcata*. An inferior product is also made from species of *Iridaea, Grateloupia, Chondrus,* and *Ahnfeldtia* [1].

Algal hazards

Though algae, are of great value to us, yet the negative aspect of these aquatic plants cannot be overlooked. Luxuriant growth of blue green algae leads to formation of algal blooms, which cause severe economic losses to aquaculture, fisheries and tourism operations, and have major environmental and human health impacts [21]. They produce colouration, off flavours and toxins, making the water unfit for drinking and even washing. Algae are important indicators of pollution in aquatic habitats, whilst, under other circumstances, they themselves become polluting agents. In recreational spots, exposure to water bloom often results in allergic skin irritation. Algae are also implicated as causative agents for several instances of dermatitis, e.g., swimmers itch. The respiratory disease, known as Tamadare fever, is related with blooms of *Trichodesmium erythraeum*. Eye and throat irritations are caused by seawater, containing large amounts of dinoflagellates [22].

The marine algae are known to obstruct free movement of ships and boats in water. Many of these get entangled on the body of ship and corrode it due to the acidic substances produced by them. Fouling on ships causes hull roughness leading to increased frictional resistance, resulting in loss of speed and increased use of fuel [23]. Corrosion of conduits and tanks is another problem posed by algae. The concrete and metals in pipes and boilers get corroded. *Oscillatoria* spp. cause pitting of metals [24].

Occasionally, algal growth becomes detrimental for the water body. Many algae produce toxins, poisonous to fish. Excessive growth of algae leads to filtration problem in water supply schemes. In water supply, reservoirs, purification plants and in sewage disposal plants, they play an important role in oxygenation and filteration.

It is impossible to assess the full ecological importance of algal growth. However, in aquatic habitats algae are part of the food chains leading to crustacea and fish. On agricultural land, they are important constituents of soil flora. In scientific work, they are used as assay organisms for vitamins, in the dating of sedimentary rocks in oil prospecting and lately as a possible source of absorber of CO_2 and provider of O_2 in space vehicles.

Recently, algae have been associated with regulating environmental pollution. These can help in removing pollutants from the surroundings, restoring contaminated sites and preventing further pollution. Algae fit well with this *'green movement'* of maintaining nature's harmony through microorganisms [25].

ALGAE IN CHANGING ENVIRONMENT

Algae hold an important position amongst microbes in indicating water quality and monitoring the same. Pollution may result in reducing or oxidizing activities in water, and can also bring about enrichment of algal nutrients in water. This may selectively stimulate the growth of a few types, producing massive surface growths or 'blooms'. The 'blooms' may in turn, reduce the water quality affecting its use for various purposes [1].

Although many kinds of algae are sensitive to large amounts of organic wastes in their environment and may disappear, others are tolerant and may be stimulated in their growth and reproduction by the presence of waste. The latter are also called pollution tolerant algae. In a study carried out on algal distribution in a water body, it was seen that in a reducing zone (polysaprobic) a few algae survived, e.g., *Oscillatoria chlorina, Spirulina jeaneri, Euglena* sp., and a few other flagellates. In the next zone (mesosaprobic), where oxygen was not completely depleted, algae could grow. This zone has two subdivisions - α-mesosaprobic zone, in which *Oscillatoria* spp., *Phormidium* spp., *Nitzschia palea, Gomphonema parvulum,* and *Stigeoclonium tenue* grow, indicating the improvement in water quality. A similar distribution is encountered in β-mesosaprobic zone. This zone is followed by oligosaprobic waters and katharobic waters, in which organic matter is minimum. Species of *Lemanea* and *Batrachospermum* are seen here [1] Table (1).

By selecting and adapting particular strains of algae, it may be possible to bring about changes in industrial effluents, which are otherwise unsuitable for release into the streams, lakes or marine waters. Algae may, thus, be put to use in the treatment of industrial wastes. Though algae may not be considered indicators in the real sense of the word [26]; yet in a few cases, they indicate water quality. Certain algae are marine pollution indicators, e.g., sewage pollution is reported to prevent the growth of the brown rock weed *Fucus*; while algae, that are stimulated, include *Blindingia minima, Enteromorpha* spp., *Ulva lactuca, Porphyra leucosticta, Erythrotrichia carnea, Acrochaetium irrgatulum, A. thuretii,* and *Calothrix confervicola.* Macrophytes concentrate metal ions from seawater, and the variation in the concentrations of the metal in the thallus is often taken to reflect metal concentrations in

Table 1
Survey of the saprobic zones and accompanying communities

Zone I	The Coprozoic Zone Algae absent. a: the bacterial community; b: the Bodo community; c: both communities.
Zone II	The α-polysaprobic zone 1: the *Euglena* community; 2: the Rhodo-Thiobacterial community; 3: the pure Chlorobacterial community.
Zone III	The β-polysaprobic zone 1: the *Beggiatoa* community; 2: the *Thiothrix nivea* community; 3: the *Euglena* community.
Zone IV	The γ-polysaprobic zone 1: the *Oscillatoria chlorina* community; 2: the *Sphaerotilus natans* community.
Zone V	The α-mesosaprobic zone a: the *Ulothrix zonata* community; b: the *Oscillatoria benthonicum* community; c: the *Stigeoclonium tenue* community.
Zone VI	The β-mesosaprobic zone a: the *Cladophora fracta* community; b: the *Phormidium* community.
Zone VII	The γ-mesosaprobic zone a: the Rhodophyta community (*Batrachospermum vagum* or *Lemanea fluviatilis*); b: the Chlorophyta community (*Cladophora glomerata* or *Ulothrix zonata*) (the type of pure water).
Zone VIII	The oligosaprobic zone a: the Chlorophyta community (*Draparnaldia glomerata*); b: the pure *Meridion circulare* community; c: the Rhodophyta community (*Lemanea annulata, Batrachospermum moniliforme* or *Hildenbrandia rivularis*); d: the *Vaucheria sessilis* community; e: the *Phormidium inundatum* community.
Zone IX	The Kathabrobic Zone a: the Chlorophyta community (*Chlorotylium cataractum* and *Draparnaldia plumosa*); b: the Rhodophyta community (*Chantransia chalybea* and *Hildenbrandia rivularis*); c: the encrusting algal communities (*Chamaesiphon polonius* and different *Calothrix* species).

a, b, c ... as alternative possibilities.
1, 2, 3 ... as graduations of pollution.

the surrounding seawater. On this basis, macroalgae, especially the Phaeophyta, frequently have been used as indicators of heavy metal pollution [27].

Lackey [28] reported two classes of algae, which could be used as indicators of clean waters : (i) the olive green flagellates (or Cryptophyceae) and (ii) the yellow green flagellates (or Chrysophyceae). These algae tend to be present in moderate to great abundance in clean waters and react adversely to pollution.

Over the years, some common indicators of water quality have been worked out by various workers. These are :

Navicula accomoda - indicator of sewage pollution.

Oscillatoria rubescens - first acute indicator of a lake that is undergoing unfavourable development.

Stigeoclonium tenue - present in foulest part.

Nitzschia palea and *Gomphonema* - in mild pollution zone.

Cocconeis and *Chamaesiphon* - in unpolluted regions [29].

Peridinium triquetum - in highly contaminated water [30].

Trachelomonas - in H_2S pollution.

Also, abundant growth of *Cladophora* in fresh water indicates that the water is not subjected to repeated pollution by heavy metals. On the other hand, abundant *Stigeoclonium tenue,* but no *Cladophora* spp., should be treated as a indicative for metal pollution. There are certain other algae, which flourish in water polluted with organic wastes and play an important role in self purification of the water body. The selective types of algae, that exist in polluted waters, are used as *pollution indicators* [31].

Another field, where algae have been put to use, is the metal toxicity studies and metal prospecting. The study of algal-metal ion interactions is particularly important principally because the algae and other simple plants are basic in the food chain. Thus, the accumulation of trace levels of metals, which may not be sufficiently toxic to affect the primary plant metabolism, can be magnified by passage up the food chain, thereby posing a potential threat to higher organisms. It is well known that the algae, grown for cattle fodder using sewage as the source of nutrients, can easily transfer toxic metals present in the sewage to the feeding animals. Well known cases of metal poisoning have been recorded in Japan, where the local inhabitants around Minimata Bay suffered neurologic illness after consuming seafish and shell fish contaminated with methyl mercury. Equally infamous is the Itai-itai disease, which was caused by chronic cadmium poisoning.

The upcoming area of wastewater treatment stands to gain maximum from the intrinsic capability of microorganisms to 'clean-up' waters. The conventional physical and chemical means of removing soluble metal waste, such as precipitation, ion-exchange, electrochemical treatment and evaporative recovery, are generally very expensive, especially when the contaminant concentrations are in the low range of 10-100 ppm. Microorganisms have been commonly reported to be capable of sequestering heavy metals and concentrating them up to several thousand times over their environment. With the growing scarcity and increasing value of some metals, this inherent ability of the microorganisms has also received prominence from the viewpoint of possible metal recovery. Some species are also known to have the ability to concentrate specific metals, and their relative abundance has been used as indicators for *'botanical metal prospecting'* [32]. Algae appear to offer the most easily accessible biological system for extracting phosphorus from domestic sewage.

BIOREMEDIATION

Pollution of the biosphere with toxic metals and radionuclides has accelerated dramatically since the beginning of the industrial revolution. In Canada, waters in and around Quebec and Montreal have been contaminated for over 20 years primarily due to pesticides, sewerage runoff and pollutants from the St. Lawrence River. In 1990, approximately 1,400 beaches in the United States were closed due to pollution and toxic waste hazards with another 1,700 threatened as reported by Centres for Disease Control (CDC). Also, in another study, toxic medical supplies discarded from Mexico were found washed up the shorelines of Corpus Christi, Texas, resulting in the closing of some beaches in Texas.

Off and on in the past, alarmed scientists have been concerned about 18,130 km^2 in the Gulf of Mexico, called the 'dead zone', because hypoxia occurs in this area. It is due to the presence of lethal combination of agricultural fertilizers, sewage run off and diminishing O_2. The process kills the food chain from the first trophic level, rendering the area virtually lifeless.

The Indian aquaculture industry also passed through rapid growth and has reached a near collapse situation as a result of pollution and water quality deterioration. Presence of mercury, lead, chromium, and manganese in quantities above permissible limits has been reported in some river estuaries, locations very close to alkali or rayon plants in the costal waters adjacent to Mumbai and Cochin, in the working area of some welding shops, electroplating and thermal processing units and the effluents of chrome tanneries. Relatively high levels of manganese have been detected in the ground water in certain parts of the Ganga-Yamuna basin [33].

The primary sources of this pollution are the burning of fossil fuels, mining and smelting of metalliferous ores, municipal wastes, fertilizers, pesticides, and sewage. Metals with specific gravity higher than 5 g.cm^{-3}, defined as heavy metals, have generated enormous concern amongst environmentalists. A variety of heavy metals are found naturally in freshwater and seawater environments due to weathering of rocks. Certain others are, however, added through unmindful anthropogenic activities. Heavy metals, such as iron (Fe), manganese (Mn), zinc (Zn), copper (Cu), molybdenum (Mo), nickel (Ni), cobalt (Co), and vanadium (V) are considered to be essential for plants, but may cause harm if available in excess. Other metals are categorized as non-essential, e.g., lead (Pb), cadmium (Cd), aluminium (Al), chromium (Cr), mercury (Hg), silver (Ag), arsenic (As), and tin (Sn). Those, which do not enter into the primary metabolism, are detrimental to plant growth and development even in traces. Metals in aquatic environment may exist in dissolved or particulate forms. These may be dissolved as free hydrated ions, complex ions, chelated with inorganic ligands, such as OH$^-$, Cl$^-$ or CO$_3^{2-}$, or complexed with organic ligands, such as amines, humic acid, fulvic acid, and proteins. Certain chemicals, that cannot be degraded, can accumulate in the environment to levels that threaten human health or environmental quality.

Deleterious effects of some metals have been studied in detail. The biological effects of metals are complicated by their interactions with other metals. Cadmium has received widespread attention because of its accelerated release into the environment as a result of industrial utilization and the resulting pollution [34]. Cadmium is a relatively rare element with no known biological function. It is ranked amongst the most hazardous heavy elements in the environment and is highly toxic to all components of aquatic communities [35]. Cadmium exerts its toxic effect(s) over a wide range of concentrations, and amongst microbes, algae and cyanobacteria are most sensitive organisms. It causes severe inhibition of many physiological processes, leading to retardation of growth and inhibition of photosynthesis and nitrogen fixation at concentrations less than 2 ppm. Loss of motility is seen in *Euglena gracilis*. Cadmium stress is known to alter nitrate uptake and metabolism in *Thalassiosira fluviatilis* and *T. aestivalis* [36]. It can induce cell elongation [37]. Cadmium also causes pronounced morphological aberrations in these organisms, which are probably related to deleterious effects on cell division. The impact may be direct or indirect, as a result of Cd effects on protein synthesis and on cellular organelles such as mitochondria. Mitochondrial swelling is seen in various diatoms. Development of large areas of intrathylakoid spaces, thylakoid disorganisation and degradation, increase in the number and size of polyphosphate bodies and appearance of large cyanophycin granules, and cell lysis are observed.

According, to WHO guidelines, the recommended permissible levels of metals in drinking water for Cd is 0.005 μg/ml. In Tamil Nadu, samples were collected from wells, lakes and springs in Salem, Coimbatore and Chengalpattu districts. The water samples showed a range 0.015-0.05 μg/ml, which exceeds the limits set by WHO quality guidelines [33]. Although Cd is toxic to algae, its specific mode of action is far from being understood. It may express toxic effects either indirectly by adhering to cell wall sites and preventing the transport of nutrients, or directly by becoming localized intracellularly and replacing essential divalent metal ions in enzymes. Lead (Pb) in wastewater is typical of a heavy metal environmental pollutant derived from industrial processes, mining, and automobile exhaust. Lead is used as an industrial raw material for manufacture of storage batteries, printing ink, fuel additives, photographic materials, matches, and explosives. The storage battery industry is the largest consumer of lead, followed by petroleum industry. Lead is also a common constituent of plating wastes, although it is not as frequently encountered as Cu, Zn, Cd, and Cr.

In Tamil Nadu and Kerala, more than 60%, and in Andhra Pradesh and Karnataka, 30-45% of the samples analyzed showed non-detectable levels of Pb. In all other states, lead was detectable in almost all the samples collected and the highest concentration was around in 1 µg/ml in West Bengal and Bihar [33]. Chromium in the aqueous environment can exist in either a hexavalent or a trivalent form. Of the two, former has the greater chronic toxicity for fresh water fish. Once inside the body, chromium becomes more or less immobilized as trivalent chromium and, therefore, tends to accumulate. High median values were seen in drinking water samples of Bihar, Haryana, Himachal Pradesh and Andhra Pradesh. High mean values were reported from West Bengal [33].

Ground water in Bangladesh and in eastern parts of India has high level of Arsenic, leading to a major health hazard and drinking water problem. Cobalt is not generally in the lists of heavy metals which are harmful to the environment. Although essential for microbial growth, it can cause adverse effects at high concentrations. Mercury exists in aquatic environments as inorganic divalent ion, which may be hydrated or complexed, or as organic monomethyl mercury. Lack of iron, an essential element, results in many morphological changes in plants. At higher concentrations in microalgae, heterocyst formation, branching, and hair formation are affected.

Adverse effects are also seen due to increase in Ni and Sn concentrations. It is quite clear that toxic metal contamination of soil, streams and ground water poses a major environmental and human health problem, which is still in need of an effective and affordable technological solution. The precise impact of pollution on species and genetic composition of aquatic ecosystems is difficult to fathom. In most of the countries, the polluted water is often used for irrigation of crops, leading to accumulation in the food. It is, therefore, important to control aquatic pollution through biological methods. One of the potential methods of bioremediation is through growth of algae, which can degrade or consume pollutants from water.

MEETING THE CHALLENGES THROUGH BIOREMEDIATION

Bioremediation, which refers to the use of living organisms (primarily microbes) to degrade environmental pollutants or to prevent pollution through waste treatment, is emerging as one of the several alternative technologies for removing pollutants from the environment. It is potentially a simple, low cost and self sustaining option for amelioration of wastewaters. Both, the natural strains and genetically modified microbes may play a role. The objective of bioremediation is to decontaminate polluted soil, water or air, using microbes or other organisms. In this endeavour, microbial biomass has emerged as an option for developing economic and ecofriendly wastewater treatment processes.

The metal uptake by microbes can occur by one of the following two different mechanisms:

(i)   *Biosorption*: It is a generic term for the passive attachment (sorption) of heavy metals to biological molecules based on complexation, chelate formation, or ion exchange. It can be characterized as the reversible and fast reaction of metal ions with functional groups of cell wall polymers. Adsorption phenomenon in extracellular biopolymers is often classified as biosorption. As a rule, the term biosorption is understood as the whole of the accumulation procedure in respective living or dead biological cell material [38].

(ii) *Bioaccumulation*: It is defined as the active uptake of metals by vital organisms into living cells. For this purpose, the microorganisms consume energy and the process is slow. Bioaccumulation is a growth-dependent process.

Algae as bioremediators

In many ways, living plants can be compared to solar driven pumps, which can extract and concentrate certain elements from their environment. All plants have the ability to accumulate from soil and water those heavy metals, which are essential for their growth and development. These metals include Fe, Mn, Cu, Mg, and possibly Ni. Also, certain plants have the ability to accumulate heavy metals, which have no known biological function. These include Cd, Cr, Pb, Co, Ag, Se, and Hg. However, excessive accumulation of these heavy metals can be toxic to most plants. The ability to accumulate these metals to unusually high concentrations has evolved both independently and together in a number of different plant species.

There are a number of reports on the feasibility of developing technology for removal of metals from polluted waters with photoautotrophs. These photoautotrophs are generally less resistant/tolerant to heavy metals for growth, but can be grown cheaply on minimal nutritional medium without sugars [32,39]. These organisms are capable of removing metals mainly by three phenomena: (i) biosorption, (ii) extracellular precipitation, and (iii) binding by biopolymers. Algae as biosorbents of heavy metals offer a cost effective, potential and alternative to conventional methods for decontamination of water bodies at global, regional, and local levels.

Microalgae (green and cyanobacteria) have been used to remove heavy metals from aqueous system, since they have a high capacity to accumulate dissolved metals [32]. It is also known that phytoplanktons affect trace metal chemistry in natural waters, not only by surface reactions, but also by metal uptake and by production of extracellular organic matter with metal complexing properties. The exudates of these algae may be important in keeping the free metal ion concentrations at low levels in natural waters, thus decreasing the toxic effects. It has been observed that polyhydroxamate siderophores, which are strong metal complexing agents, are released by cyanophytes under certain conditions [40]. These are also discharged by some species of eukaryotic algae [41,42]. With respect to macro marine algae, there has been a particular interest because of their biomass. Brown algae are considered efficient and cheap material for biosorption, e.g., *Sargassum* and *Ascophyllum*. Thus, various groups of algae have potentials of bioremediation.

Recent applications of metal-algae interactions reported include the use of algae as biosorbents for recovery of metals from industrial solutions, either to sequester toxic metals or to recover precious metals. And algae have also been proposed as monitoring organisms for Cu and Hg in estuaries. The popularity of algal based systems lies in the fact that these systems have several advantages over currently available chemical technologies. These are:

(i)   Versatility and flexibility for a wide range of applications.

(ii)   Robustness.

(iii)   Selectivity for heavy metals over alkaline earth metals.

(iv)   Ability in some cases to reduce metal concentration to drinking water standards (through biosorption).

(v)   Cost-effectiveness.

(vi)   Sustainability.

(vii)   Ability to remove the pollutants without themselves contributing any harmful substances to water bodies.

(viii)   Easy to filter out.

BIOSORPTION

Biosorption can be defined as the removal of metal or metalloid compounds and particulates from solution by biological material. Algae are known to be effective biosorbents.   Living and dead algal biomass as well as cellular products, such as polysaccharides, can be used for metal removal. The algal biomass is known to bind passively large amounts of metal(s). Algal biomass can be used in natural form, free in solution or immobilized by various techniques on to a solid support or to produce granules for a metal removal/ recovery process.   However, on prolonged contact with the metal-bearing solution, the living biomass is also capable of sequestering metal intracellularly by an active process called bioaccumulation.

Biosorption is possible by both living and non-living biomass. However, bioaccumulation is mediated only by living biomass. Thus, microbial biomass can be used and exploited more effectively in sorption rather than for accumulation (Table 2).

Important features of the algal biosorption process are:

(i)   Algal biosorption techniques/processes can be used to remove toxic metals and/or radionuclides from liquid effluent before its discharge into water bodies. In addition, the algal biomass can be harvested and utilized for metals of value.

(ii)   Biosorption is a rapid phenomenon involving passive metal sequesteration by the non-growing biomass.

(iii) Biosorption mainly involves adsorption, cell surface complexation, ion exchange and/or microprecipitation. The physio-chemical phenomena, besides being rapid, are reversible.

(iv) Biosorption is a *growth-independent* phenomenon.

(v) The sorption technology has advantages of low operating cost, effectiveness in dilute solutions, and in generating minimum effluent.

Table 2
Metal biosorption by 'living' microalgae free in solution

Organism	Metal	Accumulation (% dry weight)
Chlorella vulgaris	Cadmium	0.20
	Lead	8.50
	Zinc	0.13
	Gold	10.00
Chlorella regularis	Uranium	0.39
	Copper	0.40
	Zinc	2.80
	Nickel	
	Cobalt	0.19
	Manganese	0.80
	Molybdenum	1.32
Chlorella salina	Cadmium	0.01
	Cobalt	0.67
	Zinc	0.02
	Manganese	0.01
Chlorella homosphaera	Cadmium	0.55
	Zinc	0.40
Chlorella emersonii	Technetium	0.01

Table 2 continued

Organism	Metal	Accumulation (% dry weight)
	Zirconium	0.24
Chlorella sp.	Mercury	0.01
	Uranium	0.02
Scenedesmus obliquus	Cadmium	0.30
	Technetium	0.01
	Zirconium	0.21
Scenedesmus sp.	Molybdenum	2.30
	Uranium	0.34
Chlamydomonas reinhardtii	Technetium	0.19
	Molybdenum	2.10
	Cadmium	0.35
Chlamydomonas sp.	Zirconium	0.04
	Uranium	0.07
Dunaliella tertiolecta	Uranium	0.01
Asterionella formosa	Cadmium	2.20
Fragilaria crotonensis	Cadmium	0.60
Ankiistrodesmus sp.	Uranium	1.00
Selenastrum sp.	Uranium	1.00
Euglena sp.	Aluminium	1.50
	Zinc	0.01
	Manganese	0.02
	Copper	0.01
	Lead	0.03
Thalassiosira rotula	Nickel	0.06
	Cadmium	0.09
Cricosphaere elongata	Copper	0.07
	Cadmium	0.01

Algal-based systems

Based on the promising biosorption phenomenon, systems have been developed utilizing actively growing algae in ponds or lagoons for wastewater treatment and metal removal, where metal concentrations encountered are not toxic to algal biomass [43]. In addition, artificial meander systems with algae have been developed to treat effluent from mining operations. For these systems no biomass is produced as such, once the algae are established, these grow within the system, photosynthesizing and utilising the nutrients in the effluents. Thus, the cost of biomass production in such systems is zero. In UK, with a temperate climate, large amounts of microalgae, such as *Chlorella* sp., have been grown with no additional lighting or heating. Biotechna, a UK-based company has developed a reactor 'biocoil' consisting of coiled translucent tubes. It is a system capable of producing large yields of macroalga throughout the year, even when located on a roof top [44].

Filamentous algae, such as *Cladophora* and *Spirogyra*, and cyanobacteria-*Rhizoclonium* and *Oscillatoria* have been shown to contribute significantly to the removal of heavy metals, such as Cu, Zn, Pb, and Mn, that are discharged from a lead mine [44,45]. Other algae, including species of *Chlorella* and *Scenedesmus*, have also been used for removal of Cd and Zn [46]. However, so far species of only eight genera of microalgae have been grown on a commercial scale. These genera are *Spirulina, Chlorella, Scenedesmus, Phaeodactylum, Botryococcus, Chlamydomonas, Dunaliella,* and *Porphyridium.* These are usually grown for the production of animal feed, chemicals, biochemicals or fertilizers or as a source of food for humans. Some Cyanobacteria, *Chlorella* and *Spirulina*, have also been grown for use in a metal removal process [47]. This is not because of their metal biosorption qualities, but because these are cheapest to grow and give the maximum yield. For marine environments, macroalgae have become the candidates of interest due to bulk availability of their biomass from water bodies. Many processes have been developed for the removal of metals, using algal biomass derived from macroalgae, e.g., *Sargassum natans, Ascophyllum nodosum,* and *A. vaucheria.* The biomass has shown high biosorptive capacities for various metals (Table 3). Macroalgae are found in marine waters in all parts of the world and are, thus, potentially an extremely cheap source of biomass for a metal removal process. However, in most places, macroalgae have only been cultivated and regularly harvested as a food and source of biochemicals, e.g., alginates. The cost of harvesting a particular macroalga, in the amounts required for an *industrial biosorption* process, has not been estimated.

Metals like uranium, technetium, molybdenum, arsenic, copper, plutonium, polonium, zirconium, platinum, lead, silver, thorium, and radium show only one phase of uptake with algae due to passive, metabolism-independent and rapid phenomenon of biosorption [48]. Generally, biosorption of a single metal by the algal biomass takes place. However, studies by Carvalho et al. [49] and Chong and Volesky [50] have modelled the biosorption of two metals in a 2-metal system (e.g., Zn/Cd, Cu/Cd, Cu/Zn) by biomass derived from the brown macroalga, *Ascophyllum nodosum.*

Role of immobilized biomass (living/dead)

As the success of bioremediation for the *in situ* removal of heavy metals from contaminated substrates/sites is mainly limited to their immobilization by precipitation or reduction, processes have been developed for biosorption of metals, using algal fragments immobilized in a matrix. The immobilized biomass offers many advantages, including better reusability, high biomass loading, and minimal clogging in continuous flow systems. A number of matrices have been used in metal recovery by both viable and non-viable cells. A common practice is the entrapment in the matrix of insoluble Ca-

Table 3
Biosorption of metals by macroalgal biomass

Organism	Metal	Accumulation (% dry weight)
Sargassum natans	Gold	25.00
	Lead	8.00
	Silver	7.00
	Uranium	4.50
	Copper	2.50
	Zinc	2.00
	Cobalt	6.00
	Cadmium	8.30
	Lead	21.10
Sargassum fluitans	Lead	21.60
Sagassum vulgaris	Lead	14.90
Ascophyllum nodosum	Gold	4.00
	Cobalt	15.00
	Cadmium	10.00
	Lead	20.10
Palmaria palmata	Gold	12.50
	Lead	1.10
Chondrus crispus	Gold	7.50
	Cobalt	4.50
	Lead	6.5
Porphyra tenera	Gold	15.00
	Cobalt	2.50
Halimeda opuntia	Cobalt	8.00
	Cadmium	5.20
Vaucheria	Strontium	1.98

Table 3 continued

Organism	Metal	Accumulation (in % dry wt.)
	Copper	3.20
Fucus vesiculosus	Cadmium	5.00
	Lead	17.40
Padina gymnospora	Lead	5.90
Codium taylori	Lead	13.00

alginate. An important matrix being used for immobilization for metal removal processes is silica, e.g., AlgaSORB™ [51].

Two types of immobilization strategies can be adopted with microalgae, either active or passive entrapment. *Active entrapment* involves the culturing of required biomass before its entrapment within a polymeric matrix, whereas passive methods depend upon growth to invade the matrix [52]. Natural polymers used for entrapment are Kappacarrageenan or alginates. These have the advantage of not being toxic toward the algal cells, which remain viable within the matrix, whilst artificial polymers and matrices, such as acrylamide, polyurethane or silica, may be toxic to the cells (Table 4). *Passive immobilization* methods include the use of polyurethane foam matrices, china clay particles, glass beads or inert matrices, which are colonized by the algae [52].

Chemically-reinforced biomass

Since biomass is the basis of bioremediation, its quality needs to be maintained, if one is expecting good results. Studies have been carried out where marine algal raw biomass has been chemically processed in order to reinforce it for sorption process applications and also to enhance the sorption process [4]. Metal sorption was observed to be related to the particle size. In general, big particles had a higher metal uptake than small particles.

Algal biomass material can be chemically modified in three different ways, in order to reinforce it and provide for its improved mechanical stability and pressure resistance required for sorption column applications. The biomass may be subjected to: (i) formaldehyde cross-linking (FA), (ii) glutaraldehyde cross-linking (GA), and (iii) polyethylene imine embedding (PEI). Formaldehyde modification is a chemical cross-linking between adjacent

Table 4
Biosorption of metals by treated or immobilized algal biomass systems

Algal system	Metal	Accumulation (% dry weight)
(a) Lyophilized biomass		
Chlorella pyrenoidosa	Copper	1.50
	Cadmium	0.75
	Manganese	0.25
	Zinc	0.65
Chlorella saccarophila	Copper	1.10
	Aluminium	5.03
Chlorella vulgaris	Silver	0.32
	Copper	0.17
	Mercury	2.10
Scenedesmus quadricauda	Silver	0.21
	Copper	0.27
Chlamydomonas reinhardtii	Copper	1.20
	Cadmium	0.60
	Manganese	0.20
	Zinc	0.59

Table 4 continued

Algal system	Metal	Accumulation (% dry weight)
Stichococcus bacillaris	Copper	1.50
	Cadmium	0.70
	Manganese	0.30
	Zinc	0.75
(b) **Cross-linked biomass**		
Ascophyllum nodosum	Lead	20.1
	Copper	3.2
	Cadmium	4.0
	Zinc	2.2
Fucus vesiculosus	Lead	13.1
Sargassum fluitans	Lead	19.3
Chondrus crispus	Lead	14.2
(c) **Immobilized systems**		
Chlorella vulgaris (in silica)	Gold	30.00
Anacystis nidulans	Nickel	0.48
Chlorella salina	Cobalt	0.24
	Zinc	0.20
	Manganese	0.11
Nostoc calicola (alginate beads)	Copper	0.16
Chlorella vulgaris (ethyl acetate/	Copper	0.35
ethylene glycol dimethyl acrylate	Cadmium	0.03
co-polymer)	Zinc	0.02

chemical groups, preferably hydroxylic groups in sugars of the cell wall. Macroalgae are often dried before use, homogenized to a particular size, cross-linked with formaldehyde and actively immobilized before use. Dry algal biomass before use, either powdered or in chunks or granules, tends to swell on wetting, breaking up and making handling difficult. Thus, a support matrix is often essential to such processes that utilize it. Such systems are efficient on a small scale, but may have diffusional limitations existing in large processes. Glutaraldehyde cross-links chemical groups, which are more distant from each other because of prolonged carbon chain. Glutaraldehyde and polyethylene imine modifications are likely to introduce NH_2 groups to the biomass. These groups alter the metal sorption behaviour because of ionic interactions.

The chemical modifications generally result in altering the particle swelling behaviour. After modification, biomass is expected to have a lower expansion during the swelling process because of its more rigid structure. Thus, the expansion during the swelling process is inversely proportional to the degree of cross-linking. The studies indicate that though support matrix is often essential, yet no particular procedure can be generalized for modification. This is because of differences in the structure and composition of cell walls in various taxa. Thus, modification procedures have to be evaluated and worked out for each biomass.

Desorption-sorption cycles

Processes in which the algal biosorbent can be stripped of metal cheaply and the biosorbent reused, are considered to be more viable as commercial metal removal systems. This part of the process would be similar to ion-exchange, where metals are eluted from the biosorbent by an appropriate solution to give a small concentrated volume of metal-containing solution. However, such recovery of metals must be *non-destructive* so that the regenerated biomass can be reused in multiple biosorption-desorption cycles [53].

Biotechnological exploitation of biosorption technique depends on the efficiency of the regeneration of biosorbent after metal desorption. Therefore, non-destructive recovery by mild and cheap desorbing agents is desirable for regeneration of biomass for use in multiple cycles. For metal ions, which show a marked pH dependence in binding to the microbial cells, stripping of bound metal(s) can be accomplished by pH adjustments. In *Chlorella vulgaris,* for example, desorption of Cu(II), Cr(II), Ni(II), Pb(II), Zn(II), Cd(II), and Co(II) can be achieved by lowering down the pH to 2.0. Dilute mineral acids have been used in various studies to remove metal(s) from the loaded biomass. The physico-chemically sequestered metal to the cell surface can be easily desorbed by chelating agents.

Combustion of the metal-laden algal biosorbent material to produce ash is an alternative to desorption and recycling. The combustion of the algae would produce an ash with a high metal concentration. It would also provide a method for disposing of the spent biomass, after optimum cycling of the biomass has been achieved.

Role of cell wall

The mechanism of biosorption is based on a number of metal-binding processes taking place with components of the cell wall. Various types of interactions occur between metals and cell wall biomolecules. Electrostatic attraction (ion exchange phenomena are based on such attraction) occurs with some metals (e.g., Ca, Na), covalent type bonding with others (Cu) and redox reactions with certain noble metals (Au). The algal cell wall can reversibly biosorb metals and, thus, functions in a way similar to an ion-exchange resin [51]. Thus, the biosorption mechanism can be considered as being dependent on the composition of the algal cell wall. Mostly, algal cell walls are made up of cellulose, polysaccharides, such as mannans, xylans, and chitin. These components, along with the proteins present, can provide acid-binding sites, such as amino, amine, hydroxyl, imidiazole, phosphate, carboxylates, thiols, and thioesters and sulphate groups. Fucoidin and alginic acid of brown algal walls offer anionic sites to which metals bind readily. Carboxylic and sulphate groups have been identified as the main metal-sequestering functional ionic groups in marine algal cell wall.

Amino acids in the proteins could provide functional groups such as:

| amine | amide | carboxyl | imidiazol |

The amino and carboxyl groups, the imidiazole of histidine and the nitrogen and oxygen of the peptide bond could be available for characteristic coordination bonding with metallic ions like Cu^{2+}. Such bond formation could be accompanied by displacement of protons dependent, in part, on the extent of protonation, as determined by the pH. Metallic ions could also be electrostatically bonded to unprotonated carboxyl oxygen and sulphate [54]. Both ionic charge and covalent bonding are involved in the biosorption process. Stary and Kratzer [55] working on microalga *Scenedesmus obliquus* proved that the cell wall behaves like a weak acidic cation exchanger containing various cell wall ligands with different exchange capacities. In Cyanobacteria, walls are murein containing diaminopimelic acid, muramic acid, and N-acetyl glucosamine. Much of the metal deposition occurs at polar regions of the constituent membranes or within the peptidoglycan layer [56].

Metal binding appears to be at least a two-step process, where the first step involves a stoichiometric interaction between the metal and the reactive groups in the cell wall; and the second step is an inorganic deposition of increased amounts of metal(s). All the metal ions, before gaining access to the plasma membrane and cell cytoplasm, come across the cell wall.

Metal-binding has been studied in several algae. Green et al. [57] demonstrated the binding of Cu^{2+}, Pb^{2+}, Zn^{2+}, Ni^{2+}, Cd^{2+}, and Cr^{2+} in *Spirulina platensis* to be accompanied by the liberation of protons, suggesting an ion-exchange reaction. Similar results were obtained in Cd^{2+}, Cu^{2+}, and Zn^{2+}-binding by previously protonated biomass of *Sargassum fluitans,* where the metal binding was coupled with the release of H^+. In *Oscillatoria anguistissima,* copper and zinc binding was accompanied by release of large amounts of magnesium ions. The studies on *Oscillatoria* sp. showed its excellent copper biosorption qualities. It could absorb Cu^{2+} even at lower concentrations, indicating a great affinity for the metal. The uptake was found to increase with the increase in concentration of Cu^{2+}. It could absorb about 268.45 mg/g dry wt. of Cu^{2+} at a residual concentration of 23 mg/l, and it increased to about 1g/g Cu^{2+} adsorption at a residual concentration of 89.33 mg/l. Thus, *Oscillatoria* sp. showed a high sequestration of copper at lower equilibrium concentrations. Further, although the amount of Cu^{2+} adsorbed per unit dry weight decreased, the total amount of Cu^{2+} adsorbed increased, indicating the effect of increasing concentration of biomass on metal uptake. It was also found that 90% of adsorption was complete within just 15 minutes of initial contact with the metal-bearing solution and adsorption of the metal was found to increase with increase in temperature from 25 to 45°C [39]. The dominant mechanism of Cu^{2+} biosorption in *Ecklonia radiata* is ion exchange mechanism, involving exchange of Cu^{2+} and Mg^{2+} present in their cell walls [58]. Williams and Edyrean [59] have observed a concomitant release of Ca^{2+} during the biosorption of Ni^{2+} by the brown marine alga *Ecklonia maxima.* In brown algae, alginates present in the cell wall occur as natural salts of Na^{2+}, Ca^{2+}, and/or Mg^{2+}. These metallic ions can exchange with the contact ions, such as Co^{2+}, resulting in the biosorptive uptake of metal. Thus, biosorbents can be viewed as natural ion-exchange materials, that contain weak acidic and basic groups.

Factors affecting biosorption

Certain factors, that affect biosorption, are cell size and morphology, pH of the medium, and physiological state of algal biomass. Greater the cell surface area to dry weight ratio, more is the amount of metal biosorbed by a cell surface per unit weight. The pH of the external medium also affects biosorption of metals. The solution pH affects the solution chemistry of the metals, the activity of functional groups in the biomass as well as the competition of metallic ions for binding sites. Since the pH is related to the net charge on the cell surface, which determines the extent of the cellular sites occupied by protons and other ions of the medium [35], the competition between heavy metals and H^+ for same cellular binding sites can result in a decrease or an increase in cellular heavy metal uptake and toxicity, depending on pH.

The amino and carboxyl groups as well as nitrogen and oxygen of peptide bonds are also available for coordination bonding with metal ions, such as lead (II), copper (II), or chromium (IV). Such bond formation could be accompanied by displacement of protons and is dependent, in part, on the extent of protonation, which is determined by the pH. The isoelectric point of most algal cell walls lies between pH 3 and 4; thus, the net overall charge on the cell walls under low pH conditions promotes easier access of anions to positively charged binding sites, as the pH is decreased below the isoelectric point.

Green and Darnall [60] classified metal ions into three classes based on dependence on pH of their biosorption by algae. Those in the first class are tightly bound at pH > 5 and can be desorbed at pH < 2. The metal ions, which fall in this class, are Al^{3+}, Cu^{2+}, Pb^{2+}, Cr^{3+}, Cd^{2+}, Ni^{2+}, Co^{2+}, Zn^{2+}, Fe^{3+}, Be^{2+}, and Uo_2^{2+}. For these positively charged ions, at increased pH, there is higher biosorption of metals. The protons, thus, compete for the same active binding sites on the algal cell walls and reduce the amount of metal biosorbed at low pH (high proton concentration). In the second class, with anionic metal species such as TcO_4^-, $PtCl_4^{3-}$, CrO_4^{2-}, SeO_4^{2-}, and $Au(CN)_2^-$, the situation is reverse. In these cases, at decreased pH values, there is increase in biosorption. This effect is due to increased binding of protons to their active binding sites, which, in turn, increases anionic binding. The third class belongs to metals like Hg^{2+}, Ag^{2+}, and $AuCl_4^-$, whose biosorption is independent of pH. These preferentially form covalent complexes with ligands, containing nitrogen and sulphur.

The biosorption is also greatly affected by the physiological state of algal biomass. There is a large difference between the metal biosorbed by dead and living algal biomass. The dead microbial biomass may passively sequester metal(s) and, thus, provide a *cost-effective solution* for industrial wastewater management. It functions as an ion exchanger by virtue of various reactive groups available on the cell surface, such as carboxyl, amine, imidiazole, phosphate, sulfhydryl, sulphate, and hydroxyl. In most studies, biosorption of metals is greater than or equal to 'dead' biomass, as compared to an equal amount of living biomass. When algal biomass is in a 'dead' state, the cells are permeable and allow metals to enter and bind to internal components and surfaces of the cell as well as on the external surface, thus increasing the metal uptake or biosorption. Moreover, dead biomass tends to sequester and retain the metals better, probably due to the absence of active transport mechanisms. There is also evidence that when cells are immobilized and still 'living', e.g., in alginate beads, there is an increase in metal biosorption, which can be related to changes in cell physiology due to immobilization [61].

Considering the above factors, when developing an industrial algal biosorption system, biomass is to be selected on the basis of ease of handling and rate of uptake in the conditions prevalent at the aquatic site.

Commercial biosorbents from algae

The first algal-based commercially used system, was based on benthic algal mats contained in a meander system. These were successfully used by New Lead Bett, Missouri (USA) to remove lead and other metals from its mining and milling effluents. Gale and Wixon [44] have reported that the levels of lead in the effluent decreased from 3 ppm to 500 ppb within 3 hrs when algal-based system with benthic algal mats were used in a meander system. The advantage of this system was that, after the initial cost of construction, there was no further cost as long as the algae continued growing. The disadvantage of the system is that the metals are removed from the effluent but remain in the environment, immobilized in the algal mats. In addition to this, some other examples of use of algal-based biosorption systems are:

(i)   Bio-recovery Systems Inc. (Las Cruces, New Mexico) developed a biosorbent based on immobilized, *Chlorella vulgaris* in silica or polyacrylamide gel. This is called AlgaSORB^{TM}, and it efficiently removes metal from concentrations of 100 mg/l. The biosorbent resembles an ion-exchange resin and can undergo more than 100 biosorption-desorption cycles. It can efficiently remove metallic ions from dilute solutions, i.e. 1-100 mg/l, and reduce the concentrations of metal(s) down to 1 mg/l or even below. In addition, the system is not affected by concentrations of elements such as Ca and Mg [62].

(ii)   B.V. Sorbex Inc. (Montreal, Canada) have developed different biosorbents for specific metal recovery processes using all types of biomasses, including the algae *Sargassum natans, Ascophyllum nodosum, Halimeda opuntia, Palmaria palmata, Chondrus crispus,* and *Chlorella vulgaris.* These can work over a range of pH values and solution conditions and can biosorb varied metals without suffering interference from concentration of Ca and Mg. These can be used with a wide range of metal concentrations, not affected by organics, and can be regenerated easily [63].

(iii)   US Bureau of Mines (Golden, Colorado) has produced Bio-fix, a granular biosorbent consisting of a variety of biomasses, including algae, such as *Spirulina,* immobilized in porous poly-propylene beads. The biomass is blended with xanthan gum (derived from pure culture fermentation of glucose with *Xanthomonas campestris*) and guar gum (derived from *Cyamopsis tetragonoloba*) to give a consistent product and immobilized as beads using polysulfone. Zinc binding to this biosorbent is approximately 4-fold higher than the ion exchange resins. Metals can be eluted using HCl or HNO_3, and the biosorbent can be used for more than 120 extraction-elution cycles. The biosorbent is selective for heavy metals over alkaline earth metals, and the beads have been tested extensively for treatment of acid mine wastes [64].

BIOACCUMULATION

In practically all cases, microbial biomass can retain more or less significant amounts of metal ions, by means of 'passive' sorption and/or complex formation. This type of interaction called biosorption is characteristic of both live and dead cells. However, in a number of cases, it is possible to observe *'active concentration'* of metal ions by live cells, which is markedly dependent on the metabolic activity of the organism. Such a type of interaction has been given a generalized name, bioaccumulation.

Many species of algae have been reported to accumulate metals from their aqueous environment. This characteristic is expressed quantitatively by a concentration factor (CF), i.e. the ratio of metal concentration in the organism to that in the surrounding waters. This approach makes it possible to compare the metal avidity of different species and strains of algae. Several investigators have demonstrated that metal removal by algae is probably largely due to chelation and surface adsorption and constitutes an important response of aquatic flora to metal pollution [32,65].

Metal-binding at cellular levels

Accumulation of certain metals has been studied in detail in a few algae [66]. At ultrastructural level, the polyphosphate bodies in *Plectonema boryanum* are said to have the ability to sequester significant amounts of a number of heavy metals. Small amounts of metals, however, are generally sequestered in other parts of the cell. The polyphosphate bodies contain three types of components, which may bind metals. Polyphosphate, because of its very negative surface charge, could bind positive ions. Lipids may also bind metals, although, they are generally not considered to be a major binding site. Proteins are generally the cellular components, which are suggested to be sequestering maximum heavy metals. Rothstein [67] demonstrated that metals interacted with sulfhydryl groups in cell membrane to produce S-metal-S bridges. Hayward [68], Davies [69] and Cossa [70] have shown that the metabolic control of metal uptake is probably due to the indirect result of the variation in protein content of the cytoplasm. One of the most thoroughly studied metal-binding proteins is ferritin. These proteins are generally considered to be the primary metal-binding sites in cells. Thus, polyphosphate bodies, which generally function as phosphorus storage sites or as a source of energy, may also act as sites for storage of metal ions and for detoxification, if metals are present at toxic levels [66].

Studies by Conway and Williams [71] showed that the intracellular soluble fractions in *Chlorella regularis* contained Cd, that was irreversibly bound to metalloenzymes and complexed with amino acids, peptides, and proteins. Work on *Porphyra umblicalis* also suggests a specific intracellular location for bound cadmium, but with this organism the location is apparently the nucleus [72].

Mechanism involved

In most organisms the amount of metals taken up by active processes far exceeds that bound by cell surfaces [73], even though the metal sorption is related mainly to chemical or physical changes in the cell, rather than physiological activity [74]. Thus, it becomes important to study the process of bioaccumulation.

A model for metal uptake by microorganisms was developed, based on surface adsorption. According to this model, the initial uptake of Cd by *Chlorella* occurs in an extremely short-time, rapidly establishing equilibrium between the concentrations of metal adsorbed and metal in solution. At low concentrations, equilibrium is described by a linear relation between these concentrations. Over a longer period, as the algal cell density increases due to growth, it appears that some secondary mechanism, other than adsorption, causes metal uptake to be much greater than predicted by adsorption model. This indicates that a secondary uptake, which is slower, is involved. The kinetics of heavy metal uptake by algal cells thus involves two stages [75]. The first is a very rapid and short lived phase of physico-chemical binding of metal ions, occurring immediately after initial contact with the heavy metal and usually lasting for less than 5 to 10 min. This initial phase is thought to be passive, involving physical sorption or ion exchange phenomena at cell surfaces. The second phase is slow, extended and has been followed for up to 600 h in some algae. It may be separated from the last phase by a lag period and may be linear or hyperbolic in nature. The slow stage is possibly active, being related to some type of metabolic activity of the cell. The relative importance of these two stages depends upon the organism involved. For example, Geisweid and Urbach [76] studied cadmium accumulation in several unicellular green algae and noted a significant difference in sorption processes. The cobalt accumulation in *Chlorella salina* has an initial phase of biosorption in the first five minutes and it is independent of light or the presence of metabolic inhibitors such as KCN, DNP, and CCCP. In contrast, the second phase is clearly dependent on light and could be inhibited by metabolic inhibitors, indicating that it is an active process. This biphasic uptake has been described in many studies with microalgae, macroalgae and cyanobacteria with the metals zinc, cobalt, copper, manganese, cadmium, mercury, nickel, caesium, gold, and magnesium [77,78].

Ankistrodesmus braunii and *Chlorella vulgaris* evidenced very little and slow uptake of cadmium with greater than 80% of the metal being accumulated in the first five minutes of fast phase. With *Eremosphaera viridis*, the slow component contributed significantly to the amount of Cd^{2+} accumulated by the cells, so that 85% of the Cd^{2+} present was probably taken up internally within three hours. In *Spirulina* sp., the sorption process leading to the concentration of gold could be controlled by the action of energy metabolism inhibitors *in vitro* [79].

Little information is available on the genetic mechanism associated with Cd uptake and resistance. Cd resistance in *Euglena gracilis* is indicated to be determined by plasmids, which confer the alga with an ability to control Cd^{2+} uptake [80].

Role of immobilized biomass (living)

Role of immobilized living algae in *bioaccumulation* of metals has been investigated by several workers [81,82]. Such systems could be used for radionuclides and for metal removal from contaminated wastewaters and effluents, as stated in biosorption section. Immobilized cell systems possess several advantages over freely suspended cells in both batch and continuous flow systems. These include better capability of reuse and regeneration of biomass, easy separation of cells from the reaction mixture, high biomass loadings within a given bio-reactor, manipulation of biomass independent of dilution rate and minimal clogging in continuous flow systems. Work involving the use of immobilized algae for removal of metals from solution has mainly concentrated on dead cells. For example, Nakajima et al. [83] used polyacrylamide-entrapped *Chlorella vulgaris* to remove UO_2^{2+}, Au^{3+}, Cu^{2+}, Hg^{2+}, and Zn^{2+}, while Darnall et al. [84] used *C. vulgaris* cells immobilized in a silica gel matrix for removal of Al^{3+}, Be^{2+}, Cu^{2+}, Pb^{2+}, Ni^{2+}, Zn^{2+}, Cr^{3+}, CO^{2+}, Fe^{2+}, UO_2^{2+}, Ag^+, Ag^{2+}, and Mn^{2+} from solutions. Such studies have shown that immobilized dead algae can accumulate at least as much metal as freely suspended dead cells. However, relatively few studies have used living cells in immobilized systems for metal removal. One such study has been carried out on *Chlorella salina*. The algae immobilized in calcium-alginate microbeads accumulated significantly more cobalt, manganese and zinc than free living cells. This is of interest not only with respect to biological systems for removal of metals from contaminated effluents, but also in relation to accumulation of metals by algae in the natural environments. In aquatic habitats, algae are not only always free floating, but also a large number typically occur in clumps and biofilms attached to solid substrates, a condition perhaps analogous to algae immobilized in beads. The metal accumulation by immobilized *C. salina* is considerably greater than that of free cells. Therefore, it is possible that immobilized algae may have an important and

underestimated role in cycling of metals and radionuclides in aquatic habitats
[61] (Table 2).

MULTI-ION SITUATION

While much research has been carried out on the uptake of single species
of metal ions by microbes, little attention seems to have been given to study
of multimetal ion systems (Table 5). The knowledge of the toxicity of single

Table 5
Multimetal interactions with algae

Metal ion	Parameter	Effect	Algal species
Cu/Fe	Growth	A	*Chlorella pyrenoidosa*
Cu/Cd	Cell number	A	*Selenastrum capricornutum*
Cu/Ni	Uptake	S	*Scenedesmus acutiformis*
Cu/Ni	Cell number	S	*C. vulgaris*
Cd/Se	Cell number	A	*C. vulgaris*
Cu/Ni	Cell number	S	*Haematococcus capensis*
Cu/Zn	Cell number	S	*Amphidinium carteri*
Cu/Zn	Cell number	S	*Thalassiosira pseudonana*
Cu/Zn	Cell number	A	*Phaeodactylum tricornutum*
Cu/Zn	Cell number	S	*Skeletonema costatum* (Skel-0 and Skel-5)
Zn/Cu	Growth	S	*Hormidium rivulare*
Cd/Zn	Uptake	A	*Chlorella fusca*
Mn/Pb	Cell volume	A	*S. capricornutum*
Cu/Pb	Cell volume	A	*S. capricornutum*
Mn/Cu	Cell volume	S	*S. capricornutum*
Hg/Cd	Growth	A/S	*Anabaena inaequalis*

Table 5 continued

Metal ion	Parameter	Effect	Algal species
Hg/Cd	$^{14}CO_2$ uptake	S	*A. inaequalis*
	Nitrogenase activity	S	*A. inaequalis*
Hg/Ni	Growth	A/S	*A. inaequalis*
	Nitrogenase activity	A/S	*A. inaequalis*
Ni/Cd	Growth	A	*A. inaequalis*
	$^{14}CO_2$ uptake	A	*A. inaequalis*
	Nitrogenase activity	A	*A. inaequalis*
Hg/Cd/Ni	Growth	A	*A. inaequalis*
Hg/Cd/Ni	$^{14}CO_2$ uptake	S	*A. inequalis*
	Nitrogenase activity	S	*A. inaequalis*
Zn/Cd	Uptake	A	*P. tricornutum*
Zn/Cd	Uptake	A	*S. costatum* (Skel-0)
Zn/Cd	Uptake	S	*S. costatum* (Skel-5)
Cd/Ca	Growth	A	*C. pyrenoidosa*
Cd/Fe	Growth	A	*C. pyrenoidosa*
Cd/Mn	Growth	S	*C. pyrenoidosa*
Cd/Zn	Growth	S	*C. pyrenoidosa*
Zn/Fe	Dissolved oxygen	V	*C. pyrenoidosa*
			Scenedesmus carinatus
			Ankistrodesmus convulutus
			Aphanizomenon flos-aquae

Abbreviations: A: antagonism; S: synergism; A/S: antagonism and synergism; V: various

heavy metals to algae is interesting and has been well investigated in experimental systems. However, the aquatic habitats, which need amelioration, are usually polluted with several metal compounds. The multiplicity of metals at such sites often gives rise to interactive effects. In general, a mixture of heavy metals can produce three possible types of behaviour: synergism, antagonism, and non-interaction. These terms may be defined as follows:

Synergism : The effect of the mixture is greater than that of each of the individual effects of the constituents in the mixture.

Antagonism : The effect of the mixture is less than that of each of the individual effects of the constituents in the mixture.

Non-interaction : The effect of the mixture is no more or no less than that of each of the individual effects of the constituents in the mixture.

Each of these kinds of responses has been shown by a number of microorganisms, including algae (Table 6). Under situations of multiple ions, factors that affect the uptake include environmental conditions, such as pH and temperature. Uptake also depends upon species of algae, metal combination, levels of metal concentration, order of metal addition, and test criterion.

Though insufficient attention seems to have been paid to the problem of multicontrol uptake, yet several reasons have been advanced for the synergistic or antagonistic actions. In *Chlorella vulgaris*, combined effect of copper and nickel on cell population was found to be synergistic, which results probably from an increase in membrane permeability of algal cells [85]. In another instance in the same alga, the long term uptake of Zn is impeded by the presence of Cd. Increasing the concentration of the competing Cd ions leads to correspondingly less intercellular uptake of Zn. The presence of Cd ions appears to promote the reverse carrier reaction in the membrane transport of Zn, although Cd transport is not influenced by the presence of Zn ions. The difference in the behaviour of these two metals is likely to be due to the essential and nonessential nature of Zn and Cd ions respectively for cellular metabolism. Antagonistic interaction was reported in *Chlorella vulgaris* in the case of cadmium and selenium. It was suggested that screening or competition for the binding sites on the cellular surfaces had resulted in the metal ions mutually ameliorating their individual toxic effects [85].

TACKLING GLOBAL WARMING

Besides pollutants like metals, pesticides, chemicals, and radionuclides, environmental issues like global warming can also be addressed through bioremediation, using algal systems. Since algae are the largest group in terms of photosynthesis and biomass, they are considered most appropriate reservoirs to 'fix' atmospheric carbon dioxide. The technology has the potential to reverse global warming. The thrust is on bioremediation systems to remove CO_2 emitted into the atmosphere from the burning of carbon-based fossil fuels. The increased CO_2 contributes to the global warming and the removal of the 'extra' CO_2, polluting the atmosphere, would be a big achievement in the field of bioremediation. Theoretically, microorganisms can remove enough of this greenhouse gas from the atmosphere to avert threat of global warming. Some microorganisms are known to convert CO_2 to various organic compounds, and new improved strains could be produced, either by *genetic selection* from cultures grown in bioreactors, or by the use of *recombinant-DNA technology*. In future, it may be possible to earmark some portions of the sea as algal sanctuaries, where some of the carbon may be 'fixed' in the biomass of giant marine algae.

Attempts have been made at Marine Biotechnology Laboratory at Kamaishi, Japan to isolate algae that convert CO_2 to carbohydrate 10 times faster at 40°C than terrestrial green plants can. Rapidly growing alga *Chlorococcum littorale* in a continuously illuminated bioreactor outside the institute can tolerate high levels of CO_2, which is converted to polysaccharides. At the same institute, a second alga *Prasinococcus capsulatus* was observed to produce large amounts of extracellular mucilaginous polysaccharides. This microbe is being studied both for its potential to remove CO_2 from the atmosphere and for commercial applications of the polysaccharides.

It is believed that such algae could be grown in bioreactors near power plants where these would remove CO_2 from the atmosphere as it is released by fossil fuel burning at the power plant. However, the biggest drawback is that these algae convert the CO_2 into organic compounds that would subsequently be biodegraded to CO_2 and H_2O, releasing the CO_2 back into the atmosphere.

A successful bioremediation process will depend on finding microorganisms that produce lignins or other polymeric compounds that are resistant to biodegradation. Such conversions would immobilize the carbon, reducing the build up of CO_2 in the atmosphere. Microorganisms, particularly marine organisms called foraminifera, can convert carbon from CO_2 into calcium carbonate. This process occurs in coral reefs in which algae are known to have symbiotic relationship with animals. Formation removes carbon from the normal *biogeochemical carbon cycle*, that circulates carbon between organic compounds and CO_2. Such treatment, however, would consume enormous quantities of calcium, creating its own impact on environment [25].

Limitations

Bioremediation, at present level of development of technology, has its own limitations. The main limitation is the inability of microorganisms to adequately attack everything that is released into the environment. The researchers have proved that the DDT (a widely used pesticide) and PCBs (polychlorinated biphenyls) are some of the chemicals that are resistant to microbial attack. As a consequence, the bioaccumulation of these chemicals occurs in organisms at higher levels of ecological food web.

Another limitation is posed by the biomass itself. Although immobilized microalgae offer considerable potential for varied applications, they suffer from poor cell retention. Such cell leakage on a commercial scale could cause severe problems. In addition, many of the immobilized systems, e.g., alginate or polyacrylamide [83] tend not to be very robust and fall apart under certain conditions, many of which are found in industrial environments. Another practical limitation of the systems utilizing actively growing algae is that the growth is inhibited by high metal concentrations or when significant amounts of metal ions are biosorbed by algae. Such systems tend to perform best in warm, sunny climates where algal growth is encouraged [86]. Thus, complexity of *bioremediation technologies* requires that each treatment be adapted to the individual situation of the contaminated site. Therefore, prior to treatment of a contaminated site, a fundamental analysis of the biological, geological and hydro-geological facts is essential for successful realization of bioremediation technology.

Further, as yet, only laboratory-scale processes, using relatively small amounts of algae, have been developed and hence actual cost incurred on the proposed clean-up system has not been assessed. Cost effectiveness is important if algal bioremediators are to be used in green technology, as any other algal resource (like carrageenan and alginic acid) being used in industry, based on these plants. The most important point that needs to be understood is the acceptance of the technology by people all over the world. Since bioremediation is an innovative biological technology, engineers may not feel quite comfortable with it. They could perhaps relate it better to the conventional and tried chemical systems. Hence, to fully exploit this new way to clean-up the environment, it will be necessary to adopt a multidisciplinary approach in which chemists, biologists, and engineers work together.

Genetic engineering for 'Designer-Algae'

Since algae hold an important position amongst microorganisms being used in bioremediation, there is a need for improving the quality of algal biomass. To achieve this goal, gene technology is a promising option. One area, which needs to be investigated extensively, is the ultrastructural organization of the cells, particularly the organelles involved in sequestration and accumulation of heavy metals. For example, in cyanobacteria, this function is performed by polyphosphate bodies that possess the metal-binding protein, ferritin. Genetic regulation of metal-binding proteins needs to be elucidated.

As the biosorption process involves mainly cell *surface sequestration*, cell wall modification can greatly alter the binding of metal ions. A number of methods have been employed for modification of walls of microbial cells, in order to enhance the metal-binding capacity of biomass and to elucidate the mechanism of biosorption. These modifications can be introduced, either during the growth of the microorganism, or in the pregrown biomass. The latter could be given several physical and chemical treatments to tailor the metal-binding properties of biomass to specific requirements. The physical treatments include heating/boiling, freezing/thawing, drying and lyophilization. Various chemical treatments include washing the biomass with detergent, cross-linking with organic solvents and alkali or acid treatment. Some work has already been done on chemical treatment of biomass of *Sargassum fluitans* and *Ascophyllum nodosum* [4]. However, more research is required in this direction. Attempts must also be made to produce metal-binding proteins cheaply enough or in sufficiently large amounts for a viable removal process. The genetic basis for metal resistance, uptake mechanisms and tolerance also need to be worked out. Some success has been achieved with Cyanobacteria, where the role played by plasmid in Cd-resistance has been indicated. However, the genetic basis of such a phenomenon in macroalgae is still mysterious. The metal resistance genes could be transferred in this group through genetic engineering, giving a boost to biomass from macroalgae.

Working out physiological and metabolic pathways as well as the genetics of algae will enable us to develop methodologies, allowing evaluation of concentration of free metal ions and metal complexation. Besides, there is also a need to understand how combinations of metal ions affect the physiological, biochemical and ecological process of various algae.

This seemingly difficult task of catering to field conditions through 'designer algae' can be handled with ease through gene technology. It is believed that some instances of pollution can be readily bioremediated, using the existing technology. However, those involving toxic and chemically stable compounds, such as polychlorinated biphenyls (PCBs) or chlorinated dibenzo-p-dioxins, require the development of new and innovative technologies that involve designing new pathways. It may be worthwhile mentioning here that certain microbes show inadequate catabolic potential, leading to persistence of metals in the environment. Thus, if a productive metabolic route within an organism or community for a pollutant is not known, or known routes are ineffective, it would be rewarding to design and create a new effective pathway (with improved catabolic pathways). The work has already started in a few places and the results are being watched with interest.

Some institutes, for example, the University of Minnesota Biocatalysis/ Biodegradation Database would soon offer a systematic display of theoretical routes from one substrate to a specific intermediate or central metabolite. This will, in principle, not only permit multiple options to be considered and tested but also expose truly novel possibilities [87]. Hence, 'designer algae' may be tailored in the laboratory with bioremediation potential for natural or field conditions. Apart from this, improved strains of algae may be produced through mutation research or recombinant-DNA technology. Algae with an ability to take up 'extra' CO_2 from the atmosphere and convert it into polymers, which immobilize carbon, would give a great impetus to bioremediation technology.

Future of bioremediation

Based on the estimates obtained from a variety of sources, it is estimated that clean-up of hazardous wastes by conventional technologies may cost at least $ 400 billion in US alone. Sites contaminated with heavy metals can cost $ 7.1 billion, while mixtures of heavy metals and organics bear an additional $ 35.4 billion price tag. However, the clean-up of contaminated sites that have been identified and characterized to date will cost over $ 10 billion, using current treatment technologies. This overpowering cost burden has opened a path to the market place for innovative technologies like bioremediation.

Eccles and Holroyd [64], made a cost assessment of biosorption processes as compared to chemical systems. The cost of the *Bio-fix process* (partially based on algal biomass) was compared with lime precipitation for treating a waste containing zinc, manganese and cadmium in mg/l with a pH of 6.9. A three-column packed-bed system of Bio-fix beads was envisaged. The first column would be for removal of most metals, the second would scavenge for residual metals, and the third would be eluted and the metals recovered. At the end of each loading cycle, column one would be eluted, column two would become the new lead column and the previously eluted column would become the scavenger. When the cost of this system was compared with lime precipitation, it was found that costs of the two systems were similar per 1000 gallons of waste treated. However, the income from the possible sale of recovered metals from the Bio-fix and the increased cost of lime precipitation, when the waste liquor contains iron, were not considered.

Since most of the bioremediation systems use algal biomass and procuring it is not difficult, the cost incurred is expected to be low. In many places where large amounts of macroalgal growth has occurred on beaches and in the sea due to *eutrophication* caused by pollution, the macroalgae is considered a 'nuisance'. A large amount of money is spent in removing such algae from tourist beaches, and the subsequent transport and disposal of the algal biomass is expensive. Such algae could easily be used in bioremediation. In the long run, through location-specific research, it may be possible to utilize wastewater bodies for growing a mix of algal species in an equilibrium, and harvesting a certain amount of biomass on a periodical basis. With improvement in metal extraction technologies, it may also become feasible and economical to extract a variety of metals and other products from the harvested biomass. The success of any bioremediation technology will depend upon its ability to yield profit, or atleast to pay for itself.

The wealth of information and studies on such algal-based processes and, in many cases, the availability of abundant and relatively inexpensive algal biomass for use in future biosorption processes is tremendous. Despite technical and conceptual difficulties, this research effort may succeed in attacking global pollution problems.

CONCLUSIONS

Algae, which include both prokaryotic and eukaryotic organisms, constitute a diverse group of plants found in freshwater and marine ecosystems. Considerable variation is seen in their cellular structure, cell wall composition, pigmentation, food reserves, and reproduction. They account for more than half the total primary productivity on earth. Most of the aquatic animals depend on food chains, which begin with these phototrophs. While algal growths may lead to discoloration of lakes, reducing their appeal for recreational use, their significance in changing global environment is undebatable.

Many algae are used as food, fodder, fertilizers and as a source of hydrocolloids. Their utility as potential bioremediators in aquatic ecosystems is being assessed and algal application in 'green movement' is under investigation the world over. Resistance of algae to metal ions, their metal-uptake ability and capacity to accumulate dissolved metals have made it possible to have algal-based systems that can clean-up water bodies of heavy metal and radionuclide pollution. Various types of algae, living, dead or immobilized, and algal-derived products are used in bioremediation processes like biosorption and bioaccumulation. The cell walls are the main sites for metal-binding in cyanobacteria (blue green algae), micro- and macroalgae. Because of their photosynthetic potential, algae are also implicated in removing 'extra' atmospheric CO_2 and, thus, check the global warming.

World wide surveys indicate that the water bodies all over have been converted into dumping sites for all kinds of pollutants. The problem worsens when these polluted waters are put to uses, like irrigation. This sets in a chain of reactions, which ends up in *biomagnification*. Pollutants are also known to affect the growth of primary producers adversely, thus breaking the food chain at the first trophic level or at grass-root level. However, the ease with which various algae can be cultured and the variety of biomasses rendered by them as living, dead or immobilized, have made algae one of the most potent bioremediators. Though the bioremediation paradigm using algae to degrade pollutants *in situ* has lately attracted a lot of public attention, yet introducing the *'genetically engineered'* algae into the environment to enhance the process is yet to be demonstrated with success.

REFERENCES

1   Round FE. The Biology of The Algae London: Edward Arnold Ltd., 1965.

2   Hassett JM, Jennett C, Smith JE. Appl Environ Microbiol 1981; 41: 1097-1106.

3   Fernandez-Pinas F, Mateo P, Bonilla I. Arch Environ Contam Toxicol 1991; 21: 425-431.

4   Leusch A, Holan ZR, Volesky B. J Chem Tech Biotechnol 1995; 62: 279-288.

5   Fritsch FE. The Structure and Reproduction of The Algae, Vol. I London: Cambridge University Press, 1945.

6   Whittaker RH. Science 1969; 163: 150-160.

7   Stainer RY. Carlsberg Res Comm 1977; 42: 77-98.

8   Parker SP. (ed.). Synopsis and Classification of Living Organisms. Vol. 1&2, New York: McGraw-Hill, 1982.

9   Kingman AR, Moore J. Botanica Marina 1982; 25: 149-154.

10   Featonby-Smith BC, van Staden J. Botanica Marina 1984; 27: 527-531.

11   Metting B. Developments in Industrial Microbiology 1990; 31: 265-270.

12   Singh RN. Role of Blue Green Algae in Nitrogen Economy of Indian Agriculture, New Delhi: ICAR, 1961.

13   Hoppe HA. In: Hoppe HA, Levring T, Tanaka Y, eds. Marine Algae in
        Pharmaceutical Science. Berlin: Walter de Gruyter, 1979; 25-119.

14   Nisiwaza K. In: Hoppe HA, Levring T, Tanaka Y, eds. Marine Algae in
        Pharmaceutical Science. Berlin: Walter de Gruyter, 1979; 243-264.

15   Mathur HH. In: Seaweed Research and Utilization Association's Silver
        Jubilee Celebrations, 2-6 Sept., 1994; 15-18.

16   Chapman VJ, Chapman DJ. Seaweeds and Their Uses. London:
        Chapman and Hall 1980.

17   Michanek G. Ceres 1981; 14: 41-44.

18   Bonotto S. In: Hoppe HA, Levring T, Tanaka Y, eds. Marine Algae in
        Pharmaceutical Science. Berlin: Walter de Gruyter, 1979; 121-137.

19   Gross R, Gross U, Ramirez A, Caudra K, Collazos C, Berlin FW. Arch
        Hydrobiol Beih 1978; 11: 161-173.

20   Radmer RJ. Bioscience 1996; 46: 263-270.

21   Stein JR, Borden CA. Phycologia 1984; 23: 485-501.

22   Lee RE. Phycology. Cambridge: Cambridge University Press, 1989.

23   Townsin RL, Byrne D, Milne A, Svensen T. Speed, Power and
        Roughness: The Economic of Outer Bottom Maintenance. Trans R.
        INA. 123, 1981.

24   Myern HC. J Amer Water Wks Assn 1947; 39: 322-324.

25   Atlas RM. Bioremediation. C & EN Special report, 1995; 32-42.

26   Whitton BA. Phykos 1970; 9: 116-125.

27   Morris AW, Bale AJ. Estu Cstl Mar Sci 1975; 3: 153-163.

28   Lackey JB. Lectures Presented at Inservice Training Course in Sewage
        and Industrial Waste Disposal (Mimegraphed), 1948; 109-118.

29   Clausen HT. Jour Proc Institute Sewage Purif 1959; 3: 345-348.

30   Silva PC, Papenfuss GF. California State Water Pollution Control
        Board, Publ. 7. 1953; 34.

31   Palmer CM. Algae and Water Pollution: The Identification, Significance and Control of Algae in Water Supplies and in Polluted Water, Castle House Publications Ltd., 1980.

32   Ting YP, Lawson F, Prince IG. Biotechnol Bioeng 1989; 34: 990-999.

33   Krishnamurthy CR, Viswanathan P. eds. Toxic Metals in the Indian Environment, New Delhi: Tata McGraw-Hill Publishing Company Ltd, 1991.

34   Nriagu JO, Pacyna JM. Nature 1988; 333: 134-139.

35   Vymazal J. In: Toxicity Assessment, An International Quarterly, Vol. 2, New York: J Wiley & Sons, 1987; 387-415.

36   Li WKW. J Phycol 1978; 14: 545-560.

37   Thomas WH, Hollibaugh JT, Seibert DLR. Phycologia 1980; 19: 202-209.

38   Muraleedharan TR, Iyengar L, Venkobachar C. Curr Sci 1991; 61: 379-385.

39   Ahuja P, Gupta R, Saxena RK. Current Microbiology 1997; 35: 151-154.

40   McKnight DM, Morel FMM. Limnol Oceangr 1980; 25: 62-71.

41   Trick CG, Anderson RJ, Gillam A, Harrison PI. Science 1983; 219: 306-308.

42   Trick CG, Anderson RJ, Price NM, Gillam A, Harrison PI. Mar Biol 1983; 75: 9-17.

43   Hammouda O, Gaber A, Abdel-Raouf N. Ecotoxicology and Environmental Safety, 1995; 31: 205-210.

44   Gale NL, Wixon BG. Proceedings of the International Conference on Management and Control of Heavy Metals in the Environment, Edinburgh: CEP Consultants Ltd, 1970; 580-583.

45   Gupta R, Ahuja P, Khan S, Saxena RK. Curr Sci 2000; 78: 967-973.

46   Da Costa ACA, Leite SGF. Biotechnology Letters 1992; 12: 941-944.

47   Bedell GW. In: Volesky B, ed. Biosorption of Heavy Metals. Boca: Raton, Florida, CRC Press, 1990; 360-368.

48   Nakajima A, Horikoshi T, Sakaguchi T. Eur J Appl Microb Biotechnol 1981; 12: 76-83.

49   Carvalho De RP, Chong KH, Volesky B. Biotechnology Progress 1995; 11: 39-44.

50   Chong KH, Volesky B. Biotechnol Bioeng 1995; 47: 451-460.

51   Darnall DW, Greene B, Hosea M, McPherson RA, Henzl M, Alexander MD. In: Thompson R, ed. Trace Metal Removal from Aqueous Solutions. Whitstable, Kent: Litho Ltd, 1986; 1-24.

52   Robinson PK, Mak AL, Trevan MP. Process Biochemistry 1986; 122-127.

53   Tsezos M. Biotechnol Bioeng 1984; 26: 973-981.

54   Crist RH, Oberholser K., Shank N., Nguyen M. Environ Sci Technol 1981; 15: 1212-1217.

55   Stary J, Kratzer K. Toxicol Environ Chem 1984; 12: 67-71.

56   Shumate SE, Strandberg GW. In: Robinson CW, Howell JA, eds. Comprehensive Biotechnology, Vol. 4, London: Pergamon Press, 1985; 235-247.

57   Greene B, McPherson R, Darnall DW. In: Patterson J, Pasino R, eds. Metal Speciation and Recovery, Michigan: Lewis 1987; 315-332.

58   Matheickal JT, Yu Q, Feltham J. Environ Technol 1997; 18: 25-34.

59   Williams CJ, Edyvean RGJ. Biotechnology Progress 1997; 13: 424-428.

60   Greene B, Darnall DW. In: Ehrlich HL, Brierly CL, eds. Microbial Mineral Recovery, New York: McGraw-Hill, 1990; 227-302.

61   Garhnam GW, Codd GA, Gadd GM. Environ Sci Technol 1992; 26: 1764-1769.

62   Kuyucak N. In: Volesky B, ed. Biosorption of Heavy Metals, Boca: Raton, Florida, CRC Press, 1990; 372-377.

63   Volesky B. In: Volesky B, ed. Biosorption of Heavy Metals. Boca: Raton Florida, CRC Press 1990; 7-44.

64   Eccles H, Holroyd C. Proceedings of SCI Conference on Biological Removal of Toxic Metals, Preston, 1993; 1-2.

65   Gadd GM. Biosorption Chemistry and Industry 1990; 13: 421-426.

66   Jensen TE, Baxter M, Rachlin JW, Jani V. Environ Pollut Ser A. Ecol Biol 1982; 27: 119-127.

67   Rothstein A. Fedn Proc Fedn Am Soc Exp Biol 1959; 18: 1026-1028.

68   Hayward J. J Mar Biol Ass UK. 1969; 49: 439-446.

69   Davies AG. J Mar Biol Ass UK. 1974; 54: 157-169.

70   Cossa D. Mar Biol 1976; 34: 163.

71   Conoway HL, Williams SC. Journal of Fish Reserves Board of Canada 1979; 36: 579-586.

72   McLean MW, Williamson FB. Physiol Plant 1977; 41: 268-272.

73   Gadd GM, Griffiths AJ. Microb Ecol 1978; 4: 303-317.

74   Bollag JM, Duszota M. Arch Environ Contam Toxicol 1984; 13: 265-270.

75   Khummongkol D, Canterford GS, Fryer C. Biotechnol Bioeng 1982; 4: 22643-22660.

76   Geisweid HJ, Urbach W. Z Pflanzenphysiol 1983; 109: 127-141.

77   Gadd GM. In: Rehm HJ, ed. Biotechnology - Special Microbial Processes, Vol. 6B, Weinheim: VCH Verlagsgellschaft, 1988; 401-433.

78   Kuyucak N, Volesky B. In: Volesky B, ed. Biosorption of Heavy Metals, Boca: Raton, Florida, CRC Press, 1990; 173-198.

79   Karamushka VI, Gruzina TG, Ul'berg ZR. Microbiology 1995; 64: 157-160.

80   Bariaud A, Mestre JC. Bull Environ Contam Toxicol 1984; 32: 597-601.

81   Macaskie LE, Dean ACR Enzyme Microb Technol 1987; 19: 2.

82   Macaskie LE. CRC Crit Rev Biotechnol 1992; 11: 41.

83   Nakajima A, Horikoshi T, Sakaguchi T. Eur J Appl Microb Biotechnol
      1982; 16: 88-91.

84   Darnall DW, Greene CB, Hosea JM, McPherson RA, Henzl M, Alexander
      MD. In: Thompson R, ed. Trace Metal Removal from Aqueous Solution.
      London: Royal Society of Chemistry, 1989; 1-26.

85   Hutchinson TC, Stokes PM. Water Quality Parameters ASTM 1975;
      53: 320.

86   Wong PK, Chan KY. Agriculture Ecosystems Environment 1990; 30:
      235-250.

87   Timmis KN, Pieper DH. TIBTECH 1999; 17: 201-204.

Biotransformations: Bioremediation Technology for Health and Environmental Protection
V.P. Singh and R.D. Stapleton, Jr. (Editors)

Some physiological characteristics of saprotrophic and ectomycorrhizal fungi producing sporophores on the urea-treated forest floor

T. Yamanaka

Soil Microbiology Laboratory, Forestry and Forest Products Research Institute, MAFF, P.O. Box 16, Tsukuba Norin Kenkyu Danchi-nai, Ibaraki 305-8687, Japan.

INTRODUCTION

Reports on changes in occurrence of fungal fruit-body flora following treatment of forest soil with chemicals or fertilizers are documented with orchard, coniferous, and deciduous forests [1-8]. According to the reports, changes in occurrence of fungal species after chemical treatment depend on vegetation, the intensity of treatment, and the type of substances applied. Changes in occurrence of fruit-body could be used as an indicator of responses of forest ecosystem to environmental changes.

Sagara and Hamada [2] and Sagara [9] applied a wide variety of substances to forest soils in Japan to study possible effects of these chemicals on fungi. They found that a particular group of fungi sporulated or fruited on the soil after application of urea, ammonia water, or some ammoniacal substances that produced ammonium ions with alkaline condition. These fungi were, therefore, called as *ammonia fungi* [9]. The number of the species belonging to this group is about 40 [10]. The amounts of the nitrogenous substances applied to the soil were generally large. For example, urea was applied to the soil at the rate of 80-320 g N/m^2 in this study [9] while the amounts of urea for fertilization are usually 15-60 g N/m^2 [11-15]. It means that these nitrogenous substances applied are not for forest fertilization but for causing soil disturbances [16]. Therefore, a relatively small area (0.5 to 5 m^2) was used for this kind of experiment [9]. The occurrence of the ammonia fungi is divided into early phase and late phase [10,16]. In the early phase, about one to three months after application of nitrogenous compounds, mitosporic fungi (e.g., *Amblyosporium*), cup fungi (e.g., *Peziza* and *Ascobolus*), and smaller agarics (e.g., *Tephrocybe* and *Coprinus*), all saprotrophic, sporulate for a short period of time (Figure 1). In the late phase, larger agarics (e.g., *Hebeloma* and *Laccaria*), most of which are ectomycorrhizal, occur in the mushroom season for several years [16,17].

Figure 1. Fruiting of ammonia fungi after soil treatment with urea. a, *Coprinus echinosporus*; b, *Hebeloma radicosoides*.

In this article, I review some physiological characteristics of the ammonia fungi and the environmental conditions under which the ammonia fungi fruit or sporulate. I begin with the reports on the occurrence of the ammonia fungi in countries other than Japan and their occurrence under natural conditions in order to show the ubiquitous characteristics of this particular fungal group.

Ammonia fungi in the countries other than Japan

Lehmann [6] reported the occurrence of the following ammonia fungal species on the soil treated with urea in the United Kingdom: *Ascobolus denudatus, Tephrocybe tesquorum*, and *Coprinus echinosporus*, although late stage species were not noted in his study. In 1986 at U.K., Sagara [16] confirmed the Lehmann's results and added the following species to those already reported: *Doratomyces putredinis, A. botrytis, Rhopalomyces elegans* var *minor, Peziza moravecii,* and *T. mephitica*. In U.S.A., Sagara [16] reported *A. denudatus* and *D. putredinis* on urea-treated forest litters in the laboratory. In Taiwan, Wang and Sagara [18] obtained *P. urinophila* on the urea-treated forest litters.

In U.K., Hora [19], Khan [20], and Hora and Khan [21] obtained *T. tesquorum, C. echinosporus, T. mephitica* and *P. natropholia* after treatment of forest litters with K_2CO_3 or Na_2CO_3. Afterwards, Sagara [16] added *D. putrdinis, R. elegans* var *minor, A. botrytis, C. laanii, P. moravecii, A. hansenii,* and *Iodophanus carneus* to the list prepared by Hora [19]. In Denmark, treatment with K_2CO_3 produced *A. denudatus, Coprinus* sp., *T. ambusta , T. tesquorum, P. palustris,* and *Peziza* sp. [4]. Thus, the ammonia fungi have been observed after the urea treatment of forest litters in countries other than Japan. These observations indicate that this fungal group exists latently as spores or hyphal fragments in the forests with various types of vegetation and climate, and is likely to play an important role in nutrient dynamics under the conditions of high ammonium concentrations resulted from decomposition of waste materials. This ubiquitous group of fungi, therefore, may be one of the indicators of a healthy forest ecosystem [2].

The ammonia fungi under natural conditions

Sagara [9,10,16] found fruiting or sporulation of the ammonia fungi after decomposition of urine and/or dung of mammals, after decomposition of faeces of wasps, or after decomposition of dead-bodies of animals. Therefore, when the ammonia fungi are taken as an ecological group, the term *post-putrefaction* fungi [10] was used for implying natural habitats, that is, the sites in which faeces or dead-bodies of animal had been decomposed. Some of the ammonia fungi, that fruit at the early stage, belong to *coprophilous fungi* [9]. *Amblyosporium botrytis* and *T. tesquorum* were reported to occur on decayed or decaying fleshy fungi [22,23].

Hebeloma aminophilum forms fruit-bodies after decomposition of dead-bodies of animals [24] and *H. syrjense* fruits on decaying corpse [25], although these *Hebeloma* species have not been obtained after experimental application of nitrogenous substances.

Hebeloma radicosum and *H. spoliatum* form fruit-bodies on moles deserted middens (laternes) near their nests [10], where a tripartite relationship among animal, fungus (*Hebeloma*) and plant establishes [10]. In this relationship, the plants provide waste cleaning (by ectomycorrhiza) for the animal, and energy and other compounds to the fungi. The animals provide niches and nutrients to the plants and fungi. The fungi provide nutrients, growth regulators and so on, to the plants, and waste cleaning (by ectomycorrhiza) for the animals [10]. Therefore, this tripartite relationship may be viewed as a habitat-cleaning symbiosis [26].

CHANGES IN SOIL PROPERTIES AFTER UREA TREATMENT

Ammonium- and nitrate-nitrogen concentrations, pH, and water content

In the early phase after addition of urea at a rate of 700 g urea/m^2, the amount of ammonium-nitrogen in soil was 16-6.2 mg N/g dry soil, which was 840-190 times higher than that in the untreated soil (Figure 2). During the late phase, ammonium-nitrogen gradually decreased and nitrate-nitrogen temporarily increased to 0.3 mg N/g dry soil; 180 times higher than that in the untreated soil [17,27,28].

Soil pH increased to 8.0 seven days after addition of urea, and remained at 7-8 during the early phase, then declined to 4-5 in the late phase [17]. Soil water content also increased after urea treatment. Afterwards, the content decreased gradually to the level lower than that of the control [17,27,28].

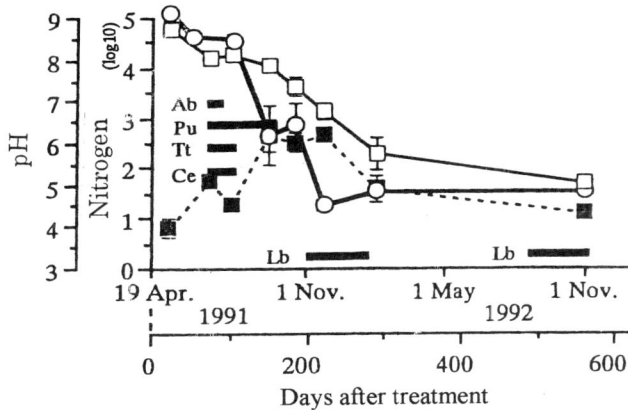

Figure 2. Changes in pH value and the concentration of ammonium- and nitrate-nitrogen in A_{01} layer after urea treatment at the rate of 700 g urea to 0.5 × 2 m plot. □, ammonium-nitrogen; ■, nitrate-nitrogen; ●, pH value. The fungal occurrences are shown within the figure: Ab, *Amblyosporium botrytis*; Pu, *Peziza urinophila*; Tt, *Tephrocybe tesquorum*; Ce, *Coprinus echinosporus*; Lb, *Laccaria bicolor*

Soil organic matter

The growth of ammonia fungi and other soil organisms after treatment with nitrogenous materials change soil organic matter quantitatively and qualitatively. The procedures of Schlesinger and Hasey [29] were adopted to analyze organic matter components of lignin, holocelullose, 50% (v/v) methanol extract, water soluble carbohydrates (WSCs) and water soluble phenolics (WSPs), petroleum ether extract and ash in urea-treated soil [27].

After urea treatment, the methanol extract increased, but subsequently decreased to the lower level as compared to the control [27]. The concentrations of WSC and WSP in the treated soil were lower than those of the control. Contents of other types of the organic matter did not change consistently after the treatment. WSC in soil decreased after urea treatment [27]. However, enzyme activities of the ammonia fungi break down high-molecular weight substances, allowing them to acquire the carbon sources. The enzymatic activities of the ammonia fungi are described later.

Upon urea treatment, the soil turned black and water extract of the soil changed from yellow to brown [9,17]. This change of the soil color was rapid after ammonia water was added to the forest soil in the laboratory [30]. When the ammonia water at pH 6 or 4 was added to the soil, color of the soil extract did not change. When buffer solution, adjusted to an alkaline pH range, was added to a forest soil, color of the soil extract changed from brown to black with concomitant increase of initial pH in the buffer solution. These observations show that the soil organic matters are dissolved partly in water under the alkaline condition stemmed from an increase of ammonium nitrogen. According to Stevenson [31], soil organic matters become soluble under alkaline conditions as acids groups are converted to soluble salts. And ammonium ion forms the soluble salts with humic acid and fulvic acids [32].

Microfungi

Ammonia fungi have been recognized by their fruiting or sporulation. Therefore, fungi other than the ammonia fungi might grow only vegetatively without fruiting or sporulating in the soil on which the ammonia fungi appeared. Furuya [33] isolated fungi from the urea-treated soil using the dilution plate technique. He obtained 15 species of fungi, most of which have not been reported as the ammonia fungi. The following species were predominantly isolated from the urea-treated sites in six different locations: *Kernia retardata*, *Lophotrichus barteletii*, *Microascus manginii,* and *Petriella setifera*. These fungi can affect the growth of the ammonia fungi in the treated soil. However, the information about the effects of these fungi on developmental patterns of the ammonia fungi is lacking.

Other soil-microorganisms

Soon after treatment of urea, high levels of ammonia released from decomposition of urea kill indigenous soil-organisms [28]. Afterwards, some soil-organisms such as bacteria and nematodes increase in the forest soil after urea treatment [17,28]. The number of bacteria in the early stage after urea treatment is well correlated with the amount of ammonium nitrogen in the urea-treated soil [17,28]. This suggests that nitrogen may be immobilized primarily by bacteria, and then released via excretion or from the decomposed bodies. Thus, these bacteria are important in nitrogen dynamics in the soil after urea treatment. The nematodes, increased after urea treatment, were bacteria feeders [17], and bacterial populations affect development of nematode population. *Rhopalomyces elegans* var. *minor,* obtained after urea treatment in U.K. [16], is noted with the nematode population. Other varieties of *R. elegans* are known as a nematode-trapping fungi, which take nutrients from the trapped nematodes [34]. A tripartite food-web, involving bacteria, nematodes, and fungi, as assumed by Barron [34], may be established here [16]. In Japan, this fungus, however, has not been obtained after treatment with any nitrogenous treatment, or the ability of the ammonia fungi to trap nematodes has not been reported. Nitrifying bacteria also increase 3 months after urea treatment, which is well correlated with increase of nitrate in soil [28].

Changes in populations of soil organisms have been well documented after use of fertilizers [11-14]. Among the reports, only Franz and Loub [11] noted changes in fruit-body occurrence, although the period of fruiting after fertilizer application was not mentioned. Soil organisms increasing together with the ammonia fungi after urea treatment influence the growth of the ammonia fungi. Effects of bacteria on spore germination, vegetative growth, and fruiting of coprophilous fungi have been investigated [35,36]

PHYSIOLOGY OF AMMONIA FUNGI

Effect of nitrogenous substances on spore germination and fruit-body formation

Ammonium ions under neutral to alkaline conditions stimulate spore germination of *C. cinereus* and *C. phlyctidosporus* [37]. Aqua ammonia promotes spore germination of *H. vinosophyllum* [38]. These results suggest that ammonia fungi may exist as a form of spore under the condition in which ammonium-nitrogen is low in acidic condition, and the fungi may start growing after ammonium concentration and pH increase.

Fruit-body formation in some *Coprinus* species of the ammonia fungi has been examined. Morimoto et al. [39] reported that addition of urea at the rate of 0.25 to 4 g urea/liter to an agar medium stimulated vegetative growth and sclerotia formation of *C. stercorarius* in darkness. Under illumination, this fungus formed fruit-bodies with urea at the rate of 0.25 to 1.0 g/liter and sclerotia or rudimentary fruit-bodies at the rate of 2.0 and 4.0 g/liter. *Coprinus cinereus* formed fruit-body primordia in a liquid medium supplemented with ammonia water to 5 days-old cultures at the concentration of 0.05 to 0.4 g NH_3/liter in darkness [40]. Some ammonium salts causing alkalinity in the medium, (e.g., $(NH_4)_2HPO_4$, NH_4HCO_3) also induced fruit-body formation of this fungus. Under illumination, this fungus formed fruit-bodies without addition of nitrogenous substances. Morimoto et al. [40] reported that ammonium seems to exert a similar effect as light on fruit-body formation for this species.

Tephrocybe tesquorum also forms fruit-bodies on an agar medium containing an extract from urea-treated forest soil, on Soytone (Difco)-glucose agar medium, and on glucose-dry yeast agar medium supplemented with urea [16,41].

Mycelial basidia formation

Tephrocybe tesquorum, an ammonia fungus, is an agaric species forming fruit-bodies, in which basidiospores (sexual spores) were produced. This fungus also forms mycelial basidia and basidiospores [42], that is, basidia and basidiospores without fruiting [2,43].

The formation of mycelial basidia and basidiospores in *T. tesquorum* was described as follows. Basidia and basidiospores developed from vegetative hyphae spreading around a disc (4 mm in diameter and 1.5 mm in depth) of a nutrient agar on a glass slide [43]. This finding indicates that this species could produce basidiospores (sexual spores) in a small amount of forest soil after release of a small amount of nutrients. Mycelial basidia and basidiospores formed also on the surface of aborted primordia of fruit-body in the forest soil collected from the field and then treated with urea in the laboratory [43]. In agar culture, mycelial basidia and basidiospore developed on the surface of primordium-like structures at the concentration of 1% of glucose in an agar medium, although fruit-bodies of this species developed at the concentration of 0.1, 0.01 and 0% of glucose [41]. These results show that this characteristic is advantageous in order to sporulate under the conditions not permitting the formation of a complete basidiocarp. The mycelial basidia and basidiospore formation in Agaricales have also been reported in *Armillaria mellea* and *Crinipellis perniciosa* [44,45], although ecological advantages in the characteristic of these species were not mentioned.

Hydrogen ion concentration

The growth of the ammonia fungi on different pH values of the agar medium was examined [46] (Table 1). The optimal growth of early phase species (EP) was generally observed in the media initially adjusted to pH 7 or 8, except for one strain of *C. phlyctidosporus,* which showed the best growth at pH 6 (Table 1). The best growth was in a broad range of initial pH from 6 to 8 for *C. echinosporus* and a strain of *T. tesquorum. Peziza urinophila* and *C. echinosporus* grew well in media initially adjusted to pH 5. Many of the EP did not grow at pH 4. However, a slight growth was observed for *T. tesquorum* at pH 3 and 4, and for *Pseudombrophila petrakii* and *C. echinosporus* at pH 4. *Coprinus echinosporus, A. botrytis,* and *P. urinophila* grew well even at pH 9. Late phase species (LP) grew well at pH 5 to 7. All but one strain of *H. vinosophyllum* grew slightly at pH 3. No growth of these species was observed at pH 9. The optimal pH for non-ammonia fungi was 5 or 6, and these species did not grow at pH above 7.

Generally, fungi grow well in acid conditions [47-49], but some exceptional species favor neutral to slightly alkaline conditions. Fries [50] reported that *Coprinus* species (e.g., *C. radiatus, C. micaceus,* and *C. ephemerus*) grew well at above pH 8. El-Abyad and Webster [51] reported that some carbonicolous species grew well at pH 6.2 to 8.2, and that the highest percentage of germination in these species was in alkaline conditions. In their studies, the substrata from which these fruit-bodies were collected were alkaline to neutral [50,51]. For the ammonia fungi, the optimal pH for EP ranges from pH 7 to 8, and that for LP is pH 5 to 6 (Table 1). The optimum pH for these fungi is generally correlated to the soil pH, where they successively fruit; EP sporulate in neutral to slightly alkaline conditions, LP fruit on acid soil [16,17,27,28]. Although the favorable condition of vegetative growth is not necessarily correlated with the condition for fruit-body formation, the pH value of the habitat could be a determinant for the developmental patterns of these fungi.

Erland et al. [52] showed that mycorrhizal fungi possessed a generally broader tolerable range of pH in symbiosis than in pure culture, and emphasized the danger of extrapolating the results from pure culture studies to symbiotic systems. Typical LP, such as *Hebeloma* spp. and *L. bicolor* are mycorrhizal [10] and, therefore, may have the ability to tolerate alkaline conditions better as a symbiont with their host trees than as a saprophyte. In field observations, however, fine roots of trees in soil were damaged by urea treatment, and did not spread again into the treated soil until fruiting of the LP of ammonia fungi, when the soil again became acidic [9].

Some ecological implications, such as possible growth forms of the ammonia fungi under the conditions unfavorable for fruiting or sporulation, have been suggested from the present results. Growth of six out of ten EP at pH 4 (Table 1), which is the value of the untreated soil indicates that even in the soil not treated with nitrogenous materials, these species may survive in the form of hyphal fragments. Growth of *Hebeloma* spp. and *L. bicolor* (LP) at pH 8 (Table 1) also shows that these species can develop their mycelia during the early phase, although these species do not form fruit-bodies until the late phase. LP with such a characteristics may infect root tips and fruit more easily than the fungi growing only in acid conditions, such as the non-ammonia species used in the present study.

Nitrogen nutrition

The ammonia fungi generally utilize a variety of nitrogen compounds as a sole nitrogen source, although there were some variations among EP or LP [53]. Many EP utilize ammonium, amino acids, urea, and bovine serum albumin (BSA). In contrast, LP grow on nitrate as well as on the nitrogenous

Table 1
Effect of pH on vegetative growth of the ammonia and non-ammonia fungi. Yields (mg dry weight/40 ml medium) were measured after culture in liquid medium initially adjusted to pH 3 to 9. Values are means calculated from five replicates for each treatment. Yields followed by different letters within rows are significantly different at $p < 0.05$ according to Scheffe's test

Species	Strain No	Initial pH						
		3	4	5	6	7	8	9
Ammonia fungi								
Early-phase species								
Amblyosporium	202	0a	0a	9a	45b	89c	57b	52b
botrytis	220	0a	1a	8a	29b	42bc	51c	34b
Peziza urinophila	165	0a	9ab	29abc	36bc	47c	100d	81d
Pseudombrophila petrakii	236	0a	0a	1a	9ab	29c	18bc	1a
Tephrocybe	212	2a	8a	9a	61b	55bc	35c	1a
tesquorum	190	1a	5a	16a	64b	64b	77b	0a
Coprinus	179	0a	2a	38b	111c	148d	135cd	21ab
echinosporus	215	0a	2a	85b	171c	172c	162c	37a
C. phlyctidosporus	NAO554	0a	0a	0a	10b	2a	0a	0a
	NAO559	0a	0a	3a	51bc	69c	24ab	2a
Late-phase species								
Laccaria bicolor	252	3ab	6abc	10cd	12d	7bcd	2a	0a
Hebeloma	110	1a	18ab	90d	102d	62c	30b	0a
vinosophyllum	135	0a	3a	69b	113c	108c	17a	0a
Hebeloma	9	1a	9ab	87c	71bc	36abc	13ab	0a
radicosoides	S610603	1a	9a	32b	45c	30b	4a	0a
Non-ammonia fungi								
Collybia dryophila	86	2a	37b	103c	17ab	0a	0a	0a
Marasmius pulcherripes	77	0a	9ab	20bc	28c	5a	1a	0a
Lyophyllum semitale	AK9301	0a	2ab	12b	2ab	1a	0a	0a
Amanita rubescens	S401801	1a	1a	3b	1a	0a	0a	0a
Suillus luteus	S710802	1ab	1ab	2bc	3c	0a	0a	0a

compounds utilized well by EP. EP and LP are characterized by the mode of nutrient acquisition, EP is saprotrophic and LP was ectomycorrhizal [10]. In general, saprotrophic fungi grow better on ammonium-nitrogen or amino acids than on nitrate [54-56]. This pattern of nitrogen utilization by saprotrophic species is also true for *mycorrhizal fungi* [57,58]. Some mycorrhizal fungi are additionally capable of utilizing nitrate [59-62]. Thus, nitrogen utilization by the ammonia fungi is generally similar to that by other saprotrophic and mycorrhizal fungi, as reported by many authors [53-62].

The ammonia fungi did not grow on the media containing amines (ethylenediamine or putrescine) as a sole nitrogen source. Amine utilization by fungi has been studied little. Ludeberg [59] reported that seven isolates of saprotrophic fungi and 31 isolates of ectomycorrhizal fungi did not grow on diethylamine and glucosamine, except for *Tricholoma pessundatum* and *Boletus elegans*, both of which grew well on glucosamine.

Amines are produced during the decomposition of faeces or dead-bodies of animals under anaerobic conditions, namely putrefaction [63]. The ammonia fungi sporulate or fruit after decomposition of dead-bodies or faeces of animal [9,10,16], although none of the ammonia fungi grew on media containing amines as a sole nitrogen source [53]. This result suggested that the ammonia fungi utilize not amines, but ammonia produced by the oxidation of amines. In field experiment, some species of the ammonia fungi sporulated after application of ethylenediamine, trimethylamine or putrescine, but these species occurred also after application of aqueous ammonia [9].

LP and EP showed different responses of the vegetative growth to a nitrogen source in pure culture, as well as to pH range. Many EP grew well on ammonium-nitrogen as a sole N source and LP grew well on nitrate as well as on ammonium-nitrogen [53]. This characteristic of these fungi is well correlated with the dominant form of inorganic nitrogen in soil during the occurrence of the ammonia fungi [27,28]. Ammonium-nitrogen in soil increase at the early phase after the treatment with urea, and nitrate-nitrogen as well as ammonium-nitrogen increase at the late phase. Harper and Webster [64] pointed out that, in the studies on coprophilous fungi, the successional occurrence of fruit-body of each species is a result of a specific minimum time required to form a fruit-body, and interspecific antagonism was an explanation for the cession of fruiting in each species. Both of the previous and the present studies [46,53], however, showed that the successive occurrence of the ammonia fungi from EP to LP was probably controlled by both pH and the form of inorganic nitrogen.

Hebeloma species, which are both ammonia fungi and non-ammonia fungi, have a potential to utilize various N sources. *Hebeloma cylindrosporoum* and *H. crustuliniforme* utilized both ammonium and nitrate as nitrogen source [61,65-67]. Two *Hebeloma* species, belonging to the ammonia fungi, grow well on ammonium, nitrate, amino acids, urea, and bovine serum albumin [53]. In addition, *H. aminophilum*, *H. radicosum*, *H. spoliatum,* and *H. syrjense* have been reported to fruit on animal wasted sites [10,24,25], and these species are also expected to have a high potential for nitrogen utilization. Besides, five species of *Hebeloma* (*H. crustuliniforme*, *H. cylindrosporum*, *H. longicaudum*, *H. mesophaeum,* and *H. velutipes*) are capable of decomposing urea [68]. All the reports indicate that *Hebeloma* spp. have a high potential to utilize various nitrogen sources. This nitrophilous character of the genus may have an advantage for the plants symbiotically associated with it at the sites after N deposition, such as intensive stock rearing, or the areas exposed to industrial and automobile exhausts.

Enzyme activity

The ammonia fungi were screened for the abilities to decompose cellulose, lignin, chitin, protein, and lipid [27]. *Pseudombrophila petrakii, Hebeloma* spp., and *L. bicolor* did not degrade cellulose. *Tephrocybe tesquorum, C. echinosporus,* and *P. petrakii* showed strong lignolytic activity assayed by laccase assay. *Amblyosporium botrytis, T. tesquorum, C. echinosporus,* and *H. vinosophyllum* decomposed chitin. All strains possessed proteolytic and lipolytic ability. Supply of glucose to the culture media resulted in weaker enzyme activities, except for lignolytic ability. The results indicate that enzymatic activity, except for lignolytic activity, is effective for obtaining carbon source, and that lignolytic ability probably functions for other reasons, e.g., acquisition of other sources, or removal of barrier for fungi to access cellulose [69]. Ectomycorrhizal species (LP) have weak abilities to decompose high-molecular substances, and are likely to get carbon sources from associated roots.

Enokibara et al. [70] investigated optimal pH range of cellulolytic activity of the ammonia fungi. In their study, EP showed a high activity of cellulase in alkaline to neutral conditions, and LP showed a weak activity of this enzyme in acidic conditions. The results show that optimal pH range of cellulolytic ability is correlated with the soil pH range where LP and EP sporulate or fruit in the field. Activity of cellulase in EP was stronger than that in LP [70].

CONCLUSION

Ammonia fungi fruit or sporulate after experimental treatment of forest floor with some nitrogenous substances and also after the decomposition of dead-body or faeces of animals [9,10,16,26]. The abilities of these fungi to grow well under alkaline conditions [46], to utilize a variety of nitrogenous compounds including ammonium, nitrate, amino acids, etc. [53], and to decompose organic matters [27] are unique and suggestive of their role in *nutrient dynamics* in forest ecosystems.

Hebeloma species, including both ammonia fungi and the non-ammonia fungi, utilize a variety of nitrogen compounds [61,65-67]. This characteristic feature may be valuable for the *bioremediation* processes, using the mycorrhizal association between fungi and plants in nitrogen-polluted areas. Further studies are necessary on the mycorrhizal association between fungi and plants under various nitrogen conditions, the impact of other soil organisms to the association, and on their benefits to the survival and productivity of forests.

ACKNOWLEDGEMENTS

I wish to express my thanks to Dr C. Y. Li, Forestry Sciences Laboratory, USDA Forest Service, Pacific Northwest Research Station for reading the manuscript, and to Dr N. Sagara, Graduate School of Human and Environmental Studies, Kyoto University, Ms K. Akama and Mr T. Akema, Forestry and Forest Products Research Institute, for kindly providing isolates of some of the fungi used.

REFERENCES

1   Hora FB. Nature 1958; 181: 1668-1669.

2   Sagara N, Hamada M. Trans mycol Soc Jpn 1965; 6: 72-74.

3   Laiho O. Acta For Fenn 1970; 106: 1-73.

4   Petersen PM. Bot Tidsskr 1970; 65: 264-280.

5   Bond TET. Trans Br mycol Soc 1972; 58: 403-416.

6   Lehmann PF. Trans Br mycol Soc 1976; 67: 251-253.

7   Ohenoja E. Ann Bot Fennici 1978; 15: 38-46.

8   Kirsi M, Oinonen P. Karstenia 1981; 21: 1-8.

9   Sagara N. Contr Biol Lab Kyoto Univ 1975; 24: 205-276.

10   Sagara N. Can J Bot 1995; 73 (Suppl 1): S1423-S1433.

11   Franz H, Loub W. Cbl ges Forstwesen 1959; 76: 129-162.

12   Roberge MR, Knowles R. Soil Sci Soc Am Proc 1967; 31: 76-79.

13   Mai H, Fiedler HJ. Zentbl Bakteriol 1979; 134: 651-659.

14   Arnebrant K, Bååth E, Söderström B. Soil Biol Biochem 1990; 22: 309-312.

15   Miller RE, Reukema, DL. USDA Forest Service Res. Note PNW-291. Portland: Pacific Northwest Forest and Range Expt. Stat, 1977.

16   Sagara, N. In: Carroll GC, Wicklow DT, eds. The Fungal Community, Its Organization and Role in the Ecosystem 2nd ed. New York: Marcel Dekker, 1992; 427-454.

17   Yamanaka T. Bull Jpn Soc Microbiol Ecol 1995; 10: 67-72.

18   Wang Y-Z, Sagara N. Mycotaxon 1997; 65: 447-452.

19   Hora FB. Mycopathol Mycol Appl 1972; 48: 3-42.

20   Khan AZMNA. Trans Br mycol Soc 1976; 67: 540-543.

21   Khan AZMNA, Hora FB. Trans Br mycol Soc 1976; 67: 358-360.

22   Pirozynski KA. Can J Bot 1969; 47: 325-334.

23   Smith AH. Contr Michigan Univ Herbarium 1941; 5: 1-73.

24   Miller OK Jr, Hilton RN. Sydowia 1986; 39: 126-137.

25   Lincoff GH. The Audubon Society Field Guide to North American Mushrooms, New York: Alfred A. Knopf, 1981.

26   Sagara N. In: Yokohata Y, Nakamura S, eds. Recent Advances in the Biology of Japanese Insectivora. Hiwa, Hiroshima: Hiwa Society of Natural History and Hiba Museum for Natural History, 1999; 33-55.

27   Yamanaka T. Mycoscience 1995; 36: 17-23.

28   Yamanaka T. J Jpn For Soc 1995; 77: 232-238.

29   Schlesinger WH, Hasey MM. Ecology 1981; 62: 762-774.

30   Yamanaka T. unpublished data.

31   Stevenson FJ. Humus Chemistry: Genesis, Composition, Reactions. New York: John Wiley & Sons, 1982.

32   Norman RJ, Kurtz LT, Stevenson FJ. Soil Sci Soc Am J 1987; 51: 809-812.

33   Furuya, K. Sankyo Kenkyusho Nempo 1990; 42: 1-31.

34   Barron GL. The Nematode-Destroying Fungi. Guelph: Canadian Biology Publications, 1977.

35   Anson AE, Fisher, PJ, Kuthubutheen AJ. Trans Br mycol Soc 1985; 85: 161-164.

36   Safar HM, Cooke RC. Trans Br mycol Soc 1988; 91: 73-80.

37   Suzuki, A, Motoyoshi N, Sagara N. Trans mycol Soc Jpn 1982; 23: 217-224.

38   Suzuki A. Trans mycol Soc Jpn 1978; 19: 362.

39   Morimoto N, Suda S, Sagara N. Trans mycol Soc Jpn 1982; 23: 79-83.

40   Morimoto N, Suda S, Sagara N. Plant Cell Physiol 1981; 22: 247-254.

41   Yamanaka T. Mycoscience 1994; 35: 187-189.

42   Singer R. The Agaricales in Modern Taxonomy, 4th edn, Koenigstein: Koeltz Scientific Books, 1986.

43   Yamanaka T, Sagara N. Mycol Res 1990; 94: 847-850.

44   Kniep H. Z Bot 1911; 3: 529-553.

45   Bastos CN, Andebrahan T. Trans Br mycol Soc 1987; 88: 406-409.

46   Yamanaka T. unpublished data.

47   Dix NJ, Webster J. Fungal Ecology. London: Chapman & Hall, 1995.

48   Hung L-L, Trappe JM. Mycologia 1983; 75: 234–241.

49   Myers DF, Campbell RN. Phytopathol 1985; 75: 670–673.

50   Fries L. Svensk Bot Tidskr 1956; 50: 47-96.

51   El-Abyad MSH, Webster J. Trans Br mycol Soc 1968; 51: 353-367.

52   Erland S, Söderström B, Andersson, S. New Phytol 1990; 115: 683-688.

53   Yamanaka T. Mycol Res 1999; 103: 811-816.

54   Hacskaylo J, Lilly VG, Barnett HL. Mycologia 1954; 46: 691-701.

55   Fries L. Svensk Bot Tidskr 1955; 49: 475-535.

56   Niederpruem DJ, Hobbs H, Henry L. J Bacteriol 1964; 88: 1721-1729.

57   Mikola P. Acta For Fenn 1965; 79: 1-56.

58   France RC, Reid CPP. Can J Bot 1983; 61: 964-984.

59   Ludeberg G. Studia For Suecica 1970; 79: 1-95.

60   France RC, Reid CPP. Microb Ecol 1984; 10: 187-195.

61   Finlay RD, Frostegård Å, Sonnerfeldt A-M. New Phytol 1992; 120: 105-115.

62   Keller G. Mycol Res 1996; 100: 989-998.

63   Alexander M. Introduction to Soil Microbiology. New York: John Wiley & Sons, 1961.

64   Harper JE, Webster J. Trans Br mycol Soc 1964; 47: 511-530.

65   Scheromm P, Plassard C, Salsac L. New Phytol 1990; 114: 227-234.

66   Quoreshi AM, Ahmad I, Malloch D, Hellebust JA. New Phytol 1995; 131: 263-271.

67   Littke WR, Bledsoe CS, Edmonds RL. Can J Bot 1984, 62: 647-652.

68   Hutchison LJ. Can J Bot 1990; 68: 1522-1530.

69   Cooke RC, Rayner ADM. Ecology of Saprotrophic Fungi. London: Longman, 1984.

70   Enokibara S, Suzuki A, Fujita C, Kashiwagi M, Mori N, Kitamoto Y. Trans mycol Soc Jpn 1993; 34: 221-228.

Biotransformations: Bioremediation Technology for Health and Environmental Protection
V.P. Singh and R.D. Stapleton, Jr. (Editors)
© 2002 Elsevier Science B.V. All rights reserved.

Bioremediation of contaminated water bodies

Balwant Kumar Singh[a], Ved Pal Singh[a] and Mahendra Nath Singh[b]

[a]Department of Botany, University of Delhi, Delhi - 110 007, India

[b]Department of Botany, S.S.N. College, Alipur, Delhi - 110 036, India

INTRODUCTION

Pollution of the biosphere with toxic elements has been accelerated dramatically since the beginning of the industrial revolution keeping pace with the progress of science. Toxic metal pollution of water has become a major environmental problem. Due to their non-biodegradable nature and *biomagnification* through the food chain, heavy metals are adversely affecting the human health. The primary sources of these pollutants are the burning of fossil fuels, mining and smelting of metalliferous ores, municipal wastes, including sewage, fertilizers, pesticides, and industrial effluents. Most conventional remediation approaches do not provide acceptable solutions and there is need for an effective and affordable technological solution.

The basic idea that both lower as well as higher plants can be used for environmental remediation is certainly very old. For example, the knowledge that aquatic and semi-aquatic vascular plants such as water-hyacinth (*Eichhornia crassipes*), penny worth (*Hydrocotyle umbellata*), duckweed (*Lemna minor*), and water-velvet (*Azolla pinnata*) can take up Pb, Cu, Cd, Fe, and Hg from contaminated solution has been around for a long time. Recently, other aquatic macrophytes such as *Hydrilla verticillata, Ceratophyllum demersum, Pistia stratiotes,* etc. have been reported as hyperaccumulators of Cr, Cd, Mn, Hg, Pb, Fe, and Cu. These aquatic plants can tolerate high levels of heavy metal concentration by sequestering them through either phytochelation or phytometallothioneins. Similarly, lower organisms, including blue-green algae (cyanobacteria) have been shown to accumulate heavy metals from polluted/contaminated waters [1].

Most plants are able to accumulate those heavy metals from soil and water which are essentially required for their growth and development. Such heavy metals are Fe, Mn, Zn, Cu, Mg, Mo, and possibly Ni. Certain plants have the ability to accumulate those heavy metals which are not needed for their biological functions. These include Cd, Cr, Pb, Co, Ag, Se, and Hg. However, excessive accumulation of these heavy metals can be toxic to most of plants. The ability of certain plants to tolerate increased levels of toxic elements and their accumulation at high concentrations from various polluted environments can be exploited for removal of toxic substances, including heavy metals from water. The use of plants for remediation of toxic substances is known as *"Phytoremediation"*.

MECHANISMS INVOLVED IN BIOREMEDIATION

There are various mechanisms involved in removal of organic and/or heavy metal pollutants from the environment by plants. It is important to understand the mechanism of metal tolerance in aquatic plants.

Aquatic plants play a significant role in maintaining water quality and *biogeochemical cycling* of elements in water bodies. They possess tremendous ability to concentrate metals in their tissues at levels much higher than the surrounding water. Plants capable of absorbing substantially greater amounts of metal than other species are termed *'hyperaccumulators'*. Because of their ability to concentrate metals, they have been exploited extensively in monitoring the concentration of pollutants and abatement of water pollution. However, mechanism of accumulation of metals and their toxic effects on the metabolism of submerged plants are not fully understood [2].

The mode of metal uptake by various forms of aquatic plants is a matter of concern to plant scientists. In submerged species due to poor development of root and transport system, it was considered that such plants received nutrients directly from the surrounding water and the roots served merely to anchor the plants to substrate. Studies showed that uptake of trace metal depends on the chemical form in which they are present in the system and on the life-form of macrophytes (floating, free-floating, well rooted, rootless submerged and emergent species). It was reported that free floating plants such as *Lemna, Eichhornia* and *Pistia* take up elements from water by root and/or leaves, whereas rootless *Ceratophyllum* takes up elements through its finely divided leaves.

A number of metal (Cu, Zn, etc.) ions are essentially required by proteins involved in metabolic functions. There also exist some other toxic metal (Cd, Pb, Hg, Cr, etc.) ions, which are non-essential for plant and animal life. All metal ions are toxic if the concentration of the free ion exceeds a specific threshold, thus causing physiological and biochemical changes. In a changing aquatic environment, there are some organisms which are capable of shifting internal levels of these *bioreactive* metal ions. Various hypotheses have been developed to explain metal tolerance by plants; these include exclusion of metal from plants, compartmentalization of metal tolerant enzymes [3,4]. However, the physiological basis of toxicity is not confined to a single susceptible site within the plant and tolerance cannot be explained by a single mechanism.

Study of *metal bioaccumulation* and *detoxification* in metallophytes and aquatic plants experiencing metal stress is of much significance due to likely transfer of toxic metal ions from metal loaded plants to the organisms at higher trophic levels in the food chain. Plants have divergent mechanisms for modulating metal levels to adapt to changes in concentration of metals. Many plants and algae tolerate heavy metals by sequestering them through either phytochelation (PCs) [5-7] or phytometallothioneins [8,9]. However, induction of PCs as metal sequestration systems appears to be a predominant mode of metal detoxification for a variety of heavy metals such as Cu, Cd, Pb, Hg, Zn, etc. in lower and higher plants.

REMOVAL OF HEAVY METALS FROM WATER

The intrinsic ability of various micro- and macrophytes to sequester and concentrate metals and other chemicals from the water bodies has long been recognized. Various green and blue-green algae possess high metal uptake capacities and, thereby, they decontaminate the water from pollutants. However, only a few studies have been made to overcome metal toxicity by aquatic macrophytes such as *Eichhornia crassipes* [10,11], *Pistia stratiotes*, and *Hydrilla verticillata* [12].

Potential of *H. verticillata* for *bioamelioration* of wastewater, containing the toxic metals, Cr, Cd, Mn, Hg, Pb, Fe, and Cu has been demonstrated [13,14]. These studies have shown high accumulation and tolerance of the macrophyte to elevated metal concentrations.

Microorganisms can also concentrate, absorb or adsorb heavy metals in the cells or cell walls. Some microfungi show positive response towards heavy metals, as they grow very well along heavy metal gradient [15]. Interestingly, application of fungal biomass to decontaminate metal polluted water or the recovery of economically important metals from natural or industrial water is worth mentioning. For example, some of the common fungi, found to grow in iron-ore tailings, can accumulate iron [16]. The mixed cultures of *Aspergillus niger* and *Trichonderma harzianum* were found to oxidize sulphides of heavy metals [17]. The former oxidized the sulphides of copper, lead and zinc to their respective sulphates. Except cadmium sulphide, the sulphide particles in the medium were adsorbed on to the surface of the mycelium. On the other hand, the oxidation of sulphides by *T. harzianum* was less. The successful removal of heavy metals has been achieved by using a *Penicillium* sp. [18]. Mushrooms can also concentrate toxic metals [19]. Various other fungi, such as *Candida humicola, Saccharomyces cerevisiae, Penicillium digitatum, Trichonderma cutaneum*, and *Poxillus filamentous*, can remove cations, including toxic and radioactive metal ions from aqueous solutions by bioabsorption [20,21]. The potential of using fungal biomass especially that of *Rhizopus* sp. is particularly effective and looks promising for practical exploitation for removal of metal ions [21].

Copper and nickel

Copper and Nickel are the chief metals, which are discharged into water bodies through mining and smelting, and other industrial operations. Copper is an essential micronutrient for algae but is toxic at higher concentrations. Nickel is not generally considered to be an essential element for growth but is toxic to most plants and animals.

One of the isolates, identified as *Scenedesmus acutiformis* var. *alternans* (referred to as B-4 *Scenedesmus*), can accumulate these metals to a substantial degree from its environment. Algae isolated from the lakes are able to grow in media containing copper upto 1.0 ppm and nickel upto 3.0 ppm [22].

Arsenic, zinc and cadmium

The essential trace metallic element, zinc, and the non-essential metal, cadmium, are accumulated by the freshwater alga, *Chlorella vulgaris*, which also show high resistance to arsenic. *Chlorella vulgaris* has a capability to tolerate and accumulate arsenic to the extent of 5,000 As mg.L^{-1} and 50,000 mg As.Kg^{-1} of dry cell, respectively [23] and has the ability to biotransform toxic inorganic arsenic to less toxic methylated arsenic compounds. The bioaccumulation and methylation of arsenic by the freshwater algae are applicable for the removal and detoxification of environmental arsenic. Arsenic contaminated waters are usually polluted by the other heavy metals, such as zinc and cadmium. The mechanisms of algal resistances to zinc and cadmium and the bioaccumulation should be different from those for arsenic. Maeda et al. [23] reported that zinc and cadmium are accumulated in *C. vulgaris* due to the physico-chemical adsorption by cell components, while arsenic accumulation was mediated largely by metabolic processes. Also in *Penicillium digitatum*, uptake of high amounts of zinc, and cadmium along with nickel by its mycelium has been reported, which is sensitive to inhibition by acidic pH [24].

Biosorption of Cd^{2+} generally comprises: (i) binding of cations to negatively charged groups on the cell surface, and (ii) energy-dependent cellular Cd^{2+} uptake. Recently, *Ceratophyllum demersum* has been demonstrated as Cd^{2+} hyperaccumulator species under field conditions [13]. *Hydrilla verticillata* and *Pistia stratiotes* are the other aquatic plants, which have been reported to be hyperaccumulators of Cd, and have demonstrated the ability to remove toxic metals, including Cd from wastewater. It is hypothesized that cadmium hyperaccumulating ability of the macrophyte is associated with induction of the metal chelating peptides, the phytochelatins (PCs), to cope up with high cellular Cd levels. Chromium and Cadmium bioaccumulation and toxicity studies in *H. verticillata* and *Chara corallina* during single and mixed treatment revealed that Cd uptake was enhanced in combined metal treatments, and the combined metal toxicity was more pronounced as compared to that of individual ones [25].

Lead, chromium and mercury

Lead, a cumulative poison, released from various industries, contributes significantly towards Pb pollution of both lentic and lotic water bodies. Lead has been found to be accumulated by a large number of aquatic plants. Recently, substantial accumulation and high tolerance of Pb has been reported in *H. verticillata* [14]. Although reasonably high concentration of Pb (1000 µM) has been found to induce synthesis of phytochelatins (PC_2, PC_3, and PC_4) in the cell suspension culture of *Rauwolfia serpentina* [6], Pb-binding complexes were absent in *Rhynocostegium riparioides* growing in flowing water contaminated with Pb.

A unicellular green alga *Glaucocystis mostochinearum* was studied for Cr uptake potential and toxicity bioassays. The alga was sensitive to various photosynthetic and nitrogenous metabolites and seems to be a suitable organism for monitoring Cr in polluted water bodies. Wastewater treatability potential of aquatic macrophytes (removal of heavy metals from metal contaminated pond water) was indicated for *C. demersum, Hydrilla reticulatum* and *Sporodela polyrrhiza* for removal of chromium.

Both free (*Hydrilla verticillata*) and rooted (*Vallisneria spiralis*) macrophytes showed high potential to accumulate mercury, maximum being in the roots of *V. spiralis*. Mercury stress induced the synthesis of various species of *phytochelatins* which bind to Hg (II) and detoxify it. Phytochelatin induction was accompanied by a decline in the levels of cellular glutathione, although this decline was only observed at high concentrations and long duration of mercury exposure, indicating the involvement of glutathione in PC synthesis. Root showed less concentration of PCs than leaves.

REMOVAL OF PESTICIDES FROM WATER

The amount of a pesticide residue accumulated by an organism by adsorption and absorption resulted in an increased concentration in cells, or specific tissues of microscopic plants and higher aquatic flora, and toxicants taken up by primary producers passed along the food chain to higher trophic levels. Detailed information on the *biodegradation* or detoxification of pesticides by blue-green algae is not available, although their effects have been reported on growth and nitrogen fixation [26,27]. The preliminary reports have indicated the degradation of lindane, accumulation of 2,4-dichlorophenoxyacetic acid (2,4-D) by green algae and detoxification of alicides by blue-green algae [28].

Benzenehexachloride (BHC)

The nitrogen fixing blue-green algae *Anabaenopsis raciborskii* and *Anabaena aphanizomenoides* were employed for detoxicating the toxic effect of pesticide, BHC in nutrient medium. A gradual loss in the toxicity of BHC was noticed, when repeated inoculation and removal of the algae was ensured [26]. Therefore, the pesticide could be detoxified by repeated growing and removing blue-green algae from the pesticide containing medium. It has also been reported that the green algae *Chlorosarcina* sp. and *Ankistrodesmus braunii* were capable of absorbing 2-8% of herbicide ^{14}C simazine.

DDT, dieldrin and photodieldrin

The extensively used chlorinated hydrocarbon insecticides, DDT and dieldrin, are common microcontaminants of the environment. The residues of these insecticides in aquatic environments are picked up by food-web and food-chain organisms and can become concentrated through successive trophic levels in organisms at top of the food chain. This *bioconcentration* can result in ecological hazards to fish, birds and other wild life.

DDT is slowly converted to DDE by various marine algae [29]. Dieldrin is extremely resistant to biodegradation. It is slowly isomerized in environment by sunlight and by marine algae [30]. Photodieldrin, which is the *"terminal residue"* of dieldrin, is also extremely persistent in the

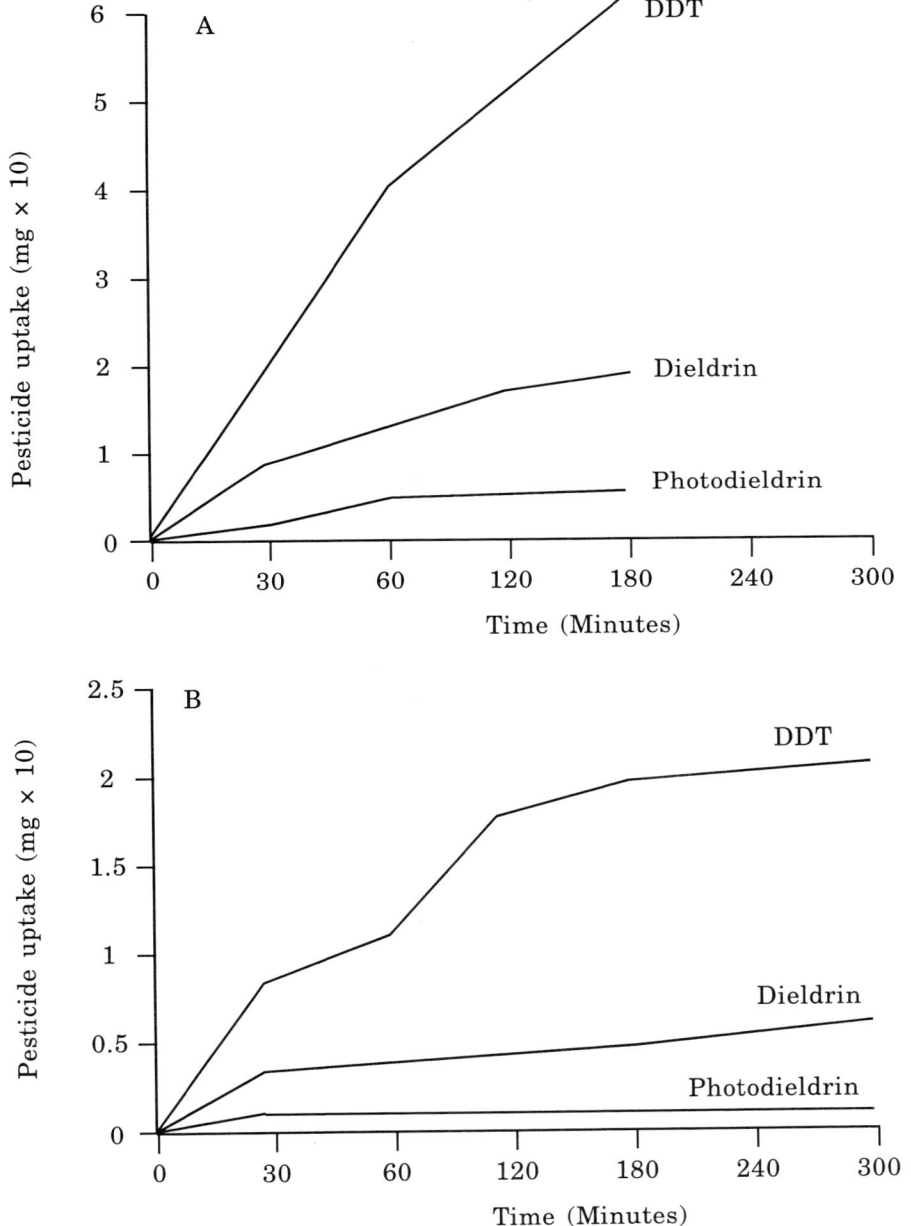

Figure 1. Absorption of DDT, dieldrin and photodieldrin by freshwater alga, *A. amalloides* at 100 cells/ml (A) and 1,000 cells/ ml (B) concentrations.

environment. Neudorf and Khan investigated the absorption/adsorption and metabolism of DDT, dieldrin and photodieldrin in the fresh water alga, *Ankistrodesmus amalloides*.

The amounts of DDT, dieldrin, and photodieldrin adsorbed/absorbed by the freshwater alga, *Ankistrodesmus amalloides* are shown in Figures 1A and 1B. The total pick-up of DDT during 1-3 hours (100 algal cells/ml) was 2.5 times higher than that of dieldrin and 10 times higher than that of photodieldrin.

CONCLUSIONS

Bioremediation of heavy metals is designed to concentrate metals in living organisms, including plant tissues, thus, minimizing the amount of solid or liquid hazardous wastes, which need to be treated and deposited at hazardous waste sites, with the ultimate goal of developing an economical method of reclaiming metals from plant residue. This will completely eliminate the need for costly off-site disposal. *Bioremediation* is certainly very old field, but it holds great potential. In order to realize this promise, it will be necessary to build a greater understanding of the many and varied processes that are involved. This will require a multidisciplinary approach, spanning fields as diverse as plant biology, agricultural engineering, agronomy, soil science, microbiology, and genetic engineering.

REFERENCES

1   Pubbi P, Singh PK. In: Rai B, Upadhyay RS, Dubey NK, eds. Trends in Microbial Exploitation. Varanasi: International Society for Conservation of Natural Resources, 1998; 27-37.

2   Guilizzoni P. Aq Bot 1991; 87-109.

3   Baker AJM. New Phytol 1987; 106: 93-111.

4 Reddy GN, Prasad MNV. Environ Exp Bot 1990; 30: 251-264.

5 Grill E, Winnacker EL, Zenk MH. Science 1985; 230: 674-676.

6 Grill E, Winnacker EL, Zenk MH. Proc Natl Acad Sci USA 1987; 84: 439-443.

7 Steffens JC. Ann Rev Plant Physiol, Plant Mol Biol 1990; 41: 533-574.

8 Robinson NJ, Tommey AM, Kuske C, Jackson PJ. Biochem J 1993; 295: 1-10.

9 Zhou J, Goldsbrough PB. Plant Cell 1994; 6: 875-884.

10 Fujita M, Kawanishi T. Plant Cell Physiol 1986; 27: 1317-1325.

11 Fujita M. Plant Cell Physiol 1985; 26(2): 295-300.

12 Tripathi RD, Rai UN, Gupta M, Chandra P. In: XVI Int Congr Biochem Mol Biol, New Delhi, 1994; Abst. 94.

13 Chandra P, Tripathi RD, Rai UN, Sinha S, Garg P. Wat Sci Technol 1993; 28(3/5): 323-326.

14 Gupta M, Chandra P. J Environ Sci Health 1994; 29: 503-516.

15 Nordgren A, Baath E, Soderstrom B. Can J Biol 1985; 63: 448-455.

16 Wong MH. J Environ Sci Health 1978; 13: 33-46.

17 Wainwright M, Graylson SJ. Trans Brit Mycol Soc 1986; 86: 269-272.

18 Golum M, Golum E, Siegel BZ, Keller P, Lehr H, Siegel SM. Water Air Soil Pollut 1987; 33: 359-371.

19 Allen RO. Chemosphere 1978; 7: 371-378.

20 Cullen WR, Mcbride BC, Reimer M. Bull Environ Contam Toxicol 1979; 21: 157-161.

21 Volesky B. Biotechnology and Bioengineering Symp No 16, John Wiley & Sons Inc 1986; 121-126.

22 Stokes PM, Hutchinson TC, Kranter K. Water Pollution Research in Canada 1973; 8: 178-201.

23 Maeda S, Mizoguchi M, Ohki A, Takeshita T. Chemosphere 1990; 21(8): 953-963.

24 Babich H, Stozky G. Appl Environ Microbiol 1977; 33: 681-695.

25   Rai UN, Tripathi RD, Sinha S, Chandra P. J Environ Sci Health A 1995; 30: 537-551.

26   Das B, Singh PK. Arch Environ Contamin Toxicol 1977; 5: 437-445.

27   Dasilva EJ, Henriksson LE, Henriksson E. Arch Environ Contamin Toxicol 1974; 3: 197-204.

28   Fitzgerald GP. Water and Sewage Works 1975; 82-85.

29   Rice CP, Sikka HC. J Agric Food Chem 1972; 21: 148.

30   Matsumura F, Patil KC, Boush GM. Science 1970; 219: 965.

Biotransformations: Bioremediation Technology for Health and Environmental Protection
V.P. Singh and R.D. Stapleton, Jr. (Editors)
© 2002 Elsevier Science B.V. All rights reserved.

Biotransformations and biodegradation in extreme environments

Anthony V. Palumbo[a], Jizhong Zhou[a], Chuanlun Zhang[b], Raymond D. Stapleton[a], Barry L. Kinsall[a] and Tommy J. Phelps[a]

[a]Environmental Sciences Division, P.O. Box 2008, Oak Ridge National Laboratory, Oak Ridge, TN 37830-8002, USA

[b]Department of Geological Sciences, 101 Geological Sciences Building, University of Missouri Columbia, MO 65211, USA

INTRODUCTION

The manner in which we view our world naturally defines normal environments as those that best support human life. However, it becomes more evident every year that the range of environments which are hospitable to microbial life far exceeds that portion of the environmental conditions which support human, mammalian, vertebrate, or even eucaryotic life. This recognition has long been evident in the microbial literature with the recognition of bacteria able to exist at very low and high temperatures and at extremes of pH. During the past half century, recognition of the ability for life to thrive under extreme conditions has extended the range in which we expect life to exist under wide ranges of temperature, pressure, and pH. We must now consider the ability of microorganisms to at least survive in extremes of radiation exposure. This microbial capability to thrive under extreme conditions has practical implications in applied microbiology, hence providing the justification of this volume. One of its areas is the *bioremediation* of organic contaminants and transformation of metals, and the purpose of this paper is to present an account of some of the researches that have been carried out in recent years.

METAL TRANSFORMATIONS

Interest in bacterial metal transformations at extreme temperatures has derived from the appreciation of pure sciences and their applied aspects, such as bioremediation [1] and leaching of ores [2]. The literature related to iron and other metal reduction has been largely driven by the interest in bioremediation, as metal reduction often results in the precipitation of less mobile forms of the metal. Interest in microbial metal oxidation is often developed by problems associated with the leaching of iron. In this case a more soluble form is produced, which can either be a problem in an uncontrolled situation or a resource in systems designed for *bioleaching*. The bulk of the following discussion focuses on metal reduction, as interest in this field has been relatively more intense in recent years. Metal oxidation and metal volatilization in extreme conditions are briefly discussed.

The cycling of organic mater and iron in the environment is profoundly influenced by microbial iron reduction [3-6]. The cycling of other metals is likely similarly influenced by biological reduction, as the ability to reduce iron is generalized to other metals in many iron-reducing bacteria. Thus, cycling of manganese, chromium, cobalt, and uranium may be at least influenced by metal reduction, as these metals are also commonly reduced by naturally occurring bacteria [4,7-9]. The examination of microbial iron reduction has increased in intensity, as there has been an appreciation for its potential role as an early form of respiration and for the implications of metal reduction for the evolution of life on Earth [1,3,4,11].

Dissimilatory Fe(III) reduction by mesophilic (20 to 35°C) bacteria has been extensively studied [3,7,12-16]. For example, *Geobacter* and *Shewanella* species were isolated from diverse sedimentary environments, including marine and freshwater aquatic sediments, pristine deep aquifers, and a petroleum-contaminated shallow aquifer [6,12-15]. Phylogenetic analysis showed that most of the mesophilic iron-reducing bacteria belong to the *delta* and *gamma* subdivisions of the *Proteobacteria* [4,16,17]. In general, iron reduction among mesophiles, thermophiles, and psychrophiles is represented throughout a wide range of prokaryotes (Figure 1).

THERMOPHILES AND IRON REDUCTION

It has been speculated that thermophilic metal reduction is a primitive form of respiration that could have been common early in the evolutionary history of biology on the planet [10]. Part of the evidence is that the ability for thermophilic iron reduction is widespread in both a spatial and biological diversity sense and that the extant *hyperthermophiles* appear closely related to the last common ancestor [10]. Bacteria capable of iron reduction at high temperatures have been reported from geothermal environments, such as the deep subsurface [18-20] and continental hot springs [20].

Vargas et al. [10] reported that in hyperthermophiles, both Archaea and Bacteria, a common property is the ability to reduce Fe(III) to Fe(II). The authors found that *Pyrobaculum islandicum, Thermotoga maritima, Pyrodictium abyssi, Methanopyrus kandleri, Archaeoglobus fulgidus, Pyrococcus furiosus,* and *Methanococcus thermolithotrophicus* all exhibited Fe(III) reductase activity. Slobodkin et al. [21] have reported similarly the widespread ability of hyperthermophiles to reduce iron. Most or all of these are capable of iron reduction, when grown on H_2. The temperature range for observation of iron reduction has been extended up to 100°C in studies of *Pyrobaculum islandicum* [22]. Thermophilic iron-reducing bacteria obtained from the deep subsurface produced magnetite particles (Figure 1) that are distinct from those formed by mesophilic iron-reducing bacteria [23].

552

Figure 1. Magnetite produced by thermophilic bacteria.

Some groups of moderately thermophilic bacteria also exhibit the ability to reduce iron. For example, Kieft et al. [20] have isolated a *Thermus* species that is able to grow with iron as an electron acceptor at 65°C. Boone et al. [18] isolated a strictly anaerobic *Bacillus* species, which they designated as *Bacillus infernus,* that was capable of Fe(III) and Mn[IV] reduction. It exhibited growth at 61°C. Interestingly, this *Bacillus* strain is unique in that it is strictly anaerobic. There are also acidophilic thermophilic iron-reducers and thermophilic iron oxidizers, which are covered under acidophiles.

PSYCHROPHILES AND PSYCHROTROPHS IN METAL TRANSFORMATIONS

Recent studies have revealed the diversity of *cold-adapted* metal-reducing bacteria. The study of bacteria adapted to growth at low temperatures has a long history [24]. However metal reduction by bacteria at low temperature has remained relatively unexamined. The diversity of psychrophilic bacteria has been examined in samples from Antarctica [25-27]. The ability of enrichment cultures from Pacific Ocean sediments and Alaskan tundra permafrost samples to reduce iron at low temperatures was described [28]. Only recently studies have been done that clearly demonstrate the low temperature metal reduction capabilities of individual organisms. Both iron [25] and manganese reduction [29,30] at low temperatures has now been described.

Zhang et al. [28] have examined psychrotrophic enrichment cultures for their ability to reduce metals and found that iron reduction at low temperatures [4-10°C] was exhibited by most of their enrichments from deep Pacific marine sediments and from Alaskan tundra permafrost. Both iron reduction and biomass production peaked at 10°C for most of the enrichments. Although solid phase ferric oxyhydroxide was reduced at a much slower rate than the soluble Fe(III)-EDTA, both were reduced. The lack of growth on yeast extract without iron provides evidence that cellular growth in the enrichments was the result of iron reduction coupled to the oxidation of the organic acids and hydrogen.

Many of the iron-reducing psychrophiles described to date have been or are closely related to *Shewanella* sp. This may be an artifact of the selective media used and the environments examined, but the pattern is clearly observable in the literature. Bowman et al. [25] defined two groupings of strains, which they designated *Shewanella gelidimarina* and *Shewanella frigidimarina* isolated from Antarctica. The strains differed in temperature optima and substrates they utilized. However, both groups could grow by dissimilatory Fe(III) reduction. DeLong et al. [26] described *Shewanella benthica* as barophilic, psychrophilic, and capable of Fe(III) reduction. Many of the cultures isolated were also most closely related to a variety of known *Shewanella* strains (Figure 2). A grouping of isolates, most closely related to a *Shewanella* sp., was designated as ANG, and another grouping most closely related to *Shewanella* sp. as SQ10. A third grouping was similar to the species *Shewanella gelidimarina,* isolated from Antarctic sea ice [25]. Single isolates were most closely related to *S. oneidensis* MR-1 [31] and *Shewanella algae* BRY. Interestingly, only one of these psychrotrophic strains grouped with another cold-adapted strain, *S. frigidimarina*. Thus, cold tolerance must be widely distributed within the genus *Shewanella*. This is not surprising as the vast majority of the world's surface is covered by oceans, environments that are characteristically cold (~4°C), representing a significant portion of potential microbial habitats. The ocean surface may harbor even colder environments, where sea ice bacteria can thrive [32,33]. Oceans play key roles in driving biogeochemical cycles and serve as important carbon and nutrient sinks [34]. Thus, cold-adapted microorganisms isolated from these environments provide important information to evaluate key biogeochemical reactions important to global nutrient cycling.

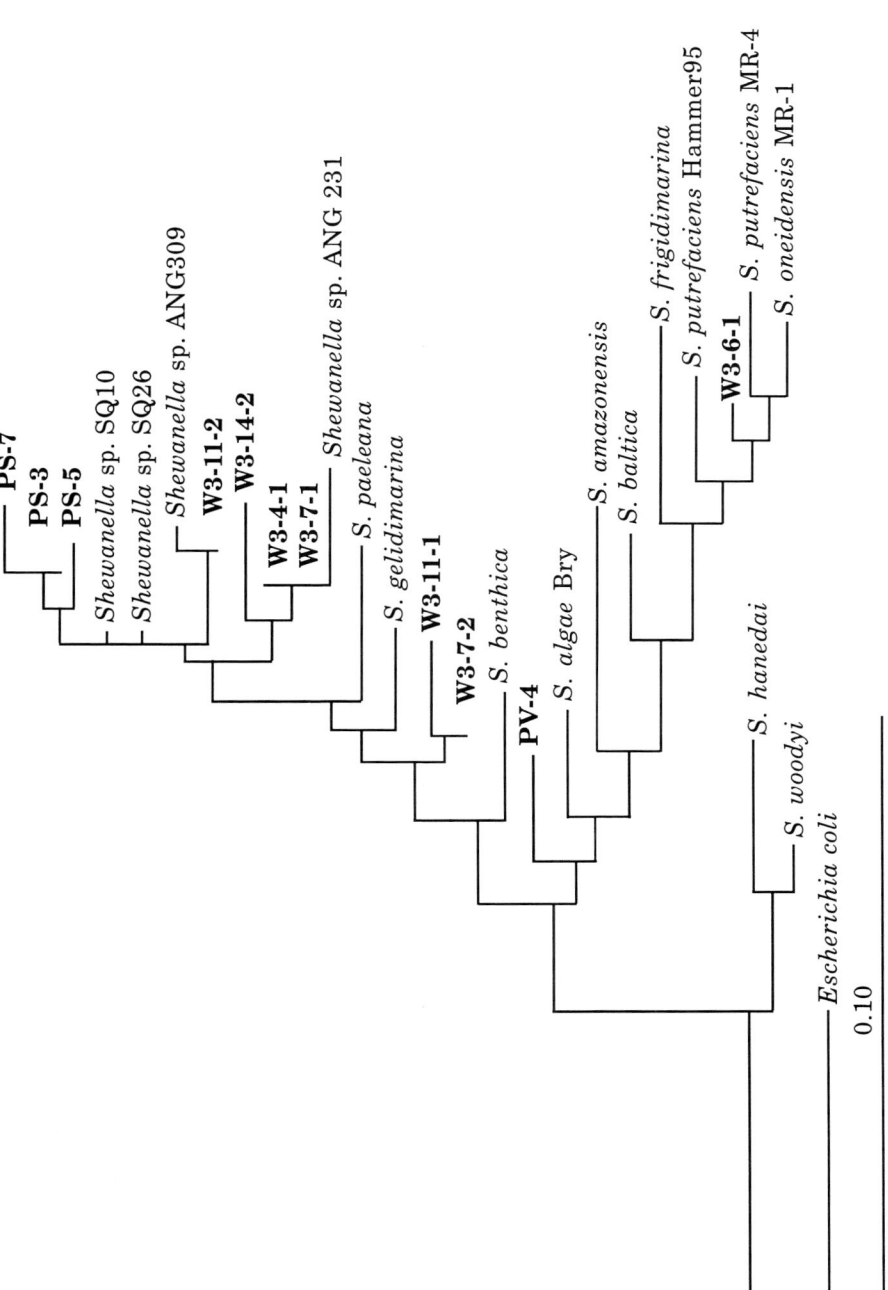

Figure 2. Phylogenetic comparison of bacterial isolates from marine sample (Stapleton et al., submitted) with other described members of the genus *Shewanella*.

Further consideration of manganese reduction may extend the diversity of bacteria capable of low temperature metal reduction. Bratina et al. [29] described a group of manganese oxidizers isolated from a lake in Antarctica. They found many strains similar, based on 16S rRNA gene sequence, to *Carnobacterium* in these extremely *oligotrophic* waters. The authors estimated 1 to 5 $\times$ 10^3 bacteria per ml and suggested that the bacteria apparently produce a diffusible compound responsible for manganese reduction. They also found a strain most closely related to *Aerococcus*. Temperature studies were not performed and the microbial cultures were incubated at 15°C. Thus, it is likely that these were either psychrotolerant or mesophilic strains.

ACIDOPHILES AND METAL TRANSFORMATIONS

Often, iron oxidation is associated with acidophilic conditions and can occur at both mesophilic and thermophilic temperatures [35]. Takai et al. [36,37] isolated a strain of iron-oxidizing bacterium, that exhibited a temperature optimum of 55°C and a pH optimum of 3.0. They found evidence that cytochrome a, and not cytochromes b and c, was involved in the iron oxidation. The moderately thermophilic *Sulfobacillus acidophilous* and *Sulfobacillus thermosulfidooxidans* are capable of growth on both ferrous iron and elemental sulphur [38]. Stoner et al. [39] have defined conditions of pH 1.8 and 45°C as the most favorable for iron oxidation in enrichment cultures of moderately thermophilic acidophilic mining bacteria.

A recent review of the diversity of acidophilic microorganisms [40] illustrates the point that there are iron-oxidizing acidophilic prokaryotes representing mesophiles, moderate thermophiles, and extreme thermophiles. There are consistent phylogenetic differences among these groups of acidophiles. Redox constraints at these low pH levels would seem to limit the ability of organisms, for which ferrous iron is the only energy source, to use any other electron acceptor except oxygen. Thus these organisms are expected to be obligate aerobes, e.g., *Leptospirillum ferrooxidans*. However, by coupling their metabolism to elemental and reduced sulfur compound some, e.g., *Thiobacillus ferrooxidans* (now *Acidithiobacillus ferrooxidans*) chemolithotrophic and mixotrophic prokaryotes can grown as facultative anaerobes [41,42].

Most of the mesophilic acidophilic iron-oxidizers isolated and characterized are eubacteria. The well known *A. ferrooxidans* , for which the literature is too extensive to cover in detail, heads this group, which also includes *T. prosperus* (which is also halotolerant), *Leptospirillum ferrooxidans*, and *Ferromicrobium acidophilus* [43]. *Leptospirillum ferrooxidans* may have a higher temperature optimum than *A. ferrooxidans* [43]. There has also been a mesophilic iron-oxidizing *Sulfobacillus* strain isolated [43].

There are apparently specific groups of thermophilic acidophilic procaryotes involved in iron oxidation with archaea dominating the extreme thermophiles [40]. Moderately thermophilic iron-oxidizers represent a somewhat more diverse group both phylogentically (*Sulfobacillus acidophilus*, *Acidomicrobium ferrooxidans,* and *Leptospirillum thermoferrooxidans*) and metabolically [40,42]. The extreme thermophiles are represented by several *Acidianus* species (e.g., *Acidianus brierleyi*, *A. infernus*, and *A. ambivalens*). Also included are *Metallosphaera sedula* and *Sulfurococcus yellowstonii* [40].

Thermophilic anaerobic oxidation of metals by bacteria is also a relatively rare phenomenon, but has been demonstrated with *Methanobacterium thermoautotrophicum* [44]. Oxidation is coupled to H_2 utilization by this methanogenic bacterium. This organism was apparently able to oxidize a variety of metals to support growth, including aluminium, cobalt, iron, magnesium, manganese, and tin.

As with many other groups of bacteria, examination of acidophilic communities using molecular techniques has indicated very wide diversity of bacteria, living in these environments with the capability to oxidize metals. Examination of slime layers from an *"extreme acid mine drainage site"* has revealed a diverse assemblage of acidophilic bacteria with some having apparent capabilities for metal reduction [45]. The authors used 16S rRNA techniques to look at the diversity of these communities and found that sequences related to *Leptosprillum* were very common. However, based on these results, there may be a wide diversity of yet unknown groups of eubacteria and archaea. The authors detected apparent relatives of known iron-oxidizing acidophiles, and many 16S rRNA sequences they found were most closely related to *Thermoplasmales* and *Acidimicrobium* among others.

There is a less extensive literature on acidophilic iron-reducers than on the iron-oxidizers probably due to the interest on the later related to mining. Some species of the mesophile *Acidiphilum* and some strains of the moderate thermophile *Alicylobacillus* reduce iron [40]. If it is true that all extreme thermophiles are capable of iron reduction [22], then the acidophilic extreme thermophiles, e.g., several *Sulfolobus* species, should be capable of iron reduction. Bridge and Johnson [46] have described iron reduction resulting in dissolution of many iron containing minerals and amorphous ferric hydroxide at pH 2 by *Acidiphilium* SJM. Interestingly, there is evidence that, in some cases, a bacterium grown under different conditions can be either an iron oxidizer or an iron reducer [42].

BIODEGRADATION OF ORGANIC POLLUTANTS

With the increased interest in remediation of the environment related to oil spills, gasoline leaks, and industrial waste sites, there is a vast literature on biodegradation of organics. The complete literature cannot be described in this limited format and only the most recent findings will be highlighted.

There is also an extensive literature on high temperature organic degradation dealing with composting. However, only the literature that pertains to organic contaminants will be considered.

The abilities of bacteria to degrade hydrocarbon compounds were appreciated early in evaluating environmental contamination by these potentially hazardous substances. Many conventional culture techniques were used to isolate numerous bacteria that exhibited capabilities to degrade alkanes and alkenes. More complex molecules, such as heterocyclic compounds were difficult to deal with, but eventually bacteria with capabilities to degrade many of these were also isolated. In the last 20 years, the abilities of bacteria to degrade chlorohydrocarbons have been demonstrated due to the use of more exotic isolation and culture techniques that allowed for co-metabolic and halo-respiration activities to be detected and demonstrated. However, relatively little of this extensive body of research has been focused on biodegradation in extreme environments.

Common sources for isolation of organic degrading thermophiles are compost, sewage sludge, and hot springs [47-49]. Interestingly, some of the research on specific enzymes involved in organic degradation has indicated that these enzymes are not particularly thermostable but are produced at high rates so that enzyme activity is maintained at high temperature [50]. Other enzymes, which appear to be more thermostable at temperature up to 100°C, have surprisingly little structural difference from less stable enzymes [51], leading to speculations on the capability to engineer enzymes with greatly enhanced thermostability.

THERMOPHILIC ORGANIC DEGRADATION

Degradation of aromatic and polycyclic hydrocarbons by thermophiles has been demonstrated to occur in a broad group of prokaryotes, and significant findings on degradation pathways have been made using thermophilic bacteria. Phenol degradation in some thermophilic *Bacillus* species has been demonstrated to be by the *meta*-pathway and to occur at 60°C [47]. There are now examples of phenol degradation by many *Bacillus* species such as *B. thermogucosidasius* and *B. thermoleovorans* [52] with temperature optimum up to 70°C. *Thermus* species, including *T. aquaticus,* that are capable of BTEX degradation at temperatures up to 70°C, have also been isolated [53]. There is some evidence that degradation under anaerobic thermophilic conditions can proceed, using pathways that differ from those seen in mesophilic organisms. Karlsson et al. [54] reported a higher sensitivity of phenol transformation to H_2 partial pressure. Also, no benzoate was observed under the thermophilic conditions. The authors attributed this to either a high turnover of benzoate or an alternative degradation pathway. Benzoate degradation at elevated temperatures was observed by An et al. [55], using thermophilic *Streptomyces*. They reported the presence of a thermostable catechol-1,2-dioxygenase in the *S. setonii* culture they used in the studies. Annweiler et al. [56] isolated a *Bacillus* species that could grow on naphthalene at 60°C. The authors reported that bacteria produced a number of metabolites that were typical of mesophilic degradation. However, presence of some more exotic metabolites indicated that thermophilic degradation by this organism might have occurred using pathways not yet demonstrated in mesophilic bacteria. It appears that, for aromatic and polycyclic compounds, the investigations on thermophilic organisms can lead to a greater understanding of degradation pathways and the discovery of new ones.

Degradation of long chain hydrocarbons at elevated temperatures is important in several oil-producing regions (e.g., southern Mexico and the Persian Gulf Region) and in degradation in compost/bioreactor applications. Degradation of long-chain hydrocarbons at 40°C has been demonstrated with bacterial isolates from the United Arab Emirates [57]. These bacteria have maximum growth rates at 60 to 80°C and have been used successfully in a bioreactor. Degradation of mineral oil and grease in weathered soil within a reactor-based composting system has been shown to proceed at 50°C [58], indicating the activity of thermophilic bacteria. Obligate thermophilic bacteria utilizing alkanes have also been isolated [59,60].

Thermophiles and hyperthermophiles may be very active in deep oil fields, which appear to be good sources for isolating thermophilic hydrocarbon-degrading bacteria. Bacteria degrading a range of organic substrates, including *n*-alkanes with a temperature optimum of 55-60°C, have been isolated from Kazakhstan and Siberia oil fields [61,62]. Orphan et al. [63] described the thermophilic assemblages obtained from high temperature oil reservoirs by enrichment samples. From heterotrophic enrichments at 100°C, they isolated a wide range of *Thermococcus*, *Thermoanerobacter*, *Desulfothiovibrio*, *Anaerobaculum*, *Thermotogales*, *Methanobacterium*, *Deferribacter*, and *Methanococcus* as well as others, that had been previously reported. However, they also found other species not described in other petroleum studies including *Acidaminococcus*, *Methanosarcinales*, and *Methanoculleus*.

Chlorine-substituted compounds, such as halophenols, polychlorinated biphenyls (PCBs), TCE, and carbon tetrachloride, can be *troublesome environmental pollutants*. However, some recent work has resulted in isolation of thermophiles, including many *Bacillus* species, that are capable of transforming and degrading many of these compounds. Reinscheid et al. [64] isolated two *Bacillus* spp. that transformed a variety of halophenols at a temperature of 60°C. Even though PCB contamination is a significant problem at mesophilic temperatures, there have only been a limited number of strains which have been found to degrade PCBs. Shimura et al. [48] have isolated a *Bacillus* sp. from compost that can degrade several PCB congeners, including some tetra- and penta-PCBs. A *Bacillus* species (*B. pallidus*) has been found to degrade compounds, such as pinene and limonene, which have been suggested as substitutes for chlorinated solvents [65].

PSYCHROPHILIC AND PSYCHROTROPHIC ORGANIC DEGRADATION

Hydrocarbon degradation at low temperatures is important in remediation of many contaminated sites. Subsurface contamination often reaches groundwater with year round temperatures of 12-18°C. Low temperature remediation is also important in surface spills at sites of high latitude or high elevation, where temperatures are characteristically low. Alpine soils have been shown to have considerable potential for bioremediation with cold-adapted bacteria [66], and numerous studies have shown potential for degradation in cold climates [67-69]. However, as pointed out by Mohn and Stewart [69], most of these hydrocarbon-degrading communities appear to be psychrotolerant and not psychrophilic. They also pointed out that many of the hydrocarbon-degrading isolates were also psychrotolerant.

Hydrocarbon-degrading *Pseudomonas* species [70] and other common soil microorganisms [72,73] are often described in low temperature environments. However, there are limited examples of the isolation of cold-adapted strains, capable of the degradation of organic contaminants, have been identified upto the species level. Most often field, microcosm, or other studies with mixed cultures are described. Thus, characteristics or identifications of individual isolates were not made. One psychrotrophic inoculum [73-75], capable of degrading diesel oil, was thought to contain a *Pseudomonas* sp. and an *Arthrobacter* sp. Aromatic hydrocarbon-degrading bacteria have also been isolated from a JP8 contaminated site in Antarctica [70]. The authors identified *Sphingomonas* spp. and *Pseudomonas* spp. *Pseudomonas* strains capable of degrading alkanes, toluene, and naphthalene were also isolated from petroleum-contaminated arctic soils [71]. A psychrotrophic *Rhodococcus* capable of growth on alkanes at low temperatures assisted by specific *physiological adaptations* has been described [76,77]. A psychrotrophic *Acinetobacter* strain was isolated by MacCormack and Fraile [72] from contaminated Antarctic soil. The widespread presence of these cultures has lead to optimism regarding the potential for low temperature bioremediation of hydrocarbons [78].

PCB degradation at low temperatures is important due to widespread distribution of PCBs and its presence in aquatic sediments where temperatures are often below 10°C. However, information on low temperature degradation of PCBs is extremely limited. Mohn et al. [79] and Master and Mohn [80] examined PCB degradation using Arctic soil bacteria. They isolated several psychrotolerant as well as psychrophilic strains which grew on biphenyl. Two psychrotolerant strains could degrade some congeners up to trichrolobiphenyl. Several species of *Pseudomonas* have also been identified as being important in this metabolic activity.

TCE degradation by psychrophilic or psychrotolerant ammonia oxidizers and methanotrophs may contribute to aquifer remediation [81]. The authors found that degradation rates by methanotrophs dropped to a greater degree with a temperature drop from 24 to 12°C than did that of ammonia oxidizers. TCE degradation by five of eight enrichment cultures was shown to be severely limited at 10°C [83]. As pointed out by Moran and Hickey [81], the widespread distribution of TCE in low temperature aquifers makes these observations critical to the potential for biostimulation for remediation at these sites.

ACIDOPHILIC ORGANIC POLLUTANT DEGRADATION

Degradation of organic contaminants by bacteria at extreme pH's has not been well documented. In fact, most often in field studies, low pH has been cited as a limiting factor in the degradation rates [83-85]. In limited cases there are reports of maximum degradation rates being observed for specific compounds at low pH [86]. The pathway for degradation and the specific compounds examined in these studies seems to play a role in determining the effect of low pH on degradation rates. For example, one report of increased degradation rate at low pH was for quinoline [86] and the authors point out that the degradation pathway for quinoline may involve a spontaneous conversion [87,88] at low pH which assists in the degradation. The limitation of pH on degradation has been observed for styrene [83] and PAH degradation [84,85].

Reports on acidophilic organic pollutant degrading bacterial isolates are also limited. Gemmell and Knowles [88] have reported the isolation of acidophilic bacteria, capable of degradation of aliphatic compounds at pH 3.0; but the isolates were not identified. Stapleton et al. [89] have reported that enrichments from a coal pile storage area and nearby environments are capable of PAH degradation. The authors isolated a number of bacteria with 16S rRNA similarity to *Acidocella* sp. as well as *Acidphilium facilis* and an isolate that was not very similar to any organism in the database. These could grow on salicylate at pH 3.0 but not on toluene, naphthalene, or phenanthrene. The authors speculated that either uncultured bacteria or the combination of these bacteria and fungi, were capable of the degradation observed in the enrichments.

DEGRADATION OF OTHER ORGANIC COMPOUNDS BY EXTREMOPHILES

A number of other relatively recalcitrant compounds, not considered as organic pollutants, are also degraded by thermophiles. Of primary interest in this area is the considerable literature on xylanase and cellulase activities. For example, a xylanase has been isolated from a thermoalkalophilic *Bacillus* [91]. Moderate thermophiles from hot springs have been shown to degrade agar at temperatures of approximately 40-60°C [49]. These Gram-positive anaerobes were considered by the authors to represent a new genus, *Alterococcus* with a high *G+C content* (65.5 to 67.0 mol%). Interestingly, these bacteria were also halotolerant with optimal growth occurring at 2.0 to 2.5% NaCl.

Bioplastics and biodegradable plastics are of increasing interest in this era of environmental concern, and thermophilic degradation of these compounds has been demonstrated. Many of these materials may be degraded by composting and, thus, thermophiles will be involved. Kleeberg et al. [91] have looked at the degradation of aliphatic-aromatic copolyesters at elevated temperatures. They found that numerous strains of thermophiles could be isolated from compost, that would degrade the terephthalic acid component of these copolyesters. Actinomycete strains played an important role, and the authors proposed that a *Thermomonospora fusca* strain, they isolated, exhibited high terephthalic acid degradation rates and could be used in comparative *biodegradation* studies.

Mixed cultures of thermophiles may have broad ranges of organic degradation capabilities. These capabilities can potentially be exploited in thermophilic treatment of wastewater [92]. Potential advantages of using thermophiles include high biodegradation rates and low sludge yields [92].

CONCLUDING REMARKS

Interest in *extremophiles* for applied microbiology is growing, and the possible uses of these organisms for metal transformation and organic pollutant degradation is being evaluated by numerous researchers. Thermophiles are amongst the many organisms for which the whole genome sequence has been determined [93], and the potential for genetic engineering of traits from these extremophiles is obvious. This may even involve a different type of "extremophile" such as *Deinococcus radiodurans* who's resistance to radiation could be considered as an extremophile trait. *Deinococcus* does show some metal reducing traits [94], that could be useful in bioremediation and it also may be a psychrophile [95,96]. Traits, such as radiation resistance, are being combined with biodegradation traits to target remediation of mixed wastes [97,98]; and this should be even more feasible in the future, as the whole genome sequence for *Deinococcus* has been published [99]. Other extremophile traits, such as adaptations to high temperatures, low temperatures, and low pH, could potentially be transferred into other strains for remediation or other purposes. There is very limited information on alkalophilic metal reduction or organic biodegradation [100]. Thus, if there is a need for remediation at high pH sites, the need could foster research in the *genomics of alkalophiles*. Much work remains to be done in defining how all these organisms adapt to extreme environments.

ACKNOWLEDGEMENTS

Support for this work was provided by United States Department of Energy, Office of Science, Microbial Genome Initiative and the Marine Biotechnology Program. Additional support was provided by the Laboratory Directed Research and Development Program through Oak Ridge National Laboratory.

REFERENCES

1   Gadd GM. Current Opinion in Biotechnol 2000; 11: 271-279.

2   Norris PR, Burton NP, Foulis NAM. Extremophiles 2000; 4: 71-76.

3   Lovley DR. Microbiol Rev 1991; 55: 259-287.

4   Lovley, DR. Annu Rev Microbiol 1993; 47: 263-290.

5   Nealson KH, Saffarini D. Annu Rev Microbiol 1994; 48: 311-343.

6   Lovley DR, Anderson RT. Hydrogeol J 2000; 8: 77-88.

7   Myers CR, Nealson KH. Science 1988; 240: 1319-1321.

8   Gorby YA Lovley DR. Environ Sci Technol 1992; 26: 205-207.

9   Zhang C, Liu S, Logan J, Mazumder R, Phelps TJ. Applied Biochemistry and Biotechnology 1996; 57/58: 923-932.

10   Vargas M, Kashefi K, Blunt-Harris EL, Lovley DR. Nature 1998; 395: 65-67.

11   Nealson KH, Myers CR. Am J Sci 1990; 290: 35-45.

12   Lovley DR, Phillips EJP. Appl Environ Microbiol 1986; 57: 751-757.

13   Caccavo Jr. F, Blakemore RP, Lovley DR. Appl Environ Microbiol 1992; 58: 3211-3216.

14   Roden EE, Lovley DR. Appl Environ Microbiol 1993; 59: 734-742.

15   Coates JD, Phillips EJP, Lonergan DJ, Jenter H, Lovley DR. Appl Environ Microbiol 1996; 62: 1531-1536.

16   Lonergan DJ, Jenter HL, Coates JD, Phillips EJP, Schmidt TM, Lovley DR. J Bacteriol 1996; 178: 2402-2408.

17   Nealson KH, Little B. Appl Microbiol 1997; 45: 213-239.

18   Boone DR, Liu YT, Zhao ZJ, Balkwill DL, Drake GR, Stevens TO, et al. International J Systematic Bacteriology 1995; 45: 441-448.

19   Liu SV, Zhou JZ, Zhang CL, Cole DR, GajdarziskaJosifovska M, Phelps TJ. Science 1997; 277: 1106-1109.

20   Kieft TL, Fredrickson JK, Onstott TC, Gorby YA, Kostandarithes HM, Bailey TJ, et al. Appl Environ Microbiol 1999; 65: 1214-1221.

21   Slobodkin AI, Zavarzina DG, Sokolova TG, Bonch-Osmolovskaya EA. Microbiol 1999; 68: 522-542.

22   Kashefi K, Lovley DR. Appl Environ Microbiol 2000; 66: 1050-1056.

23   Zhang C, Liu S, Phelps TJ, Cole DR, Horita J, Fortier SM, Elless M, Valley JW. Geochimica et Cosmochimica Acta 1997; 61: 4621-4623.

24   Monita RY. Bacteriol Rev 1975; 29: 144-167.

25   Bowman JP, McCammon SA, Nichols DS, Skerratt JH, Rea SM, Nichols PD et al. International J Systematic Bacteriology 1997; 47: 1040-1047.

26   DeLong EF, Wu KY, Prezelin BB, Jovine RVM. Nature 1994; 371: 695-697.

27   Gosink JJ, Staley JT. Appl Environ Microbiol 1995; 61: 3486-3489.

28   Zhang C, Stapleton RD, Zhou J, Palumbo AV, Phelps TJ. FEMS Microbiology Ecology 1999; 30: 367-371.

29   Bratina BJ, Stevenson BS, Green WJ, Schmidt TM. Appl Environ Microbiol 1998; 64: 3791-3797.

30   Thamdrup B. Advances in Microbial Ecology, 2000; 16: 41-84.

31   Ziemke F, Hofle MG, Lalucat J. Rossello-Mora R, International J Systematic Bacteriology 1998; 48: 179-186.

32   Staley JT, Gosink JJ. Ann Rev Microbiol 1999; 53: 189-215.

33   Sullivan CW, Palmisano AC. Appl Environ Microbiol 1984; 47/4: 788-795.

34   Parkes RJ, Cragg BA, Bale SJ, Getliff JM, Goodman K, Rochelle PA et al. Nature 1994; 371: 410-413.

35   Norris PR, Barr DW. FEMS Microbiol Letters 1985; 28: 221-224.

36   Takai K, Horikoshi K. Appl Environ Microbiol 1999; 65: 5586-5589.

37   Takai M, Kamimura K, Sugiot T. Bioscience Biotechnol Biochem 1999; 63: 1541-1547.

38   Norris PR, Clark DA, Owen JP, Waterhouse S. Microbiology-UK 1996; 142: 775-783.

39   Stoner DL, Miller KS, Fife DJ, Larsen ED, Tolle CR, Johnson JA. Appl Environ Microbiol 1998; 64: 4555-4565.

40   Johnson BD. FEMS Microbiology Ecology 1998; 27: 307-317.

41   Pronk JT, de Bruyn JC, Bos P, Kuenen JG. Appl Environ Microbiol 1992; 58: 2227-2230.

42   Bridges TAM, Johnson DB. Appl Environ Microbiol 1998; 64: 2181-2186.

43   Atkinson T, Gairns S, Cowan DA, Danson MJ, Hough DW, Johnson DB, et al. Extremophiles 2000; 4: 305-313.

44   Loworwitz WH, Nagle DP, Tanner RS. Environ Sci Technol 1992; 26/8: 1606-1610.

45   Bond PL, Smriga SP, Banfield JF. Appl Environ Microbiol 2000; 66/9: 3842-3849.

46 Bridge, TAM, Johnson, DB. Geomicrobiol J 2000; 17: 193-206.

47 Ali S, Fernandez-Lafuente R, Cowan DA. Enzyme Microbial Technology 1998; 23: 462-468.

48 Shimura M, Mukerjee-Dhar G, Kimbara K, Nagato H, Kiyohara H, Hatta T. FEMS Microbiol Lett 1999; 178: 87-93.

49 Shieh WY, Jean WD. Canadian J Microbiol 1998; 44: 637-645.

50 Milo RE, Duffner FM, Muller R. Extremophiles 1999; 3: 185-190.

51 Daniel RM. Enzyme Microbial Technology 1996; 19: 74-79.

52 Duffner FM, Reinscheid UM, Bauer MP, Mutzel A, Muller R. Systematic Appl Microbiol 1997; 20: 602-611.

53 Chen CI, Taylor RT. Biotechnol Bioeng 1995; 48: 614-624.

54 Karlsson A, Ejlertsson J, Nezirevic D, Svensson BH. Anaerobe 1999; 5: 25-35.

55 An HR, Park HJ, Kim ES. J Microbiol Biotechnol 2000; 10: 111-114.

56 Annweiler E, Richnow HH, Antranikian G, Hebenbrock S, Garms C, Franke S et al. Appl Environ Microbiol 2000; 66: 518-523.

57 Al-Maghrabi IMA, Bin Aqil AO, Islam MR, Chaalal O. Energy Sources 1999; 21: 17-29.

58 Beaudin N, Caron RF, Legros R, Ramsay J, Ramsay B. Biodegradation 1999; 10: 127-133.

59 Zarilla KA, Perry JJ. Archives Microbiol 1984; 137: 286-290.

60 Zarilla KA, Perry JJ. Systematic Appl Microbiol 1987; 9: 258-264.

61 Nazina TN, Ivanova AE, Mityushina LL, Belyaev SS. Microbiol 1993; 62: 359-365.

62 Nazina TN, Ivanova AE, Borzenkov IA, Belyaev SS, Ivanov MV. Geomicrobiology J 1995; 13: 181-192.

63 Orphan VJ, Taylor LT, HaFenbradl D, DeLong EF. Appl Environ Microbiol 2000; 66: 700-711.

64 Reinscheid UM, Bauer MP, Muller R. Biodegradation 1997; 7: 455-461.

65   Margesin R. International Biodeterioration and Biodegradation 2000;
      46/1: 3-10.

66   Agosti, JM, Agosti TE. In: Proceedings of US Symposium on impact of
      Oil Resource Development on Northern Plant Communities. Institute
      of Arctic Biology, Fairbanks, Alaska 1972; 80-85.

67   Bradley PM, Chapelle FH. Environ Sci Technol 1995; 29: 2778-2781.

68   Braddock JF, Lindstrom JE, Brown EJ. Marine Pollution Bulletin 1995;
      30: 125-132.

69   Mohn WW, Stewart GR. Soil Biology Biochem 2000; 32: 1161-1172.

70   Aislabie J, Foght J, Saul D. Polar Biology 2000; 23: 183-188.

71   Whyte LG, Bourbonniere L, Greer CW. Appl Environ Microbiol 1997;
      63: 3719-3723.

72   MacCormack WP, Fraile ER. Antarctic Science 1997; 9: 150-155.

73   Margesin R, Schinner F. J Chemical Technology Biotechnol 1997; 70:
      92-98.

74   Margesin R, Schinner F. Appl Environ Microbiol 1997; 63: 2660- 2664.

75   Margesin R, Schinner F. Chemosphere 1999; 38: 3463-3472.

76   Whyte LG, Hawari J, Zhou E, Bourbonniere L, Inniss WE, Greer CW.
      Appl Environ Microbiol 1998; 64: 2578-2584.

77   Whyte LG, Slagman SJ, Pietrantonio F, Bourbonniere L, Koval SF,
      Lawrence JR et al. Appl Environ Microbiol 1999; 65: 2961-2968.

78   Whyte LG, Greer CW, Inniss WE. Canadian J Microbiol 1996; 42:
      99-106.

79   Mohn WW, Westerberg K, Cullen WR, Reimer KJ. Appl Environ
      Microbiol 1997; 63: 3378-3384.

80   Master ER, Mohn WW. Appl Environ Microbiol 1998; 64: 4823-4829.

81   Moran BN, Hickey WJ. Appl Environ Microbiol 1997; 63: 3866-3871.

82   Broholm K, Christensen TH, Jensen BK. Water Research 1993; 27:
      215-224.

83   Fu MH, Alexander M. Environ Sci Technol 1992; 26: 1540-1544.

84   Durant ND, Jonkers CAA, Bauwer EJ. Biodegradation 1997; 8: 77-86.

85   Durant ND, Wilson LP, Bauwer EJ. J Cont Hydrol 1995; 17: 213-237.

86   Thomsen AB, Henricksen K, Gron C, Moldrup P. Environ Sci Technol 1999; 3: 2891-2898.

87   Schwartz G, Bauder R, Speer M, Rommel TO, Wingens F. Biol Chem Hoppe-Seyler 1989; 1183-1189.

88   Gemmell RT, Knowles CJ. FEMS Microbiol Lett 2000; 192: 185-190.

89   Stapleton RD, Savage DC, Sayler GS, Stacy G. Appl Environ Microbiol 1998; 64: 4180-4184.

90   Takahashi H, Nakai R, Nakamura S. Biosci Biotechnol Biochem 2000; 64: 887-890.

91   Kleeberg I, Hetz C, Kroppenstedt RM, Muller RJ, Deckwer WD. Appl Environ Microbiol 1998; 64: 1731-1735.

92   Lapara TM, Alleman JE. Water Research 1999; 33: 895-908.

93   Nelson KE, Paulsen IT, Heidelberg JF, Fraser CM. Nat Biotechnol 2000; 18: 1049-1054.

94   Fredrickson JK, Kostandarithes HM, Li SW, Plymale AE, Daly MJ. Appl Environ Microbiol 2000; 66: 2006-2011.

95   Carpenter EJ, Lin SJ, Capone DG. Appl Environ Microbiol 2000; 66: 4514-4517.

96   Nandakumar R, Mattiasson B. Biotechnol Tech 1999; 13: 689-693.

97   Daly MJ. Curr Opin Biotechnol 2000; 11: 280-285.

98   Brim H, McFarlan SC, Fredrickson JK, Minton KW, Zhai M, Wackett LP, Daly MJ. Nat Biotechnol 2000; 18: 85-90.

99   White O, Eisen JA, Heidelberg JF, Hickey EK, Peterson JD, Dodson RJ et al. Science 1999; 286: 1571-1577.

100   Shimao M, Onishi S, Mizumori S, Kato N, Sakazawa C, Appl Environ Microbiol 1989; 55: 478-482.

Biotransformations: Bioremediation Technology for Health and Environmental Protection
V.P. Singh and R.D. Stapleton, Jr. (Editors)
2002 Elsevier Science B.V.

573

Bioremediation of hazardous ethylenebisdithiocarbamate (EBDC) fungicides

Dileep K. Singh

Department of Zoology, University of Delhi, Delhi - 110 007, India

INTRODUCTION

Fungicides are chemicals used to prevent or minimize the crop losses caused by pathogenic fungi. They can be non-systemic or systemic compounds (Table 1) and used in crop protection through seed treatment, soil treatment, fruit and foliage application, trunk infection and post-harvest treatments.

An alkali and metal salts of alkylenebis (dithiocarbamate) acid are some of the most widely used fungicides in agriculture. The mode of action of this dithiocarbamate compound is not well defined and involves multiple target sites. One possible mode of action is the interaction of isothiocyanate and/or carbon disulfide radicals, released during the initial phase of the degradation of the parent molecule, with proteinacious sulphydryl groups, leading to deactivation/inhibition of enzymes and cellular functions [1].

Degradation of alkylenebis (dithiocarbamate) in soil, water plants and animals follows the common pathways. The disassociation of the metal complex and oxidation reactions lead to the formation of carbon disulfide (CS_2), ethylene thiuram disulfide (ETD), ethylene thiuram monosulfide (ETM) and isothiocyanate as major products [1]. Further degradation yielded ethylenethiourea (ETU), ethyleneurea (EU), and 2-imidazoline as terminal products [1].

Table 1
Fungicides used in crop protection

(A) Non-systemic fungicides

  (a)   Sulfur

  (b)   Copper

  (c)   Copper (II) oxychloride

  (d)   Cuprous oxide

  (e)   Fenthin hydroxide

  (f)   Dithiocarbamates

  (g)   Phthalimides

  (h)   Dicarboximides

(B) Systemic fungicides

  (a)   Organophosphates

  (b)   Benzimidazoles

  (c)   Carboxanilides

  (d)   Phenylamides

  (e)   Phosphites

(C) Miscellaneous fungicides

  (a)   Propamocarb

  (b)   Isoprothiolans

  (c)   Pyroquilon

  (d)   Hymexazol

  (e)   Tricyclazole

  (f)   Ethirimol

  (g)   Bupirimate

  (h)   Probenazole

  (i)   Carproparnid

  (j)   Spiroxamine

  (k)   Fenpropidin

  (l)   Famoxadone

PERSISTENCE AND RESIDUES OF EBDCs AND ETU

The IUPAC commission on pesticide chemistry has defined persistence as the residence time of a chemical species (parent compound or a metabolite) in a specifically defined compartment of the environment [2].

Persistence is the net result of many interacting factors. Chemical properties of pesticides such as volatility, solubility, stability to ultraviolet irradiation, tendency to adsorb onto or dissolve into tissue surfaces, ease of hydrolysis, sensitivity to humid atmospheres, potentiality to polymerize with or without UV irradiation, excitation, possible isomerization or other rearrangement, etc. are all important properties that play a significant role in persistence. In considering persistence, it is necessary to take into account the metabolism or alternation products of a pesticide besides the study of the residues of the parent pesticides itself.

RESIDUES OF EBDCs AND ETU IN THE ENVIRONMENT

After the patenting in 1934 by Tisdale and Williams, the use of dithiocarbamic acid derivatives as fungicides (EBDCs) has been increasing with a tremendous speed throughout the world. Till 1977, the EBDCs were registered for the use against 1296 diseases and for 271 crops constituting the largest group of commercial fungicides [3]. Their extensive use in agriculture leads to the residue problem, as posed by other pesticides. A major proportion of any *agricultural pesticide*, no matter how and where applied, finds its way eventually to the soil [4], from where if enters air, water or biota [5]. Residues may enter the living organisms by absorption, inhalation, and ingestion of resides present in food and in the environment [6]. A number of workers and organizations have been engaged in the determination of EBDC and ETU residues in the environment since the confirmation of *teratogenic* and *carcinogenic* potentials of ETU.

RESIDUES OF EBDCs AND ETU IN SOIL

Besides being used as foliar protectants, EBDCs are also used in seed dressing. During foliar spray, a considerable amount of these fungicides finds its way to the soil. Similarly, when the fungicide is used in seed dressing its most of the amount reaches the soil. According to Rich [7] zineb has a negative charge similar to the leaf surface, hence the electro-kinetic force necessary for adsorption is lacking between zineb and leaf surface. Hence the weathering becomes more prominent here. It may be same for all other EBDCs due to their similar structures. The deposits of EBDCs on leaves are likely to leach from the upper leaves to the lower leaves and ultimately to the soil along with run off due to rain and irrigation. Both EBDCs and ETU added to the soil in various ways are likely to further degrade into other compounds due to microbial as well as certain physical and climatic factors. Blazquez [8] found only 30 per cent of the initial residues of ETU to be present in soil after 13 days. However, only a trace amount of ethylene thiuram monosulfide (ETM) could be detected at that time when applied at the rate of 80 ppm of soil. No Dithane M-45 could be detected after 13 days, when applied at a rate of 360 ppm in soil. Using [^{14}C] ETU, Rhodes [9] found that the residue level of ETU and its degradation products came down to 48 per cent of the initial amount of ETU present in 4 weeks, however, the residue level for intact ETU was much less. This loss in overall amount of ETU and its degradation products may be attributed to its conversion into CO_2 by soil microorganisms [10]. Lyman and Lacoste [11] have also observed this process of *mineralization* in non-sterile soils. The residues of ETU in the upper 2.5 inches of soil near the plants treated with maneb were observed to be 0.475 ppm, 15 days after spray [12].

METABOLISM AND DEGRADATION IN SOIL

The pesticide residues present in soil, either as a result of their direct application to soil or as a result of their movement due to various factors

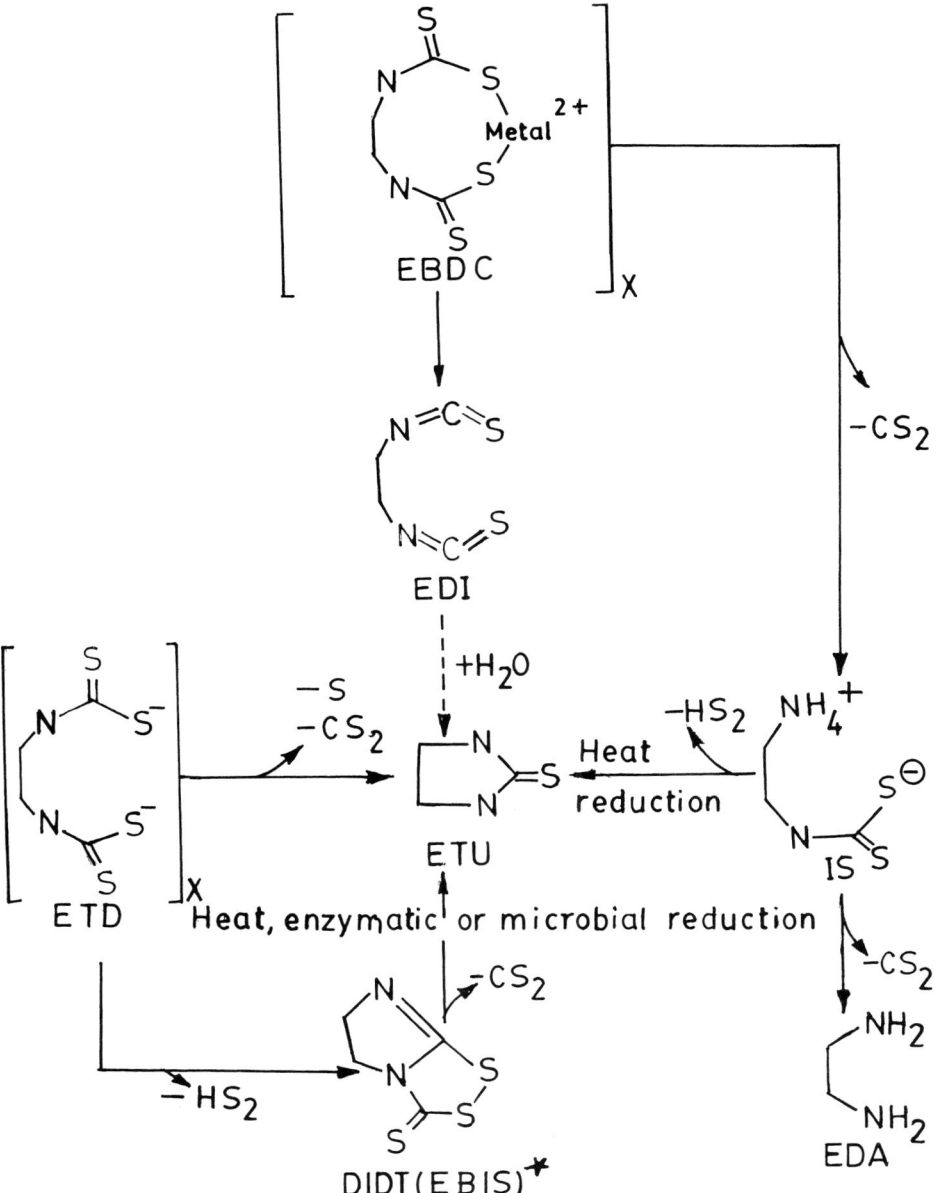

Figure 1. EBDC degradation routes for ETU formation [13,14]. EDA = Ethylenediamine, DIDT = 5, 6-dihydro-3H-imidazo (2, 1-C)-1, 2, 4-dithiazole-3-thione, EU = Ethyleneurea, EBDC = Ethylenebisdithiocarbamate, EBIS = Ethylenebisisothiocyanate sulfide, EDI = Ethylene diisothiocyanate, ETD = Ethylene thiuram disulfide [15,16], ETU = Ethylenethiourea, IS = Inner salt, X = "Polymeric" material, * = Original name: ethylene thiouram monosulfide (ETM) [17], ... = Possible route.

578

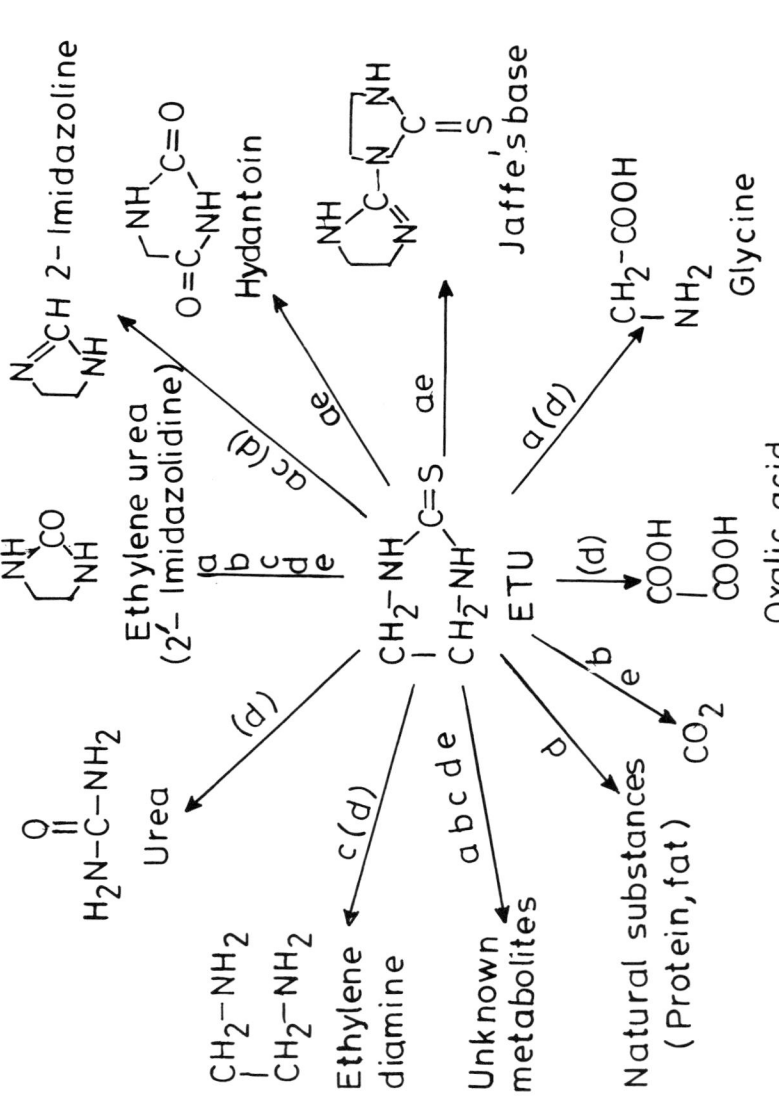

Figure 2. Reaction products of ETU in biological and non-biological systems [13]. a = photodecomposition, b = chemical oxidation, c = plants, d = animals, e = soil. Letters in parentheses indicate proposed pathways.

such as rain, leaching and winds, undergo both chemical as well as *microbial degradation*. The chemical transformation of pesticides is a wide spread phenomenon occurring in soil by various reactions such as oxidation, hydrolytic, and non-hydrolytic nucleophilic reactions [18,20]. Reactions with free radicals present in soil are also reported to occur [21]. These reactions may be catalyzed by clay surfaces, metal oxides, metal ions, organic surfaces [20] and, in some cases, by soil enzymes [22,23]. All these reactions are mediated through water which functions as a reaction medium and as a reactant or both [20].

Besides the chemical constituents of soil, bacteria, actinomycetes, fungi, algae and protozoa, which constitute the major groups of microorganisms in soil, play major role in the degradation and metabolism of pesticides in soil [24]. Population densities of bacteria may be as high as 10^9 per gm soil, and the length of fungal hyphae may be hundreds of thousands of meters per gm soil [25]. Microbial metabolism is one of the most important biotic factors, influencing the fate of pesticides in soil. Microbes may treat pesticides as substrates and, thereby, derive energy or may use them as raw materials for synthetic reactions [26]. Conversely pesticides may be metabolized incidental to normal substrates in the process of cometabolism. In cometabolism, the microorganisms derive benefit by metabolizing the chemical; however, the compound is biodegraded [27, 28]. Microbial degradation of pesticides is known to occur by various types of reactions namely hydrolytic, reductive, oxidative, dehydrochlorination, isomerization, polymerization and synthetic metabolism [29]. Zhang et al. [30] reported rapid dissipation of EBDCs from soil. The half-lives estimated in case of mancozeb and zineb were 7.6 and 16 days respectively. Gennari et al. [31] reported the degradation of mancozeb to ETU in soil. In case of unsterilized soil, 50 per cent degradation of initial deposit was observed after 15 days; however, the degradation was slower in case of sterilized soil. Kaufman and Fletcher [10] studied the fate of ETU in both non-sterile and autoclaved soils. They observed the formation of EU in both conditions. However, the *biologically active* soils were found to degrade ETU to carbon dioxide, hydantoin, Jaffe's base, and two more unidentified degradation products. Lyman and Lacoste [11] also reported the mineralization of ETU in non-sterile soils. Rhodes [9] reported rapid dissipation of ETU from Keyport silt loam soil with a half-life of less than four weeks. However, the half-life of (^{14}C) maneb in the same soil was 4 to 8 weeks. In case of Hagerstown silt loam soil, the half-lives of ETU and mancozeb were estimated to be 22 and 90 days respectively [11]. Since the formation of (^{14}C) carbondioxide did not parallel the disappearance of ETU from the soil [10] and the above value of half-life of ETU was based on (^{14}C) carbondioxide formed, it may be too high from the exact value. (^{14}C) ETU in soil treated at 2 ppm was converted to ethyleneurea in 2 days and 43 per cent was degraded to (^{14}C) carbondioxide within 4 days after treatment [10]. Rhodes [9] also reported that 79 per cent of (^{14}C) ETU present in soil was converted to (^{14}C) EU after 1 week of exposure to field conditions. Doneche and Segmin [32] also studied the degradation of mancozeb in vineyard soils. Blazquez [33] found slower degradation of ETU on soil surface as compared to mancozeb.

EBDCs have also been detected in the form of bound residues. The bound residues of zineb in soil increased with elapsing time [30]. Pazmino et al. [34] reported pronounced binding in case of maneb treated soil and the bound residues exceeded the extractable residues.

REFERENCE

1   Robberts T, Huston D. In: Metablic Pathways of Agrochemicals. Cambridge, UK: Royal Society of Chemistry, 1999; 937-956.

2   Greenhalgh R. Pure Appl Chem 1980; 52: 2565.

3   Brandes GA. DOC 09596. Philadelphia: Rohm and Hoas.

4   Harvey J Jr. Rev 1983; 85: 149.

5   Woodwell GM, Craig PP, Johnson HA. Science 1971; 174: 1101.

6   Robinson J. In: Edwards CA, ed. Environmental Pollution by Pesticides. New York: Plenum Press, 1973; 459-493.

7   Rich S. Phytopathol 1954; 44: 203.

8   Blazquez CH. J Agric Food Chem 1973; 21: 330.

9   Rhodes RC. J Agric Food Chem 1977; 25: 528.

10   Kaufman DD, Fletcher CL. Abstracts of the 165th National Meting of the American Chemical Society, Dallas, Taxes, 1973.

11   Lyman WR, Lacoste RJ. Proc Int IUPAC Congr Pest Chem, "Pesticides", Stuttgart: Georg Thieme, 1975; 67.

12   Newsome WH, Shields JB, Villenenue DC. J Agric Food Chem 1975; 23(4): 756.

13   IUPAC Publication, 1977.

14   Marshall WD. J Agric Food Chem 1977; 25: 357.

15   Hylin JW. Bull Environ Contam Toxicol 1973; 10: 227.

16   Engst R, Schnaak W, Rattba H. Nachrichtenbl. Dtsch Pflanzenschutzdienst (Berlin) 1968; 22: 26.

17   Vonk JW, Sypestejn AK. Appl Biol 1970; 65: 489.

18   Kearney PC, Helling CS. Residue Rev 1969; 25: 25.

19   Goring CAI. In: Goring CAI, Hamker JW, eds. Organic Chemical in the Soil Environment Vol. II, New York: Marcel Dekker, Inc, 1972; 569: 631.

20   Goring CAI, Laskowshi DA, Hamaker JW, Meikle RW. In: Haque R, Freed VH, eds, Environmental Dynamics of Pesticides, New York: Plenum Press, 1975; 135: 172.

21   Helling DS, Kearney PC, Alexander M. Adv Agron 1971; 23: 147.

22   Skuins JJ. In: MC Clarran AD, Peterson GH, eds. Soil Biochemistry: Dekker 1967; 371.

23   Burns RG, Edwards JA. Pestic Sci 1980; 11: 506.

24   Alexander M. In: Introduction to Soil Microbiology, 2nd eds., New York: John Wiley and Sons, 1977; 467.

25   Torstensson L. In: Hance RJ, ed. Interactions Between Herbicides and Soil. New York: Academic Press, 1980; 159-173.

26   Anderson AC. J Environ Sci Health 1986; B21: 41.

27   Alexander, M. (1979). In: Pritchard PH, Bourguin AW, eds. Microbial Degradation of Pollutants in the Marine Environment. EPA – 600/9 – 79 – 012, 1979; 69-75.

28   Alexander M. Science 1981; 211: 132.

29   Matsumura F. In: Krishna Murthy CR, Matsumura F, eds. Biodegradation of Pesticides. New York: Plenum Press, 1982; 67-87.

30   Zhang L, Shi G, Mo H, An F, Wang G. Report to IAEA for Research Contract No. 4277/R1/RB, Res Cent for Eco-environ Sci Academia Sinica, 1988.

31   Gennari M, Cignetti A, Crosa M, Negre M. Def Veg 1988; 42: 19.

32   Doneche B, Segmin G. CR Hebd Acad Sci Ser D 1975; 280: 2265.

33   Blazquez DH. J Agric Food Chem 1973; 21: 330.

34   Pazmino O, Balanos M, Espinosa L, Moran M, molineres J, Merino R. Final FAO/IAEA Research Coordination Meeting at Beytepe-Ankara, Turkey, 13-17 March, 1989.

INDEX